Agribusiness Supply Chain Management

Agribusiness Supply Chain Management

N. Chandrasekaran ▪ G. Raghuram

CRC Press
Taylor & Francis Group
Boca Raton London New York

CRC Press is an imprint of the
Taylor & Francis Group, an **Informa** business

First published in paperback 2024

First published 2014
by CRC Press
2385 NW Executive Center Drive, Suite 320, Boca Raton FL 33431

and by CRC Press
4 Park Square, Milton Park, Abingdon, Oxon, OX14 4RN

CRC Press is an imprint of Taylor & Francis Group, LLC

© 2014, 2024 Taylor & Francis Group, LLC

Library of Congress Cataloging-in-Publication Data

Chandrasekaran, N. (Nagarajan)
  Agribusiness supply chain management / by N. Chandrasekaran and G. Raghuram.
    pages cm
  Includes bibliographical references and index.
  ISBN 978-1-4665-1674-8 (hardback)
    1. Agricultural industries--India--Management. 2. Business logistics--India. I. Raghuram, G., 1955- II. Title.

  HD9016.I42C465 2014
  630.68'7--dc23   2013036689

ISBN: 978-1-4665-1674-8 (hbk)
ISBN: 978-1-03-291775-7 (pbk)
ISBN: 978-0-429-25369-0 (ebk)

DOI: 10.1201/b16704

Visit the Taylor & Francis Web site at
http://www.taylorandfrancis.com

and the CRC Press Web site at
http://www.crcpress.com

Contents

SECTION II INTRINSIC ISSUES

Case Studies

Preface

Agribusiness is one of the key economic activities in India that links the agricultural economy, and its demand and supply management is aimed at the well-being of the society. An agribusiness supply chain includes a number of processes such as supply management, production management and demand management to ultimate customers through a competitive distribution channel. India leads in production of many of the agricultural commodities but is still widely criticized for improper management of distribution. This is due to diversity of production and demand bases and inefficient logistics management. Further, complexity of agribusiness supply chains is due to bulkiness of produce, perishability and seasonality. It is imperative to have a well-designed supply chain.

It is important to understand issues from the perspectives of various actors in the agribusiness supply chain. This includes the input sector, farms as production centres processing units as value-adding production centres, and delivery to ultimate customers through retailers and directly from the higher echelons of the distribution channel. The supply chain effectiveness also depends on intrinsic issues like perishability, quality and risk.

An agribusiness supply chain is dependent upon the functioning of a support system that includes infrastructure like cold storage and warehouses, adoption of information technology and a well designed and implemented financial system to overcome the vagaries of informal financial systems affecting cost and responsiveness of the supply chain.

Finally, an agribusiness supply chain depends upon government and its bodies at the central (federal), state and local levels to function effectively. There are a number of laws and regulations that impact this sector. Agribusiness supply chain role agents, analysts and policy makers must be able to relate to this in decision making.

Agribusiness supply chain management focuses on how a firm organizes receipts of material and components, adds value and delivers to the ultimate customer through planning, sourcing, manufacturing and distribution processes engaging multiple stakeholders in the supply chain. With increased proliferation of products and services, and customers becoming ever demanding and expecting differentiation at low cost, supply chain challenges are going to be aplenty for firms. Supply chain management impacts a company significantly for achieving a competitive edge in any market. Managing fertilizer production and distribution and, in the same way, managing processing and distribution of commodities like tea are highly competitive, and supply chain design matters most.

In today's world, companies compete on supply chains. Those who have the capability to establish a distinctive supply chain and create it as a strategic asset are leaders in their business and in fact emerge as the best in class across industries and markets.

Some of the examples of such distinctive agribusiness supply chains in India include those established by Gujarat Cooperative Milk Marketing Federation (GCMMF), brand owner of AMUL, ITC Ltd—Foods Division, PepsiCo India and HUL, to name a few. Similarly, in the input sector, companies in fertilizers, seeds and machinery compete through their supply chains.

This book is an attempt to provide readers a comprehensive perspective of agribusiness supply chains. It covers issues across various business and government roles in the agribusiness supply chain domain. The book also provides different cases for readers to relate to decision-making situations. The aim of this book is to provide a treatment of agribusiness supply chain management that is clear, well-structured and interesting. It brings out inter-relations within its drivers and across functions in the agribusiness supply chain. The text provides a logical approach to key activities of agribusiness supply chain management and relates principles and practices predominantly with examples from India.

N. Chandrasekaran
G. Raghuram

Introduction

This book addresses issues that help a reader to systematically approach decision making in agribusiness sector. It focuses on actors in supply chains, intrinsic issues that would impact the actors and then the support systems that are essential to make the supply chain achieve its effectiveness. This Introduction provides the structure with which the authors approached the subject.

The structure of the book is as follows:

Introduction

- Nature of Agribusiness Supply Chain: Chapter 1

Section I: Actors

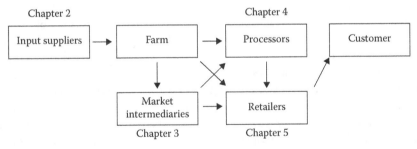

Section II: Intrinsic Issues

- Perishability: Chapter 6
- Quality: Chapter 7
- Risk: Chapter 8

Section III: Support Systems

- Infrastructure: Chapter 9
- Information Technology: Chapter 10
- Financial Systems: Chapter 11
- Role of Government: Chapter 12

Chapter 1 discusses "Nature of Agribusiness Supply Chain." The complexity of agribusiness has two components: farm output for direct demand and farm-based intermediate products for final demand. Then, three flows—physical, financial and information, and their interrelation for an effective agribusiness supply chain—are discussed. An introduction to contemporary issues like quality, security and challenges for society, and the role of supply chains is provided next. Then discussions follow on different drivers and their role in managing effectiveness of the agribusiness supply chain.

Section I: Actors

This section discusses actors in agribusiness supply chains, which include input providers, farm output market intermediaries, processors and retailers.

Chapter 2 is on "Input Suppliers" with the objective of understanding specificity of each input and importance of a driver for supply chain effectiveness. An input is taken up with reference to a case highlighting demand and supply management, and the way supply chain processes like planning, sourcing, value addition and distribution to users are critical. The chapter addresses inbound material and services at farm level operations.

Chapter 3 is on "Farm Output: Market Intermediaries" with the objective of understanding distribution options and forward linkages aimed at reaching ultimate customers. The coverage includes: farm gate procurement, local procurement through intermediaries, role of *mandis*/auction centres in farm output distribution, role of government agencies and challenges of organized players in a supply chain.

Chapter 4 is on "Processors." The objective of the chapter is to understand the challenges faced in this sector by processors when organizing their production and when reaching markets with their products. A number of issues like planning processes, sourcing, facilities, distribution management and waste management are discussed here.

Chapter 5 is on "Retailers" with the objective of understanding the complexity of retailing agricultural produces, both farm fresh and processed, and the need for efficiency and effectiveness of the supply chain. Both organized and unorganized markets are covered. Also, discussions are presented on the role of intermediaries and a short disguised case on challenges of managing a retail store.

Section II: Intrinsic Issues

After having looked at the major actors in the agribusiness supply chain, we move to discussing three intrinsic issues namely perishability, quality and risk.

Chapter 6 is on "Perishability." The objective is to understand the challenges faced in this sector with respect to distribution of produce, short life, lack of temperature control for storage and transportation, wastages and pricing pressures.

Chapter 7 is on "Quality" and how it is important for agribusiness supply chain effectiveness. The discussions include appreciating the importance of grading and quality management; current practices and brief overview of different policies that may impact on the quality of goods especially from trade perspective.

Chapter 8 is on "Risk" in agribusiness with the objective of understanding a number of risks that could hamper supply chains and the ways to mitigate them.

Section III: Support Systems

After having looked at the major actors and intrinsic issues in the agribusiness supply chain, we analyse the support systems required for managing effectiveness of agribusiness supply chain. These include: infrastructure, information technology, financial systems and government policies.

Chapter 9 is on "Infrastructure," which discusses the importance of transport connectivity, availability of storages including cold storages, communication network and so on.

Chapter 10 discusses importance of "Information Technology" with the objective of understanding the scope of deploying technology in agribusinesses at the farm level and in the different stages of agribusiness supply network. A number of instances of success in Indian agribusiness are cited in discussions.

Chapter 11 is on "Financial Systems" with an objective to enumerate the various financial challenges in this sector that impact the supply chain. The discussions are on: informal credit operations; role of commodity exchanges for efficiency and on awareness and new systems approach such as *e-choupal* and community communication centres. Here again instances of various policies and systems are cited.

Chapter 12 discusses the "Role of Government" with the objective of understanding various initiatives taken by the government to improve efficiency in this sector. Discussions include trade negotiations and subsidy; domestic subsidy schemes and their impact on supply chains and direct interventions through measures like buffer stock creation, releases, price and distribution control, Essential Commodities Act, food credit, farmers credit support system, cooperatives and so on, and their impact on the supply chain.

Each chapter starts with learning objectives. The subject is then presented with illustrations and examples. Agribusiness supply chain management is a practical application subject based on theoretical concepts. Most of the chapters contain a framework, and issues in handling the supply chain with real-life examples.

These 12 chapters are followed by 8 cases listed below.

Sr. No.	Case	Context
1	Bayer CropScience: Science for a Better Supply Chain	Supply chain improvement possibilities for a pesticide firm
2	Marico: Disintermediating the Copra Supply Chain	Supply chain disintermediation strategy for a coconut oil manufacturer
3	Hasmukhbhai K. Nakum: Cold Storage Entrepreneur	Market choice for an independent cold storage entrepreneur
4	Chilli in Soup (A)	Choices for the Spices Board of India to improve the image and quality of chilli exports in the context of detection of banned substances in the product by EU
5	Chilli in Soup (B)	Choices for Spices Board of India in the context of product recall in UK involving chilli exported from India, including testing for quality.
6	Adani Wilmar Limited	Distribution logistics and mode choice for an edible oil processor
7	Woolworths Limited, Australia	Warehouse consolidation strategy for a large food retailer
8	Akshaya Patra, Gandhinagar	Integrated view of the issues in the supply chain of a midday meal provider

Each case begins with a brief of the case context, followed by the case, questions for discussion and an approach to analysis.

A table on relevant issues that drive decisions versus the cases (Table 0.1) and a table on linkages between the chapters and cases (Table 0.2) are provided for readers to relate cases and discussions in each of the chapters.

Given the vast nature of agricultural production and economic system in the country, the book focuses on issues from the perspective of a supply chain analyst to enable different role agents in agribusiness to fulfil supply chain objectives of customer effectiveness and cost efficiency. The book is targeted at students studying agribusiness supply chain management in a management course or a specialization course in other disciplines and for working professionals and policy analysts.

The 12 chapters are authored by Chandrasekaran and the eight cases are coauthored by Raghuram with various others.

Table 0.1 Issues That Drive Decisions vs. Cases

Issues	Bayer CropScience: Science for a Better Supply Chain	Marico: Disintermediating the Copra Supply Chain	Hasmukhbhai K. Nakum: Cold Storage Entrepreneur	Chilli in Soup (A)	Chilli in Soup (B)	Adani Wilmar Limited	Woolworths Limited, Australia	Akshay Patra, Gandhinagar
Supply Chain Mapping	*	*	*	*	*	*	*	*
Competitive Forces	*	—	*	—	—	—	—	—
Regulation	—	—	—	*	*	—	-	*
Customer Service/Effectiveness	—	—	—	*	*	—	*	*
Quality Management	—	*	—	*	*	—	—	*
Pricing Policy	—	—	*	—	—	—	—	—
Cost Efficiency	—	*	*	—	—	—	*	*
Demand Management	*	—	*	—	—	—	—	—
Distribution Management	*	*	—	—	—	*	—	—
Production/Value Addition	*	—	—	—	—	—	—	—
Sourcing	—	*	—	—	—	—	—	—

(Continued)

Table 0.1 (Continued) Issues That Drive Decisions vs. Cases

Issues	Bayer CropScience: Science for a Better Supply Chain	Marico: Disintermediating the Copra Supply Chain	Hasmukhbhai K. Nakum: Cold Storage Entrepreneur	Chilli in Soup (A)	Chilli in Soup (B)	Adani Wilmar Limited	Woolworths Limited, Australia	Akshay Patra, Gandhinagar
Network Design	–	*	*	–	–	–	*	–
Facilities: Retail	–	–	–	–	–	–	*	–
Facilities: Distributions Centres	–	*	–	–	–	*	*	–
Facilities: Processing Unit	–	–	–	–	–	*	–	*
Facilities: Market Yards	–	*	–	–	–	–	–	–
Transportation Network	–	–	–	–	–	*	*	*
Inventory and Warehousing	*	–	–	–	–	–	–	–
Information Technology	–	*	–	*	*	–	*	–
Investment Criteria	–	–	*	–	–	–	–	–
Third Party Services	–	–	*	–	*	–	–	*
Returns Management	–	–	–	*	*	–	–	*
Responsible Supply Chain	–	–	–	*	*	–	–	*

Table 0.2 Linkages between Chapters and Cases

Chapters	Bayer CropScience: Science for a Better Supply chain	Marico: Disintermediating the Copra Supply Chain	Hasmukhbhai K. Nakum: Cold Storage Entrepreneur	Chilli in Soup (A)	Chilli in Soup (B)	Adani Wilmar Limited	Woolworths Limited, Australia	Akshaya Patra, Gandhinagar
Nature of Agribusiness Supply Chain	*	*	*	*	*	*	*	*
Actors								
Input Suppliers	*	—	—	—	—	—	—	—
Farm Output: Market Intermediaries	—	*	*	—	—	—	—	*
Processors	—	*	—	*	*	*	*	*
Retailers	—	—	—	—	—	—	*	-

(Continued)

Table 0.2 (Continued) Linkages between Chapters and Cases

Chapters	Bayer CropScience: Science for a Better Supply chain	Marico: Disintermediating the Copra Supply Chain	Hasmukhbhai K. Nakum: Cold Storage Entrepreneur	Chilli in Soup (A)	Chilli in Soup (B)	Adani Wilmar Limited	Woolworths Limited, Australia	Akshaya Patra, Gandhinagar
Intrinsic Issues								
Perishability	*	*	*	-	-	-	*	*
Quality	*	*	-	*	*	-	-	*
Risk	*	*	-	*	*	*	-	-
Support Systems								
Infrastructure	*	*	*	*	-	*	*	*
Information Technology	*	*	-	*	*	*	*	-
Financial Systems	—	—	*	*	*	—	—	*
Role of Government	—	*	*	*	*	*	—	zq

Acknowledgements

I thank Mr. H. R. Srinivasan, vice chairman and managing director, Take Solutions Ltd., for his continued support and encouragement on all my academic endeavors. Without his understanding and open-minded approach, this initiative would not have been accomplished. I thank Rev. Fr. Christie and Rev. Fr. Peter Xavier, director, LIBA, Chennai, for their continued support in pursuing this work.

I would like put on record my sincere gratitude to Dr. T. V. Subramaniam, Prof P. S. Anantha Narayanan, Mr. P. G. Subramanaiam, Dr. T. K. Nathan, executive director, KKID and Mr. K. Krishnamurthy, publisher from Chennai for their guidance and comments on various issues connected with this work. I would like to thank Prof. P. Chandiran, Prof. M. Ramasubramaniam, Mr. H. Sai Sridhar and Ms. M. Rammyaa, research scholar from LIBA, who have been encouraging and supportive at different stages of this work. I would like to thank Mr. K. Venkatesh who has provided editorial support.

I thank Oxford University Press India and NABARD for giving permissions for reproducing some of their material.

I am grateful to Prof. Raghuram and his family, my mother Mrs. Annapoorani Nagarajan, my spouse Prabha and daughter Sangeethaa for their support and understanding.

N. Chandrasekaran

I thank the following coauthors of the cases for permitting use of the material for the publication of this book:

Tathagata Bandyopadhyay
Faculty, IIM-A

N. Vijaya Baskar
(then) PGP ABM Student, IIM-A

Sanjay Choudhari
(then) Academic Associate, IIM-A

Sarang Deo
(then) PGP Student, IIM-A

Atanu Ghosh
(then) Faculty, IIM-A

G Kuberkar
(then) Research Associate, IIM-A

Santosh Kumar Mishra
(then) PGP ABM Student, IIM-A

Shravanti Mishra
(then) Research Associate, IIM-A

Saral Mukherjee
Faculty, IIM A

Sanjay Kumar Singh
(then) PGP Student, IIM-A

Neeraj Sisodia
(then) Research Associate, IIM-A

The organizations that have rendered direct support in the material development efforts include:

The Akshaya Patra Foundation
Mother Shree Cold Storage Private Limited
Marico Industries Limited
Spices Board of India
Woolworths Limited
Bayer CropScience Limited
Adani Wilmar Limited

I acknowledge the research assistance provided by:

Ramesh Reddy Amereddy
Anjali Dave
Vishal Kashyap
Kruti Mody
Shivani Shukla
Anju Singla
Niti Sirohi

I also thank Tata McGraw-Hill for permitting us to print the cases, "Chilli in Soup (A)" and "Bayer CropScience: Science for a Better Supply Chain," and Asian Case Research Journal for the case "Adani Wilmar Limited (A)."

G. Raghuram

We thank Ganesh Nachiappan, Rachna Rana and Niraja Shukla for content and data validation, and administrative aspects of editorial work.

We are grateful to Taylor & Francis for coming forward with this publication. We sincerely thank Lara Zoble and Laurie Schlags for their continued support and encouragement. We also thank copy editors of Taylor & Francis in helping us in standardizing and improving the content. We are also grateful to all those who would use and come back with feedback on this book.

N. Chandrasekaran
G. Raghuram

Authors

N. Chandrasekaran holds a PhD in financial management (1990) from Institute for Financial Management and Research (IFMR), University of Madras, and has about 25 years of corporate experience in areas relating to supply chain management, corporate planning, strategic management, mergers, acquisitions, corporate restructuring and has also worked with start-ups. He is also a Certified Supply Chain Professional (CSCP) awarded by Association of Operations Management, USA.

He is currently with Take Solutions Ltd as vice president, Corporate Affairs. He is also director, Centre for Logistics and Supply Chain Management, Loyola Institute of Business Administration, Chennai. Dr. Chandrasekaran has worked for agro-based business for organizations like National Dairy Development Board, leading groups engaged in the business of sugar manufacture, and with IT and knowledge-based companies as head of strategy, Human Resources Management and Policy and Systems.

He has published a number of articles in leading professional journals and brought out a first annual publication named *Indian Supply Chain Network*. He has authored a book *Supply Chain Management* published by Oxford University Press India in 2010. He is also coauthor of *Strategic Management*, published by Oxford University Press India in 2011. He led a team of four researchers on a project "Ethical Supply Chain Management in India," which was later published by Shroff Publishers in 2012.

G. Raghuram has been a faculty member at IIM Ahmedabad since 1985. He is the dean (Faculty) since September 2013. He was the vice chancellor of the Indian Maritime University from July 2012 to March 2013 and was the Indian Railways Chair Professor from January 2008 to August 2010. He specializes in infrastructure and transport systems and logistics and supply chain management. He conducts research on the railway, port, shipping, aviation and road sectors. He has published over 30 refereed papers in journals and written over 140 case studies. His fifth coauthored book is in the press now. He was awarded "Academician of the Year" by the Chartered Institute of Logistics and Transport in 2012. He is a fellow of the Operational Research Society of India, and Chartered Institute of

Logistics and Transport. He has teaching experience at universities in India, the U.S., Canada, Yugoslavia, Singapore, Tanzania and the UAE.

He has offered consultancy services to over 100 organisations including multilateral agencies. He is on the board of eight companies in the fields of infrastructure and logistics. He has served on various government policy making and advisory committees for the Ministry of Railways, Ministry of Shipping, Ministry of Civil Aviation, Planning Commission, Comptroller and Auditor General and the Cabinet Secretariat.

Prof. Raghuram has a BTech degree from IIT, Madras; a postgraduate diploma in management from IIM, Ahmedabad; and a PhD from Northwestern University, U.S.

ACTORS

1

Chapter 1

Nature of Agribusiness Supply Chain

> **OBJECTIVE**
>
> This chapter introduces the concepts of supply chain and logistics and their relevance to the agribusiness system. Furthermore, it discusses the role of the supply chain and the importance of supply chain drivers, namely, facilities, inventory, transportation, sourcing, pricing, and information technology.

1.1 Supply Chain

We may understand supply chains from the perspective of how an end consumer of tea gets it in his or her tea pot. This tea consumer is the ultimate customer of say, Lipton, a Unilever brand. Lipton tea blends are selected from many different plantations around the world in well-known source countries such as India, Sri Lanka, Kenya, and China. Lipton sources tea both from its own and from other plantations, processes the tea, and moves its products across borders and through distribution networks to reach the ultimate customer. The entire process, which runs into months, involves different role players, with responsibilities discharged by each of them under the guidance of Unilever, all aimed at satisfying the ultimate customers. The planning and managing of demand, and the organization of supply, involving the holding and moving of stock, with value addition at each step across the chain, constitute supply chain management (SCM) activities. While performing these activities, several other managerial disciplines, such as production, marketing, finance, operations, and information management come into play.

The supply chain focuses on managing a network of organizations and their activities to fulfil the demands of the ultimate customers of a focal firm in a dynamic environment.

SCM is the integration of key business processes for serving customers. During these processes, value is added to goods and services right from the original suppliers to each manufacturer and other intermediaries in the chain until they reach the end customers. While managing these processes, the focus is also on providing value to all stakeholders. These business processes are not limited to buying, movement, storage, and their integration. Some interesting terms such as original suppliers and adding value to customers and stakeholders also have to be considered.

According to Monczka et al. (2002),

> a supply chain encompasses all activities associated with the flow and transformation of goods from the raw material stage (extraction) to consumption by the end users, as well as the associated information flows both up and down the supply chain. Therefore, it comprises a physical element (the strategic partnering of various market-focused, responsive organisations involved in the transformation of specific goods) and an information element (controlled sharing of business data and processes). A supply chain can, therefore, be viewed as a value chain network consisting of individual functional entities committed to the controlled sharing of business data and synchronized coordination of processes for optimizing supply chain profit.

Hence, the term SCM is used to describe the management of the flow of materials, information, and funds across the entire supply chain, starting from suppliers and going to component producers, final assemblers and distributors (warehouses and retailers), and ultimately, to the consumers. This description talks about managing three flows—physical, informational, and financial—across the chain, and also about the importance of the customer. In contemporary business practice, a fourth flow, namely, reverse physical goods flow or reverse logistics, is gaining importance.

Thus, the objective of a supply chain manager is to manage a network of organizations that are involved, through upstream and downstream linkages, in the different processes and activities that produce value in the form of products and services in the hands of the ultimate customer (Christopher, 1998). This connotes that a supply chain involves two or more organizations, and exists by articulating the flow of material, finance, and information with the purpose of serving the ultimate customer. It may be worth reiterating here that the key to the existence of a supply chain is the presence of an ultimate customer and its value to him.

An example of a network for processing and marketing of edible oil in India, where it is a regulated business, is discussed here. The schematic representation is given in Figure 1.1.

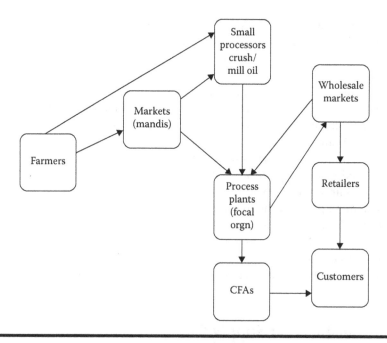

Figure 1.1 Schematic representation of an edible oil processor supply chain network. CFA, carrying and forwarding agent.

In this example, the focal organization, whose supply chain is discussed, is an edible oil processing and marketing plant, say ABC Ltd that processes and sells refined groundnut oil. The company in its supply chain network has other entities, namely, farmers, who directly sell to the company and the small processing units who buy the produce from the farmers and crush it. The company may procure directly from the farmers through its own procurement teams, or can outsource the procurement work to other partners. All these entities would be a part of the sourcing/procurement process components of the supply chain.

ABC Ltd does the milling, solvent extraction and refining, and sells refined groundnut oil. The company may have expellers/crushers for separating oil from the groundnut kernels. The next stage involves solvent extraction, and then comes the refining stage. The company may buy expeller oil and/or extracted solvent (edible grade) and refine them. These activities are part of the internal supply chain processes.

ABC Ltd would sell refined oil through their carrying and forwarding agents (CFAs)/depots/distributors. This would be a part of their process of primary distribution. From this node, secondary distribution and last mile reach to customers happen as CFAs sell through retail networks. ABC Ltd would also move its products through wholesale markets, as many small- and medium-sized retailers prefer to buy from wholesale markets. Another channel for ABC Ltd is selling directly to institutional buyers, who would be bulk consumers. A few situations may arise where the CFAs also sell to wholesale market players on a need basis; this is not depicted in Figure 1.1, as this is more an exception rather than a normal practice. All these are parts of the distribution network, which is a part of the customer relationship management process in a supply chain network.

Hence, ABC Ltd is the supply chain focal organization whose refined groundnut oil supply chain is relevant for decision making. The participants in the network could be CFAs and others in the distribution network, as well as procurement agencies and institutions that constitute the supplier network. ABC Ltd would work towards maximizing its profit while processing groundnut kernels and marketing groundnut refined oil. It would refrain from being intra-firm focused, and would not like to make profits at the cost of the role agents across its supply chain. On the other hand, the supply chain framework, which is responsive and pegged towards value for money, would help to improve profits of all the players across the chain and enhance customer satisfaction.

This explains the typical operation and relevance of a supply chain network.

1.2 Logistics and Supply Chain

There has been a tendency to use these two terms interchangeably, even by management professionals. It may be useful to understand the focus areas of both of these terms so that we are able to identify their distinguishing and common features. The evolution of each of these terms has been an advantage over the other; logistics functions have been in existence since time immemorial and have contributed to public life, military operations, and business.

> Even today, it is amazing to study the logistics operations for milk marketing, as it is happening in India. States like Gujarat have a surplus of milk, and have successfully driven cooperative movements. The success of AMUL and the initiative of the National Diary Development Board have led to the success of the Gujarat Cooperative Milk Marketing Federation Ltd. (GCMMF), the brand owners of AMUL. It is one of the largest diary companies in India, selling milk and milk products across various states in India. GCMMF/AMUL works through its milk unions in Gujarat to sell milk across deficit regions in bulk. Just consider the operations involving a perishable product like milk, which is collected from remote villages, processed in dairy plants and moved across India in wagons for augmenting supplies in domestic markets. (Chandrasekaran, 2010)

The entire supply chain needs to be handled with an emphasis on high quality and hygiene in a temperature-controlled environment. These logistical operations are very extensive and are critical for the success of the dairy supply chain network.

At this point, we should understand some definitions of logistics management.

In 1991, the international Council of Logistics Management (CLM) defined logistics as "the process of planning, implementing, and controlling the efficient, effective flow and storage of goods, services, and related information from the point of origin to the point of consumption for the purpose of conforming to customer requirements." This definition focuses on logistics as management of movement and storage of goods and services from the point of origin to the point of consumption, based on consumer needs. The definition sounds more operational, focuses on a single entity, and is customer focused. Typically, in circumstances where competition is restricted, or operations are highly focused, like military operations, such definitions clearly depict the functions that constitute logistics management. (Chandrasekaran, 2010)

This definition was later modified to read "Logistics is the process of strategically managing the procurement, movement and storage of materials, parts and finished inventory and the related information flows through the organization and its marketing channels for the cost effective fulfilment of customers' orders." This encompasses, apart from storage and movement, key aspects like strategic focus, by covering both product and information flows. This definition is more realistic for commercial organizations and businesses. Many organizations, over the decades, have focused on the logistics aspect of business. (Chandrasekaran, 2010)

A number of firms operating in this sector, especially in processing, realized that their growth depended on their ability to efficiently manage transportation of inbound and outbound material and stocking, which in many cases may be influenced by characteristics such as bulkiness, perishability (requiring, for example, temperature control), and seasonality in demand and supply. These characteristics require specialized temperature-controlled handling, procurement management, contract manufacturing, and the maintaining of high quality standards. Such complex requirements can be handled only by those with lots of expertise in logistics operations, such as managing transport operators, distribution centres, service providers, and facilities.

Terms such as logistics, inbound logistics, materials management, physical distribution, and supply chain management seem to be used interchangeably. Very briefly, inbound logistics covers the movement of material, components and products received from suppliers. Materials management describes the material handling part of the movement of goods and components within a factory or firm. Physical distribution refers to the movement of the finished goods outward from the end of the plant operations to the shipping or dispatch department. (Chandrasekaran, 2010)

Logistics describes the entire process of material and products moving into and through process centres, and out of a firm. The facets of logistics management include

- Order management
- Outbound transportation and distribution management
- Inventory management
- In-plant logistics, such as stores and movements towards lines and shops
- Inbound transportation
- Procurement
- Information management

From the above-mentioned features, it can be said that logistics management is oriented towards process optimization, that is, cost minimization for the focal organization. It gives a sense of internal SCM. But current thinking on SCM and experience has evolved to a broader perspective, where one needs to go beyond logistics operations (Chandrasekaran, 2010).

1.3 Agribusiness

Agriculture in India has a long history, and today India ranks second worldwide in agricultural output. Agriculture and allied sectors, such as forestry and logging, accounted for 16.6% of the GDP in 2007. The sector employs about 60% of the total workforce and plays a significant role in the overall socio-economic development of India. India is the largest producer of milk, coconut, tea, ginger, turmeric, and black pepper in the world. It has the world's largest cattle population, at about 281 million (in 2009). India is the second largest producer of wheat, rice, sugar, groundnut, and inland fish. Furthermore, India accounts for 10% of the world's fruit production, ranking first in the production of banana and *sapota* (*marmalade plum*). One of the challenges for India is that its population is growing faster than its ability to produce rice and wheat.

It may be noted here that apart from agricultural and allied sector production as farm outputs, a country needs to have capability to process farm or unit level output through processing sector. Agribusiness sector considers all commercialization of farm raw outputs and processed output. Here, we may note that processing is the largest sub-sector in agribusiness.

Apart from providing food, the agribusiness sector is important because it allows us to undertake routine activities similar to those in science, government, and education, which are important for developing economies. Agriculture has traditionally been one of the key sectors contributing to the national income, and is very important socially, as the country has a huge population and different strata of economic groups that are dependent on agriculture for food and on agricultural products for living.

Agriculture relates to farming, animal husbandry, and aqua-/marine-related activities. "Agribusiness" is a complex system reaching beyond the farm and commodities or produce. It includes everything required to bring food to the consumer, even the challenges of providing inputs to farms for producing food. For example, if we consider the production of wheat to be a part of agriculture, agribusiness relates to all associated activities, from procurement of wheat and conversion into flour and/or processed food, till the end products reach the customers. Similarly, all aspects of the input side of the business, namely, capital equipment, and inputs such as fertilizers, are part of the agribusiness sector.

As in manufacturing, the agribusiness supply chain also depends on various role agents and their activities. If we look at the case of the branded wheat flour supply chain in India, there could be a number of actors, whose roles need to be clearly understood for the various activities performed. For example, the company that owns the brand Pureatta, which sells nationally, procures wheat mainly from western Uttar Pradesh and Punjab. The produce is seasonal, whereas consumption is year round. Thus, the company will be serving a huge market, and therefore setting up processing facilities across regions would not make sense. Similarly, centralized production and redistribution throughout regions would also not be effective. Thus, the company has adopted the strategy of engaging contract manufacturers, who procure wheat and convert it as per standards set by the company, and sells the end product through them in the form of Pureatta branded wheat flour. In this process, there are procurement agencies that buy wheat; convertors; brand owners who release production schedules, distribution plans, and controlled marketing; and a host of intermediaries such as regional distribution centre operators, transport operators, CFAs, and retailers. All these players are aligned towards optimizing resources and efforts for fulfilling Pureatta customers' needs in the supply chain network. Each player has clearly defined roles and responsibilities, and is remunerated for the services rendered. The orchestration of such supply chain activities is what makes for the challenge, especially in agribusiness, where seasonality, bulkiness, and perishability are involved.

Resources are another key aspect in agribusiness supply chains. Often, production centres and demand centres are widely spread out. Typically, production centres would be in clusters, whereas demand centres would be spread across the geography. This requires a number of role agents and activities to be structured to serve the demand. This would require resources such as processing units, storage centres, and transport operators, including, possibly, temperature-controlled trucks. Similarly, in terms of time, agricultural produce is bunched in certain time windows, based on seasonality, whereas consumption is all year round. This again requires very effective resource management through storage, handling, and even processing. In a country like India, which has a large geographical spread, with

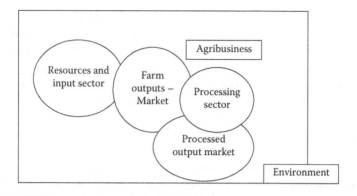

Figure 1.2   The agribusiness system.

varying climatic conditions, developing infrastructure, and a large population to be served, resource management is a challenge. Hence, resources, role agents, and activities in the agribusiness supply chain must be in synchronization to establish successful supply chains.

Thus, the agribusiness system includes not only those who produce at the farm gate but also those who

- Provide inputs (e.g., seeds, chemicals, nutrients)
- Process the output (e.g., processing plants)
- Manufacture goods and commodities using produce
- Transport/sell the products (e.g., retail grocery stores, wholesalers)

Figure 1.2 depicts the agribusiness system.

1.4   Challenges Faced

It may be noted that there has been a constant mismatch between demand and supply. India is a vast country with a population of about 1.2 billion. Although India has a large and diverse agricultural sector and is one of the world's leading producers of agricultural products, it is also a major consumer because of the sheer size of its population. The pressure on demand and supply can be attributed to geographical spread. Moreover, India's diversity in terms of socio-economic factors imposes additional pressure on demand and supply.

India's GDP growth rate has been next to China's during the first decade of this century and millennium. "High growth rates have significantly reduced poverty in India. However, India's per capita national income is still low, estimated at US$ 820 in 2006. Based on this, the World Bank still classifies India as a low-income

Table 1.1   Food Consumption—Daily Calories Per Capita

	1990–1991	*1998–2000*	*2003–2005*
Brazil	2860	3001	3223
China	2696	2917	2957
India	2396	2463	2512

Source: European Commission. India's role in world agriculture. *Monitoring Agri-Trade Policy* newsletter 03-07, December 2007.

country, and the World Development Report 2008 estimates that one-third of the population still lives below the poverty line" (European Commission, 2007).

Table 1.1 gives basic statistics on food consumption.

India is among the world's leading producers of paddy, wheat, buffalo milk, cow's milk, and sugarcane. It is either the world's leading or second largest producer in eight out of its top ten products. External trade is driven by surplus over domestic demand.

India exports products such as tea, coffee, rice, wheat, cotton, tobacco, spices, fruits and vegetables, oil meals, processed fruits, juices, marine products, sugar and molasses, meat, and other agricultural products. The total value of exports of these products was around US$ 24.7 billion for the year 2010–2011. Apart from these, the agribusiness system would include more product categories such as cotton yarn, fabrics and garments, and other similar products that are not mentioned here.

The composition of exports includes commodities to the extent of one-third of total agricultural exports, intermediate products over one-fourth, and final products for the remaining 40%. This indicates that some of the goods are produced far in excess of domestic demand.

Similarly, we can look at the import data and understand that products for which demand was higher than domestic production were imported to augment the supply. Indian agricultural imports are focused mainly on intermediate products, which account for 56% of the total imports. Final products account for 31%, and commodities account for 13%. The bulk consumption goods include cereals and cereal preparations, edible oils, pulses, and sugar, which accounted for US$ 8.7 billion out of the total imports of US$ 61.41 billion during 2010–2011. Bulk consumption items accounted for 14% of total imports. The agribusiness sector also imports fertilizer, capital equipment, transport equipment, cotton, and other items, which are not included in the above figures.

Thus, the demand and supply situation influences trade and the domestic economy in terms of the agribusiness sector.

1.4.1 Production

As may be observed from the data on food consumption, India rates among the lowest. In spite of achieving high growth rates, India is still categorized as a low-income economy. Moreover, the data on imports reveal that edible oils and pulses, which are important dietary inputs in India, are imported in large quantities.

India had its first green revolution in the 1960s. However, the population has more than doubled since then. This requires improving productivity with better input and resource management. Furthermore, we need to understand market sensitivities and, if required, to work on the demand pattern for optimal use of available resources. Hence, there is a need to stimulate production or induce socially desirable standards for production functions, instead of allowing autonomous production functions to operate. This calls for active policy orientation by agricultural economists and supply chain specialists, who need to be involved in decision making.

1.4.2 Labour and Inputs

Agriculture is the key employer in India, accounting for 60% of the labour force, down from 70% in the 1990s. The movement of the labour force has not kept pace with the development of the economy. Agriculture accounted for about 44% of the labour in China in 2002, and for 21% in Brazil in 2004. What do these numbers mean? Probably, low value added per unit of agricultural labour.

In India, agricultural value added per worker has grown by only 15% in real terms from 1990 to 2004, whereas it rose by 60% in China and more than doubled in Brazil. Apart from labour productivity issues, India suffers from the problem of low per farmer holding, which has an average size of less than 1.4 ha. Small farm holdings are fit just for subsistence, with low investment and productivity. Because of this, subsidies on farm inputs, including fertilizers, electric power, and irrigation water, have led to inefficient use of inputs. Among the challenges, a prominent one is the vagaries of nature, such as floods, droughts, and cyclones in the same year, all of which lead to supply side issues in Indian agriculture.

1.4.3 Supply Chain Network

The management of post-harvest losses is one of the major challenges faced by the supply chain network. Agricultural products are perishable in nature and losses occur due to various reasons such as lack of knowledge, improper handling, lack of storage facilities, and the inability of the various supply chain partners to work in unison.

Farmers are focused on producing a crop, purchase a number of inputs. They are capable of handling production in given conditions. A vast proportion of

Indian agriculture is dependent on two monsoons, namely, the south-west and the north-east. Although cereals such as paddy and wheat benefit from irrigational support from rivers in catchment areas, a large proportion of the land is arid, and crops such as edible oilseeds and pulses that are grown here are mostly dependent on the monsoons. Farmers are exposed to high levels of production risk, and investments in inputs in the early stages of farming are not covered appropriately. There is a tendency to go to moneylenders or intermediaries who typically charge high interest. Also, farmers are not fully aware of crop insurance and other support provided by the government. They approach middlemen for their needs. Thus, by configuration, the supply chain becomes weak, with frequent breaks.

After the produce leaves the farm gate, processing and food manufacturing become the centrepieces of the supply chain network. There are a number of logistics and supply chain challenges that have to be overcome. The first and foremost issue from the supply chain profits perspective is whether there are fair price mechanisms and distribution management functioning in this sector. In a well-orchestrated supply chain network, firms and participating role agents are rewarded based on activities and performance, and there is a fair distribution of profits. There is great transparency, as all players are in partnership mode. However, in agriculture, individuals and disaggregated farmers are unable to keep current with the markets, and thus are often exploited by the intermediaries.

The growth of organized corporate involvement in the fast moving consumer goods (FMCG) and the retail sectors has helped the farming community. Corporates such as ITC (a subsidiary of British Tobacco) have started engaging at the farm level, effecting better dissemination of information on market trends and reducing the levels of intermediaries through *e-choupals*, which are discussed later in Exhibit 1.1. Over the coming years, this situation is likely to improve efficiency in this part of the supply chain.

The next stage in the supply chain network is the conversion of raw products to forms that are more convenient to consumers. This stage would involve processing, grading, and even converting to products for consumption. Examples include the making of sugar from sugarcane, dairy products from milk, a host of beverages for adults and children, and jams and juices with improved preservation, and so on. The challenge here is procurement dynamics, susceptible to seasonality, and the carrying forward of inventory, given regulatory conditions in India.

The following lists provide quick summaries of the input, production, and processing subsectors.

The input subsector includes

- seed, nutrients, credit, equipment, fuel and chemicals, all of which are needed for operation (i.e., production)
- level of inputs and challenges

- management of the physical flow, as in the case of improved seeds
- management of financial flows
- roles of intermediaries
- efficiency in terms of cost, availability on time and trade-offs
- other input change; use of energy
- due to efficiency, energy usage has not changed much on farms, just the type of input
- relatively few input businesses compared to production or processing (look at feed manufacturing vs. the number of farms!).

The production subsector includes

- the challenge posed by the average size of farms in India
- the trend towards increased specialization of production, taking advantage of new technologies
  - genetically improved growth/survival
- the involvement of large corporations
- increased production efficiency brought about by specialization (increased production in the face of decreased or constant levels of input)
- production economics
  - production costs increase every year due to increased input costs
  - but costs of inputs are not related to commodity prices; when commodity prices drop, gross farm income falls, but the amount spent on inputs does not.

The processing-manufacturing subsector includes

- all businesses that turn raw materials into finished (or partially finished) products
- aquaculture-related activities, largely accomplished by processors/processing plants
- those who package, distribute, and sell at times, places and in forms desired by consumers (the marketing "lot")
- distribution, which represents a major share of the total amount spent by consumers on food.

1.4.4 Corporate and Organized Business

In India, agriculture and agribusiness are highly regulated in terms of procurement price, minimum guaranteed output price, distribution control under The Essential Commodities Act, land holding, and so on. Corporate involvement and organized approaches to agribusiness face challenges, although a few entities have successfully established their credentials.

Exhibit 1.1 Highlights of the Success of ITC Limited in India in Food and Agribusiness

ITC was incorporated on August 24, 1910, under the name Imperial Tobacco Company of India Limited. In 1970, the Company's ownership was made largely Indian, and the name of the company was changed from Imperial Tobacco Company of India Limited to India Tobacco Company Limited and then to I.T.C. Limited in 1974. The company holds a multi-business portfolio encompassing a wide range of businesses—cigarettes and tobacco, hotels, information technology, packaging, paperboards and specialty papers, agribusiness, foods, lifestyle retailing, education and stationery, and personal care. Since 2001, the company has been called ITC Limited.

In its first six decades of existence, the company's activities were primarily devoted to the growth and consolidation of the cigarettes and leaf tobacco businesses. This was based on intense relationships with tobacco farmers. The abilities to procure, process, and market cigarettes were the key core competencies required. During the 1970s, the company witnessed the beginnings of a corporate transformation, maybe because of an ability to foresee opportunities it could exploit, and the limitations of tobacco-based businesses.

In 1979, ITC entered the paperboards business by promoting ITC Bhadrachalam Paperboards Limited, which today has become the market leader in India. Bhadrachalam Paperboards amalgamated with the company effective March 13, 2002, and became a division of the company, Bhadrachalam Paperboards Division. This business also manages natural resources and works closely with local communities.

In 1990, ITC set up the Agri Business Division for export of agri-commodities by way of leveraging its agri-sourcing competency. According to company sources, this division was one of India's largest exporters in 2011. ITC's unique initiative in agribusiness was setting up the *e-choupal* initiative in 2000. This started with soya farmers in Madhya Pradesh; steps were taken to cut out intermediaries and facilitate improved realizations for farmers. *e-choupal* has been conceived as a more efficient supply chain aimed at delivering sustained value to customers around the world. The *e-choupal* leverages information technology (IT) to virtually cluster all the value chain participants. Thus, it delivers the same benefits as vertical integration.

The *e-choupal* is a blend of click and mortar capabilities. The village-level internet kiosks are managed by farmers—called *sanchalaks*. This enables the agricultural community to access ready information on the weather and market prices in its local language, and to benefit from knowledge on scientific farm practices and risk management, facilitating the sale of farm inputs and the purchase of farm produce from the farmers' doorsteps.

Real-time information and customized knowledge provided by the *e-choupal* enhance the ability of farmers to take decisions and align their farm output with market demand. The farmers benefit through enhanced farm productivity and higher farm gate prices, whereas ITC benefits from the lower net cost of procurement, having eliminated costs in the supply chain that do not add value.

ITC Limited has extended *e-choupal* to 10 states, covering over 4 million farmers. ITC's first rural mall, christened "Choupal Saagar," was inaugurated in August 2004 at Sehore, Madhya Pradesh. On the rural retail front, 24 "Choupal Saagars" are operational in the three states of Madhya Pradesh, Maharashtra, and Uttar Pradesh.

ITC Limited entered into the food processing business by blending multiple internal competencies to create a new driver of business growth. It began in August 2001 with the introduction of ready-to-eat Indian gourmet dishes. In 2002, ITC entered the confectionery and staples segments with the launch of the brands mint-o, Candyman confectionery and Aashirvaad atta (wheat flour). In 2003, the company entered the biscuits segment, which was highly competitive. ITC Limited has managed a high degree of market penetration with its Sunfeast brand of biscuits. In 2007, ITC entered the fast-growing branded snacks category with its brand Bingo! In eight years, ITC's foods processing business has grown to a significant size, featuring over 200 differentiated products under six distinctive brands, with an enviable distribution reach, a rapidly growing market share, and a solid market standing.

Source: ITC Limited, Our profile. 2012.

It can be seen here that the ability of corporations to be innovative and flexible is critical for growth and success. Also, we may infer from the ITC Limited example that establishing conglomerates and a multi-divisional diversified business would also help in a big way. Hindustan Unilever Limited (HUL) also has a similar track record. Many large companies in other sectors have divisions in the agribusiness along with their main line of activities. An example is the Jubilant Group, which primarily had a base in organic chemicals and pharmaceuticals and later entered the ready-to-eat-food segment. Household-name brands such as Amul, which is focused on dairy products, have grown horizontally across India integrating a large number of players under the umbrella of the Gujarat Cooperative Milk Marketing Federation Limited (GCMMF). A company such as Marico has stayed focused on its core area of operation involving branded oils and spices, and has become a market leader. Similarly, a brand such as Mother Dairy has been successful in the dairy and fruits and vegetables sectors in a focused geography. There are numerous examples of such businesses that have established themselves as lead players under challenging circumstances.

1.5 Role of Supply Chain

As in any supply chain network, SCM in agribusiness involves the strategic management of a network of relationships by a focal organization, aimed at catering to the demands of its ultimate customers, and in the process achieving efficiency and competitive advantage through its role agents and their individual and joint discharge of activities.

1.5.1 Players and Their Activities

As depicted in Figure 1.3, the following are the different players and their activities in the agribusiness supply chain.

1. *Resource owners:* Resource owners are the trigger point in any supply chain. Most of the resources are naturally endowed and need to be subjected to a production process. In an agrarian economy, resources owners need not necessarily engage in production, but can lease their resources for third parties to produce. For example, owners may lease their land for others to raise crops. Furthermore, satellite farming is increasingly being practised in India where corporates are involved. For example, a sugar manufacturing company may have in its command area registered plots of farmers—either their own or leased—where crops are raised under the management guidance of the factory. The farmers harvest and sell the produce to the factory. Thus, resource management is a key aspect of the agribusiness supply chain. There are situations where seed producers lease farmers' land for raising and harvesting their crop, and sell their produce for high profits. Farmers lend their resource in such cases.

2. *Input providers:* The agricultural production function requires a number of inputs, namely, seeds, fertilizers, pesticides, irrigation and equipment, capital equipment such as tractors, threshers, and so on. Labour is another critical

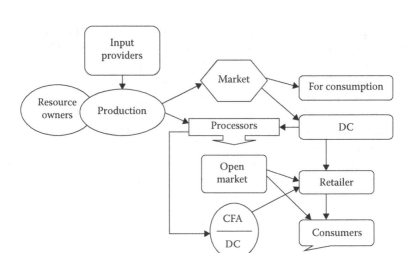

Figure 1.3   The agribusiness supply chain. DC, Distribution channel; CFA, carrying and forwarding agent.

input. The efficiency with which the input market functions would impinge on the efficiency of the agribusiness supply chain. A disconnect between input providers and their value creation could adversely affect the supply chain network, not only in terms of price efficiency but also in terms of production efficiency. Timely availability of improved seeds, fertilizers, and pesticides is important for efficient raising of crops. Similarly, the capital equipment market and its evolvements, such as rentals and leased services, are also vital for land preparation and harvesting. Thus, input providers play an important role in the agribusiness supply chain.

3. *Production:* Resources and inputs are deployed for harvesting agricultural produce such as cereals, pulses, cash crops, spices, and oil seeds. This is an important stage in agribusiness, as many of the products are moved to markets as commodities for direct consumption, or are procured by processors for value addition and conversion into food products. Production is defined by resource endowment and factor conditions, including agro-climatic conditions. Production risk is high in India because of its dependency on the monsoons, with all their vagaries. As demand is more stable and predictable except during times of volatility, it is the supply side, as measured by the volume of output, which influences the market for exchange.

4. *Market for direct consumption:* There is a huge market for direct consumption of agricultural produce as food. This is mainly with respect to products such as cereals, pulses and other food grains, fruits and vegetables, salt crystals (especially among low-income groups in India), and allied sector products such as dairy and marine products. We may have to understand the dynamics of direct

consumption and the channels through which direct consumption products reach the ultimate customers. After keeping a small amount of produce for their own consumption, farmers sell their output through the market, or to those who come for direct procurement. The government has established regulated markets for agricultural products. This helps farmers to sell in markets where there is a fair play of market forces. An alternative channel available to them is direct procurement agents, who act as intermediaries between the farmers and the direct consumption markets. Farm produce are also traded through *taluk*-wise agriculture producers' cooperative societies, which act as intermediaries and make sure that all farmers get the right prices, and that payment is made within defined times (examples of such produce include cotton, groundnut, and turmeric). Whether through intermediaries or direct markets, the produce moves through wholesalers and retailers before being bought by the ultimate customers. The ultimate consumers could be individuals or institutional buyers who serve individual consumers, such as large hotel chains and canteens.

5. *Processors:* Processors are of paramount importance in agribusinesses. Much of the agricultural produce gets converted into food and other products as it reaches the final consumers. For example, India is the leading producer of sugarcane. Yet, only less than 1% of sugarcane is directly consumed. Almost all of the sugarcane grown is processed as plantation white sugar, *gur*, and *khandasari*, which are common sweeteners in India. Similarly, cotton, jute, and edible oil seeds need to be processed before reaching forms and shapes that are of value to consumers. Even food grains go through some processing, such as the conversion of wheat into flour for direct consumption and for the manufacture of products such as biscuits. Direct consumption items are also branded, and increasingly, consumers prefer small lot sizes over bulk buying, due to the conveniences of modern living. Processors' business dynamics are influenced by the margins between input prices and output prices, conversion costs and margins on transactions, and value additions. Processors play a vital role in the supply chain network.

6. *Open market:* The open market has a key role to play in distribution channels. As mentioned in Fig. 1.3 and "market for direct consumption", produce for direct consumption moves through open markets. Similarly, processed agricultural goods also get sold through open markets, and directly from processors. For example, products such as edible oil, coffee, tea, and spices are widely traded in open market operations. Not only are physical flows enabled, but also the existence of commodities markets, where spots and futures and other options are available. Open market operations play a vital role in a supply chain, defining its strength and the sharing of profits across the network.

7. *Retailers:* Retailers number in millions across the country, enabling the movement of products in the agribusiness supply chain. Traditionally, India has had the "Mom and Pop" store format of retail, and organized retail is just picking up. Although organized retail would benefit supply chain efficiency,

local governments are protective of small traders, and hence, additional inter-mediaries are being encouraged. It is likely that with increasing experience of organized retail served by domestic firms, and with improvement in overall standards with technology breakthroughs, retail formats are likely to become more efficient.

8. *Consumers:* Consumers are the final point at one end of the spectrum of a supply chain network. In India, agribusiness is regulated with consumers' interests being protected, as the socio-economic structure calls for such regu-latory measures. Consumers as a group exercise a lot of competitive pres-sure. There are certain areas where farmers' interests and consumers' interests are directly pitched against each other, resulting in competitive pressure and stress for policy makers.

9. *Financial agents and impact on physical flows:* In India, a majority of farmers are weak due to poor holding sizes. The ability to raise funding for operations from the organized financial system is limited. Hence, unorganized money-lenders take advantage and set low procurement prices, exploiting the plight of farmers. Due to poor information, poor knowledge, and lack of drive to break out from this cobweb, farmers are affected, and supply chain profits are usurped by the middlemen. This is the classic weakness in the system.

In early 2002, the government set up nation-wide multi-commodity exchanges, using modern practices for clearing and other activities. The system is yet to mature fully, although it has attracted huge volumes of trade. The ability to go beyond selective cash crops and the scope to regulate effectively on payments and deliveries will help improve financial and physical flows, and thereby, supply chain benefits will flow across the system. At this point, it is important to note that synchroniza-tion of agribusiness supply chains to enable distribution of supply chain profits among role agents and actors based on their time, effort, and risk assumption is vital. This would require effective management of financial, informational, and physical flows within, out of, and into supply chain networks.

1.5.2 Needs of Producers and Consumers

Producers, processors, and intermediaries work in silos to maximize their profits. Most of the time, producers are exploited by intermediaries in inverse proportion to their ability to reach the markets. Producers who have committed to resources look at long-term profits and the ability to recover a decent return on capital. Unfortunately, in the prevailing situation, this group carries most of the produc-tion risks in the supply chains.

As we can observe from Figure 1.4 on the conflicting needs of producers and consumers, consumers demand a higher level of satisfaction on a limited budget, and hence, demand favourable prices. Furthermore, they buy very small quantities of many products. Hence, we observe conflicts between consumers and producers

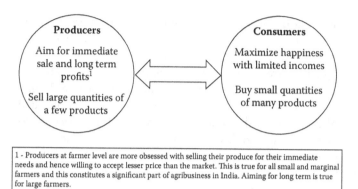

Figure 1.4   The conflicting needs of producers and consumers.

that complicate supply chain networks, which are ideally needed to synchronize the chain and make it relationship and partnership driven. Setting up an efficient, equitable agribusiness supply chain can be a challenge of the highest order.

1.5.3   Four Utilities

There are four forms of utility, namely, form, place, time, and possession, which would gain/deliver value as produce moves from the producers to the ultimate customers of the focal firm. These are described in the following sections.

1.5.3.1   Form

This relates to processing of the products into forms desired or needed by consumers. It could be fish in the form of rounds or nuggets. There are a host of products made out of each agricultural produce, and the forms these products take are decided by demand and by their ability to create new demand. For example, the Indian snacks market was less developed a decade back, and there were very few chips brands. Today, the market has a host of brands, with multinationals leading the field. The same is true with respect to the ready-to-eat-food segment and a number of processed food products. The ability to create form utility and newer markets and products improves value realized, but complicates the supply chain network, as every activity involved, from demand estimate to last mile reach, becomes a challenge.

1.5.3.2   Place

This relates to transportation of the product to locations desired by the consumers. It can also refer to movement of goods for value addition during intermediary processing. For example, tea is produced in the north-eastern and southern states of India. But a large quantity of tea is consumed in the western and northern states. Hence, tea needs to be transported in the most cost-efficient manner, with quality being maintained in the process. Place utility plays a major role as demand centres and consumption centres are two different concepts. Baskin-Robbins, a leading ice cream brand in India, has a centralized manufacturing plant at Pune and distributes its products across India by outsourcing to a cold chain operated by a third party, namely, Snowman, the cold chain company. Thus, place utility is accomplished through this method. We can consider a number of such examples. Onion is produced largely in Maharashtra, but consumed across the length and breadth of India. Transportation facilitates place utility for customers.

1.5.3.3 Time

This relates to storing of raw material, products in process, and end products until they are needed, that is, time utility refers to the matching of demand and supply in terms of any time lag between their occurrences. For example, milk is produced in plenty during the flush season, between October and March, but demand is low during this period. Milk demand peaks during April to September, which is a lean period for milk production. Milk being a perishable product, the dairy will have to separate butter oil (fat) and other milk derivatives (milk solids other than fats) and store. During the lean season, when demand is high, it is possible to recombine milk derivatives along with fresh milk to enhance supply volumes. Thus, time value is created. Similarly, mango is procured during the summer months in India, namely, April to July, but the juice concentrate is consumed in the form of beverages throughout the year. Thus, time utility plays an important role in overcoming issues associated with seasonality in production where demand exists throughout the year.

1.5.3.4 Possession

This relates to the gaining of ownership or the right to use a resource for production, or a product for consumption. For example, land resource owners need not necessarily be producers. Many times, they lease out their lands to producers who raise crops. Similarly, in the case of processing, common facilities are created through cooperative organizations, and processing could happen through sharing of fixed resources. More commonly, this is seen in storage, where common facilities are shared. In all these cases, the right to use is what provides possession utility, which is important from the supply chain perspective.

1.6 Supply Chain Drivers

Supply chain drivers are key to the performance of supply chain constituents and role players across the network. A network is operational only when supply chain drivers are in place. The drivers are components of the supply chain structure that lead to effectiveness of the supply chain. In a way, supply chain drivers are to be viewed as operating tools for implementing supply chain strategies and carrying out operations. Drivers act as pillars of the supply chain on which blocks are laid and operations dwell. Drivers could be logistical drivers such as warehouse facilities, inventory, and transportation, or cross-functional drivers such as pricing, sourcing, and information. Apart from these drivers, certain external factors such as regulatory systems, international agreements, tax systems, and infrastructure may impact the performance of supply chains indirectly, through configuration of the drivers. Hence, it is important to understand the role of drivers in supply chain performance, their components, and how they influence and impact competition.

Supply chain drivers ensure a balance between responsiveness to customers and efficiency in the supply chain, which enables a company to be competitive in its chosen strategic arena. It may be noted that there is a need for synchronization of the supply chain strategy with functional strategies such as product, marketing, and finance and with competitive strategies. The supply chain structure defines the strategic focus for the supply chain that needs to be accomplished through effective deployment of supply chain drivers. It may be important here to define these drivers in terms of facilities, inventory, transportation, sourcing, pricing, and information. Then each of the drivers would be taken up and each of its components and their importance in the supply chain would be analysed.

1.6.1 Facilities

Facilities in a supply chain refer to the physical location of the nodal organization or its partners in the supply chain; the locations where a product or service is being fabricated, assembled, produced, processed, or stored. Theorists developed propositions based on weight-losing material and volume-based raw material tending to be located near raw material sources. These are products such as sugarcane, cotton, edible oil seeds, and many other agro-based goods, which are located near the sources of supply of the raw material. FMCG such as staple food, biscuits, and other similar products have a normal tendency to be located near markets. They engage contract manufacturers to serve the market.

The role of production facilities in a supply chain is derived from the supply chain strategy on whether to achieve responsiveness for a given level of efficiency or to achieve efficiency at a desired level of responsiveness.

Facilities are of two types as shown in Figure 1.5.

The set-up for a warehousing facility varies from industry to industry and across firms. It also changes over time, as can be seen in modern times, where corporates follow just-in-time and vendor-managed inventory practices and use third-party logistics services. However, in agribusiness, the main purpose is to manage demand and supply gaps across time and regions. Furthermore, temperature control is also another important factor in warehouse decisions.

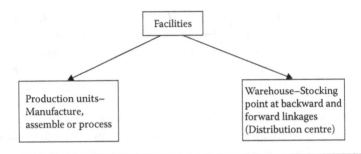

Figure 1.5   Types of facilities.

The importance of facilities in supply chains varies depending on the industry structure, industry practices, nature of goods, buyer characteristics, and the strategic intent of the focal organization. "These are the typical competitive factors that may influence the role and importance of the supply chain drivers" (Chandrasekaran, 2010). With respect to facilities, a number of competitive forces may impact the choice of location. Here, the forces and approaches are discussed in the indicative list below.

1. *Seasonal supply:* When production is to be based on seasonal inputs, the process capacity needs to be high and the ability to carry large stocks of near-finished goods is important. Edible oil seeds are available during certain months of the year as a post-harvest crop, and production peaks immediately after the harvest. Edible oil can be easily carried over as stock in hand to cater to the demand over the different months of the year.
2. *Bulkiness:* Wherever the material is of a bulk nature, it is convenient and efficient to locate processes closer to the farm gates than to the consumption points. For example, the milling of paddy gives about 80% rice, and hence, processing happens closer to the resource origins. Similarly, in the case of fruits and vegetables that are processed, the processing plants are again close to the production centre because of bulkiness. Oilseeds like groundnut are also processed near the procurement centres, because of their weight-losing nature.
3. *Perishability:* Marine products, dairy products, and sugarcane are highly perishable and extraction value is at a peak only when they are processed immediately. Sugarcane is both bulky and perishable in nature, so any facility location strategy must consider such parameters.
4. *Input industry location considerations:* Input factors such as fertilizers, say ammonia, are located closer to the port where raw material (naphtha) is imported or near plants from where raw material is procured. There are other input factors such as capital equipment, where economies of scale are decided by volume as they are assembly plants. Inputs such as seeds, pesticides, and so on are a trade-off between source and demand centres.

Apart from the above competitive factors, competitive structures influence facility location. In the case of agriculture, the market is imperfect, with many suppliers and buyers. A small number of firms that are in the branded market, with focus on selling processed goods and establishing branding in a commodity market, operate in oligopolistic conditions. Imperfect conditions with a large number of small processors may have inefficiencies, but the structure is difficult to alter for the better. The large players are few in number, and operate under oligopolistic conditions, where product and market differentiation are key. Facility decisions would also be determined based on these factors.

Key facility decisions, whether it is a manufacturing unit or a warehousing facility, would include location, capacity sizing, and operations methodology. As explained above, location decisions are driven by a host of factors such as competitive conditions, which include rival firms' behaviours, resource availability, buyer behaviour, and entry and exit conditions. Many of these are strategic in nature. The fundamental trade-off is between cost efficiency and responsiveness to customers in most of the situations.

> Capacity sizing decisions would be driven by both strategic and planning level decisions. At the time of capacity creation, it would be more strategic, looking at the long-term demand and supply conditions, plant cost and the economies of scale in production and other managerial functions. But once capacity is created, it is the near-term business cycle that influences capacity utilization, depending on demand and supply conditions. There is a need to have flexibility built into the capacity created, to enable adjustments to changing market scenarios. (Chandrasekaran, 2010)

Production choices and manufacturing and process designs are again driven by many competitive and technological factors. They are driven by the level of maturity of the technology deployed, the degree of associated capital that provides flexibility, the ability to handle flexibility, and its impact on operating cost efficiency. Changes in the snack foods market, sugar manufacturing along with co-generation of power, and increasing acceptance of refined edible oil are some of the examples of intensified competitive conditions impacted by choice of technology and product features in India.

It is interesting to note here that there have been developments in using mobile processing facilities at the farm level in India. These developments have happened through learning from developed countries and customizing to local requirements. Small and medium farmers can optimize returns by using such technology for value enhancement. For example, there is a company that markets mobile seed cleaning and grading plants, which help to process seeds directly in the fields. "The machine is capable of cleaning and grading of all cereals, pulses, oil seeds, spices, forestry seeds, vegetable seeds, fodder crop seeds etc., including wheat, maize, barley, millet, gram, soya, sunflower, mustard and jute" (http://www.agrosaw.com/smp.html).

In the case of warehouses and distribution centres, facility designs relate to location, the number of warehouses and the level of investment in the warehouses. The trade-off here is decided by responsiveness, which leads to decentralization and increased number of warehouses compared with centralization of facilities with one large warehouse, which focuses on cost efficiency.

Overall, facility decisions depend on a number of competitive factors, and from a supply chain perspective, the choice depends on which matters more: frequency and level of service to customers or cost of providing the same. Resource endowment and its influence on supply chain decisions in the case of farm output to direct markets and intermediary processes and market sectors of agribusiness are key decision areas.

1.6.2 Transportation

Transportation plays a key role in the SCM of a nodal organization, as the movement of goods across supply chains to ultimate customers is an important value-creating activity. From the available array of transportation options, the nodal organization opts for a specific choice at every stage based on cost efficiency and responsiveness as the supply chain focus. When a firm operates on a supply chain focus of efficiency, then it chooses low-cost transportation, which may not necessarily be responsive.

On the other hand, when a firm operates on responsiveness as the supply chain focus, then speed of delivery would be the criterion, and cost is given lesser importance, as customers would be able to absorb the incremental cost for the service levels desired. The choice of transportation is driven not only by cost and responsiveness but also by a host of other factors within the framework of decision making. For example, highly perishable products are moved through air cargo, which would be expensive. Thus, responsiveness could be due to factors such as the criticality, the value, or the perishable character of the product.

> For a supply chain manager, the ability to balance the transportation cost with responsiveness would become important. Also, the spread of customer groups to be served and the efficiency of operations could be influencing factors in transportation decisions. It may also be noted that in a supply chain situation, the transportation decision is influenced by the location of facilities, inventory and information efficiency, like in the case of a national brand of processed agricultural produce such as wheat flour and edible oils. (Chandrasekaran, 2010)

The effectiveness of transportation as a driver depends upon the competitiveness forces which are operating in the industry in which firm is into. They are briefly discussed here.

Cost leadership: When a firm adopts cost leadership as a strategy, transportation decisions may involve working out the most cost-effective options. More importantly, transportation can facilitate cost leadership through structuring of the network for effective planning of routes and schedules of vehicles, for both procurement and distribution. For example, effective procurement management of inbound milk collection in the dairy business achieves cost leadership through transportation. Similarly, on outbound transportation, managing multiple chamber temperature-controlled vehicles for distribution could be an issue for a firm handling marketing of ice creams, dairy products such as flavoured milks and chocolates, and dry products such as biscuits.

Differentiation: When a firm adopts product and market differentiation as a strategy, transport decisions impact the choice of strategy. For example, when a plant decides to be a national brand or even a global brand in selling processed agricultural produce such as tea, transportation decisions play a vital role, as mode selection and assured placement of vehicles are critical for sustaining brand share.

Focus: When a firm adopts a focus strategy, transportation decisions can play a vital role in achieving competitive advantage. Let us take an agribusiness input industry such as fertilizer. The firm focuses on selling fertilizers of different nutrient values based on agro-climatic conditions. Consumption patterns and timings could be variants, and the firm needs to structure its distribution plan by designing transportation based on local market conditions.

Alliance and co-options: Alliances and co-options among firms could be an important transportation strategy for a firm, especially when it moves bulk material across long routes. Furthermore, when there is a need for deploying dedicated vehicles such as temperature-controlled and containerized trucks, backhaul support through co-option among user non-competitive companies can provide a competitive advantage. In the case of containerized trucks, alliances with truck operators for deducted trucking again provide a competitive advantage.

Apart from the firm level and strategic business unit or product group-wise transportation strategy deployed for supply chain efficiency and effectiveness, industry-led practices can also facilitate improvements. For example, transport operators in India have strong associations that constantly work on improvements for goods carriage. Furthermore, they have effective methods of exchanging information in order to improve asset utilization and overall economy of business, thus contributing to cost efficiency in transportation.

Similarly, there are regional and district-level operators' associations that procure orders for movement from major industries and distribute demand and supply to streamline operations. Such initiatives over the years are bound to strengthen practices, and a number of competitive advantages are likely to come up. For example, for inbound transportation of sugarcane, which is an intense seasonal activity, transportation cannot be resolved without the involvement of local associations. In the same way, manufacturers, especially those who move products in specialized vehicles such as temperature-controlled and containerized trucks, promote good practices through co-operation among themselves, enabling better utilization and investment in assets and bringing about overall improvements in the efficiency and effectiveness of all players in the supply chain network.

It may be noted that the key transportation decisions that a supply manager may look into would include selection of the transportation modes and the design of a transportation network. Generally, there are a number of options available, involving alternative modes of transport, for movement of cargo. However, the choice is determined by the cost of transportation and the ability of the network to pass it on to the customer.

A regular raw material item, which is an input in manufacturing, would have a standard flow, and its transportation mode would be fixed on the basis of cost and flow time factors. For example, in the case of edible oil refining processes where the brand owner buys material from the open market and processes, brands, and markets its products, production is defined by a production plan and schedule, and the arrival pattern is defined based on a given stocking plan. Transportation is defined by the bulk of material and distances involved, as also on the options available for the nodal organization. In this case, network selection would also be involved, as material may come from multiple points or vendors, and aggregation through routes is possible. On the other hand, where decisions involve issues such as supplementing inputs with import of raw material from abroad, the cost of imports and transportation, including handling, would decide the selection of the final option.

Thus, transportation as a driver plays a critical role in supply chain network decisions, and is an important competitive factor that could influence the business success of the nodal organization as well as the participating players. Transportation as a driver is key to structuring inbound and outbound transportation options for farms outputting to both direct markets and intermediate processors. Disaggregation, cost management, and network design factors are some of the managerial issues that would need focus.

1.6.3 Inventory

Inventory plays a significant role not only in the supply chain but also in competitive strategies. A company's competitive leadership could be impacted positively in certain industries because of effective inventory management practices. If responsiveness is the supply chain strategy, the competitive situation must be one where the customer is keen on product variability and quick response rather than cost. Leadership here would depend on the ability to accomplish this with the highest degree of responsiveness, and the ability to be perfect or near perfect in fulfilment of orders. For example, in the marketing of processed agricultural outputs such as juices, inventories across the chain, ensuring freshness, would decide the success of the brand.

Inventory as a force in competition is also influenced by business cycles and by the need of a firm to adjust inventory, facility management, and production in response to movements in business cycles. For example, during the flush production season perishable products would be handled under an inventory policy of liquidating stocks and implementing an appropriate inventory carrying policy.

Inventory management policies are interwoven with stages of the product life cycle. During the introduction stage, product availability is the key factor in capturing market share, and the firm would focus on responsiveness, not on cost. The high level of demand uncertainty of the introduction stage does not lend itself to effective demand planning. At this stage, the firm is far from finalizing its supply chain strategy, as customer acceptance and growth plans are yet to be crystallized. In such conditions, holding high levels of inventory in the emerging product supply chain network assures product availability. Failing this, a firm not only incurs lost sales opportunities but also loses a valuable intangible asset, namely, its image in the market. Therefore, overstocking is the preferred tactic.

During the growth stage, inventory becomes a less important factor in ensuring responsiveness than at the introduction stage. As demand uncertainty decreases, the accuracy of demand forecasting improves, which in turn helps assure product availability by improving demand planning. The approach could still be responsive, as the firm may not like to lose sales opportunities during the growth phase, and this would be critical for building up the brand image. The competitive strategy would be to augment market share, and cost may not be the focus. Because it is the growth phase, the ability to pass on cost would also be available.

At the product maturity stage, efficiency is the dominant supply chain strategy. At this stage, the product becomes standardized, and growth rate is stable and predictable. A firm would focus on economies of scale to lower unit cost. In competitive situations, inventory becomes a cost factor in the profit and loss account and a liability in the balance sheet. The focus would be to reduce inventory to an optimal level, given demand predictability. Production focus would also move on to increasing commonality among products in a group and achieving a configuration close to the demand point so that the value of the inventory held is low.

During the decline stage, falling sales volumes and product prices are the primary features. A firm would focus on liquidating stocks in order to reduce losses due to inventory obsolescence. Cost would be the main driver in such conditions. This situation need not necessarily apply to a strategic business unit or a firm in a declining phase. It could also be applicable to a product in a product category or group.

> Thus, inventory management becomes crucial, and is related to a firm's strategic intent and the competitive situation in which it operates. Apart from product lifecycle, many industry factors such as stage of industry and competitive responses from stakeholders may also influence inventory policies, which need to be understood by the supply chain manager. (Chandrasekaran, 2010)

Inventory could include raw material, intermediary or in-process goods, and finished goods. These forms of inventory decide every stage of the supply chain, and how it intertwines with its role players and creates value for the ultimate customers. Further, such categories of inventory form the basis for inventory decisions relating to order phases, uncertainty handling, and seasonality factors. (Chandrasekaran, 2010)

Goods that are used as inputs in a process are classified as raw material. They form a part of the total value in goods produced in a supply chain at every stage of the supply chain as it progresses towards the ultimate customers. For a sugar manufacturer in India, the raw material is cane and the output is plantation white sugar, whereas in Brazil. In India, beet is used to produce raw sugar or refined sugar. In the refinery process, raw sugar becomes the input. In the manufacture of chocolates or syrups, sugar or refined sugar could be the input for different product category outputs. It is important to understand the different classifications of inventory in any production process, and their criticality in designing inventory holding policy. Although it is a rudimentary concept, it is basic process mapping and plays a vital role in policy making.

To decide on the balance between efficiency and responsiveness, one may have to understand cycle inventory, seasonal inventory, and safety inventory. Cycle inventory is the average amount of inventory used to satisfy demand between receipts of shipments. The choice is between a large inventory with a longer cycle time and less inventory with frequent offering, thereby reducing average lot size. Seasonal inventory refers to the beefing up of or slashing of inventory for peak and slack inventory seasons, respectively, based on experience and predictability. Seasonality could be due to demand or supply or both factors. The challenge for management would be to decide on the quantum of seasonal inventory based on the efficiency level for a desired level of responsiveness.

Safety inventory is basically built up to counter uncertainty in demand. If demand is certain, managerial decisions on inventory are handled rather firmly, based on experience and customer research. However, sudden factors can unexpectedly affect demand. Firms may not like the opportunity to sell and corner market share during such spurts in uneven demands. Any firm will have to carefully balance between the loss of likely sales for want of stocks and excess stocks due to overestimation and attempting to be responsive.

It may be noted that inventory plays a key role in the supply chain and in the competitive position of firm in its industry, irrespective of the form of inventory and the value creation arising from the utility it brings to all players across the supply chain network. The overall objective of balancing responsiveness and cost in terms of efficiency has to be achieved. Overcoming the challenges of managing stock and making the right managerial decisions with respect to stock and releases are key to success in the competitive agribusiness market. (Chandrasekaran, 2010)

1.6.4 Sourcing

Sourcing has a key role in the supply chain, and would be considered as one of the drivers in many businesses, especially in agribusiness systems. Sourcing refers to a number of strategic activities that go into supplier relationship management right from selecting a vendor, engineering and configuring products along with vendors, to making strategic investments for committing with the vendor as a competitive differentiator to being a single vendor who has a proprietary commitment on resource and technology.

In the supply chain, sourcing becomes important, as it is the set of business processes required to purchase goods and services to make products for the ultimate customers. Supplier development and synchronization for production, product development, and cost management are going to be challenges. If we consider a manufacturer of capital machinery, say tractors, the demand for tractors for new capacity creation, expansion, and replacement is a derived demand. New capacity creation and expansion especially depend on the performances of agriculture and the monsoon. Replacements could be a function of the age profiles of the machinery deployed, technology advancement, and the industry cycle. However, tractor manufacturers will have to keep their suppliers engaged as per the requirements of their business. During a recessionary trend, derived demand may not be high. But critical component suppliers have to be taken into confidence and kept engaged. A supplier at this stage would expect transparency and commitment to get involved in future growth plans, when agriculture turns around. Hence, it may be noted that suppliers need to be categorized based on their positions with respect to bargaining power, criticality, and the value of items procured from them. Investing time, effort, and at times even financial inputs is important for effective SCM, as these would decide long-term cost efficiency and responsiveness.

The effectiveness of sourcing as a driver depends upon the competitive forces of the industry in which a firm is operation. They are briefly discussed here.

Cost leadership: Sourcing as a driver in supply chain focus may have conflicting influences when it comes to cost leadership and responsiveness. Generally, economies in production would be facilitated to reduce costs of manufacture and at the same time, optimize sourcing efficiency by minimizing inefficiencies in buying, such as wastages due to quality, adjustment costs in input, number of defects, and so on. "Long term contracts in volume, with a commitment on purchase, are more price efficient" (Chandrasekaran, 2010). However, in agriculture, factors such as seasonality and perishability adversely influence sourcing strategies for cost advantage. Because production is lumped, sourcing also needs to be managed when output reaches markets, as farmers in India do not have holding size power. Cost effectiveness in sourcing is determined by negotiating power, size, and so on. Nevertheless, cost leadership can be achieved if appropriate procurement policies are in place.

In India, there is an informal market for futures and options and there are formal, officially regulated commodities markets operating, which helps in managing cost. When we mention informal markets, we are talking about a system wherein the intermediary, who is typically a moneylender funding agriculture operations, ties up the procurement of produce at a predetermined price. Invariably, this leads to exploitation of the suppliers and poor synchronization among the components of the agribusiness supply chain network.

To overcome this, the government has permitted futures and options in selected commodities, so that market operations determine the values of the produces and improve supply chain benefits. Organized players can gain cost leadership by participating in this market.

Differentiation: Although a firm addresses differentiation as a sourcing initiative, responsiveness and customer satisfaction play a major role. Differentiation could be in the market, the product, or both. Multiple stock keeping units and varied customer groups would be the normal trend while deploying differentiation. Flexibility in supply as per the demand of the nodal firm's production plan is the key. It may involve incremental costs due to low volume and change over. Customers could be in a position to absorb the costs. Sourcing strategies must enable the serving of different markets and product segments; for example, a dairy company must be able to sell in all regions and all through the year, although the production of milk may go through peaks and troughs. Similarly, the firm must be able sell differentiated products, as one market area may demand bulk milk, whereas another may demand dairy products, and so on.

Focus strategy: A firm may adopt a low-cost focus strategy for a large segment or a focus strategy with differentiation for a market or a combination of both for nation-wide territories. In such cases, balancing between efficiency and responsiveness is important for success. The low-cost focus would be suitable for volume-based businesses, where customers could be price sensitive and product groups could be the focus of the nodal organization. Sourcing options in such cases should focus on efficiency, and the low-cost mode is preferred. Many products in agribusiness, which serve staple consumption markets, would belong to this category. Alternatively, "differentiation to build customer loyalty with broad product groups with narrow segmentation is possible. This may serve responsiveness to customers and sourcing for product variability or speed of service would require a customized approach and increased cost" (Chandrasekaran, 2010). Examples of this market could be ice creams, health quotient-driven high-end edible oils, and so on.

Thus, sourcing forms play a key role in managing responsiveness and cost efficiency in any supply chain network, depending on the chosen focus of the nodal firm. Certain business sourcing could play a larger role as product development and the proprietary nature of resources and engineering may decide industry competition; suppliers may have strong bargaining power, compelling the nodal firm's supply chain manager to work in a partnering mode.

There are certain country-based factors that give some firms a competitive advantage as potential sourcing partners. These advantages bring value to the firms, and also drive industry to higher levels of maturity. Similarly, foreign firms look at India for sourcing agricultural produce in processed forms for competitive advantage. Typical produce items are tea, coffee, and spices.

Domestic firms in agribusiness will have to work closely with farmers and intermediaries to plan and utilize their production capacities. Although time-consuming, it is important to conduct negotiations and work closely with these supplier segments. It requires lot of maturity and time to work fruitfully on such relationships. There are a number of leading companies such as Amul, ITC, and HUL that have established well-refined processes and demonstrated success.

1.6.5 Pricing

An efficient supply chain moves the right product at the right time, the right place, the right quantity, and the right quality to the right customer, and most importantly, at the right price for everyone in the supply chain network. Hence, it is clear that price would be one of the drivers of the supply chain. With one's understanding of supply chain focus, in the trade-off between efficiency and responsiveness in a supply chain structure, price has connotations for both. An efficient supply chain is one where customers would prefer the best price at close to market efficient levels. Price would be one of the key factors in deciding demand. Although simple economics states that price is determined by demand and supply factors, in real life, business decides on supply under normal conditions keeping in view the demand and realizable price. Therefore, prices would necessarily be a factor in defining and configuring a supply chain.

However, when the supply chain manager focuses on responsiveness, where customer service is of paramount importance, it is believed that the customer would be able to pay a premium price, and price setting decisions would take this into consideration. "Supply chain network players reckon on this while defining and executing their activities, so that the incremental cost and margin could be borne by customers. In this case, price setting happens based on the level of customer service, and monopolistic practices such as price fixation, product differentiation and so on would be based on the realizable price" (Chandrasekaran, 2010). However, in agribusiness, farm output creates imperfect market conditions, where producers would not be able to exploit market conditions to their own benefit. In the case of the processed output segment, price is determined by supply conditions in terms of volume, demand factors, and sensitivities—for example, pressure built up in public forums and policy-making circles.

Thus, price would be one of the drivers of the supply chain network, whether the focus is on efficiency or responsiveness. The other drivers would influence and be influenced by the price factor.

We can consider the Indian biscuits market, which is worth more than ₹5000 crores, a large share of which comprises the unbranded and rural markets. Until 2003, Britannia and Parle were the two companies with brand names that dominated the branded biscuits market. For any company to be successful in this arena, it must be able to create a deep impact on the "bread-and-butter low margin, high volume, popular and low-priced small-pack glucose biscuit segment," apart from the high-margin urban segment for cream biscuits. A typical customer is a member of a low-income group/daily wage earner who looks at energy boosting bouts during breaks in work. The selling points here are price and availability. Brand loyalty and recall matter least when price and availability are ignored. In this market, because the product is standardized for a branded player, price is a key differentiator. The price, coupled with supply chain efficiency (especially in distribution management), is a competitive force that determines success.

> Another area which is of interest is the sale of perishables in supermarkets in India. The perishables segment, especially in the case of fruits and vegetables, is one of the segments that determine the success of a store. Perishables are sensitive to price and freshness factors. Price is actually a function of freshness. The challenge for the supply chain manager is to ensure that fresh stocks keep rolling over the counter at competitive prices. When a customer walks into a store and is unable to find perishables of the right quality and price, the customer moves on to the next store, which means the store manager loses a potentially large sale. Thus, to overcome this problem, stores work on daily low-price time windows, when they liquidate stocks of perishables at low price points, increasing footprints into the store during non-peak hours, boosting sales volumes and allowing replenishment with fresh stocks for peak hour sales, thereby improving on price realization. Thus, price plays a key role in the ability to stay competitive. We may note that this strategy of price discounting and liquidation is not limited to perishables alone, but is prevalent for many other commodities also. (Chandrasekaran, 2010)

The role of pricing as a supply chain driver is crucial in supply chain effectiveness, as every intermediary completes a transaction at a price that forms the cost for the next stage. The ultimate customer, for whom the focal organization works to provide value, pays for it. Competition and industry structure have a role in pricing. Traditionally, economists have proposed various theories on how price fixation happens in markets, and on how non-price competition can influence business. There are situations where non-price factors may attract customers. We must realize that all such factors are influenced and balanced by way of cost and price. We may have to delve into the fundamental concept of cost efficiency, which is primarily price sensitive and opt between functional product groups and responsiveness, or customer service-driven factors where customers are willing to pay premium prices for the desired levels of service. Hence, price is an aspect that a supply chain manager has to reckon with in order to achieve the overall business objective of synchronizing the contributions to the value chain of all the players linked to the nodal organization. Typical issues would be, given the government policy and incentives structure, is it a boon or bane for agribusiness efficiency for farm gate sector and for processing sectors? It looks like that the choices are between the ability to cushion margins by passing on inefficiency to consumers or the inability to manage margins and get stuck between farm gate and consumers through intermediaries.

1.6.6   Information

In today's business world, information and IT are paramount, and it needs a broad conception to encompass all the information that businesses create, and to use it. A wide spectrum of increasingly convergent and linked technologies helps to process information to collaborate, partner, and co-perform with entities and players across the supply chain, creating value for customers, suppliers, intermediaries, and the nodal organization. All such linkages are usually managed by the nodal organization. Earlier, companies were competing through IT as a differentiator. However, the evolution of technology and reduced costs have made it a threshold competence for an organization.

IT changes industry structures and the fundamentals of competition with the application of new generation technology and process speeds. This is true for the supply chain domain as well. Some examples in business are the application of technology platforms for business processes such as sourcing and the facilitation of buying and selling through cyberspace.

ITC's *e-choupal* clearly shows the competitive advantage that technology can bring to a company and its supply chain. IT can also create competitive advantages by giving companies new ways to outperform rivals in growth strategy and in the ability to create a new business model for existing business operations.

We could describe information as a driver along with the product life cycle of a focal organization. During the introduction stage, product-specific activities and transactions are limited to the focal firm, its closely located divisions, and a limited number of trading partners. The underlying business processes of the introduction stage tend to be less complex with respect to the number of entities. The focus here would be to ensure that the product meets customer acceptance and that the model is scalable. Hence, the range of IT capability requirements is limited to this ambit, which makes for a simple structure. Thus, during the introduction stage the focus would be on a reliable, cost-efficient information system to expedite an efficient supply chain, although the predominant supply chain focus would be responsiveness.

As sales and market share increase during the growth stage, the focal firm establishes more trading partners and suppliers with increased interaction across the network. Therefore, information becomes a critical driver, as it helps coordinate activities across the supply chain. At this stage, information systems such as Enterprise Resource Planning (ERP) and domain-based SCM systems are deployed to improve supply chain functioning.

The mature commodity product's supply chain may extend over a wide area in terms of geographical locations. Efficiency is the dominant supply chain strategy. A stable market share and mature relationships with trading partners facilitate the establishment of complex transaction systems across the supply chain. The supply chain becomes more complex and highly networked, with multiple tiers and partners. The reach requirements include the ability to interact over the entire supply chain network, including customers, irrespective of the technologies at specific network locations. The range requirements can be extensive, because the capability is needed to execute complex, demanding transactions. Investment in IT could have reached the maturity level, but needs to be responsive to market requirements based on technology obsolescence and the needs of the supply chain network.

During the decline stage, as the focal firm loses trading partners, the supply chain shrinks. The focus would be to be cost efficient, and it may not be possible to update technology and keep it contemporary, but what exists needs to be very effective.

> Apart from these, technology can be a key input in many strategic situations involving the supply chain. The fundamental trade-off is to prioritize between efficiency and responsiveness, which could be related to the push and pull strategy of the supply chain. Investment in technology is largely driven by budgets and ease of implementation, without the need for change of business process time and again. Though process improvements and changes are welcome, frequent interventions irk the partners, and learning time requirements need adjustment along the entire supply chain network. Hence, while deploying enabling technologies, a supply chain manager would consider such soft issues in terms of acceptability and ease of implementation. (Chandrasekaran, 2010)

Although there might have been scepticism about IT applications in India a decade back, a large number of firms have since revolutionized their operations and their industries with technology applications, as in the case of ITC, Marico, Britannia, and Amul (GCMMF). Proliferation in the use of handheld devices would further improve the pace of technology adoption.

1.7 External Factors

These drivers are influenced by certain external factors such as regulatory systems, international agreements, tax systems, and infrastructure. For example, the lack of cold storage availability at a reasonable cost is a limiting factor in India, and farmers with small- and medium-sized holdings expect government support for this facility. For many reasons, establishing and operating cold storages would not be the core competency of the government, and hence, the system suffers.

Regulatory systems, especially with respect to procurement price, output price ceiling, and interventions through export–import policies impact agribusiness. Agricultural input industries are heavily dependent on government support in the form of subsidies and duties exemptions. Price and distribution control also impact the input segment. Infrastructure issues can badly hurt production, especially when there are power shortages or withdrawals of subsidies to farmers.

1.8 Conclusion

To conclude, the agribusiness supply chain hinges on two criteria, namely, how well goods and services flow from businesses to consumers, which is referred to as efficiency, and how the SCM system meets the needs of consumers, which is referred to as responsiveness. Although supply chain decisions in manufacturing and services are driven by these two criteria through six drivers, agribusiness becomes complex because of two groups within the system, namely, the farm gate produce market and the processed goods market, and the play of a host of operators who tend to have divergent business objectives, with a narrow focus on short-term gains and decision unit level profit making. Farm gate produce markets are mostly dominated by traders, and are biased towards traders, not farmers. Consumers also do not benefit, and the inefficiency in the system is only helping intermediaries.

References

Chandrasekaran, N. *Supply Chain Management*, Oxford University Press, India, 2010.

Christopher, M. *Logistics and Supply Chain Management: Strategies for Reducing Cost and Improving Service*, Financial Times Pitman Publishing, London, 1998.

CSCMP Supply Chain Management. Available from: http://cscmp.org/aboutcscmp/definitions.asp

Engle, C. R. and Quagrainie, K. *Aquaculture Marketing Handbook*, Blackwell Publishing, London, 2006.

European Commission. India's role in world agriculture. *Monitoring Agri-Trade Policy* newsletter 03-07, December 2007.

ITC Limited, Our profile. 2012. http://www.itcportal.com/about-itc/itc-profile/itc-profile.aspx

Mobile Seed Cleaning Grading Plants. Available from: http://www.agrosaw.com/smp.html

Monczka, R., Trent, R., and Handfield, R. *Purchasing and Supply Chain Management* (2nd ed.), South-Western, Cincinnati, 2002.

Reserve Bank of India. *Handbook of Statistics on Indian Economy*. 2013. Available from: http://www.rbi.org.in/scripts/AnnualPublications.aspx?head=Handbook%20of%20 Statistics%20on%20Indian%20Economy

Chapter 2

Input Suppliers

OBJECTIVE

This chapter discusses each agri input in specific detail and importance of drivers in terms of their effectiveness. The discussion goes further to supply chain processes such as planning, sourcing, products/value addition, distribution and demand management. The chapter also addresses inbound material and services for farm level operations.

2.1 Introduction

The inputs for an agricultural produce are seeds, fertilizers, crop protection products, farm machinery and equipment, veterinary products, agri-biotech, and environment-friendly products. These inputs are important for producing agricultural output and allied sector products and commodities in the required quantity and at the desired quality. Agricultural inputs play a crucial role in determining yield levels, and in turn, in the augmentation of the levels of production in the long run. It is well appreciated that improvement in the yield depends on the application of the right technology, the use of quality seeds, fertilizers, pesticides and micronutrients, and the availability of adequate irrigation resources. These are aspects stressed by agricultural domain experts and inter-disciplinary professionals.

In this chapter, we take a supply-chain-centric approach to optimum management of agricultural inputs for achieving overall efficiency in the agribusiness sector. The focus of discussion would cover the passage from farm gate to production centre.

2.2   Categorization of Inputs

During production, the inputs consisting of all material, machinery, men, efforts, and capital are converted through various means to produce the required output. Various disciplines define the input–output relationship based on their industry characteristics. One can classify agricultural inputs as follows:

■ Natural resources such as land, soil, water and monsoon effects
■ Technological resources, including inorganic fertilizer, insecticide, improved seed varieties and equipment
■ Human resources (including labour, extension services, and life-cycle concerns)
■ Access to capital through credit and financial services, with farm produce as collateral, other forms of credit and microfinance, and farm subsidies and grants
■ Facilities that can count as being logistics/supply chain drivers, including seed multiplication, fertilizer plants, and so on, transportation of inputs by road, rail, or other means, inventory policy with respect to procurement of inputs for use and for holding for re-use
■ Social, political, capital and government policies that influence the input industry.

A wide range of dynamics influence the agribusiness inputs industry. For example, international agreements such as patent laws, Trade-Related Intellectual Property Rights (TRIPS), and other multilateral agreements regulate the industry. Similarly, certain market structures in the agricultural inputs industry are evolving, impacting the concentration and behaviour of firms and thereby influencing supply chain integration. This is mainly because of ongoing changes in power relationships, which jeopardize supply chain profits, though a few players are able to increase their profitability. Some initiatives have the potential to transform industries, such as the National Dairy Development Board (NDDB) in India organising the input industry in the dairy sector. The NDDB pioneered transformative activities by working continuously with its member cooperatives on using best practices, including animal health, reproduction through biotechnology, mobile veterinary support services and more importantly producing and marketing of balanced feed. The natural result of these activities led to an increase in overall efficiency of the sector because of improved supplier, focal firm, and customer linkages, which in turn enhanced profitability through the supply chain.

2.3   Growth of Inputs Manufacturing Sector

After independence, and with the onset of the planning era, India witnessed rapid growth in the agro-processing sector. The first phase of development started with the Green Revolution, which resulted in increased agricultural production and emphasized the need for post-harvest management. The Green Revolution refers to a series of research, development, and technology transfer initiatives, based on activities that happened between the 1940s and the late 1970s. These activities increased agriculture production around the world, and significant changes took place in the late 1960s. Norman Borlaug, the father of Green Revolution, created a number of initiatives, including the development of high-yielding varieties of cereal grains; expansion of irrigation infrastructures; modernization of management techniques; and distribution of hybridized seeds, synthetic fertilizers, and pesticides to farmers. In 1961, Borlaug was invited to India by Dr M. S. Swaminathan, who pioneered Green Revolution in India, thereby earning the title Father of the Green Revolution in India, at a time when the country was experiencing the worst-ever famine in modern history. Agronomists developed high-yielding varieties of rice, wheat, and maize, which led to phenomenal production, as may be observed in Table 2.1. The average yield of crops also increased with the adoption of the Green Revolution in India. These yield growth patterns can be observed in Table 2.2.

Inferences regarding the impact of the Green Revolution on agro-based manufacturing and processing are as follows:

1. Growth in agricultural output increased demand for agricultural inputs such as certified seeds, fertilizers, tractors, and so on. These challenges are briefly discussed here to enable both the input and the output sectors of agro-based industries to be viewed in the right perspective.
2. The growth in agricultural output also gave opportunities for agro-based primary, secondary, and tertiary processing of goods. Primary processing mostly included sorting, grading, and storage before industrial processing. Secondary and tertiary processing comprised the use of capital equipment and machinery for conversion of raw material into processed foods and other products that would come under the industry classification. This kind of demand also poses innumerable supply chain challenges.
3. The growth in output and processing also increased the scope for managing recovery of produce and handling waste generated in production systems. We have attempted to provide examples of a few developments in these areas, which increased value generation and contributed to improving the economics of agricultural produce processing and the manufacturing of goods based on agriculture.

2.4 Manufacturing and Processing

The growth in agricultural output with the advent of the Green Revolution was possible with fertilizer application and better irrigation practices. It may be observed from Table 2.3 on fertilizer production, imports, and consumption that the production of all NPK fertilizer increased from 1,059 thousand tonnes of nutrients in

Table 2.1 Production of Major Crops (in Million Tons)

Group/Commodity	1970–1971	1980–1981	1990–1991	2000–2001	2006–2007	2007–2008	2008–2009	2009–2010	2010–2011[a]
Food Grains	108.4	129.6	176.4	196.8	217.3	230.8	234.4	218.1	241.6
Rice	42.2	53.6	74.3	85	93.4	96.7	99.2	89.1	95.3
Wheat	23.8	36.3	55.1	69.7	75.8	78.6	80.7	80.8	85.9
Cereals	96.6	119	162.1	185.7	203.1	216	219.9	203.4	223.5
Pulses	11.8	10.6	14.3	11	14.2	14.8	14.6	14.7	18.1
Oilseeds[b]	9.6	9.4	18.6	18.4	24.3	29.8	27.7	24.9	31.1
Sugarcane	126.4	154.2	241	296	355.5	348.2	285	292.3	339.2
Cotton[c]	4.8	7	9.8	9.5	22.6	25.9	22.3	24	33.4
Plantation Crops									
Tea[d]	0.4	0.6	0.7	0.8	1	0.9	0.9	0.9	0.9
Coffee	0.1	0.1	0.2	0.3	0.3	0.3	0.3	0.3	0.3
Rubber	0.1	0.2	0.3	0.6	0.9	0.8	0.9	0.8	0.8
Potato	4.8	9.7	15.2	22.5	22.2	28.5	34.4	36.6	36.6

Source: Government of India, *Economic Survey 2011–2012.*

[a] Fourth Advance Estimates as on July 19, 2011.
[b] Includes groundnut, rapeseed and mustard, sesame, linseed, castor seed, safflower, sunflower, and soya bean.
[c] Bales of 170 kg.
[d] Calendar year.

Table 2.2 Yield per Hectare of Major Crops

Group/Commodity	(kg/ha)									
	1970–1971	1980–1981	1990–1991	2000–2001	2005–2006	2006–2007	2007–2008	2008–2009	2009–2010	2010–2011[a]
Food Grains	872	1023	1380	1626	1715	1756	1860	1909	1798	1921
Rice	1123	1336	1740	1901	2102	2131	2202	2178	2125	2240
Wheat	1307	1630	2281	2708	2619	2708	2802	2907	2839	2938
Cereals	949	1142	1571	1844	1968	2020	2151	2183	2075	2247
Pulses	524	473	578	544	598	612	625	659	630	689
Oilseeds[b]	579	532	771	810	1004	916	1115	1007	958	1159
Sugarcane (tonnes/ha)	48	58	65	69	67	69	69	65	70	69
Cotton	106	152	225	190	362	421	467	403.3	403	510
Plantation Crops										
Tea[c]	1182	1491	1794	1673	1500	1666.7	1500	1500	1500	1500
Coffee[c]	814	624	759	959	1000	1000	750	750	750	750
Rubber[c]	653	788	1076	1576	2000	1800	1333.3	1285.7	1142.9	1142.9
Potato (tonnes/ha)	10	13	16	18	17	15	18	19.1	20.3	20.3

Source: Government of India, *Economic Survey 2011–2012.*

[a] Fourth Advance Estimates as on July 19, 2011.
[b] Includes groundnut, rapeseed and mustard, sesame, linseed, castorseed, nigerseed, safflower, sunflower, and soya bean.
[c] Calendar year.

Table 2.3 Production, Imports and Consumption of Fertilizers

(Thousand Tonnes of Nutrients)

	1970–1971	1980–1981	1990–1991	2000–2001	2005–2006	2006–2007	2007–2008	2008–2009	2009–2010	2010–2011	2011–2012[a]
Nitrogenous fertilizers											
Production	830	2164	6993	11004	11354	11578	10900	10870	11900	12156	12576
Imports	477	1510	414	154	1385	2689	3677	3844	3447	4492	3883
Consumption	1487	3678	7997	10920	12723	13773	14419	15090	15580	16558	na
Phosphatic fertilizers											
Production	229	842	2052	3748	4221	4517	3807	3464	4321	4222	4432
Imports	32	452	1311	396	1122	1322	1391	2927	2756	3802	3471
Consumption	462	1214	3221	4215	5204	5543	5515	6506	7274	8050	na
Potassic fertilizers											
Imports	120	797	1328	1541	2747	2069	2653	3380	2945	4069	2248
Consumption	228	624	1328	1567	2413	2335	2636	3313	3632	3514	na
All fertilizers (NPK)											
Production	1059	3006	9045	14752	15575	16095	14707	14334	16221	16378	17008
Imports	629	2759	2758	2090	5254	6080	7721	10151	9148	12363	9602
Consumption	2177	5516	12546	19702	20340	21651	22570	24909	26486	28122	na

Source: Government of India, *Economic Survey 2011–2012.*

Note: Import figures for 2011–2012 are from April to November 2011.

Abbreviation: na, not available.

[a] Estimated.

1970–1971 to 17,008 thousand tonnes of nutrients in 2011–2012. For the same period, imports increased from 629 thousand tonnes of nutrients to 9,602 thousand tonnes of nutrients. It is interesting to note that all N/P/K component imports also went through similar increases. Further, consumption of all NPK fertilizers increased from 2,177 thousand tonnes of nutrients in 1970–2071 to 28,122 thousand tonnes of nutrients in 2010–2011. The trend over a period of 40 years shows a compound annualized growth rate in production, import, and consumption in the range of 6.5%–7%.

It may be interesting to analyse the growth of industries manufacturing agricultural inputs such as pumps and tractors, which are increasingly used for irrigation and land preparation. One would expect the impact of the Green Revolution to significantly influence the growth of the agricultural inputs production sector. It can be observed from Table 2.4 that pump production increased from 35 million units in 1950–1951 to 3,139 million units by 2010–2011. This is a compounded, annualized growth rate of about 8%, which is significant. However, while analysing the data, one might note that the statistics for pumps include pumps produced for both agricultural and industrial usage, as the data pertains to the growth of mechanical engineering industries in India. Based on the interactions at the field level, we observe that modernization of agriculture required more water. Land usages, constraints in irrigation, improved coverage of supply of electrical power, and distribution of fuel and economic rationale encouraged usage of pumps in agricultural sector. Similarly, tractor manufacturing industry, which belongs to the agricultural input sector, has experienced growth with the advent of the Green Revolution in India. It may be observed from Table 2.4 that the production of tractors in India in 1980–1981 was around 71,000 units and went up to 465,400 units by 2010–2011. This means a compounded annualized growth rate of 6.5%. This can be seen as a highly successful phase of growth in agribusiness, especially because of the fact that a number of farmers in India have fragmented land holdings, and the propensity to invest on capital machinery in agriculture is low.

One would expect other agriculture inputs industries such as those producing and distributing hybrid seeds, manufacturing machinery for other activities like harvesting and so on, would have had similar issues and opportunities.

2.4.1 Supply Chain Challenges

The challenges associated with supply chain management of manufacturing processes in the agribusiness inputs sector would include the following:

1. Demand management: Fertilizers, hybrid seeds, and other inputs are in high demand with short lead times and high peaks immediately following the monsoons. Unless the production system and the supply chain are agile and able to cater to such demand, the supply chain is bound to fail. The spikes in

Table 2.4 Pumps and Tractor Industries in India

Industry Unit	Pumps (Power Driven) in Millions	Tractors in Thousand Units
1950–1951	35	na
1960–1961	105	na
1970–1971	259	na
1980–1981	431	71
1990–1991	19	142.2
2000–2001	481.9	284.4
2004–2005	1549.8	187.7
2005–2006	1726.1	236.4
2006–2007	1954.7	300.5
2007–2008	2089.3	295
2008–2009	2140.3	293.6
2009–2010	2891.7	373.7
2010–2011	3139.1	465.4

Source: Government of India, Economic Survey 2011–2012.

Abbreviation: na, not available.

demand could go both ways, and the ability to absorb wild variations in the supply chain network is crucial for the success of individual firms and the sector as a whole.

Similarly, manufacturers of capital equipment like tractors and irrigation and harvesting equipment depend upon the realization of produce during good times to get farmers interested in making additional investments in modern farming practices. Demand management must predict and capture such opportunities and ensure product availability and quick reach to the markets. More specifically, from the manufacturing process perspective, demand management is a key input for resource, material, manufacturing, and distribution planning.

2. Supply chain planning: As may be observed from the demand management perspective, the planning process assumes greater importance for supply chain network partners in achieving targeted agriculture input movements.

The production systems of supply chain partners need to be synchronized with demand patterns, storage needs, and transportation networks at efficient costs. This is mainly to ensure that the farmers who are the end customers in this network realize value from different role agents. The big challenge with the planning process is synchronization of roles and responsibilities among players so that farmers get whatever they need on time at the level of service expected.

3. Supplier relationship management: Sourcing process is an enabler for effectively serving the customer demand, the customer in this context being the farming community and the government. Sourcing assumes greater importance as a number of input materials such as fertilizer is regulated. In fact, inadequate water availability is one of the key constraints in India, and demand for irrigation pumps is a function of water distribution in different territories in the federal system. Supplier relationship management in certain cases, such as tractors and capital equipment for the agriculture sector, is more in line with the automotive industry, where the suppliers are tapped for strategic, critical, and bulk components and accordingly, relationships get defined. When it is a strategic component supplier, the relationship is such that interaction is more frequent, and is sharing- and partnering-oriented for mutual benefits. When it is a question of critical components, then again, the relationship is highly interactive and proactive, resulting in sharing of technology and growth plans. When the purchases are for bulk values, the firm encourages competitive buying so that the value realized is the highest for the customers.

   In most of the cases pertaining to agricultural inputs, sourcing requires intense plan synchronization and the ability to work closely with the sourcing agents.

4. The manufacturing function involves supply chain activities in the agribusiness input sector, which would critically influence the performance of the sector.

   a. One activity is the choice of location of the plant and of regional distribution centres for the produce, so that the desired service level (responsiveness) and costs can be managed. For example, a fertilizer or tractor plant that requires huge capital investment is normally located in a centralized location, and the farmer community is served through regional distribution centres. On the other hand, hybrid seed plants can be spread in scattered locations and distributed through carry and forward agents.

   b. Contract manufacture of inputs and delivery of service inputs are other important strategies deployed by firms and individual entrepreneurs, especially those who offer agricultural input services and products like organic manure, certified seeds, and so on. This strategy requires ensuring of the uniform quality of products provided by all players, and localization to the demands of the environment.

c. Manufacturing of inputs like fertilizer and machinery like tractors will have to be carried out throughout the year, whereas demand, as mentioned earlier, arises in spikes. This requires effective management of inventory and optimization of inventory, transportation, and production costs.

5. Distribution management of the input sector is another key challenge for processors and manufacturers. For example, a fertilizer plant may sell its output in over an 800 km radius in India. This being a bulk commodity that has to be sold in interior lands across villages, distribution management poses a challenge. The Government of India (GOI) has fertilizer under regulatory management, where prices as well as freight cost equalization pools are in practice. If a company has to earn normal profits as determined by the GOI, it must operate efficiently at or above the level dictated by the norms prescribed by the regulatory agency. This is true for any agriculture input, because of the widely disbursed consumption centres and poor infrastructure.

6. Returns management in the agriculture input sector poses unique challenges. Returns in case of machinery is more to do with post-sale service and the ability to provide quick spares and service support to keep equipment operational at locations far away from service centres. Similarly, in case of inputs like fertilizers and pesticides, redistribution by withdrawal from one location to the other is possible. Further, withdrawal of certain other inputs from the market is also possible, as in the case of certified seeds.

Thus, input processing and manufacturing in the agribusiness input sector offer opportunities for a supply chain analyst to constantly assess the goals of efficiency of the supply chain in terms of cost for a desired high level of responsiveness.

It may be useful to mention here the use of mechanical equipment in harvesting and the resulting growth of the industry. There has been increasing usage of mechanical equipment like tillage equipment, harrows, sowing equipment, plant protection equipment and irrigation equipment for different crops and fields; we are not discussing these issues, as they are still evolving in India. Further, these issues require technical discussions rather than economic discussions from the supply chain perspective. We encourage readers to refer to agricultural engineering papers for this purpose. With regard to harvesting equipment, it is increasingly being tried out in the harvesting of paddy, wheat, cotton, tea, sugarcane, and other crops. In Maharashtra, the relevant government departments and the sugar industry have collaborated in attempts to use mechanical harvesters for sugarcane due to the decreasing availability of cane harvesting labour, and also because of the economy associated with using mechanical harvesters.

Manual cane harvesting involves hard labour, with high energy demand. A group of 15 labourers deploying their best efforts can harvest about six tonnes a day, and thus, for an acre of land, which could yield about 40 tonnes, they may take about six to seven days. The sugarcane harvester is a large mobile machine weighing about eight tonnes. It can cut about 120 tonnes of sugarcane in 16 hours.

It costs about ₹265 to cut one tonne of sugarcane with a harvester, while the manual cost is about ₹322 per tonne (Exhibit 2.1). It is more important to note that the time efficiency of harvesters and the non-availability of labourers determine the push towards mechanization. The exhibit below highlights these factors.

Exhibit 2.1 Mechanization of Sugarcane Harvesting Encouraged

The Maharashtra government has proposed to give subsidies to cooperative sugar mills and farmers from the fiscal year 2011–2012 for purchase of sugarcane harvesters. This is mainly to tide over acute labour shortage. In 2006, about 10 lakh labourers were available. However, in 2011–2012, it was reported that their number had dwindled to about five lakhs. Cutting sugarcane is a very strenuous task, and therefore, not many labourers opt for it.

There are positive economics to deploying mechanical harvesters. A harvester can cut about 120 tonnes of sugarcane in 16 hours. It costs about ₹265 to cut a tonne of sugarcane with a harvester, while the manual cost is about ₹322 per tonne. It is estimated that by 2014, about 450 harvesters costing over ₹450 crore will be acquired in the state. The state government has already approached the central government with a ₹100-crore subsidy plan proposal. However, under the Rashtriya Krishi Vikas Yojana of the Union Ministry of Agriculture, about ₹50 crores has been sanctioned. The cost of a harvester is about ₹1 crore, and if the government subsidy is given for 50%, a farmer has to invest only ₹50 lakhs. For the machines, the normal payback period is about eight years, but due to the subsidy, the payback time has been halved.

Source: Wadke, R., *The Hindu Business Line*, Aug. 25, 2011.

One may note that paddy and wheat harvesting are also attracting mechanization due to labour shortages. Increasingly, a lot of investment is going into mechanization of irrigation (using drip irrigation mechanisms), especially for cultivating grapes and oranges. Again, Maharashtra, which is the leading producer of both crops, is leading in investments. Thus, mechanization for production and harvesting is likely to continue to go up in India in the years to come. This will encourage manufacture of such equipment.

2.4.2 Supply Chain Goals

Efficiency is primarily measured by the internal measures of supply chains, namely, cost factors and asset utilization; responsiveness is the main supply chain goal, which is expected to be realized across the supply chain network. In the agribusiness input industry, these goals assume paramount importance. Responsiveness is the key for input management, especially for farm level production-driven supply chains. This is more so in a monsoon-dependent agrarian economy with large tracts of arid land that depend on rain. Inputs such as seeds, pesticides, fertilizers, human effort and capital equipment are likely to create pressure in terms of demand for these inputs and expectations regarding responsiveness. However, the input providers will have to consider cost dynamics and raise produce under certain logical estimates. The challenge in the industry is to balance the divergent demands of supply chain goals.

In India, dry land farming activities for produce such as oilseeds and cotton increase depend upon the abundance of the monsoons. India frequently experiences volatile monsoon behaviour, impacting crop plantation and growth of produce. Unfavourable monsoons have an adverse effect on supply chains, especially on input providers, as demand for inputs such as fertilizers, pesticides and other agrochemicals drops.

Exhibit 2.2 reflects a case study in 2011 in Gujarat, where land areas under oilseed cultivation increased substantially, increasing demand for farm inputs and throwing supply chains out of gear. Only those input provider firms which are agile and able to respond adequately to such unexpected developments can be successful. Many small players and intermediaries act without perspective, and work on the silo principle of maximizing their own local profits. Thus, the agricultural inputs sector faces the problem of delicately balancing its supply chain goals of efficiency and responsiveness.

Exhibit 2.2 Groundnut Production Estimated to Surge in 2011 Summer in India

A council consisting of the Indian Oilseeds and Produce Export Promotion Council (IOPEPC) and Junagadh Agriculture University (JAU) conducted a crop survey in various parts of Gujarat between May 7 and 13, 2011, and estimated that the summer groundnut production in Gujarat would be around 2.71 lakh tonnes, as against 1.29 lakh tonne in 2010—an increase of 110%. The production of sesame during the period was estimated to surge by 37.5% to 55,000 tonnes. Total area under groundnut cultivation in Gujarat during the 2011 *kharif* season increased from 68,000 hectares to 1.21 lakh hectares due to sufficient availability of irrigation water. The survey also mentions that groundnut yield increased to 2,430 kg per hectare from 1,900 kg the previous year.

Source: Dave, V. Groundnut production to rise this summer. *Business Standard,* May 16, 2011.

The experience with agricultural inputs for the processing sector is similar. The challenges of meeting efficiency and responsiveness targets are compounded by the vagaries of nature and uncertainty in produce markets. For example, as shown in Exhibit 2.2, when groundnut sowing is extensive, operators like process firms plan for bumper harvests and boost their capacities for procurement, processing, and working capital to handle the larger volumes anticipated. If one considers the figures, the system needs to show a flexibility of more than 20% in resources, which is quite a challenge. With such a degree of adjustment required, efficiency may not be ideal. The focus would be on handling the increased volumes; otherwise, it would have an adverse effect on the farm produce economy.

2.4.3 Push versus Pull

In supply chains, the push strategy is one where the supply chain is organized in anticipation of demand or customer orders. Such a situation mainly depicts the speculative character of a supply chain, where production and stocking happen on the basis of demand forecasts. This process would be driven more by economies of scale in operations and for low-value items. For example, popular models of tractors are produced in India on the basis of demand estimates. The tractors produced lie in stock, awaiting customer orders. This is also the case with respect to seeds production and distribution. In the case of agricultural processing, the push strategy is more prevalent. For example, a sugar manufacturing company will have to intensively plan all its activities based on the push strategy. Sugar is produced during six to eight months of the year, and output is released for sale over non-productive months, too. Hence, the supply of inputs, namely cane, needs to be planned well ahead.

In contrast to the push strategy, the pull strategy in supply chains is organized to serve customers on the initiation of demand. The concept of "pull" in supply chains is about the execution of processes in response to customer orders. This concept is reactive, but enables value creation to customers by allowing flexible configuration of products. The advantage to the focal organization in this process is that it reduces its finished goods inventory by postponement. For example, the demand for seeds for plantations, especially in the case of dry land farming, is based on the pull strategy, which comes into effect after the arrival of monsoon. Similarly, the demand for capital equipment services as well as labour and inputs could be based on cropping and growth patterns. Normally, a pull-based supply chain allows the focal organization to provide customers with flexibility in selecting inputs. At the same time, the pull factor works by enabling the focal organization to manage its inventory better.

However, it is important to understand the boundaries of push and pull in any supply chain network. Push and pull processes are useful in reaching strategic decisions relating to supply chain design, as the planner gets a more global view of how supply chain processes relate to customer orders. Based on such a global view, facility locations, which could be split or consolidated, and other facets of the supply chain can be designed.

2.5   Supply Chain Drivers

In this chapter, we have so far discussed the role of supply chain drivers namely facilities, inventory, transportation, sourcing and information technology impacting the performance of inputs industry.

2.5.1   Facility

Facility management for the agricultural inputs sector is an important aspect of supply chain drivers. When one looks at any facility, its various forms need to be analysed. One aspect is with respect to manufacturing, process, and resource endowment, and the other is about creating storage spaces such as warehouses and regional distribution centres, especially with respect to input management.

In the case of a seed multiplication plant for cereals and other large crop bases where seed processing is to be centralized, the facility-related decision would hinge on whether to locate centrally or be driven by location economics and distribute low weight/volume material like seeds through distribution networks. The situation would remain true in the case of crop protection chemicals like pesticides. However, centralized locations would not necessarily be the criteria for industries dealing in certain other inputs, such as fertilizers and tractors. Though these plants are expected to use economies of scale in production, they tend to be located where there is maximum cost economy from perspectives of capital and operations. One may observe that some of the plants have skewed locations for a variety of reasons.

It may be useful to highlight an agricultural equipment manufacturing company and its facility. Tractors and Farm Equipments Ltd (TAFE), a tractor manufacturer, has four plants in India at Mandideep (Bhopal), Kalladipatti (Madurai), Doddaballapur (Bangalore) and Chennai, as well as a new overseas plant in Turkey. The Mandideep plant came up in 2005, when TAFE acquired Eicher's tractor business. The company manufactures the Eicher brand of tractors at this plant, diesel engines at its plant at Alwar, and transmission components at Parwanoo.

TAFE's success can be measured terms of its performance, in terms of revenue growth and production volumes. TAFE's turnover, at ₹86 crore in 1985, had risen to ₹5,800 crore by 2010–2011. TAFE is the second biggest tractor company in India, after Mahindra Tractors. It is the third biggest in the world, after Mahindra and the US-based John Deere. From around 4000 tractors a year in 1985, the group now manufactures 120,000 units annually (*Business Today*, September 2011).

TAFE holds a market share of 22% excluding the Eicher brand; with Eicher, it holds about 28%. If one looks at India's tractor market, northern states such as Uttar Pradesh, Punjab and Haryana, western states such as Rajasthan, Gujarat and Madhya Pradesh and southern states like Andhra Pradesh dominate sales. With its major plant location in the southern part of India, namely Madurai in Tamil Nadu, TAFE was able to rise to the position of being second in market share. This was possible in spite of its factory locations being skewed, mainly because of its dealer network and distribution management. A well-designed distribution network that is part of its facilities has helped it achieve success in business. Thus, one may conclude that the management of manufacturing locations and distribution networks, especially nodes and flows to distribution points, can define the success of a supply chain.

2.5.2 Inventory

Another important supply chain driver is inventory management. In the agribusiness input industry, inventory would play a significant role, as inputs are used for producing agricultural produce during the season. In India, agriculture is carried out during two seasons, namely, the *kharif* and the *rabi* seasons. *Kharif* season is after first rains of south-west monsoon in India, which is in June–July and harvest happens in September and October. *Rabi* season starts after the *kharif* season and more popularly termed winter crop with plantation in November and harvest in February/March. Cultivation is triggered by the onset of the monsoon. Inputs such as seeds and capital equipment and preparatory farm equipment would be required at the initiation of crop farming. As the crop firms up, farmers would need fertilizers, crop nutrients, and pesticides. Later, farmers would need farmer extension support services, harvesting support, and support in reaching the markets. All these activities require an effective inventory management policy across the supply chain. The players deploy inventory concepts such as cycle inventory, safety inventory, seasonal inventory, and pipeline inventory. Inventory could be in the form of raw material, work in progress, or finished goods.

Cycle inventory is the average inventory that builds up within the supply chain, because a supply chain stage either produces or purchases in lots that are larger in volume than the demands of customers. Raw materials, components, and parts are required for production. The cycle plays a crucial role in keeping the production process continuous. Raw materials and work-in-progress inventories are a major part of the production-related inventory. This works even in the agribusiness input sector. Fertilizers typically work on building cycle inventories, as they are moved along a distribution network before they reach the farming community. Similarly, many other farm inputs such as seeds, micronutrients, and crop protection chemicals move through a distribution network requiring effective planning and management of cycle inventory.

Cycle inventories are "held primarily to take advantage of economies of scale in the supply chain. Supply chain costs are influenced by lot size. The primary role of the cycle inventory is to allow different stages to purchase products in lot sizes that minimize the sum of material, ordering, and holding costs. Ideally, cycle inventory decisions should consider costs across the entire supply chain, but in practice, each stage generally makes its own supply chain decisions, increasing total cycle inventory and total costs in the supply chain" (Chandrasekaran, 2010). The problem is compounded by multi-level intermediaries, with varying sizes and significant differences in negotiating power.

On the other hand, the safety inventory is an inventory carried for the purpose of satisfying demand that exceeds the amount forecast for a given period. Inventory is carried based on forecasts of demand, and forecasts are rarely accurate. In the agriculture sector, as noticed in Exhibit 2.1, demand for inputs like seeds surges under exceptional monsoon conditions. At such times, when demand is higher, there are lost sales if the product is not available along the distribution network.

Similarly, in the agricultural inputs equipment industry, it is important to stock critical components, as failures cannot be scientifically predicted. Breakdowns during peak operating seasons can have negative consequences. In order to avoid customer service problems and the hidden costs of unavailable components, companies hold safety stocks. This provides a cushion against uncertainties in demand, lead time and supply, thereby ensuring that operations are not disrupted.

The term, "seasonal inventory" refers to inventory that is used to absorb uneven rates of demand or supply that agribusinesses face. Manufacturers of seeds, fertilizers, nutrients, and crop protection chemicals in India experience seasonality in sale, with peaks around monsoon time—either the south-west monsoon or the north-east monsoon. If there is no stock in stock-keeping units (SKUs) during this period, the firm may lose market share. Hence, holding seasonal inventory helps in evening out volatility in demand and supply. A focal firm may stock up on SKUs to meet seasonality-related fluctuations.

> In contrast to the seasonal inventory, the pipeline inventory, or inventory in transit, refers to inventory moving from one node to another in the materials flow system. Materials move from suppliers to a plant, from the plant to a distribution centre or customer, and from distribution centre to a retailer. Pipeline inventory consists of orders that have been placed but not yet fulfilled. This can be reduced by reducing stocking locations, improving materials handling and avoiding delays in distribution. However, such a strategy is difficult in many consumables input segments of the agribusiness sector (Chandrasekaran, 2010).

Farm sector inventory issues are more to do with estimating demand and managing stocks across the distribution channel for effecting sales at appropriate times. Another critical challenge is the availability of the right SKUs in the required quantity, especially in case of seeds and crop protection. The agribusiness input sector has a varied and proliferate demand for SKUs in different locations, with a tendency to support its ultimate customers in the chain, namely, the farming community.

Another aspect of inventory management in the agribusiness input sector is the obsolescence of input material because of lapse of recommended useful period of life, loss of demand, and variance in demand estimation. These issues are quite normal, and business practices require building of inventory across the chain. The issue is: how does one handle unsold inventory or manage returns with respect to the focal firm? Generally, firms rely on the demand estimates and use extension staff to closely coordinate the information flow. Diverting products to relevant locations and quickly reprocessing before expiry are some of the practices deployed. There is a substantial cost incurred, as loss of goods due to expiry of useful life is provided for.

This leads to a discussion on the cost of managing inventory in the agribusiness sector. Discussions on inventory cost also include the holding cost, namely, financing costs such as interest on capital, obsolescence, warehousing at the institutional level, and so on. Depending upon the nature of the input industry, these costs would vary. For example, in the case of fertilizer, the inventory holding cost could be significant, as firms need to carry substantial stock to meet surges in seasonal demand.

In the case of seed crops like sugarcane, the work-in-progress inventory is critical, as the seed crop takes almost seven months to harvest. Cane cultivation is coordinated by cane department of sugar plants, which promotes cropping for registering acreage under commercial production. At the farm level, farmers hold work-in-progress inventory in the field, incurring substantial commercial risk. There must be a commercial crop plantation plan under which seed cane must be harvested and inventory needs to be monetized.

On the other hand, many other crops may have different approaches. A firm generally produces its crop seeds one year ahead. *Kharif* seeds for a given season are produced in the previous *kharif* season, and the same practice applies for *rabi* seeds. However, there have been discussions exploring the feasibility of producing seed during the *kharif* season and selling it for the *rabi* season if the seed variety permits cropping in both seasons. Further, the possibility also exists of producing seed in the early (*kharif* or *rabi*) season and selling it in the late (*kharif* or *rabi*) season. This may be possible in regions where monsoon movements allow enough time to organize the processing of seeds and logistics. Though such innovative practices would reduce inventory costs of input goods, traditional practices of carrying inventory for a year are more common in the seed industry.

Inventory management in godowns, processing plants, and other handling locations is very critical. As seeds are perishable, following proper inventory management practice is important. For example, a company may have a policy to follow the first-in first-out (FIFO) method, though one may not be sure whether it would be implemented effectively, considering the informal and unorganized nature of the sector.

Thus, inventory management in the agribusiness sector, both at the farm gate level and the processing level, brings several unique aspects and challenges.

2.5.3 *Transportation*

Transportation management is another supply chain driver that could impact the agribusiness input sector. As mentioned in the previous chapter, agricultural inputs have the characteristics of bulkiness, perishability, and seasonality, which create challenges for transport management. Over and above these characteristics, the agribusiness input sector suffers often from the need to be cost-effective, as Indian farmers are afflicted by low income and poor savings. Farming is undertaken as a compelling economic activity, which may not necessarily pay back decent returns. In fact, apart from the rising issues with respect to the socio-economic aspects of agriculture input management, one will have to probe matters from the supply chain perspective, namely, are firms in the supply chain truly working on an equitable partnering mode, such that that every player's time, efforts, and resources are fairly compensated for? This would be an important perspective while understanding transportation management aspects relating to the agribusiness inputs sector.

The transportation network is influenced by the distribution network for the product. Here, one may be looking more at outbound challenges that become inbound issues at the next stage of the supply chain. In the case of the seeds market, the transportation issue is not that challenging when farmers buy local seeds, except in the case of products that are bulky and perishable. Such products may include seed plants for plantations or cane. More often, farmers manage these challenges through better planning and use of local market vehicles.

Branded and labelled seeds are distributed through the channel partners. In such a network, these movements can be handled in the form of small parcels up to the last mile. Between the manufacturing unit and the warehouse, they will move in full truck loads. Then, between the warehouse and distribution points such as preferred dealers and stockists, they may move in trucks on the milk run. Only in case of sudden unexpected demand they would be moved in quantities lesser than truck loads, or through parcel services.

If one looks at transportation network challenges for a bulk product like cane for the sugar industry and fertilizer, the situation may be different.

In the case of the sugar industry, seed cane movement is mostly localized, as the material for planting has to procured from within a geographical area of radius 50–100 km. The challenge lies in planning and scheduling of seed cane for plantation. Farmers mostly use their own transport like tractors for this purpose. For requirements involving large volumes and long distances, they also use trucks. Coordination with seed cane farmers, plantation fields, labourers, and transport operators becomes critical in this operation. Any delay in handling caused by the failure to by any of the parties to show up can lead to a variance in the planned schedule, and to activity that may lead to poor supply chain network performance. Apart from seed cane, other transportation and distribution management of other inputs such as fertilizers, crop nutrients, and crop protection also plays a significant role.

In the fertilizer business, delivering the product through the right mode to the right customer at the right time, right cost, and right quality matters most. Appendix 2.1 explains in detail, the transportation challenges of a fertilizer manufacturing firm, using an assumed name.

Transportation problems associated with choice of mode, location of primary movement points, and management of normative and actual transportation costs are common challenges in any manufacturing set-up that has to market its products to the interior parts of the country. In the case of a number of fertilizer companies in India, transportation costs play a very significant role in the profitability of the firms. Further, there has been increasing pressure on the country's economic policy makers to reduce subsidies on fertilizer. However, at the firm level, a logistical trade-off needs to be achieved for improving operational efficiency, coupled with warehouse and inventory costs. Decisions based on the results of an evaluation of transportation options would also have a considerable impact on the procurement process.

In any agribusiness input sector, the supply chain manager would constantly be evaluating the cost comparisons of the various options for transporting goods from stock points to consumption points. The manager would also be giving thought to the extent of spill over of stock possible during the peak period, and its associated cost.

If one looks at an input like pesticide, companies may not have an extended retail presence; instead, they may depend upon on a network of dealers, distributors, and preferred dealers to service the demand. The sales organization is made up of independent operations in different zones (typically, the country is divided into four zones), with states divided into regions and districts into territories. SKUs are transported from the manufacturing plants to warehouses (regional distribution centres), which are at times managed by carrying and forwarding agents. Much of the transportation from the manufacturing locations to warehouses is done on a single source–single destination–multiple product–full truckload basis. The rest involves movement from a single source to multiple destinations on a full truckload basis. From the warehouses, the material is usually moved to the distributors in full-truckload milk runs. Occasionally, materials are sent as parcels to a single destination. Distributors manage the movement of material to dealers using smaller vehicles such as auto-rickshaws and vans.

Thus, transportation management for the agribusiness input sector is challenging, and varies widely from localized to centralized distribution involving volumes ranging from small parcels to bulk material. It is important to understand the nuances of each supply chain and optimize the transportation system taking into consideration all relevant inventory and facility decisions.

After having discussed the facility, transportation, and inventory drivers of the logistics part of agribusiness, we may analyse other drivers, namely, procurement, pricing, and information.

2.5.4 *Sourcing*

Procurement is also referred to as the sourcing and purchasing of goods and services for commercial purposes. Every decision unit, be it a farmer or a firm, sets procurement policies that govern its choice of suppliers, products, and the methods and procedures of buying. Depending upon the size of the decision unit and the complexity and level of formality of the agribusiness concern, procurement practices are evolved in various segments.

At the farm level, sourcing decisions to many inputs and services are decided by push factors, which is based on the availability of input to be sourced off the shelf. However, mismatch between what is required at the field and when may not be available.

For an individual farmer, inputs that are involved in procurement decisions include seeds, fertilizer, pesticides, plant nutrients, extension services, and labour. Though one would think that such procurement decisions are mostly driven by the fundamental economics of agrarian principles, group decisions also play a major role. For example, procurement chain decisions should focus on the micro level aspects of the geography in which the farm is located. Certain parts of one district sometimes get more rain, while neighbouring districts or *taluks* of the same district may not get as much rain. The supply of inputs should be planned at a micro level by getting enough information from the field. Farmers under such circumstances are driven by group decisions to achieve a balance between demand and supply, and also to enjoy the benefits of contiguous farming.

The need for contiguous farming can be better appreciated in wet crop lands and in some dry land irrigated cropping patterns, like with sugarcane. Since farmers are expected to support community goals in their villages, fields that are further away from access roads will have to go along with those that access the roads. Given the fact that there is a need for sharing water and optimizing efforts in crop management, including those involved in receiving inputs to and sending outputs from the farm, contiguous farming is a compulsion. This influences procurement decisions at the individual farm level, since decisions are made jointly with others in the peer farming community. Peer groups build and improve upon relationships with many stakeholders, such as input suppliers and extension support system providers. This helps better handling of procurement challenges such as identifying the needs of customers (farmers) and suppliers, choosing tools and processes to communicate with suppliers, policies for evaluating proposals, and quotes and suppliers (pricing aspects). It is important to note that group behaviour and systems dominate the procurement process at the farm level. This is explained in Figure 2.1.

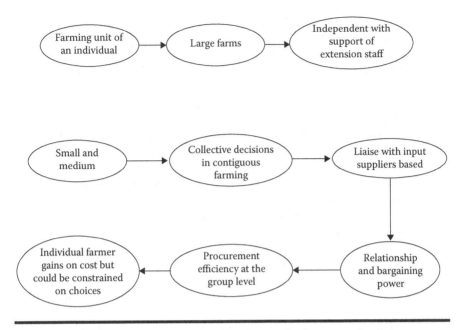

Figure 2.1 Procurement decision mechanism at the farm gate level.

2.5.5 Pricing

After having discussed procurement as a driver, one may look at pricing as a supply chain driver in the agribusiness input sector. At the farm gate level, seeds are fairly priced on the basis of market conditions. Fertilizer is a product that in India is highly price-sensitive and subsidized. The GOI has initiated nutrient-based subsidy schemes from 2010, and is expected to reduce fertilizer subsidy over the years. From the supply chain perspective, changes in central and state government support for fertilizer prices in form of subsidy would impact demand and supply.

Exhibit 2.3   Government Cuts Subsidy on Fertilizer

The Union Cabinet on March 1, 2012, approved the proposal to bring down the quantum of subsidy on decontrolled fertilizers—phosphatic (P) and potassic (K)—for 2012–2013. The proposal is to reduce subsidies on P&K fertilizers under the Nutrient Based Subsidy (NBS) policy. However, it is unlikely to have any adverse impact on fertilizer prices.

Due to the strengthening of the rupee and bearish global prices, the Department of Fertilisers had recommended reductions in subsidies on nitrogen (N) and potassium (K), which will be ₹24 a kg each for 2012–2013, with phosphate (P) at ₹21.8 a kg. For 2011–2012, the subsidy on NPK had been fixed at ₹27.15, ₹32.33, and ₹26.76 a kg, respectively, under the NBS policy. The reduction in P&K fertilizer subsidies are expected to bring down the government's total subsidy bill by 20% in the fiscal year 2012–2013.

The subsidy bill on P&K fertilizers is estimated to be ₹52,000 crore, with the overall subsidy bill touching ₹90,000 crore for the year 2011–2012. Under the NBS regime, introduced from April 1, 2010, retail prices of 22 varieties of P&K fertilizers were freed. The government fixes subsidies on nutrients such as NPK, linking them to the import parity price of di-ammonium phosphate (DAP) and muriate of potash (MoP).

The subsidy is reimbursed to fertilizer firms for selling indigenous or imported crop nutrients at lower prices to farmers. This comprises a reduction in subsidy on DAP by ₹4,763 a tonne and on MoP by ₹1,554 a tonne, a reduction of 27% and 10%, respectively. Since India imports almost its entire requirement of non-urea fertilizers, this reduction could help bring down the fertilizer subsidy bill substantially.

Source: The Hindu. Govt cuts subsidy on key fertilizers. Mar. 1, 2012.

Other inputs at the farm gate level include pesticides and nutrients, which are predominantly market driven. Though irrigation and water resources are national assets, use of which is based on water sharing and availability conditions, at the farm level, water trading, especially of groundwater, has been taking place. Pricing is based on local demand and supply conditions and the cost of power, which is based on the efficiency of generators. Given the constraints in water availability, implementing a supply chain–centric approach could be difficult, and decision units are expected to take a silo approach.

Another important agribusiness input is agricultural labourers. In India, supply of agricultural labourers is becoming a problem, as majority of the rural population is no longer willing to work on demanding jobs in the field. This situation could be the result of the impact of urbanization, with increased work opportunities in construction and industries. Apart from these, developments in infrastructure are also absorbing agricultural labourers.

There is an increasing tendency among rural people who offer their services as agricultural labourers to reduce the time given by them to agricultural work, as they are get the support of the GOI through the National Rural Employment

Guarantee Act (Exhibit 2.4). Though assessment studies have mentioned the positive contribution the scheme makes to the social fabric, one cannot ignore changing cultures and the increasing apathy towards laborious agricultural labour. This hampers agribusiness input supply chain efficiencies, and results in a lack of supply chain orientation in the sector.

Exhibit 2.4   National Rural Employment Guarantee Act, 2005. No. 42 of 2005 [5th September 2005]

An Act to provide for the enhancement of livelihood security of the households in rural areas of the country by providing at least one hundred days of guaranteed wage employment in every financial year to every household whose adult members volunteer to do unskilled manual work and for matters connected therewith or incidental thereto.

Source: Government of India. *Gazette of India,* Sept. 7, 2005.

Moving forward with the discussion on pricing, we can see that agribusiness sector farm equipment pricing is mostly decided by market dynamics. The challenge has been more severe for small and marginal farmers in using equipment that is available on lease for tasks. These services are mainly handled by intermediaries, and therefore, an exorbitant pricing mechanism is often followed. Farmers are compelled to agree to these prices because of their compulsions to proceed with their farming duties on critical farm calendar days.

2.5.6 *Information*

Information is another important driver impacting agribusiness input decisions. It can be observed from the preceding discussions that a lot of inefficiencies occur in the farm gate sector because of improper and inadequate information, handled by intermediaries through local grapevines. Further, these intermediaries, who are well entrenched in the system because of the credit mechanism, could distort information at the farm level to suit their own purposes.

The government has been initiating a number of initiatives to promote information dissemination over the decades. Agricultural extension service support by governments and universities provide for the improvement of farming economies and the aggressive posture of supply chain networks, as various parties in the network are involved in developing and promoting new varieties, practices, and market reach.

Agribusiness process units work closely with the government and universities to promote commercial farming. There is a win-win approach when stakeholders collaborate to transfer information. Though farmers respond well to advice on farming and practices, transactional information is what is critical. AMUL's ability to provide information built on strong cohesiveness leading to competitive advantages is an example of a successful cooperative mechanism.

Not only large corporate bodies, but also small- and medium-sized firms who have taken up agribusiness drive such information dissemination activities on extension services. Exhibit 2.5 provides an example of such an initiative by a company named eFresh in Andhra Pradesh. It is important to understand the role of information in supply chain efficiency in the agribusiness sector, which has just as much importance as in case of manufacturing or assembly businesses like motor cars. In fact, information is more significant in agribusiness, as it affects the economy and livelihoods more than operational transactions like buying and selling.

Exhibit 2.5   The Farmer Outreach Service

Srihari Kotela, a chartered accountant, has launched a service to reach out to farmers in Andhra Pradesh. Farmers can make a simple phone call seeking a visit by the *rythu ratham*, a comprehensive mobile information source on all aspects of farming, be it fertilizers, crop diseases, or government schemes.

According to Mr. Kotela, the concept would encourage good agricultural practices and raise farm incomes. The company owned by him, namely, eFresh Portal, has tied up with the Acharya N. G. Ranga Agricultural University, the University of North Bengal, and the Food Safety and Knowledge Network for this initiative.

Typically, the *rythu ratham* vehicles are equipped with soil-testing kits, pesticide sprayers, protective gear, an LCD projector, a computer system, training videos, and so on. The material is shown in villages, typically in the shade of trees, on *gram sabha* platforms and at *panchayat* offices. Farmers typically watch the demonstrations on farm-related activities very keenly. What is really happening is that with the government having cut down on its budget for agricultural extension activities, mainly under the auspices of the Farm Information Bureau, which used to organise such field visits, private initiatives are now beginning to occupy that space.

Source: The Hindu. The farmer outreach "ambulance". Mar. 12, 2012.

Thus, the role of supply chain drivers such as facilities, inventory, transportation, procurement, pricing, and information in bringing efficiency to the agribusiness input sector is clear. There are a number of role agents such as farmers, focal firms who could be processors, intermediaries, governments and institutions. All of them jointly lead to supply chain profits and create value in the system. Any silos approach leads to the kind of inefficiency that characterized the systems of the past. Farming has been maturing over the years, but that also brings new sets of issues that need the attention of farm managers and policy makers.

The technology, process, and people trilogy plays a significant role in defining supply chain maturity. Technology and process improvements are impacting people in different walks of life, including agribusiness. At agribusiness process unit levels, a number of initiatives with information systems and the use of handheld devices have been supporting positive trends in supply chain efficiency.

In a few situations, the lack of availability of agricultural labourers and the negative trend impacting labour availability in some areas due to the implementation of the National Rural Employment Guarantees Act have driven farmers to consider use of advanced technology-based capital equipment for farming activities such as sowing, fertilizing, harvesting, and even threshing of grains. This is an area of interesting developments in input management, as it economizes usage of labour by reducing dependence. Farmers need not necessarily buy the equipment, which in any case would be out of the reach of most of them. They can avail of the services of intermediaries who are dealers and operators of the required equipment in defined territories, and provide the equipment as a service on a use and pay model. There are cases where the dealers themselves come and perform the necessary activities as an outsourced process. This is certainly a welcome development in terms of the application of technology and process to overcome people-related issues. However, the disadvantages here are the inefficient clubbing of demand for services and the inadequate numbers of equipment in use, which deters usage at times by creating uncertain waiting times and loss of productivity due to loss of moisture in produce, and so on. One would expect this part of supply chains to pick up steam in the years ahead, improving efficiency and reducing certain value losses that are prevalent in the current frame of operations.

2.6 Conclusion

In agribusiness, input supply chain network challenges exist both in the farm gate segment and in produce processing unit segments. It is observed that the constituents of a supply chain, together with the focal firm, need to work on planning, sourcing, processing, and distributing processes to achieve efficiency. Further, the roles of agents are to be synchronized, though there has been a tendency to work on local optimization, with drops in overall system efficiencies. Procurement drivers and information technology drivers are critically impacting efficiency, as the former is largely an operational aspect which works on markets as well as on relationships.

Behavioural parameters substantially impact farm gate to direct and factory gate procurement, and also set up a balance between the cost and responsiveness of supply chains through other drivers, namely, facilities, transportation, and inventory. Agribusiness supply chain agents' roles and activities are also largely regulated by the GOI, and one needs to follow a certain framework of operations to achieve efficiency. The technology, process, and people trilogy has a role to play, and adoption of technology and matured processes are likely to benefit all concerned in the long run.

Appendix 2.1

A2.1.1 Transportation Options and Challenges for a Fertilizer Manufacturing Firm

Source: Reproduced from N. Chandrasekaran, *Supply Chain Management*, with permission of Oxford University Press, New Delhi, India.

It was the beginning of the Indian southwest monsoon in June 2009 when Mr. J. L. Narasimha Rao was firming up his SBU level plans, committed earlier to the corporate office in Chennai, on sale of fertilizers from his Kakinada plant in Andhra Pradesh. Mr. Rao was a firm believer in efficient operational management and in delivering the product to the right customer, at the right time, the right cost, the right mode and the right quality. He believed that his logistics management team is one of the best, and wanted to have a discussion with Mr. Sunil, his outbound transportation manager.

A2.1.2 Indian Fertilizer Industry

India has an agriculture-based economy, with the sector and its associated industries contributing to nearly one-fourth of the country's GDP. The fertilizer industry is one such associated sector that has shown significant growth patterns over the last decade. This industry in particular has been instrumental in achieving self-sufficiency in food grain production. The GOI has also created an environment conducive for the fertilizer business by way of subsidies/concessions to farmers and compensation to manufacturers. The Department of Fertilizers is responsible for policy formulation, and for promotion and development of the fertilizer industry. The government's participation in the industry has ensured success in addressing capacity requirements, though the sector's efficiencies have taken backstage. However, recent changes in government policies are aimed at sufficiently improving efficiencies in the sector. The government policies are oriented more towards deregulating the sector, with less emphasis on the policy parameters and long-term goals. India is one of the largest producers and consumers of fertilizers in the world.

In 2008–2009, India became the second largest consumer of fertilizer in the world, after China. The total consumption in that year was estimated at about

25.3 million tonnes in terms of various nutrients. The total production in that year was estimated at 32.8 million tonnes of urea/phosphatic/SSP combined, while the demand exceeded about 46.7 million tonnes.

The production levels of urea have ensured near self-sufficiency, while demand for nitrogenous fertilizers is wholly met by the indigenous industry. The raw materials needed for production of phosphatic fertilizers are imported on a large scale. The Indian fertilizer industry is, however, not completely devoid of challenges. The policy parameters are not lucid, especially for nitrogenous and phosphatic fertilizers, which slows down the domestic capacity-building process. Limited availability of feedstock is also forcing companies to look into joint venture options; fluctuations in gas pricing are also posing a significant challenge for fertilizer companies. The combined cost of feedstock and fuel accounts for anywhere between 60% and 80% of the total production cost, making the efficiency improvement process an immediate imperative for the sector.

The Indian fertilizer industry earlier was dominated by public enterprises, which still hold a major share; however, the private sector is also fast emerging, with a high installed capacity for phosphatic fertilizers. The sector as a whole is looking at bridging the demand–supply gap in the country, and the major players are striving towards achieving this objective. The surge in demand in the Indian fertilizer industry on one hand, and supply uncertainty on the other, render decision making a difficult task for all the companies; the cost–benefit ratio becomes a crucial factor. Fertilizer companies are hard pressed to take difficult decisions within a short period.

A2.1.3 Freight Rate Management by Government

The fertilizer industry was regulated for a long time in India because of the dynamics of the agrarian economic issues the country faces. The freight equalization policy is essentially concerned with the equalization of freight charges, where the short distance charges subsidize the long distance movements. The policy essentially operated on rail fares, and road fares are also accommodated wherever railheads are not available. In cases of combined rail and road movements, the rail freight up to the nearest railhead is taken into consideration, and then the road transport charges up to the terminal point are adjusted and equalized. Typically, the way the freight equalization scheme works is that the weighted average expenditures on freight, based on the expected lead distances and the corresponding tariffs, are estimated in advance, to be added to the ex-factory price. This provides the basis for the allocation of freight equalization fund. In the case of the fertilizer industry, not only the freight but also the total production cost is equalized by means of a subsidy, resulting in a price equalization account based on normative costs and assured rates of return. Mr. Rao is constantly worried about operating in such a regulated environment, because if the plant fails to achieve the normative basis of costs both in production and distribution, it could become a loser.

A2.1.4   Company Background

Company XYZ is a leading company in India, manufacturing a wide range of fertilizers, pesticides and specialty nutrients. XYZ has multi-locational production facilities, markets its products all over India and exports pesticides to various countries across the globe. The company has eight manufacturing units located in the states of Andhra Pradesh, Tamil Nadu, Maharashtra, Gujarat and Jammu and Kashmir. The company is also engaged in rural retail business in Andhra Pradesh through its outlets, "Mana Gromor Centres."

XYZ is among the largest phosphatic fertilizer players in India. The company also markets phosphogypsum and sulphur pastilles. The company's fertilizers are sold under well-established brand names like "Gromor", "Godavari", "Paramfos", "XYZ Gold" and "XYZ Super." XYZ has four fertilizer plants, of which two plants are located in Visakhapatnam and Kakinada in Andhra Pradesh, and two plants are located in and around Chennai in Tamilnadu. XYZ produces and sells phosphatic fertilizers of various grades, including di-ammonium phosphate (DAP) and single super phosphate (SSP), and has a combined installed capacity of 22 lakh tonnes of DAP/complex fertilizers and 1.32 lakh tonnes of SSP. The company's fertilizer products are also extensively marketed and sold through a network of dealers or "Mana Gromor Centres."

A2.1.5   Demand and Distribution Pattern

The company's sales volume in the year ended on March 2009 was 21.62 lakh metric tonnes (MT), which included 1.19 lakh MT of imported DAP and 0.55 lakh MT of imported MOP, compared to 21.71 lakh MT sold during the previous year. The company has been marketing its fertilizers in 11 states, including Andhra Pradesh, Karnataka, Maharashtra and Tamil Nadu. The production and sales of the company's fertilizers are highly dependent on and regulated by the availability of phosphoric acid and other raw materials.

The movements of fertilizers are strictly governed by movement orders issued by the Indian Government, entailing higher distribution cost because of multiple destination points. XYZ has a dominant presence in Andhra Pradesh, Tamil Nadu, Karnataka, Chhattisgarh, and Maharashtra, and is a significant player in West Bengal and Orissa.

XYZ continues to invest in infrastructural facilities, including raw material warehouses and bagging and distribution facilities. This helps the company in improving its capacity to handle higher volumes of both raw materials and finished goods. The company has also made sizeable investments at its Vishakapatnam and Kakinada plants to increase production capacity and enhance operational efficiencies. The research and development facility at Vishakapatnam enables the company to develop new grades of fertilizers, including water soluble fertilizers, sulphozinc, etc.

The setting up of 407 *Mana Gromor Centres* across Andhra Pradesh has enabled the company to sell its products directly to farmers, thereby reducing its dependence on intermediaries, which has benefitted both the company as well as the farmer community. This has triggered direct interaction with the farmers, something that has helped the company in developing new products to meet the specific and customized requirements of farmers.

A2.1.6 *Transportation Challenges*

Mr. Sunil, the logistics and distribution manager at the Kakinada plant, is confronted by various operational problems and constraints. He is primarily concerned about the modes by which his company's products reach end users, including those in remote markets, and about the average times taken by those modes. Company XYZ follows a multi-tier distribution system, as mentioned above. Sunil has to ensure the product reaches two categories of intermediaries—dealers/stockists as well as village retail centres (*Mana Gromor Centres*).

The planning for distribution of fertilizers each season is handled by the government to meet the seasonal demand. The manufacturer has to keep stock in readiness at intermediate points, such as dealers and retail centres, and satisfy the variations in demand accordingly. Company XYZ also has its individual market strategy to align its sales objective with that of the macro level policies, resulting in mutually beneficial outputs.

Table A2.1 below shows the month-wise sales of ammonium phosphatic fertilizers during the year 2008–2009.

As mentioned in Table A2.1, the total sale of ammonium phosphatic fertilizers for the year was 1.4 MT. The demand went up in the months of July–September and November–January and therefore, sales are higher during these periods. During the months of July–September, the demand stems primarily from the *kharif* season, and the demand during the months of November to January is due to the *rabi* season. It is advisable to have buffer stocks during the months preceding and following the peak seasons to accommodate sudden upsurges in demand.

The fertilizer is transported from Kakinada to various places that fall within the 800 km radius. Table A2.1 above gives the month-wise distribution plan for fertilizer from the XYZ plant. Every district within the 800 km radius has at least three stocking points. This is close to the rake point in the district for fertilizer.

The practice has been to transport the fertilizer by road if the distance from Kakinada is within 400 km and by rail if the distance is beyond 400 km. The loading factor varies significantly with road and rail. A normal truck can carry up to nine tonnes. In the case of rail transportation, fertilizer movement is achieved using rakes only. A–A broad gauge covered wagon has a capacity of 2,500 to 2,700 MT, while an open wagon carries up to 3,000 MT. A metre gauge wagon carries between 700 and 800 MT. XYZ has been moving broad gauge rakes with an

**Table A2.1 Distribution Plan for Ammonium
Phosphatic Fertilizer**

Quantity of Ammonium Phosphatic Fertilizers (000 tons) Moved Month Wise	% of Qty Sold	Qty 000 Tonnes
Apr	2	29
May	2	29
June	3	44
Jul	10	146
Aug	13	190
Sep	15	219
Oct	6	88
Nov	15	219
Dec	15	219
Jan	12	175
Feb	5	73
Mar	2	29
Total	100	1460

average load of 2,600 MT. These are the current operation conditions under which Sunil has been making his transportation decisions.

A2.1.7 Movement by Rail

Sunil has to liaise with the railway officials on a continuous basis to ensure rake availability, and he also needs to address full load requirements so as to achieve optimum benefit from the nominal cost of rail transportation. The production capacity at the Kakinada plant needs to be matched with the demand of the districts around Kakinada within a 800 km radius. While road transportation is faster, more efficient and easily accessible, the trucking industry in general poses certain challenges, such as complex documentation procedures and a higher cost per tonne compared to rail transport.

It may be noted here that the GOI offers primary movement by rail at actual cost and secondary movement by road subject to certain limits in reimbursement, like ₹200–300 per tonne. This would be a case of favouring rail transportation.

However, what is more important is to peg indirect costs such as ordering time cost and coordination cost.

The demand uncertainty needs to be met by opting for road transportation even when distances are greater than 400 km. This is because with rail transportation, rake availability needs to be confirmed and booked at least one week prior to the date of loading. Full wagon load conditions also need to be considered. This makes road transport more viable for smaller shipments to be delivered on a short notice period. Road transport service providers, however, operate in an unorganized environment, and Sunil needs to choose an appropriate transporter who will provide high service quality at a nominal cost.

The time taken for supply of rakes after placing a request with the railways is between one and five days. The turnaround time for rake load is between five and six days. The same distance when covered by road transport takes only three to four days. During the peak season, both road and rail capacities are fully utilized. In case of delay, backup storage needs to be arranged with the warehouse manager, since the stocking of other finished products such as pesticides and specialty nutrients also increases during the same period. The loading of products is possible only between 6 a.m. and 9 p.m. at the factory warehouse.

A2.1.8   Trade-Offs

While reliability of supply is a crucial factor in the distribution of fertilizers, the inventory transportation balance needs to be maintained at an optimal level so as to ensure the operational efficiency of the company. Sunil is faced with the challenge of revisiting his company's existing transportation plan and schedule and reducing the cost of operations, while maintaining an on-time delivery schedule to all the accessible districts. Demand information is given to the factory on a weekly basis; loading can be consolidated on a fortnightly or monthly basis depending on the distribution pattern, availability of road and rail capacity and cost of operations.

Sunil should also address the manpower requirements for loading of trucks directly for dispatch and for transportation to the railhead loading station. Labourers work in eight-hour shifts and during peak season, they work in two consecutive shifts. Two labourers are needed to load a nine-tonne truck. Sunil needs to tie up with road transporters occasionally to deliver cargo to the railhead loading station. On an average, 25 labourers are available to work in each shift.

Sunil is also given the option of outsourcing his transportation requirements to a third party logistics service provider. The service provider has the capabilities of providing packaging, scheduling, booking of appropriate modes of transport, and delivering goods as requested by Sunil. However, the cost of these services will be higher than the cost of conducting them in-house. By developing a long-term relationship with the service provider, cost benefits could be accrued over the years to come.

The transportation trade-off problem described above is a common challenge in any manufacturing setup that has to market products to interior parts of the country. The reliability of supply and availability of fertilizers are highly critical issues, since the agricultural sector is the backbone of the country's growth, and its products are highly seasonal. Rail is generally the preferred mode of transport for bulk commodities like fertilizers, with a lower cost of transportation, though there are constraints such as specific destinations and loading problems. Road transportation is more flexible and less time consuming than rail transportation, but comes at a higher cost. Transportation decisions revolve around rake loads, as well as road versus rail transportation benefit analysis. Till recently in India, about 73% of fertilizer was moved by rail and 27% by road. The XYZ plant has a different movement plan, as the plant is located in a fertile region where demand is concentrated within an area at a less than 400 km radius. Here, about 50% of the fertilizer is equally moved by road and by rail.

Sunil's responsibility is to meticulously evaluate the various options available for transportation, including outsourcing to a third party logistics service provider, deriving cost–benefit analyses and then making modal choice decisions. He also needs to take into account storage considerations during the peak seasons. The costs of primary and secondary movement versus storage costs need to be carefully evaluated, and appropriate decisions made. These decisions should most importantly take into account the seasonal nature of demand, and should be such that an optimum solution for the entire year is scheduled. There are also considerations of sudden upsurges in demand in specific locations, and hence, buffer stock would be needed at certain storage points and retail outlets. All these costs should be taken into account while making final decisions from the available outbound transportation options.

In the case of company XYZ, transportation cost plays a very significant role in the final profitability of the firm. Coupled with the warehouse and inventory costs, the logistical trade-off needs to be achieved for improving operational efficiency. The result of this valuation of transportation options would also have a considerable impact on the procurement process.

Discussion Points

1. Month-wise and mode-wise distribution plan
2. Budget for movement at ₹3.50 per tonne per km for road and for rail at 50% of the road cost, as it is subsidized
3. Highlight distribution challenges considering demand spikes
4. What would be the benefits of using a third party logistics service provider?

Notes

1. This case is for academic discussions only, and is not fully reflective of the actual policy environment in India.
2. The case was first written about and published by N. Chandrasekaran, *Supply Chain Management*, Oxford University Press of India, 2010, and has been revised now. Reprinted with permission of Oxford University Press of India.

References

Chandrasekaran, N., *Supply Chain Management*, India: Oxford University Press, 2010.

Dave, V. Groundnut production to rise this summer. *Business Standard*, May 16, 2011. http://www.business-standard.com/india/news/groundnut-production-to-rise-this-summer/435637/

Government of India. National Rural Employment Guarantee Act, 2005 *Gazette of India*, Sept. 7, 2005. http://www.nrega.nic.in/rajaswa.pdf (Accessed on March 28, 2012).

Government of India, *Economic Survey 2011–2012*. http://indiabudget.nic.in

PepsiCo India. PepsiCo India achieves splendid success in contract farming in West Bengal. Mar. 23, 2010. http://www.pepsicoindia.co.in/media/Press-Releases/release_03_23_10.aspx (Accessed on Mar. 27, 2012).

Punj, S. Tractor queen on a roll—Mallika Srinivasan, *Business Today*, Sept. 18, 2011. http://www.businesstoday.intoday.in/story/most-powerful-women-in-business-2011-mallika-srinivasan/1/18321.html (Accessed on Mar. 7, 2012)

The Hindu. Govt cuts subsidy on key fertilizers. Mar. 1, 2012. http://www.thehindu.com/business/Industry/article2950124.ece (Accessed on March 29, 2012).

The Hindu. The farmer outreach "ambulance". Mar. 12, 2012. http://www.thehindu.com/sci-tech/agriculture/article2985487.ece (accessed on March 30, 2012).

Wadke, R. Subsidy for sugarcane harvesters in Maharashtra. *The Hindu Business Line*, Aug. 25, 2011. http://www.thehindubusinessline.com/industryand-economy/agri-biz/article2396944.ece

Chapter 3

Farm Output: Market Intermediaries*

OBJECTIVE

In this chapter, various distribution options for farm output, such as directly from farm gate to consumer, wholesalers and other intermediaries, including mandis and auction centres, cooperatives and public agencies, and processors of farm output are discussed. The role of government agencies and organized players and challenges faced by them in managing their supply chains also are covered in the discussion.

3.1 Introduction

Marketing of agricultural produce is a crucial aspect of agribusiness, as it can significantly impact the two goals of supply chains: cost and responsiveness. The responsiveness is measured in terms of the speed with which produce reaches customers. There are a number of factors that influence the distribution channel of agricultural produce. These could include product bulkiness, perishability, seasonality, inadequate infrastructure, low market links at the farmer level and multiple intermediations because of poor structuring of financial flows. In spite of these constraints, the outbound movement of agricultural produce has been closely

---

* This chapter is reprinted from Chandrasekaran, N. *Supply Chain Management*. New Delhi: Oxford University Press India, 2010. With permission.

monitored by various interested groups such as the government for ensuring food availability and its security, farmers, consumers who are members of the general public, businesses, policy analysts and the media.

In this chapter, we analyse the various channels through which agricultural produce reaches the market, the challenges faced by supply chains and the various ways to improve supply chain objectives, so that the rewards for all constituents in the system are understood and achieved.

3.2   Trends in Output

The production of agricultural produce has increased over the years. Table 3.1 provides data on growth in agricultural production, Table 3.2 gives growth data for fish production in India and Table 3.3 reflects the data on the all-India production of milk, eggs and wool.

It may be observed from Table 3.1 that production of the major crops in India is growing, but not consistently. There have been years when agricultural production has dropped significantly, as in the case of pulses and coarse cereals in 2011–2012.

In the case of commercial crops, other than cotton, crops such as sugarcane, oil-seeds, jute and mesta have had fluctuating trends. Except for sugarcane, the other crops would directly impact trade of the produce. In India, most of the sugarcane harvested goes into production of plantation white sugar. However, during years of low production, intermediaries actively try to divert cane to low-end processes like manufacture of jaggery and *khandsari*, which are used as sweetening agents, especially in North India, for beverages.

It may be observed from Table 3.2 on fish production in India that marine production has recorded a sluggish increase over the decades. During the period 2000–2001 to 2009–2010, it grew from 28.11 lakh tonnes to 29.89 lakh tonnes. However, inland fish production has grown significantly, especially between 2000 and 2011, from 28.45 lakh tonnes to 48.62 lakh tonnes.

Similarly, Table 3.3 depicts the significant growth in milk production and egg production between 2000 and 2010. However, wool production has been sluggish during this period.

This kind of agriculture and allied products growth is bound to have a significant impact on marketing channels.

3.3   Marketing Channels

It may be relevant here to mention that the output marketing system has evolved over centuries. A number of changes have happened in recent times, especially after the country's independence, with the planned development of various sectors. The government has not only focused on agricultural policies but also on financial

Table 3.1 Production of Major Crops during 2002–2003 to 2010–2011

(Million tons)

Crop	2002–2003	2003–2004	2004–2005	2005–2006	2006–2007	2007–2008	2008–2009	2009–2010	2010–2011[a]	2011–2012[b]
Rice	71.82	88.53	83.13	91.79	93.35	96.69	99.18	89.09	95.32	87.10
Wheat	65.76	72.15	68.64	69.35	75.81	78.57	80.68	80.80	85.93	93.90
Coarse cereals	26.07	37.60	33.46	34.06	33.92	40.76	40.03	33.55	42.22	30.42
Pulses	11.13	14.91	13.13	13.39	14.20	14.76	14.57	14.66	18.09	6.43
Food grains	174.77	213.19	198.36	208.60	217.28	230.78	234.47	218.11	241.56	217.85
Oilseeds	14.84	25.19	24.35	27.98	24.29	29.76	27.72	24.88	31.10	20.89
Sugarcane	287.38	233.86	237.09	281.17	355.52	348.19	285.03	292.30	339.17	342.20
Cotton[c]	8.62	13.73	16.43	18.50	22.63	25.88	22.28	24.02	33.43	36.10
Jute and Mesta[d]	11.28	11.17	10.27	10.84	11.27	11.21	10.37	11.82	10.58	11.22

Source: Directorate of Economics and Statistics, Department of Agriculture and Cooperation.

[a] Fourth advance estimates released on 19.7.2011.
[b] First advance estimates released on 4.9.2011.
[c] Million bales of 170 kg each.
[d] Million bales of 180 kg each.

Table 3.2   Fish Production in India

	In Lakh Tonnes		
Year	Marine	Inland	Total
1950–1951	5.34	2.18	7.52
1960–1961	8.80	2.80	11.60
1970–1971	10.86	6.70	17.56
1980–1981	15.55	8.87	24.42
1990–1991	23.00	15.36	38.36
2000–2001	28.11	28.45	56.56
2001–2002	28.30	31.26	59.56
2002–2003	29.90	32.10	62.00
2003–2004	29.41	34.58	63.99
2004–2005	27.79	35.25	63.05
2005–2006	28.16	37.56	65.72
2006–2007	30.24	38.45	68.69
2007–2008	29.19	42.07	71.26
2008–2009	29.78	46.38	76.16
2009–2010	29.89	48.62	78.51

Source: Department of Animal Husbandry, Dairying and Fisheries, Government of India.

systems, public policies like the Essential Commodities Act and public distribution systems, infrastructure and protection to a few agro-based industries, which would all have impacted the reach of agricultural output market.

Marketing channels for agricultural products can be grouped under the following classifications:

1. Direct to consumer and through the market
2. Through wholesalers and intermediaries
3. Through public agencies or cooperatives
4. Through processors

3.3.1   Direct to Consumer

The marketable surplus that moves directly from the farmers to consumers is a small portion of the total market. It is estimated that this segment would constitute about 1%–2% of the marketable surplus. Over the years, this figure has improved

Table 3.3   All India Production of Milk, Eggs and Wool

Year	Milk (million tonnes)	Eggs (billion units)	Wool (million kg)
1990–1991	53.9	21.1	41.2
2000–2001	80.6	36.6	48.4
2001–2002	84.4	38.7	49.5
2002–2003	86.2	39.8	50.5
2003–2004	88.1	40.4	48.5
2004–2005	92.5	45.2	44.6
2005–2006	97.1	46.2	44.9
2006–2007	100.9	50.7	45.1
2007–2008	104.8	53.6	44.0
2008–2009	108.6	55.4	42.9
2009–2010	112.5	59.8	43.2

Source: Department of Animal Husbandry, Dairying and Fisheries, Government of India.

only marginally. For marketable surplus from rural India that is directly reaching the market, we need to evaluate the following points:

1. A large number of land owners and agricultural producers are small and marginal farmers, and most of them do not generate adequate marketable surplus. Many of them have supplementary livelihood sources and benefit from many schemes for generating economic value.

   India has increased its marketable surplus of agricultural products from 11 million tonnes in 1950–1951 to more than 200 million tonnes of cereals by 2011–2012. But according to an estimate, small farm produce accounts for nearly 41% of total food grains produced. Rice accounts for 49%, wheat for 40%, coarse cereals for 29% and pulses for 27%. Over 50% of fruits and vegetables are grown on small farms.

2. Medium-sized farmers and producers do not have the kind of economy of scale that would enable them to reach markets directly, and hence, they are dependent on intermediaries. Although this trend is likely to continue in future, the number of intermediaries may be reduced, benefiting supply chains in select areas and produces.

3. Although a number of cooperative networks are operating, the more successful are those that are involved with processing and marketing processed products. Dairies may be an exception to product form change; however,

processing is important, as the produce is highly perishable. This is mainly because of the fact that Indian milk market is for bulk and fresh milk.

4. A small portion of large farms, especially in high-value products and meat and fisheries, are engaged in direct marketing.

Thus, we believe that the direct marketing segment is still a small portion of the total agricultural produce raised in India, and is not likely to change dramatically.

Then the question that arises: how would the producers and suppliers to the agribusiness network benefit with a reasonable share of profits arising from the supply chain networks?

3.3.2 Through Wholesalers and Intermediaries

There is a massive volume of produce handled by wholesalers and intermediaries in India. According to government sources, the private sector handles nearly 80% of the marketable surplus in agricultural products. The intermediaries include village traders, primary and secondary wholesalers, commission agents and also processors.

Intermediaries thrive because of the small volumes of food grain and other agricultural products that small- and medium-sized farmers produce. Other factors that influence the role and extent of intermediaries include the following:

The changing pattern of demand conditions, which is due to increased disposable income, provides scope for wider choices; price and budget are more liberal in certain cases.

1. Change in buying habits, such as greater frequency of buying, opting for buying from departmental stores and exclusive outlets, health consciousness and limited ability to store at the family level, as there is an increase in the number of nuclear families, and space is increasingly at a premium.

2. On the supply side, in an earlier era farmers did not have an understanding of the markets, had access to inadequate data, and hence, were dependent on intermediaries. The situation has changed dramatically, as the explosion of media such as radio, television, private channels and Internet and their coverage has given tremendous, easy access to data on prices, arrivals and all other relevant information. Using these media, today farmers are able to use intermediaries more effectively. Intermediaries also now understand that their survival is more dependent on the value-added services they provide, rather than their exploitation of the gifts of gaps in the market system.

3. Another important development at the supply services side is the improvement in infrastructure such as storage, transportation and market centres. These factors, along with the access to media and information technology, act as a boon to farmers when selling their output to intermediaries. This is because they are now informed producers/traders of produce, reflected in the confidence with which they can stock and sell for much better terms, rather than the hurry to monetize that was characteristic of the earlier decades of independent India.

4. There is an increase in linkages with distant and overseas markets, which intermediaries enable access for agricultural produces. The focus of intermediaries has shifted from taking produce to the nearest market to the business of reaching large and wide markets, where the advantages of volume and sustainability of business are in focus.
5. There are a number of changes in the forms and degrees of government intervention in the marketing of agricultural goods and infrastructure. This has made the intermediation profession more evolved and matured, moving towards performance and delivery rather than short-term gains. Government thinking and adoption of policies not only bring credence to the trade, but also enable evolution of competitive characteristics, shoring up the intensity of rivalry in the informal sector and making it more scientific and sensitive to value creation.

It may useful here to go through the market systems and some of the structural features of the Indian agriculture sector in relation to the above factors.

3.3.2.1 The Mandi System

Often, agricultural and allied production by an individual farmer in a field is of such meagre quantity that it is economically inadequate for the farmer to sell it directly in nearby regulated markets. This necessitates an aggregator to act as an intermediary for the farmer. Many farmers supply to one aggregator, who then sells through marketing channels that could include nearby markets, referred to as *mandis*. The produce then moves through wholesalers, before proceeding to retailers and consumers. Essentially, these are markets in small towns and cities to which farmers from nearby villages would bring the agricultural produce at harvest time and where traders would buy this produce from a large number of farmers selling their produce to a small number of traders who are the aggregators. This differentiates agricultural marketing from industrial products marketing, necessitating creation of separate special systems, institutions and infrastructure.

The roles played by brokers and commission agents include the following:

■ Collection of the produce from a number of small farmers in villages or from larger farms and carrying everything in a single lot to towns/cities for sale
■ Arranging transportation in single lots or unit loads and moving produce to the markets
■ Enabling better price realization.

Thus, commission agents aggregate produce into marketable lots, improve price realization and enable transportation. These factors created a class of brokers and commission agents, who charged a commission from the farmers.

At times when a farmer does not find the prices quoted by the trader good enough, the trader would offer to keep the goods and wait for a buyer offering a better price. This could take days, weeks or months. The farmer would regularly check the latest price offered with the trader, and once the farmer agreed to a price, the trader would sell the goods on his behalf. For keeping the goods and arranging a buyer at a better price, the trader would again charge some extra fees/ commission. This system came to be known as *aadhat*. This arrangement is similar to a consignment sale.

3.3.2.2 Informal Financial System and Intermediation

Most of the farmers in India, as mentioned earlier, are small and marginal farmers who do not have sufficient funds or financial support for raising agricultural produce. They depend on money lenders, who are also incidentally traders and commission agents. Over years of regular buying of the produce from the same farmers, a trader had reasonable assurance that for each new harvest, his farmers would bring their produce to him for sale. The trader, therefore, found it relatively risk-free to lend money to such farmers in times of their need. A large number of traders in these markets are also money lenders. However, it is believed that such traders charge exorbitant interest rates, and bind farmers to a vicious cycle of debt burdens. It is also a sombre fact that there has been a large incidence of suicides by small and marginal farmers, as they were neither able to return the debt nor lead a decent standard living for themselves and their families.

All these transactions, as is obvious, were informal and unregulated systems and practices that evolved over centuries, mostly because of the socio-economic conditions and culture practices prevailing in the society at various times in history. The farmers were always an unorganized lot, with little or no strong resources, and the traders, *aarhitia* or brokers were wealthier and more powerful. The level of maturity of the social and legal system was inadequate to establish socio-economic justice. This led to gross exploitation of the market, wherein neither the producers nor the consumers were better off, but intermediaries in the system flourished at the cost of both.

With the independence of the country in 1947, many legislative and structural reforms such as abolition of the *zamindari* system were undertaken by the government. Similarly, reform was introduced by enacting the Agricultural Produce Marketing Committee (APMC) Act, to regulate the system (Planning Commission). However, even today, there are pockets of India where intermediaries are still taking advantage of the plight of the poor farmers.

3.3.2.3 Procurement by Government Agencies and Cooperatives

The Food Corporation of India (FCI) is the nodal Government of India (GOI) agency that, along with other state government agencies, procures wheat, paddy and coarse grains directly from farmers. This is one of the government's major interventionist programmes to

1. facilitate fair price realization for farmers and
2. ensure availability of food grains to the weaker sections of Indian society under any given circumstances through buffer stock maintenance operations. This is more of a food security measure.

The FCI and state government agencies establish a number of purchase centres at various points, so that farmers can utilize the benefits of the scheme. The number of procurement centres is decided by the state governments during every season, based on a number of parameters. The focus is to ensure that the minimum support price programmes work well, and farmers are not compelled to sell to agents or processors at a price less than the minimum support price. It may be noted here that farmers are free to sell their produce at prices higher than the minimum support price, if they find any such opportunities in the market.

Typically, the government procurement and food subsidy system in India operates as explained below:

1. The GOI fixes the minimum support price (MSP) for food grains, at which procurements will be made from farmers.
2. The food grains are stocked and carried over for distribution through the Public Distribution System (PDS), while a buffer stock is maintained for managing any uncertainties.
3. By the socialist policies of the government followed, PDS distribution is heavily subsidized, as it is targeted at benefiting low-income groups, and a number of schemes for the poor are implemented using the stock carried.
4. Food subsidies in India have been severely criticized of late, as they have increased the non-plan expenditure burden. Food subsidies, which were around ₹2,850 crore in 1991–1992, have increased to ₹72,823 crore in 2011–2012.

Food production of the main staple crops, namely, rice and wheat, has grown over the years, but with an inconsistent trend, especially for wheat. This would certainly impact procurement, as during years of low production, one would expect the market to pay more than the minimum support price, and farmers would sell through the market. In years of high production, the markets would bear down and hence the FCI would procure more. Further, the FCI would be constrained by space and policy considerations, which we will discuss later. Table 3.4 provides data on recent trends in procurement of rice and wheat, and the share of procurement in total production.

Some of the observations from the above table are as follows:

1. Rice procurement has increased from 23.7% of production in 2001–2002 to 33% of total production in 2011–2012.
2. Rice procurement increased significantly in 2008–2009, and was around the same level in 2011–2012.

Table 3.4 Trends in Food Grains Procurement (Million Tonnes)

Year	Procurement			Production		Procurement as % of Total Production	
	Rice	Wheat	Total	Rice	Wheat	Rice	Wheat
2001–2002	22.13	20.63	42.76	93.34	72.77	23.7	28.3
2002–2003	16.42	19.03	35.45	71.82	65.76	22.9	28.9
2003–2004	22.83	15.8	38.63	88.53	72.15	25.8	21.9
2004–2005	24.68	16.8	41.48	83.13	68.64	29.7	24.5
2005–2006	27.66	14.79	42.45	91.79	69.35	30.1	21.3
2006–2007	25.11	9.23	34.34	93.35	75.81	26.9	12.2
2007–2008	28.74	11.13	39.87	96.69	78.57	29.7	14.2
2008–2009	34.1	22.69	56.79	99.18	80.68	34.4	28.1
2009–2010	31.46	25.38	56.84	89.09	80.8	35.3	28.1
2010–2011	34.2	22.51	56.71	95.32	85.93	35.9	26.2
2011–2012	34.46	28.34	62.8	104.32	93.9	33	30.2
2012–2013	–	38.02	38.02				
Average	27.44	18.76	46.19	91.51	76.76	29.76	23.99
Standard deviation	5.84	5.88	10.07	8.62	8.32	4.56	6.05

Source: Government of India.

Note: Rice data are for the annual periods between October of a year and September of next year, and wheat for April of a year to March of next year.

3. Wheat procurement has increased from 20.63 million tonnes in 2001–2002 to 38.02 tonnes in 2012–2013. The trend has been volatile, with drops in procurement to as low as 92.26 lakh tonnes in 2006–2007, which was around 12.2% of the total wheat production. The year 2007–2008 was also similar. However, procurement has gone up over the years since then, reaching a peak level of 38.0 million tonnes in 2012–2013.

4. The average procurement of rice was around 27.4 million tonnes, while that of wheat was 18.8 million tonnes for the years between 2001–2002 and 2011–2012, whereas the standard deviation was around 5.8 million tonnes for both produces, namely, rice and wheat, indicating high volatility in wheat procurement.

The objective of PDS management is to make food of a certain quantity and affordable price available to the poor and vulnerable sections of the society at all times, and also to provide fair returns to farmers. The most ideal situation would be to procure more from farmers during years of surplus production, when one would expect prices to be reasonable, and release more stocks during years of shortage. However, the reality is that there is massive stress on the system, for reasons that are given below:

1. The cost of management of buffer stocks and the PDS is high, as the economic costs include procurement prices, transportation and handling, cost of carrying stocks and obsolescence due to uneven stock and off-take.
2. Inflationary trends in food prices in India during the last two years have led to severe criticism of the government's food policy. The years 2006–2007 and 2007–2008 were characterized by low-cost inflation for wheat and rice. However, procurement of both grains did not move up significantly. "In fact, wheat procurement declined sharply during these periods, contrary to expectations. The year 2009–10 witnessed sharp inflation in the prices of rice and wheat, and thus, one would have expected lower procurement of rice and wheat during this period. But the trend was again contrary" (Basu, 2011). This raises doubt about the efficacy of the procurement system that is operating in India.
3. As per the current practice, FCI releases in open market operations are bought mostly by millers and rarely by traders. The millers are not allowed to resell to traders and make profits. Ideally, one would expect traders to bid in open market sales, so that they enjoy opportunities to book profit on volumes when prices are high and low surplus is available from the farming community. This does not happen, which again defies economic logic. The reasons could be the aggressive monitoring and controlling mechanisms on reselling and profiteering, which are not helpful to control prices, as releases are not adequate when they have to be!
4. There has been severe criticism again on the shortage of storage space with FCI and state governments, and on the poor quality of rice and wheat. Obsolescence again is looked upon as a crime in a country like India, where food inflation is a shocker and a substantial population of people who are downtrodden go without adequate food. Though FCI has been doing well in adding capacities, it is a challenge to rise up to the levels of procurement volumes that are ideal.

This is a vociferously debated subject in India, and enormous amount of reports and other information are available in the public domain. Our objective here is to highlight how public procurement plays a critical role in supply chain networks, though there are lots of challenges, and the desired levels of efficiency and responsiveness are not achieved.

3.3.3 Producers' Cooperative

A cooperative system is a voluntary organization established and managed by its members under a nurtured government system that enables farmers to market their farm products collectively. A cooperative marketing society's main purpose is to help members to market their produce more profitably than would be possible if they traded through the private sector. This system works when credit flow is even and fair at the market, so that farmers need not depend upon contractors and traders for credit, and can sell directly at the markets. Even for those who have borrowed from the private sector, the cooperative system helps to establish an alternate route and structure so that they may achieve fair realization on their produce.

The producers' cooperative is mainly meant for the benefits of producers, and this would indicate a bias in principle. In reality, cooperatives are a way of improving collective bargaining power, as individual farmers operate small holdings and produce quantities that individually may not reach marketable lot sizes.

In India, there is a four-tiered federal structure percolating down to the level of primary societies at village clusters. The structure is shown below.

Level	Organization
National	National Agricultural Co-operative Marketing Federation (NAFED)
State	State Co-operative Marketing Federation (SCMF)
Regional	District or Regional Marketing Society (DMS)
Village	Primary Marketing Societies (PMS)

The National Agricultural Co-operative Marketing Federation (NAFED) is a federal organization of state level apex co-operative marketing societies in India.

It was established in 1958, with its headquarters in New Delhi. Its operations cover the entire country. The objects of the NAFED are to organize, promote and develop marketing, processing and storage of agricultural, horticultural and forest produce; distribution of agricultural machinery, implements and other inputs; undertaking of inter-state, import and export trade, wholesale or retail as the case may be; and acting and assisting with technical advice in agricultural production for the promotion and effective working of its members and cooperative marketing, processing and supply societies in India (NAFED, 2012; see Table 3.5).

Highlights of operation of NAFED in 2011–2012 (NAFED, 2012):

1. NAFED procured 23,028.24. MT of various oilseeds and oils, including groundnut pods, mustard seed, soya bean, sunflower seed and assorted oils valued at ₹530.6 million, and sold a quantity of 24,274.44 MT of various oilseeds and oils valued at ₹632.2 million.

Table 3.5   NAFED Key Operations in 2011–2012

	Procurement (million tons)	*Sales (₹ million)*	*Key Produces*
Oilseeds	530.57	632.21	Mustard and soya bean
Pulses	517.80	666.20	Gram and *moong*
Horticulture	130.84	43.36	
Food grains	5829.83	6872.92	Paddy and wheat
Total	700.90	821.47	

Source: NAFED, 2012.

2. NAFED purchased 18,625.68 MT of pulses, such as *arhar,* gram, *masoor, moong, urad* and assorted pulses valued at ₹517.8 million, and sold a total quantity of 23,357.73 MT of these pulses, valued at ₹666.2 million.
3. NAFED purchased 8,542.46 MT of various horticultural commodities such as onion, potato and other vegetables and fresh fruits, valued at ₹130.8 million, and sold a total quantity of 2,223.52 MT of these items, at a value of ₹43.4 million.
4. NAFED purchased 51,1866 MT of food grains such as barley, *jowar,* maize, paddy, rice and wheat valued at ₹5,829.8 million and sold 43,3031 MT of these grains, with a total sales value of ₹6,872.9 million.
5. NAFED achieved a turnover of ₹1,063.28 crore during 2011–2012. It earned a gross profit of ₹45.68 crore, but due to huge interest liabilities on outstanding loans, there was a net loss of ₹188.42 crore.
6. NAFED continues to enjoy the status of 'star export house' as per the certificate of recognition issued by the Ministry of Commerce, GOI. Exports by NAFED on its own account during the year 2011–2012 were to the order of ₹11.52 crore.
7. NAFED continues to be one of the canalizing agencies for the export of onion. NAFED directly exported 6,147 MT of onion to different destinations, valued at ₹11.52 crore. In addition, NAFED also issued no objection certificates (NOCs) for export of 323,293 MT of onion valued at ₹411.26 crore.

Table 3.6 shows an interesting trend, that is, the decreasing turnover of NAFED, which was substantial in the last two years. Though turnover peaked in the year 2009–2010, the next two years witnessed a sharp decline. It may be highlighted that though third-party exports of a large volume of onions are given no objection certificates by NAFED, on its own account, it has exported low volumes, showing lesser patronage. Though as a structure, NAFED is critical for the supply chain networks of Indian agriculture, it faces limitations in terms of direct operations.

Table 3.6 NAFED Turnover over the Last Five Years

	₹ million				
Year	2007–2008	2008–2009	2009–2010	2010–2011	2011–2012
Total turnover	47066.5	50650.5	63733.2	20085.1	10632.8

Source: NAFED, 2012.

The efficiency of NAFED operations depends upon the role played by the state marketing societies and regional and primary societies. The apex marketing societies are expected to undertake marketing operations on behalf of affiliated societies, particularly in the field of inter-state trade, and to support farmers through distribution of agricultural inputs and other goods required by them. The regional societies are expected to coordinate the functions of the primary marketing societies, both in regard to the marketing of agricultural produce and the distribution of agricultural requisites and consumer goods.

The primary marketing societies are, by and large, located at the local markets or *mandis*. In few states, marketing societies have been organized at the headquarters of the revenue block in local administration of a district, covering the whole block. Efficiency in operations is critical for efficient and responsive supply chain networks.

In India, there is another form of cooperative, which is again successful in the case of certain produces. These are producers' and processing cooperatives. These cooperative processing and marketing organizations are driven by the following factors.

1. The large base of farmers and producers add value through processing. Some of these include large units in plantations and processing plants for tea, coffee, spices, rubber, sugar, oilseeds, cotton and jute.
2. The number of units that functions as marketing cooperatives for procurement in the hinterland facilitate and handle value addition. These are small- and medium-sized units, such as rice mill hullers, jute making units, cotton ginning and processing units and vegetable and horticultural product processing units. A number of poultry, fishery and meat processors are also set up as farming cooperatives, and share small- to medium-sized storage and handling facilities for value enhancement.
3. Another factor is government encouragement, under tiny, micro, small and medium enterprises schemes, in the form of a number of concessions and other support, and restricting the business for them so that there is decentralization of industries and employment generation takes place benefiting a large society.

Thus, cooperative units play a leading role in agricultural produce supply chains, as they improve the marketing of produce with value addition and limit the levels of intermediation.

3.3.4 Through Processors

In India, buying and selling of agricultural produce mainly take place in market yards, sub-yards and rural periodic markets spread throughout the country. There are in all 7,246 regulated markets in the country (as on June 30, 2011) and 21,238 rural periodic markets, about 20% of which function under the ambit of regulation. The agricultural marketing system in India has experienced government intervention and evolved over the years. Private sector agricultural produce processors were earlier bought through regulated markets, directly or through contractors or other intermediaries.

However, there are certain crops like sugarcane where satellite farming is allowed. This would mean that there could be an allotted geographical area around processing plants, within which the plants could develop farming activities with landowners and cultivators. The plants can procure the produce and operate their plants profitably. Farmers have the right not to sell, if they choose to do so. In most cases, the government fixes the minimum procurement price in such farming activities, so that farmers are not exploited. Further discussions on this are handled in the chapter on agricultural processors.

A typical, formal contractual farming arrangement confers benefits to both producers and purchasers. Farmers avail of remunerative marketing opportunities by way of assured procurement of their produce of desired quality. The contractor sponsor procures produce at a pre-determined price, based on certain quality parameters. Normally, the contract sponsor provides technical support and also input services to achieve the desired outcome and quality.

"One may note that the model APMC Act 2003 stipulates institutional arrangements for registration of sponsoring companies, recording of contract farming agreements, and indemnity for securing farmers land; and the Act also lays down a time-bound dispute resolution mechanism. Contract farming with a lot of informality has been prevalent in various parts of the country for commercial crops such as sugarcane, cotton, tea and coffee" (planningcommission.nic.in).

In recent times, some Indian states have amended the APMC Act to allow contract farming. There have been private sector initiatives in direct procurement from farmers, such as Pepsi Foods Pvt. Ltd., Tata Rallies, Mahindra Shubh Labh and Cargill India. The focus in contract arrangements has been towards improving farm economics through elimination of waste and productivity enhancement, thereby increasing farm income and at the same, helping the corporate entity achieve its goal of linking agriculturalist corporate activities to production systems in the agribusiness sector.

There have been some instances of appreciably successful contract farming in India, notwithstanding the criticism raised by different interest groups. ITC promoted the *e-choupal* model, which seeks to address the constraints faced by the Indian farmer arising out of small and fragmented farm holdings, weak infrastructure, supply chain intermediaries and the lack of quality and real-time information. ITC appoints *sanchalaks*, who are local leaders. The *sanchalak* sets up a kiosk with

Internet connectivity. Farmers use this facility to obtain real-time market information on prices, availability of inputs, weather data and other issues that are of relevance to them. The local leader, also on behalf of ITC, aggregates and procures the product, which is taken over by ITC for processing. ITC also enables farmers to aggregate their buying needs both for agricultural inputs and farming as well as for consumption, and provides group bargaining power.

It is estimated that ITC's intervention in agribusiness supply chains has permitted farmers to increase their sales by 10%–15%. Further, ITC has succeeded in generating procurement cost saving to the tune of 3%–4%, allowing it to incrementally improve its competitive position (Patnaik, 2011).

Another company which is engaged in corporate farming in India is Pepsi. Similarly, McDonald's has mentioned that through its supplier McCain, it has worked closely with farmers in Gujarat to produce process-grade potato varieties, which are not common in India. According to Bakshi, "These are mainly contract farmers in Deesa (North Gujarat) and Kheda (Central Gujarat). We continue to contribute in the growth of these farmers. McDonald's remains steadfast to its commitment of working with local suppliers and farmers to source all its requirements in India." (Indo Asian News Service, 2012)

Apart from these multinationals, a number of large and middle-sized companies, who have set up successful agribusinesses, are working in partnership with farmers. This list includes companies like Tata, HUL, Britannia, Nestle, AVT Group, Cavin Kare, Hatsun Agro, Marico, Kaleesuwari, K.S Oils and Heritage, to name a few.

This fact highlights the scope for private processors to improve supply chains in agribusiness, provided the prevailing atmosphere provides all the necessary support, enabling them to compete in a fair and regulated environment with less controls and more enabling polices.

3.4 Conclusion

Supply chains for agricultural output face the challenges of product bulkiness, perishability, seasonality, inadequate infrastructure and low market links at the farmer level. There has been a trend towards multiple intermediations because of poor structuring of financial flows. Agricultural output has been growing over the years. The need to manage output and prices has been a challenge. The government has to manage farming community economics with fair price realization and at the same time provide food security to the poor and vulnerable members of society in the India population. Output flow management is a key supply chain process, wherein market and regulatory forces interact through governments, private players and the farming community to achieve efficiency and responsiveness. There seems to be a lot of misgivings about intermediaries through contractors. Traders and money lenders play active roles, as most of the farmers are small and marginal, and are unable to raise marketable produce lots. Aggregation is critical. Of late, with the liberalization of the APMC Act and a few state governments' initiatives, contract and corporate farming initiatives are gathering momentum, and the use of the latest technology is bringing about more awareness. Output management practices in agribusiness supply chains in India are expected to evolve for the better and achieve maturity, improving supply chain profits.

Appendix 3.1

A3.1.1 Agricultural Crops and Intermediation Process: Pomegranate and Banana

Agricultural crops, especially of cash crops and fruits, are likely to have an intermediation process different from each other as well as in different locations for the same crop. This is mainly because of the number of stakeholders in the supply chain of such crops and wide variation among producers' size. Also, there are factors like regional conditions and business practices influencing such supply chain decisions. These arise because of clustering of production centres, role of aggregators and agricultural practices. It is important to understand the nuances of such intermediation process to improve the effectiveness of supply chain. Here we have explained such an intermediation process through two crops, namely pomegranate and banana.

A3.1.1.1 Pomegranate

The largest producer of pomegranates in India is the state of Maharashtra, which accounts for 70% of the total production. The districts in which pomegranates are cultivated in major volume are Sangli, Nasik, Ahmedanagar, Pune, Dhule, Aurangabad, Satara, Osmanabad and Latur. Nasik district alone accounts for 26% of the total production of pomegranates in the state of Maharashtra.

Pomegranate is propagated through grafting. The plant starts yielding fruit from the second year onwards. Of late, tissue-cultured saplings are also being planted in India. Around 80%–90% of the farmers in these districts use drip irrigation, as availability of water is a problem in many areas of the cluster. Pomegranate is harvested round the year in the cluster.

Fruit harvest volumes increase by 10%–15% mainly from July to September, and again from November to March. Harvesting volumes of the fruits are relatively low from April to June.

The supply chain network of pomegranates in India is depicted in Figure A3.1.

The major players involved in the trade of pomegranates are farmers, pre-harvest contractors, village level aggregators, commission agents, wholesalers, semi-wholesalers and retailers. The role played by major stakeholders and the value added at each stage are briefly captured below.

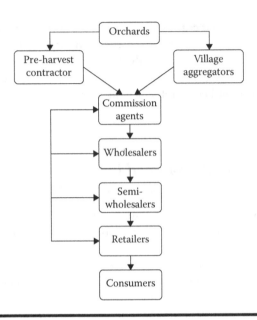

Figure A3.1 Supply chain network of pomegranates in India.

Farmers: The average landholding of pomegranate farmers is around eight hectares, spread into three to five land parcels. Around 50% of the land is used for pomegranate cultivation and the rest is used for growing other crops. The plants are sown in a square system that accommodates 750 plants in a hectare. Most of the farmers set up their orchards in different land parcels, and each orchard is of a different variety.

Pre-harvest contractors: The pre-harvest contractors are responsible for harvesting of fruits, sorting, grading, packaging and transportation to destination markets. The trader decides on the time of harvesting depending upon the market demand. When fruits attain maturity, the trader informs the farmers to start plucking. It is done by the farmers using their own family members, except in case of some of the large farms, where the local labour is employed. The contractor arranges for pickup of fruits from the farm gate and transport to the point of aggregation in smaller trucks of three metric tonne (MT) capacity. The entire cost of weighing, grading, packing, loading, transportation to destination markets, unloading and commission is borne by the trader, and the farmers are paid the farm gate price. This part of supply chain handles approximately 85%–90% of produce from the orchards. Farmers prefer to sell entire quantity irrespective of size and grade at the time of harvest. Price is linked to quality. Contractors decide on harvest and focus on aggregation of unit loads.

Village level aggregators: When the farmer sells pomegranates through a village level aggregator/transporter, the aggregator arranges for transport and brings the produce in crates to the APMC markets. The total expenses borne by the farmer includes the cost of transportation and loading/unloading charges, as well as the margin payable to the village level aggregator. Village level aggregators handle about 10%–15% of the produce from orchards. Transactions are driven by market forces and price is linked to quality. Typically, farmers decide on harvest time and village level aggregators sell in nearby APMCs.

Commission agents: They facilitate trade between the contractor and the wholesaler, for which they charge a commission of 8%–10% from the contractor. Payment to the contractor is made by the commission agent on behalf of the buyer/wholesaler.

Wholesalers: They are the bulk breakers and are responsible for distribution of pomegranates to various locations in the country. The wholesaler pays marketing cess at the rate of 1.05%, and sends the produce to semi-wholesalers or retailers.

As mentioned earlier, only 10%–15% of the produce from an orchard is sold through aggregators in APMC markets. The produce is sold by open auction through commission agents, and they charge 8% commission from the farmers. The produce is bought by traders, who pay marketing cess @1.05%. Pomegranates are graded and packed in premises provided by the commission agent, and the cost is borne by the trader.

Some of the salient features of the price build-up are mentioned below:

- There are around five intermediaries between the farmer and the consumer. The intermediaries are the contractor, the commission agent, the wholesaler, the semi-wholesaler and the retailer.
- In moving of produce from the farmer to the consumer, the price builds up by around 2.5 times.
- Since pomegranate is a hardy fruit, wastage is quite low along the value chain, that is, 3%–5% at the farm level and 1% during handling and transport. Major losses occur at the retailer level (both weight and value), that is, around 5%–10%, if the produce is not sold on the same day.

Some of the gaps widely discussed in the supply side are that

- farmers have limited knowledge about scientific crop management;
- plucking is mostly done manually and the use of equipment for harvesting is limited to a few large growers;
- pomegranates are graded and packed manually at the farm level, at which level pre-processing facilities such as pre-cooling, washing, grading and sorting are absent.

There are a number of interested parties such as state governments, central government departments, financing institutions and banks, along with infrastructure companies, that are working on improving the supply chain.

The above discussion depicts the role and responsibilities of supply chain players in the pomegranate supply chain networks of India. We will now discuss another crop, which is similar to the pomegranate in terms of economic factors.

A3.1.1.2 Banana

The banana, one of the important commercial crops, has seen production volumes steadily growing. India, which produced nearly 30 million tons of banana as of 2010–2011, is one of the largest producers of bananas globally. In India, Tamil Nadu, Maharashtra, Gujarat and Andhra Pradesh contribute nearly 65% to the production, with Tamil Nadu leading with nearly 28% of the total production in India. These four states, together with Karnataka, Madhya Pradesh, Bihar and Uttar Pradesh, contribute nearly 88% of the total production of bananas in India.

Though India is the largest producers of banana in the world, the country consumes almost all of its production, barring a small percentage that is exported mainly to the Middle East and Maldives. Table A3.1 gives the state-wise production of bananas in India.

In India, the banana is grown under different production environments, which vary with land type, irrigation facilities and so on. It is grown by both small and marginal farmers and also large farmers, as one in a portfolio of crops.

Table A3.1 Indian Production of Banana

Sr. No.	State	Production (000 ton)			Share (%)
		2008–2009	2009–2010	2010–2011	
1	Tamil Nadu	6667.00	4980.90	8253.00	27.71
2	Maharashtra	4960.00	5200.00	4303.00	14.45
3	Gujarat	3571.60	3779.80	3978.02	13.36
4	Andhra Pradesh	2804.00	2819.60	2774.76	9.32
5	Karnataka	1918.80	2132.30	2281.58	7.66
6	Madhya Pradesh	1498.00	1459.80	1719.58	5.77
7	Bihar	1373.60	1435.30	1517.11	5.09
8	Uttar Pradesh	82.73	1138.60	1346.05	4.52
9	West Bengal	954.10	982.20	1010.15	3.39
10	Assam	852.60	805.20	723.57	2.43
Total production - All India		26217.20	26469.50	29779.90	
Top four states' share		69%	63%	65%	
Top eight states' share		87%	87%	88%	
Top ten states' share		94%	93%	94%	

Source: APEDA, APEDA Agri Exchange, 2012.

Contiguous cropping could be important in cases of large commercial farming. The garden land system, with three crop cycles, is a widely adopted system of production in many parts of the country. Wetland banana cultivation with deep trenches is common in Tamil Nadu and parts of Karnataka; perennial cultivation on hill slopes is also practised in Tamil Nadu, the north-eastern states and Karnataka. The banana is grown as a subsistence crop in homesteads, and is also grown as a crop mixed with areca nut and coconut trees (Mustaffa, 2011).

In Tamil Nadu, bananas are mostly cultivated in Tiruchirappalli, Coimbatore, Erode, Karur, Dindigul, Namakkal, Madurai, Tirunelveli, Kanyakumari, Vellore and Thanjavur districts. Banana produced in the southern states of India, namely, Tamil Nadu, Karnataka, Andhra Pradesh and Kerala, are consumed in major urban markets such as Chennai, Madurai, Coimbatore, Trivandrum, Kochi, Bangalore, Mysore, Hyderabad and Vishakhapatnam, and also get redistributed to nearby markets from these cities. The banana is consumed as a ripened fruit. It is also processed as chips. Kerala gets a lot of bananas from other states, whereas Tamil Nadu is not just more self-sufficient, but also supplies surplus quantities to other states.

In another leading banana-producing state, Maharashtra, Jalgoan, Dhule and Buldhana are the major banana-producing districts. Around 80%–85% of the banana produced in Maharashtra is marketed out of the state to New Delhi, Lucknow, Kanpur, Amritsar, Jaipur and other places. New Delhi is the most important market for bananas from Maharashtra, and it acts as a major consumption as well as distribution centre serving markets in the national capital region. Rail transport is widely used for movement.

In the case of the state of Gujarat, Surat, Anand, Vadodra, Bharuch and Narmada are the major banana-producing centres and similar to Maharashtra, a substantial quantity is moved to northern India through New Delhi, Jaipur and Lucknow markets.

A3.1.1.2.1   Channels of Distribution

Farmers: Farmers raise bananas in gardens or wet land as the supplementary or main crop. They bear the risk of production, as major crop risk components are water availability, cyclones, diseases and crop economics.

Marketing through Cooperative Societies: In Maharashtra, a large volume of bananas is traded through cooperative societies. The societies handle price negotiation with the commission agents in destination markets. This involves groups of farmers with negotiating power in produce sales, facilitated by the societies. They provide marketing linkages to their members. The societies charge a fee of 2%–3% to the farmers on the selling price of the produce. This is not common in Tamil Nadu and other states. However, farmers collude to negotiate price, especially if the harvesting and contiguous planting economics force a bind to exercise more negotiating power with the commission agents. Unfortunately, lack of financial strength and borrowing through informal network weaken the supply chain networks.

Thus, cooperative societies facilitate aggregation of volume and negotiate with commission agents. Societies are voluntary group assuming power to ensure fair prices and avoid exploitation. Commission for administrative process is around 2%.

Marketing through Local Contractors: There are a number of contractors, representing commission agents, who work with the farmers both at the pre-harvest and at the post-harvest stages to procure the produce. When banana bunches are mature, local fruit merchants negotiate prices and give advance payments to the farmers. Typically, the produce is harvested at the farm and transported to the destinations by the contractors. This relieves the farmers of the burden of harvesting, especially labour management; a process of agriculture is thus taken over by the intermediary. The contractors facilitate aggregation, transportation to the destination markets and payments to the farmers from the traders/commission agents. Intermediary costs would be around 2%–3% of the value of the crop to the farmers. However, the price realized would be lower, as the contractor adjusts for harvesting, handling, transportation and risk management during this process. Nevertheless, this constitutes the most commonly used method.

Commission agents/traders: They operate at major consumption and/or consolidation markets from where bananas move down the stream. They play a significant role and control trade. Financial strengths of agents and ability to organize transport and effectively handle local contractors are critical success factors. These intermediaries command around 5%–7% of the margin. If one imputes the hidden charges as in harvesting and transportation, they may make around 7%.

Direct marketing: A small percentage of large farmers with big landholdings directly deal with traders/commission agents from the wholesale markets, and send their produce to the destination markets after fixing up the rates and quantities. The cost of harvesting is borne by the farmer. The farmer facilitates aggregation and transportation. Most of the time, the prices would be fixed as the farm gate price, and the buyer keeps the margin and bears the transportation cost.

Wholesalers: They buy through commission agents in small towns and within limited geographical area for redistribution to local trade. Typically, they handle certain geographical areas and operate out of a market or common place where many such wholesalers operate. They may have a mark-up of 5%.

Retailers: They vary from vendors with carts to organized retail who sell produce to final consumers. They assume the risks of physical damage and price discounts as the product ages. Though the post-harvest loss is estimated to be 10%, it is likely to be higher. Retailers work on a margin between 7% and 15%.

The supply chain mechanism for bananas in the region is depicted in Figure A3.2.

The banana is an important agricultural produce of wet land and irrigated regions, and an indispensable fruit for the common man in India, considering the value the fruit offers. Its availability and price would have a significant impact on markets, drawing huge attention from large sections of the society. The supply chain network needs to operate efficiently, and hence, the role of agents becomes critical.

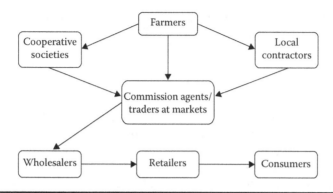

Figure A3.2 Banana supply chain in India.

Role of the intermediary in food grains and other crop marketing: Rice; wheat; coarse cereals such as *jowar, rahi* and *bajra* and other crops like oilseeds are the major crops in India. While rice and wheat are staple crops grown in wet lands and irrigated regions, coarse cereals and oilseeds are grown on rain-fed lands. The wet land farmers are comparatively rich and productive, while dry land farmers are marginal, and face huge risks in crop cultivation. However, there are a number of small and marginal farmers who do cultivate rice and wheat.

In case of rice and wheat, large acreages are involved and there is a likelihood of bunching of harvest and arrivals. This brings tremendous pressure to bear on farmers, especially those who are small and marginal, to sell the produce immediately after harvest. Often, such farmers would have also taken huge crop loans, and this means that surplus after loan repayment is meagre. There would still be pressure to retire current loans if they are obtained from organized sources, as failure to do so would affect the farmers' creditworthiness and their ability to borrow in the future. The plight of farmers who have borrowed from the informal sector and from money lenders/traders is the worst! They have no choice but to sell through traders to square off their loans and carry forward the balance, if any.

Though such a system is reprehensible, farmers' financial compulsions impact marketing of their produce. The government and its agencies are quite aware of the pitfalls and have taken a number of initiatives to help the farmers. One must understand that the rationale of such a system is derived from uneconomic holdings, low financial capabilities and limited scope for alternate employment to supplement agricultural income of the farming families. Money lenders and traders are those who understand the pressure of funding needs and provide assistance. Since these loans are unsecured and face the risk of crop failure or mismanagement, the money lenders and traders are expected to charge exorbitant rates of interest. Apart from these factors, the cost of supervision and follow-up on this lending is also high, as the lenders need to give a personal touch and attention to his portfolio of borrowers. Hence, the marketing channel for produce in such cases is determined more by crop financing than by the economics of the market system (other than funding). We conclude that the market intermediary for rice, paddy and other food grains exists because of the inadequacy of capital for a large number of small and marginal farmers, and because of the need to bunch harvest due to agricultural structural factors.

In the case of oilseeds, the intermediary exists because of his ability to support the farmer with finance and information and because he can do aggregation for market lots much more efficiently when compared to any efforts individual farmers may make. The difference between the growers of the two horticultural crops we discussed earlier, namely, pomegranate and banana, and growers of oilseeds is that the latter are economically much weaker, and thus, dependence on the intermediary is far higher. Further, the crop volume considerations, and the compulsions to sell in regulated markets to get fair prices, necessitate reaching out to distant places with small quantities.

Thus, we conclude that in India, the intermediary exists for different crops because of socio-economic compulsions, and irrespective of the government's efforts, the farming community is finding it difficult to rationalize their operations.

However, there are some positive signs in certain areas, where procurement by government agencies, cooperatives and private firms has helped to overcome the shortcomings of intermediation.

References

APEDA. APEDA Agri Exchange, 2012. http://agriexchange.apeda.gov.in/India%20Production/India_Productions.aspx?hscode=08030000

Basu, K. India's food grains policy: An economic theory perspective. *Economic & Political Weekly*, 46, 37–45. 2011.

Government of India. Report of the working group on agricultural marketing infrastructure and policy required for internal and external trade. Agricultural Division Planning Commission, 2007. http://planningcommission.nic.in/aboutus/committee/wrkgrp11/wg11_agrpm.pdf

Indo Asian News Service. We use Indian potatoes for our French fries: McDonald's. *Yahoo! News India*, Dec. 5, 2012. http://in.news.yahoo.com/indian-potatoes-french-fries-mcdonalds-103629824--finance.html

Mustaffa, M.M. Vision 2030, National Research Centre for Banana, Tiruchirappalli, 2011. http://www.nrcb.res.in/document/Vision%202030.pdf

National Agricultural Co-operative Marketing Federation (NAFED). Website. 2012. http://www.nafed-india.com/

Patnaik, G. Status of agricultural marketing reforms. *Workshop on Policy Options and Investment Priorities for Accelerating Agricultural Productivity and Development in India*. IGIDR Proceedings PP-069-11b, Nov. 10–11, 2011. http://www.igidr.ac.in/newspdf/srijit/PP-069-11b.pdf

Chapter 4

Processors

OBJECTIVE

This chapter helps to understand the challenges faced by processors in the agribusiness sector. Various problems in terms of both inputs and outputs, in organizing their production and reaching markets with their products are covered. The issues relevant for this actor in the supply chain are analysed with a few examples.

4.1 Introduction

Agro-processing is a significant activity in the agribusiness sector that contributes to value addition, product life management and the management of food availability and prices in any country. India is no exception; rather, agro-processing is more critical in this country, as the large population depends upon the agricultural sector's economic activities for employment, wages and consumption. Given the characteristics of agricultural produce, namely, perishability, seasonality and bulkiness, value addition through processing is critical for the sector. Similarly, much of the produce is consumed after value addition through form change and place and/or time of production and consumption variations. It is the processing methods and facilitation that enable a producer to bring out utility produce for consumers in such a way they are able to realize the full value of the price paid. The difference between the producer's cost and value additions through agents in the supply chain, including the channel operators, and the price paid by the consumers, makes up supply chain profits. In fact, consumer's surplus must be considered as a part of supply chain profit that is realized by the consumer, wherever such consideration is relevant for decision making.

In the post-independence era, the growth in agriculture produce was phenomenal, as mentioned in the earlier chapters. This gave opportunities for many traders to consider processing, and entrepreneurs and firms realized the potential of processed agricultural produce among people. Population growth, urbanization and growth in income and wealth, as well as the support of the government through fiscal measures and direct intervention, paved the way for increase in demand for processed agricultural produce and more so from organized business.

4.2 Primary Processing

The primary processes of sorting, grading, value addition and storage play a significant role in agro-based products business. Businesses are impacted by supply chain issues like location of processing centres in terms of closeness to the farms, clustering of units, transportation of finished goods, inventory and storage (especially with seasonality in demand and supply), level of foreign trade, length of the channel and finally regulatory aspects.

A number of agricultural commodities including cereals, pulses, grains, fruits and vegetables and products like poultry and marine go through sorting, grading and storage and then to market.

Many of these commodities and a few others fall under the categories for micro, cottage and small-scale industries sector. They are important parts of the agribusiness supply chain and value creation is achieved through activities around the supply chain processes such as planning, sourcing, processing and distribution.

One such example is cashew processing and exports, which is described here to get an understanding of how it works.

India is the second largest producer of cashews after Vietnam. India produced 613,000 MT in the year 2009–2010, as shown in Table 4.1. The crop is grown mainly in the peninsular states of India, particularly along the coastal states like Kerala, Karnataka, Goa, Maharashtra, Tamil Nadu, Andhra Pradesh and Orissa.

India is also one of the largest processors of cashew. Table 4.2 gives details on the country's cashew processing capacity as of 2005. It may be noted that India processed more cashew than it produced. Hence, trade in cashew was significant. We will first analyse cashew processing. The states like Kerala, Tamil Nadu and Karnataka handle large volumes of both indigenous and imported cashew. In this section, the processing of indigenous cashew is important. "There are two commonly followed methods of cashew nut processing, viz., the roasting process and the steam cooking process. The processes are based on the variety of raw material, location, technological mechanization and availability of secured energy supply. In the cooking process, the vegetable oil that is extracted from the shells of the cashew seeds has a market in the paint and adhesive industries" (Mohod et al., 2010).

Since roasting demands manpower that is comparatively less skilled, many of the units in the industry in India follow the roasting process. "In the cashew nut

Table 4.1 Cashew Production in India

State	2003–2004		2004–2005		2005–2006		2006–2007		2007–2008		2008–2009		2009–2010	
	A	P	A	P	A	P	A	P	A	P	A	P	A	P
Kerala	101	95	102	64	80	67	80	72	84	78	70	75	72	66
Karnataka	94	46	95	43	100	45	102	52	103	56	107	60	118	53
Goa	55	32	55	26	55	27	55	29	55	31	55	30	55	26
Maharashtra	148	120	160	174	160	183	164	197	167	210	170	225	175	198
Tamil Nadu	95	51	105	53	121	56	123	60	123	65	131	68	133	60
Andhra Pradesh	136	95	150	88	170	92	171	99	171	107	182	112	183	99
Orissa	124	71	126	74	120	78	125	84	131	90	137	95	143	84
West Bengal	9	9	9	8	10	10	10	10	10	10	11	11	11	10
Others	18	16	18	14	21	15	24	17	24	18	30	19	33	17
Total	780	535	820	544	837	573	854	620	868	665	893	695	923	613

Source: DCCD, Area, production & productivity of cashewnut in India. *About Cashew & Cocoa.* 2010.

Abbreviations: A, Area in '000 ha; P, Production in '000 MT.

Table 4.2   Cashew Processing Units in India—2005–2006

States	Processing Units (No.)	Processing			
			Utilization (000 MT)		
		Capacity	Indigenous	Import	Total
Kerala	432	600	67	320	387
Karnataka	266	300	45	20	65
Goa	45	50	21	–	21
Maharashtra[a]	2200	50	20	–	20
Tamil Nadu	417	400	294	225	519
Andhra Pradesh	175	100	92	–	92
Orissa	209	100	11	–	11
West Bengal	30	8	8	–	8
Chhattisgarh	3	5	–	–	–
North East States	22	10	15	–	15
Total	3799	1623	573	565	1138

Source: DCCD, Cashew processing units in India—2005–06. *About Cashew & Cocoa.* 2010.

[a] Includes 1850 small-scale cottage industry.

roasting process, the thick black smoke released from the rotating roasting drum through the stack has been found to be an air pollutant that affects the environment" (Mohod et al., 2010). A town called Palasa on the border between Andhra Pradesh and Orissa is one of the important clusters for cashew processing, and the government has been persuading the local manufacturers to improve their methodology or move out.

"The general steps involved in cashew processing are drying of freshly harvested raw seed for storage, soaking of seed, steam cooking or roasting, shelling, separation, kernel drying, peeling and packaging" (Mohod et al., 2010). With stringent rules on environment control, a number of units are shifting to stream processing.

It may useful to relate processing with the trade in cashew kernels from India. By corroborating data on production, processing and exports, it can be observed that India is exporting a significant volume of locally produced kernels. The import of kernels is for local consumption (Table 4.3). Hence, it can be concluded that through variety and process, India is able to produce high-value cashew nuts for export to developed countries. The leading importers of cashew nuts from India are

Table 4.3   Export of Cashew Kernel and Import of Raw Cashew Nut

	Cashew Kernel Export		Raw Nut Import	
Year	Quantity	Value	Quantity	Value
2003–2004	100828	1804.42	452898	1400.93
2004–2005	126667	2709.24	578884	2183.24
2005–2006	114143	2514.86	565400	2162.95
2006–2007	118540	2455.15	592604	1811.62
2007–2008	114340	2288.9	605970	1746.8
2008–2009	109522	2988.4	605850	2632.41
2009–2010	108120	2905.82	752894	3037.35

Source: DCCD, Cashew processing units in India—2005–06. *About Cashew & Cocoa.* 2010.

Note: (Quantity in MT) Value in ₹ crore.

the United States, the UAE, other gulf nations such as Saudi Arabia, and European nations such as the UK, France, Spain, Italy, the Netherlands and Germany (CEPCI, 2012). This shows the significance of processing for export and at the same time still satisfying local demand.

Cashew is one of the agricultural products that undergo primary processing for local consumption and exports, and also secondary/tertiary processing. Coffee beans could be another significant example. India is a leading producer, exporter and consumer of coffee. Again, coffee goes through steam processing into kernels before it is sold through marketing channels as kernels or as processed coffee powder (instant) in and outside of India. The other products in this category include marine products, a host of fruits and vegetables and basic products such as grains and spices.

Thus, the primary processes of sorting, grading, value addition and storage play a significant role in the business of agro-based products. There are a number of such examples in the sector that can be analysed. The critical factors that influence supply chain design are closeness to farms, cluster approach, transportation, storage and evolution of trade.

4.3   Food Processing

We proceed to further discuss food processing and the secondary and tertiary processing of agricultural products. The agricultural goods processing sector serves as a vital link between the agriculture and industrial segments of the economy. From the government perspective, focus is required for ensuring food security and for controlling inflation. Further, processing is important to reduce waste of agricultural raw materials, improve the value of agricultural produce by increasing shelf life and, at the same time, help fortify the nutritive capacity of the food products; processing also ensures remunerative prices to farmers as well as affordable prices to consumers. The effectiveness of these measures and the response of the agricultural and industrial sectors can be seen in export growth.

According to the Ministry of Food Processing Industry in India (2012), food processing industries would incorporate two processes:

1. "Manufacturing processes—if any raw product of agriculture, animal husbandry or fisheries is transformed through a process (involving employees, power, machines or money) in such a way that its original physical properties undergo a change and if the transformed product is edible and has commercial value, then it comes within the domain of food processing industries, and,
2. Other value-added processes—hence, if there is significant value addition (increased shelf life, shelled and ready for consumption, etc.) such produce also comes under food processing, even if it does not undergo manufacturing processes".

Thus, food processing involves primary processing that converts raw agricultural produce, milk, meat and fish into commodities that are fit for human consumption. Steps such as cleaning, grading, sorting and packing are employed in processing. Secondary and tertiary processing industries usually deal with higher levels of processing, where new or modified food products are manufactured.

The Government of India has used the Annual Survey of Industries (ASI) and National Sample Survey Organization (NSSO) data to categorize National Informatics Centre (NIC) groups, based on the International Standard Industrial Classification for aggregation of national statistics. It has been assumed that the factories listed in the groups in Table 4.4 can be summed up to constitute food processing industries.

The following are the key highlights with respect to agricultural produce processing in India (Table 4.5):

1. GDP at 2004–2005 prices in India has gone up at a compound annual growth rate (CAGR) of 8.40% (mofpi.nic.in).
2. Manufacturing, which contributes about 16% of GDP, has grown at a CAGR of 9.35%, which is slightly higher than the GDP growth rate.
3. The food processing industries sector has grown from ₹44,355 crore in 2004–2005 to ₹66,078 crore in 2009–2010 at a CAGR of 8.49%.

Table 4.4 Classification of Food Processing Industries

Sr. No.	NIC Group	Description
1	151	Production, processing and preservation of meat, fish, fruits, vegetables, oils and fats
2	152	Manufacturing of dairy products
3	153	Manufacture of grain mill products, starches and starch products and prepared animal feeds
4	154	Manufacture of other food products
5	155	Manufacture of beverages

Source: Ministry of Food Processing Industries, Data bank on economic parameters of the food processing sector, 2012.

Note: National Accounts Statistics also use the NIC groups to report the contribution of each group to the gross domestic product (GDP). About 151–155 NIC groups, as mentioned above, are added to derive the contribution of the food processing sector.

However, it may be interesting to note that registered food processing industries, which are considered to be in the organized sector, have registered a CAGR of 13%, compared to the unorganized sector, which has registered a CAGR of 1.5%.

4. Whereas the agriculture produce sector has registered a CAGR of 4.2%, the food processing industry has registered twice this growth rate, while the growth rate of the organized sector is much higher. This leads us to draw the inference that in India, agricultural produce processing has been growing significantly.

This can be validated with more analysis of the growth in the sector in terms of investments.

It may be observed from Table 4.6 that capital investments in the registered food processing sector has increased from ₹52,881.72 crore in 1998–1999 to ₹138,969.17 crore in 2007–2008 at a CAGR of more than 10%. Fixed capital has also grown at a CAGR of over 10%, whereas working capital has grown at a higher rate of 12%.

This clearly shows that the food processing industry in India is growing, with agricultural produce being processed at the primary level and further at the secondary and tertiary levels, with sophisticated investment in processing.

Table 4.5   Share of Food Processing in Manufacturing and GDP

₹ in crore

Description	2004–2005	2005–2006	2006–2007	2007–2008	2008–2009	2009–2010	CAGR%
GDP	2971464	3254216	3566011	3898958	4162509	4493743	8.4
Manufacturing	453225	499011	570436	629052	655775	713428	9.35
Manufacturing as % of GDP	15%	15%	16%	16%	16%	16%	
FPI	44355	47690	52164	57320	67122	66078	8.49
FPI as % of manufacturing	10%	10%	9%	9%	10%	9%	
Registered FPIs	22148	26780	30710	34752	43893	43910	13.16
Registered as % of FPI	50%	56%	59%	61%	65%	66%	
Unregistered FPIs	22207	20910	21454	22568	23229	22168	1.47
Unregistered as % of FPIs	50%	44%	41%	39%	35%	34%	
Agriculture, forestry and fishing	517651	543418	562316	588760	595017	638112	4.2
Agriculture as % of GDP	17%	17%	16%	15%	14%	14%	

Source: Ministry of Food Processing Industries, Data bank on economic parameters of the food processing sector, 2012.

**Table 4.6   Capital Investment in Registered Food
Processing Industries (₹ in Crore)**

Year	Invested Capital	Working Capital	Fixed Capital
1998–1999	52881.72	9138.26	26756.18
1999–2000	64544.19	13173.49	31642.34
2000–2001	70203.45	12814.78	31886.59
2001–2002	74774.26	13480.72	33907.01
2002–2003	79563.41	10919.53	37627.47
2003–2004	81308.45	17239.37	37411.68
2004–2005	84089.79	16992.79	41387.87
2005–2006	92038.54	19032.54	45356.65
2006–2007	112484.00	22055.74	57459.85
2007–2008	138969.20	28828.91	68334.83

Source: Ministry of Food Processing Industries, Data bank on
economic parameters of the food processing sector,
2012.

This clearly points to the improvements that have been occurring in supply chains in the agribusiness sector because of investment and improved economic activities in food processing sector.

Further analysis is taken up by studying the performance of certain industries based on agro-products such as sugar, tea, coffee and vanaspati over the years since 1950–1951. This compares well with the earlier tables on some secular trend analyses based on the growth of crops.

It may be observed from Table 4.7 that agro-based industries such as sugar, tea, coffee and vanaspati have grown significantly during the post-independence planning era. However, it has to be noted categorically that the sugar industry has again showed significant growth from 1990 to 1991 by doubling production, while others have also registered high growth in production.

In case of the sugar industry, our discussions reveal that apart from liberalization of government policies, the ability to invest in and operate power plants on a co-generation basis have helped to improve value creation and thus enabled significant additions to capacity. Further, it proves another important point: in agro-based industries, entrepreneurs are able to upscale technical and operating capabilities if there is scope in the manufacturing process for value enhancement.

Table 4.7   Selected Agro-Based Industries Production in India

| Year | (in '000 MT) | | | |
	Sugar	Tea	Coffee	Vanaspati/Edible Hydrogenated Oil
1950–1951	1134	277	21	155
1960–1961	3029	318	54	355
1970–1971	3740	423	71	558
1980–1981	5148	568	139	753
1990–1991	12047	705	170	850
2000–2001	18510	827	313	1445
2004–2005	13660	831	296	1898
2005–2006	19321	893	262	1193
2006–2007	28199	949	254	1285
2007–2008	26300	948	265	1380
2008–2009	14677	968	267	1532
2009–2010	18802	991	278	1122
2010–2011	24350	967	283	827

Source: Ministry of Finance, *Economic Survey* 2011–2012.

4.4   Supply Chain Drivers

It may be interesting to highlight supply chain drivers and challenges in these agro-based industries including that of sugar and plantation crops like coffee and tea, and the ways in which firms are managing growth.

4.4.1   Facilities

Location of facilities is mainly determined by factors such as bulkiness and perishability. For example, cane has to be processed close to procurement centres. Produce such as milk needs to be processed close to collection centres. Similarly, tea leaves, being perishable and of a bulky nature, need to be processed close to the estates/gardens soon after plucking. On the other hand, certain produce like wheat flour can be processed and distributed by shipping grains to local regional centres for subsequent distribution of the produce locally.

Another key aspect of facilities is the location of warehouses and distribution centres that processors manage through their network.

4.4.2 *Inventory*

Inventory management is another important perspective of supply chains that agro-based processing companies would have to handle efficiently for achieving growth and success. As mentioned earlier, much of agricultural produce is seasonal. From the facilities viewpoint, this requires the setting up of capacity for handling peak inflow. But the challenge from the inventory perspective is: how can the plant be run for a maximum time period in a year, so that idle capacity cost is minimized? A second challenge is the management of the raw material inventory, which has an uneven flow, as well as of the finished goods inventory, which would be more tapered and close to even demand? Over and above the seasonal characteristics, inventory management is complicated by regulatory measures on many of these products, as they come under the essential commodities category. The challenge for a supply chain manager is to manage this aspect, instead of leaving it to functional managers to handle. Normally, in many agro-based industries the functional managers, such as the procurement manager, the production manager and the finance manager, who directly report to the chief operations officer, drive business. This leads to the grave mistake of lack of supply chains and partnering approaches in decision making, which results in the cobweb theory taking over. Successful companies manage inventory through a partnering approach across the network.

4.4.3 *Transportation*

The transportation perspective of supply chains is another important aspect of the agro-based output sector. One may need specialized vehicles such as bulk container movement trucks. Often, less specialized trucks are operated in India because of poor economic sense, pointing to the ignorance of the cost of operation vis-à-vis the value. Further, other options like rail transport and air transport are not available widely in India because of route optimization issues, wagon/rake availability and multiple handling. In the case of some sectors like sugar manufacture, inbound issues are challenging aspects of supply chains. Cane is procured by factories from villages that are radially connected. A factory in its yard cannot plan for more than four hours of crushing. Assuming if a plant crushes 5000 tonnes of cane in a day, it has to crush approximately 220 tonnes in an hour. The minimal inflow of raw material must be maintained at that level. This means a minimum of 25 trucks an hour bringing in raw material. A factory needs to ensure that it runs 24 hours, as the processing is a continuous process, and therefore, 600 trucks per day are to be managed. If the turnaround time is two to three days for a truck from loading to unloading and returning for the next trip, then the transport manager has to manage 1500 trucks or more till the crushing season is over. On the other hand, the outbound aspects for sugar are fairly well managed, as the sales contracts are ex-factory, and buyers organize the trucks for transportation.

There are certain agro-based sectors like tea where the challenge is different. Geographically speaking, tea is procured for processing from a close radius. However, tea gardens are usually situated in mountain terrain, and tea is plucked manually (mostly by women). Since it has to be processed quickly, vehicles move quickly from aggregated load points to grading and processing centres. The challenge is tougher while moving to demand centres. Many small processors move to auction centres at a nearby location. The tea auction centres in India are located in Kolkata, Siliguri, Guwahati, Cochin, Coimbatore and Coonoor. Public tea auctions have played a key role in the primary marketing of tea in India for over a century, ever since the first tea auction centre was set up in Calcutta in 1861. However, movements from the centres to demand points are interesting, especially in the case of bulk cargoes. In India, demand centres are in the west, the north, the east and the south in the same order. The majority of tea production activities take place in the north-east. Southern states like Tamil Nadu and Kerala also contribute to production. Often, tea from the north-eastern and eastern states is moved in trucks and by rail to the western and northern parts of the country, which is a quite challenge.

Thus, agro-processing units work seamlessly with buyers and their agents in the supply chain to enable efficient movement of goods.

4.4.4 Sourcing

Sourcing issues have been discussed earlier as well. Agro-based industries have to necessarily map their procurement plans and ensure all operations are synchronized. More importantly, since agro-based industries depend upon resource endowments and their users for economic activities such as production and exchange, they have to necessarily establish partnering relationships with a strategic orientation with others in their supply chains. In the literature on supply chains, one normally comes across such recommendations for autos and auto components, high-technology industries, heavy engineering, consumer products (durable and entertainment goods), and so on. It is imperative for a large supplier-based (mostly fragmented) agro industry to have a highly supply chain-focused procurement management system.

Procurement activities for the agribusiness input sector can be seen at the farm gates as well as in process industries.

In the case of farm gate direct procurement, where relationships matter most, prices are set more on bargaining power and the value being brought in. When a process plant procures material, the price is based on market prices minus intermediation costs, including those of logistics, like moving farm produce to the market. In the current market environment, farmers' understanding of price factors is realistic, and the focal firm behind the procurement also ensures enough transparency in the system to procure adequate volumes. In the case of regulated price regimes, the government fixes a certain procurement price as the minimum support price, or the floor price. The focal firm will have to pay the minimum recommended price, and many times, farmers expect that price to be improved upon. There are cases where the central government fixes the statutory minimum price (SMP) and the state government comes up with a state advised price (SAP), which is higher than the SMP. The focal firm ends up paying SAP, as in the case of sugarcane. Group negotiating power, especially within the political system, drives pricing, rather than supply chain dynamics.

In the case of the regulated markets, prices are determined by forces of demand and supply. This would apply more to agricultural produce arrivals at the regulated markets and is linked to demand conditions during the time window of demand. With the ongoing explosion of technology and maturity of systems, market information is widely available. Farmer groups are able to effectively use mobile phones and other media to decide on arrivals, and the same situation prevails with processors as well. Price discovery and transactions are more structured. This phenomenon has brought about a considerable evolution in pricing mechanisms, leading to supply chain efficiency. One major issue here is the role of intermediaries, who could take positions and play around with stock and price. However, the evolution of the commodities market and futures trading in some selected commodities has brought further efficiency to the supply chain.

It may be interesting to look at the procurement decisions and challenges in the agribusiness input segment of process industries. Many items of farm produce are procured from the following networks:

1. Directly from the farm gate
2. From the regulated markets
3. Delivery at the units.

Figure 4.1 on procurement challenges for processing units in agribusiness highlights some of the key challenges for each of the nodes/options in the network. A firm may follow each of the options while relying predominantly on its chosen ones.

ITC's *e-choupal*, PepsiCo India's initiative for procuring potatoes and dairy processing firms like Amul have achieved huge success in directly procuring from the farm gate. The key elements of success in direct procurement lie in a company's initiative in achieving unique partnerships with local agencies, ensuring transfer of technology through well-trained employees and supply of relevant agricultural implements. At times, suppliers may have to provide farm inputs on credit, and ensure it is done in a timely fashion, with the right quality. Optimum operations call for prompt procurement and timely payments, with effective use of modern communication technology and maintenance of perfect logistics and marketing practices.

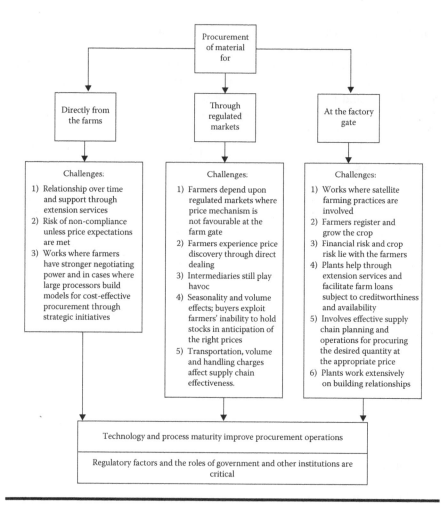

Figure 4.1 Procurement challenges for processing units in agribusiness.

The Amul model showcases long years of partnerships based on trust, supported by excellence in operating performance, with the combination leading to the success of its procurement policies. The processing unit works closely with farmers through its primary cooperative societies, which basically operate at the village level, ensuring collection of milk twice everyday. Further, farmers are paid on time, and the amounts are based on the fat content of the milk supplied. As mentioned above, extensive planning, careful organizing of logistics and the support of extension staff are critical for success. One needs to invest time and effort in understanding the local dynamics and setting up a system that delivers value to stakeholders through proper procurement strategies.

Exhibit 4.1   PepsiCo India's
Success in Contract Farming

West Bengal farmers supplying potatoes to PepsiCo India farmers registered a phenomenal growth of 150% in the harvest for 2010. The company's potato supplies were derived from a total acreage of 2,600 acres in 2009. PepsiCo India imparted to its 6,500 contract farmers knowledge of the correct geometry and chemical kits required for optimal harvesting. Further on, PepsiCo India provided better quality seeds and ensured timely irrigation by maintaining day-to-day follow-up on all relevant agricultural practices. Post-harvest, the company provided the required transport and storage facilities as well. This phenomenal growth resulted in the farmers earning a profit of between ₹25,000 and 40,000 per acre, as compared to ₹15,000–25,000 in the previous year.

According to Nishchint Bhatia, Executive Vice-President, Agro, PepsiCo India, "Being the largest procurer of potato in India, PepsiCo India has not just given more opportunity to potato farmers but also worked in close partnership with the farming community with a vision of creating a cost-effective, localized agro-base in India by leveraging its access to world class agricultural practices. A unique feature introduced by FritoLay India is the 'weather insurance' offered to the farmers. It also guarantees a 'buy back' price for the farmers which insure a regular income for them. During the course of this partnership, PepsiCo has transformed the lives of thousands of farmers by helping them refine their farming techniques and raise farm productivity."

PepsiCo India runs the country's largest contract farming operation in potatoes. The company works with nearly 15,000 farmers and procures 70,000–75,000 tonnes of potatoes. Through contract farming, total procurement is about 150,000 tons. PepsiCo India's Direct seeding initiative saves 30%–40% of water in paddy cultivation by doing away with the traditional method of flooding the fields. In 2009, PepsiCo India incorporated 6,500 acres and saved nearly 5.7 billion litres of water.

Source: PepsiCo, PepsiCo India achieves splendid success in contract farming in West Bengal, Mar. 23, 2010.

Thus, direct procurement is effective only when the bargaining power of both farmers and processing units is high, and requires a support system to scale up value in the supply chain network by the elimination of intermediaries. There may be cases where a firm makes efforts for procurement, but a farmer resorts to non-compliance for short-term gain. These instances are more aberrations than a regular practice.

One may also look at some retailing companies such as Reliance, Spencer's, and Food Bazaar, which procure produce directly from farm gates. Direct procurement from farmers can be done only in states that have amended their Agricultural Produce Marketing Committee (APMC) Act to permit buyers to purchase directly from producers and farmers, in line with the Model Act 2003 proposed by the central government. Reliance had to face a lot of negative publicity. Small traders organized agitations against the practice of direct procurement from farmers.

Agribusiness processing units alternatively procure from regulated markets as well. Often, they use a combination of direct procurement and procurement from regulated markets. For example, the oilseeds processing plant of the Karnataka Co-operative Oilseeds Growers Federation Limited (KOF) has a processing plant at Hospet in Bellary district, with a crushing capacity of 250 tonnes per day (TPD) for groundnuts and 50 TPD for sunflower seeds, as well as a 100 TDP solvent extraction plant and a 50 TDP continuous refinery. The plant usually tries to procure oilseeds for processing directly from growers/cooperative societies. But due to the dynamics of price and production, farmers may choose to sell their produce in regulated markets. Further, there could be scope for extending the operational period if a plant can supplement its procurement levels with purchases from market operations. More importantly, refinery operations would require their own feedstock from expellers and also from the markets to utilize maximum capacity for longer periods of operations.

The plant operations manager may have to work on an effective production plan incorporating optimum use of plant capacity with a mix of direct procurements and material from market operations. Since market operations are based on market prices, which could be volatile, the ability to understand parity in price with input–output yield parameters and margin on operations is mandatory. Further, market operations require the ability to stock or hold goods by buying produce cheap and processing it while capacity is available and prices are conducive for output. This makes it necessary to balance inventory policy and procurement policy; further, synchronization should also be achieved with respect to transportation costs. Thus, market operations challenge the supply chain functions of procurement, price management, inventory, facility management and transportation.

It may be noted that the network structure for the factory gate procurement option requires a certain ecosystem to work with. This approach works in the case of satellite farming, as demonstrated by sugarcane factories. The plant is a command area allocated by the state government to develop and nurture cane farming. The sugar factory is run by procuring cane from the command area. The firm that has been granted the license promotes registration of the land, wherein the farmer contracts to produce cane and offers it for procurement by the factory. The plant will support the farmer with extension services during the growth phases, and will monitor crop growth.

Though it looks rather simple, it is quite a complex operation, with detailed procurement planning, policy and process set in place. For a 5,000-tonne per day cane crushing (TCD) capacity plant to be operational for 210 days in a year, around 10.50 lakh tonnes of cane are required. With an average yield of 35 tonnes per acre, the plant should have a minimum of 30,000 acres under sugarcane cultivation. In reality, it must provide for a 20% variation and, therefore, should target 36,000 acres. With average holding in certain parts of the country as low as one acre, a plant of this size may have to manage up to 36,000 farmer units. Further, there must be sequencing of plots in such a way that the procurement feeds plant operations for a period of eight months. One may note that the process in a sugar plant of this size has an ability to handle varying feeds of 3,750–6,750 TCD depending upon climatic conditions and the imbibition levels that can be handled while milling cane.

It is just not operations that challenge procurement but organizing harvest labourers for cutting cane from the field and loading is another challenge. Cane harvesting and loading cannot be easily done by regular agricultural labourers, as it is laborious and difficult. There are specialized labour groups that do this, and they are in high demand. The plant has the responsibility of organizing labour on behalf of farmers, as individual farmers do not have the wherewithal for the same. Hence, there must be a team that coordinates production planning, procurement and operations at the field. The other challenge in field operations is that of organizing transportation of cane from the fields to the factory. Some of the rich farmers have their own means of transportation. Typically, cane is shipped to the factory from the fields in carts, trailers and trucks. Trucks are the most popular mode of transportation, but a 5000 TCD plant needs to organize approximately 500 trucks a day, which is a daunting task. Generally, the factory enters into agreements with local truck associations to ensure smooth operations. Transportation costs are paid in terms of tonnes carried.

Thus, at the factory gate, procurement for operations is a planned and well-structured activity involving relationships with many supply chain stakeholders. With this, one can see that the complexities of procurement operations in the agribusiness sector are wide ranging, and require meticulous planning and strong operational capabilities.

4.4.5 Pricing

Pricing is another important perspective which supply chain managers need to focus on in agro-based industries. Input prices are sensitive to crop output and seasonal characteristics. Many times, the government regulates the input price through a minimum floor price stipulation (support price schemes). In the case of agro-based industries, output prices were highly regulated till the 1990s. Since then, there have been a number of liberalization policies like decontrolling of sugar, dairy processing and edible oils. However, one still needs to understand that output prices in India are not fully under market conditions, as the prices of sugar, edible oil and other outputs are very sensitive. In adverse situations, the government quickly imports these commodities or comes in with direct intervention. Another important aspect in the supply chain networks of agro-based industries that affects prices is the presence of a number of intermediaries in the system, and even more, their inability to relate reward to responsibility, which affects pricing and profits. Agro-based industries improve distribution management and trying to reduce intermediation. However, fragmented markets and the vast reach required make this a challenging task. The progress of organized retail in India and customer demand for quality would help to improve price realization. Customer preferences for recent products, irrespective of long shelf life, also deter pricing. Firms are compelled to use discounting and markdown pricing to ensure freshness of stocks. Thus, pricing plays a significant role in agro-based produce processors.

4.4.6 Information

Finally, information, which is now considered as one of the key inputs to supply chain decisions, enables efficiency in those decisions. Agro-based processors are today heavily deploying information systems and also looking for avenues for benefiting by aligning systems and processes around breakthrough technological developments that have surfaced in the market.

As may be seen in the tea industry, the newly designed e-auction system is being used by all the six auction centres. The e-auction system integrates the process, starting from the creation of invoices till the delivery of tea from the warehouses, enabling better control and avoiding duplication of work, thus reducing the time taken. The benefits of e-auctions are aplenty for all supply chain network operators, as summarized in Table 4.8. However, processors also have immensely benefitted in the form of streamlining of information, financial and physical flows in the system, enabled by technology upgrades. Though it is believed that there was an initial reluctance to move to e-auction, the success of it is now reflected in the increased patronage.

This is not only true of tea auctions in India, but also in many other trades where technology can be integrated. We would be discussing this in the chapter on IT for agribusiness efficiency, wherein a few case studies would be discussed. ITC, as mentioned in one of the earlier chapters 1 & 2, is a classic case. We believe that companies like Amul have adopted state-of-the-art technology applications for deriving supply chain efficiencies.

Thus, we observe the growth of the agricultural produce sector and processing industries in India impacting supply chain efficiencies. There is another which is gaining greater attention in contemporary agro-based business is of wastages and linkage to supply chain management.

While processing agricultural products, wastes are generated because of the fibre content after extraction or by process additions such as water and chemicals.

Table 4.8 Benefits of e-Auction System

Sr. No.	Benefit	Comment
1	Direct participation by each auction stakeholder, with access possible from anywhere, since system is Web based, and with anonymity maintained	Improving market reach
2	Flexibility in setting up auction sessions on any auction day; option for parallel sessions and parallel bidding	Improving market reach
3	Facilities for participation of a large number of buyers in the bidding process, which is not possible in manual auctions due to the limitations of auction hall space	Improving market reach
4	Scope for sharing of any lot with the next highest bidder	Buyer coordination
5	Quick dissemination of auction sale information, compared to manual auctions	Information efficiency
6	Reduction in transaction time and cost	Supply chain goals
7	Planning tool in the hands of buyers and other stakeholders, as bid history and analyses are easily available	Better analytics
8	Dynamic knock down process as, "reserve price" can be changed by auctioneers even during the time of bidding by the buyers	Real time process
9	Facility for manufacturers to upload tea dispatch information on the e-platform directly, in addition to viewing the auction prices during the auction sessions	Information efficiency
10	Facility for warehouses to upload Advance warehouse Receipt (AWR) directly on the e-platform	Improving financial process
11	Integration of pre-auction and post-auction activities of different stakeholders with the auction activities in the e-auction platform	Overall process improvement and better planning

Source: Ministry of Commerce & Industry, Tea sold through e-auctions crossed 300 million kgs, *Tea File,* Apr.–Sept. 2010., p. 3.

We also observe that the sector faces a number of challenges, and has still been able to take advantage of attractive growth opportunities successfully. It also shows the unleashing of the large potential for processing goods, as the market widens with changing customers' preferences and urbanization.

This can be better explained with an industry situation. In the case of sugar, cane is crushed using a lot of water in the crushers. The imbibition level is one such measure of performance. The more water is used, the better is the extraction of the juice from the cane. However, this necessitates the demand for more steam to process the juice. At this stage, bagasse is the residual from the milling section. The sugar mills use this bagasse for combustion in the boilers, which produce steam for processing, while electricity is also generated for use in the factory. Traditionally, earlier low-pressure boilers and turbines were reduced, which meant low efficiency, with the electricity produced just adequate for the factory. Over the last two decades, sugar plants have installed high-pressure boilers and matching turbines to generate electricity. With the same quantity of bagasse, now they are able to export substantial amounts of electricity to the grids. The increased recovery through this system has changed the fortunes of sugar manufacturers.

Moving further in the sugar manufacturing process, pressed mud is generated during the purification of juice process, and is sold as organic manure to farmers. Farmers pay for it and clear the load from the sugar factory; thus, sales are ex-factory, and transportation is through the farmers' vehicles. After this, the crystallization process takes place through vacuum pans. The process generates molasses at this stage. Storage of molasses is a tough process, especially during the peak season, and can impact the environment, as the processes are rigid. Molasses stocks are sold, and in cases where there is manufacture of industrial alcohol attached to sugar plants, the process generates spent wash and solids. Disposing of spent wash as per the demands of environmental regulation is another major challenge.

Thus, as one may note, different types of materials are generated during processes, raising opportunities to create by-products, and residue management becomes important. A similar kind of waste management is necessary in every other agriculture process business. Often, the generated materials are used for preparing organic manure or by-products like bricks and land fillers. Of course, there are industries like those that process raw leather from animal skins, which are facing challenges because of their inability to handle effluents, and thus, are being asked to close. At times, this forces India to import processed raw skins for further value addition and re-export to developed nations.

4.5 Industrial Production

We move further to discuss on the use of agro-based products as input in the processing and manufacture of goods classified under Standard Industrial Classification (SIC) codes.

It may be interesting to look at the performance of a few more industries that would have been placed in various SIC categories, but use agricultural produce for secondary and tertiary processing. This may include industries for producing cotton, paper and capital goods like textile machinery.

The following are certain observations based on Table 4.9 on the performance of industries for cotton textile and paper-related products in India. The performance and inferences on supply chain issues are drawn here more as illustrative of the issues faced by this sector.

1. The corrugated and other similar paper industry has demonstrated a fluctuating trend in production for the period 2004–2005 to 2010–2011. The industry has been plagued with problems of over-capacity, manual operations and low productivity. Further, transport constraints and high freight costs lead to setting up of small- and medium-sized plant near consumption points.

2. The craft paper and writing and printing paper industries have registered CAGRs of 7% and 6%, respectively, for the period 2004–2005 to 2010–2011.

3. Though the paper industry has grown at an average growth rate comparable with the manufacturing sector, the industry has specific challenges, especially in terms of the availability of forestry resources and sustainability. One of the chief concerns of the supply chain managers in these industries is to ensure raw material availability and replantation of trees for sustainability of resources. Action must be undertaken in accordance with a well-designed plan to ensure resource availability. On the downstream of the supply chain network, managers have to handle distribution management, as large paper plants could be located close to raw material sources (paper raw material is weight losing in character), and the end products need to be moved across the network.

4. Cotton cloth and mixed/blended cloth manufacturing units use an agricultural produce, namely cotton, which is ginned. In fact, ginning is a large industrial activity which is mainly confined to tiny (cottage) and small- and medium-sized industries in India. The ginned cotton is then moved for making yarn and then cloth.

5. It can be observed from Table 4.9 that cotton cloth registered a CAGR of 7.5% from 2004–2005 to 2010–2011. The industry has grown phenomenally from 1950 to 1951 onwards. It may be noted that the cotton cloth industry has two components, namely, the mill sector and the decentralized sector. From 1990 to 1991 onwards, the decentralized sector has grown significantly compared to the mill sector mainly because of the labour problems and cost factors in the mill sector. The decentralized sector units are able to handle growth, as they are small in size and a lot of fiscal concessions are available. There are a number of factors which led to the fall in the mill sector, and units in certain mill sector centres like Mumbai and Ahmedabad have started switching over to other businesses because of the attractive real estate pricing

Table 4.9 Performance of Cotton Textile and Paper-Related Industries in India

Industry	Textile Machinery	Corrugated & Other Paper	Craft Paper	Writing & Printing Paper	Cotton Cloth	Cotton Cloth-Mill Sector	Cotton Cloth-Decentralised Sector	Mixed/Blend Cloth	Mixed/Blend-Mill Sector	Mixed/Blend-Decentralized Sector
Unit	₹ Million	₹ Crore	Th tonnes	Th tonnes	Million sq. m.	Million sq. m.	Million sq. m.	Million sq. m.	Million sq. m.	Million sq. m.
1950–1951	Na	Na	Na	Na	4215	3401	814	Na	Na	Na
1960–1961	Na	Na	Na	Na	6738	4649	2089	Na	Na	Na
1970–1971	Na	Na	Na	Na	7602	4055	3547	170	107	63
1980–1981	Na	Na	Na	Na	8368	3434	4934	1270	730	540
1990–1991	Na	Na	Na	Na	15431	1859	13572	2380	698	1682
2000–2001	125296	Na	Na	Na	19718	1106	18612	6351	332	6019
2004–2005	14382	266	859	1821	20655	1072	19583	6032	243	5789
2005–2006	17115	270	899	1974	23873	1192	22681	6298	252	6046
2006–2007	24101	261	977	2066	26238	1305	24933	6882	330	6552
2007–2008	27447	214	1014	2144	27196	1249	25947	6888	422	6466
2008–2009	16890	223	1060	2288	26898	1259	25639	6766	426	6340
2009–2010	15742	226	1184	2387	28914	1465	27449	7767	482	7285
2010–2011	24572	269	1260	2577	31742	1604	30138	8279	526	7753
CAGR	9.50%		7%	6%	7.50%	7%	7.50%	5.50%	14%	5%

Source: Ministry of Finance, *Economic Survey* 2011–2012.

their lands attract. From a supply chain perspective, poor cost conditions and the scope for realizing the real estate potential raised the scope for revisiting supply chain configurations in the decentralized sector.

6. Finally, a look at the growth of the mixed/blended cloth sector shows that it has registered a CAGR of 5.5%. Here again, the decentralized sector has a larger share compared to the mill sector, but has grown at a lower CAGR compared to the mill sector.

7. In the cloth manufacturing industry, cotton assumes a significant part, which is determined by cotton prices, availability and trade policies.

8. Here, we have not discussed the situation regarding man-made fibre, filament yarn, spun yarn and so on as our discussions are focused on challenges faced by the agro-based processing sector rather than the cloth sector.

Thus, agro-based secondary and tertiary processing industries face innumerable challenges in designing supply chains in order to achieve the overall supply chain goals of efficiency and service levels. This is because of the characteristics of agricultural produce and the number of market intermediation levels. The latter is being addressed through application of technology and spread of information and data for decision making. But the former would continue to be a challenge.

4.6   Conclusion

To conclude, agro-based processing can include primary, secondary and tertiary processing industries. The challenges are in planning network operations and synchronizing all stakeholders to optimize the supply chain goals of cost efficiency and responsiveness, while still achieving maximum supply chain profits shared across the network players, based on their respective roles and responsibilities. Because of the typical characteristics of agricultural products, namely, their seasonality, bulkiness and perishability, certain role agents may skew relationships and achieve increased importance, reaping more benefits. Processing firms need to establish partnering relationships and synchronize operations like any other industry's evolved supply chain network designs.

References

CEPCI. Cashew Export Promotion Council of India. http://www.cashewindia.orgDCCD (Directorate of Cashewnut and Cocoa Development). Area, production & productivity of cashewnut in India. *About Cashew & Cocoa*. 2010. http://dccd.gov.in/stat. htm#Areahttp://dccd.gov.in/stat.htm#Area

DCCD. Directorate of Cashewnut and Cocoa Development. Cashew processing units in India—2005–06. *About Cashew & Cocoa*. 2010. http://dccd.gov.in/stat1a. htm#PROCESSING

Ministry of Commerce & Industry, Tea sold through e-auctions crossed 300 million kgs, *Tea File*, Apr.–Sept. 2010, p. 3. http://www.teaboard.gov.in/pdf/Tea%20File%20PDF

Ministry of Finance, *Economic Survey* 2011–2012. http://indiabudget.nic.in

Ministry of Food Processing Industries, Data bank on economic parameters of the food processing sector, 2012. http://mofpi.nic.in

Mohod, A., Jain, S., and Powar, A.G. Pollution sources and standards of cashew nut processing. *American Journal of Environmental Sciences* 6(4), 324–328, 2010.

PepsiCo, PepsiCo India achieves splendid success in contract farming in West Bengal, Mar. 23, 2010. http://pepsicoindia.co.in/media/Press-Releases/release_03_23_10.aspx

Tea e-Auction. https://www.teaauction.gov.in/Home.aspx

Chapter 5

Retailers

OBJECTIVE

This chapter helps in understanding the complexity of retailing agricultural produce, such as farm fresh and processed produce, and the need for addressing challenges of achieving efficiency and effectiveness in supply chains. The discussions include those of unorganized and organized retailers and role of intermediaries.

5.1 Introduction

Typically, the term retail is used to refer to sales for final consumption, in contrast to sales for processing or resale (i.e., wholesale). A retail sale is a *sale to the ultimate consumer*. Retailing can be said to be the interface between the producer and the individual consumer buying primarily for personal consumption. This definition of retail excludes direct interface between manufacturers and institutional buyers such as the government and other bulk customers. It may be appropriate to mention here that retailing is the last link that connects the individual consumer with the manufacturing and distribution chain.

"A retailer is involved in the act of selling goods to the individual consumer at a margin of profit. Retailing can be classified as organized or unorganized, based on the structure or format of business. Typically, organized retailing refers to trading activities undertaken by licensed retailers, that is, those who are registered for sales tax, income tax, etc. This category includes corporate-backed hypermarkets and retail chains, and also privately-owned large retail businesses. On the other hand, unorganized retailing refers to the traditional formats of

Table 5.1 Advantages of Organized and Unorganized Retail

Unorganized	Organized
Proximity to customer	Better quality and grading
Excellent customer relationship	Economies of scale
Long operating hours	Better loyalty programmes
Favourable price points	Appropriate standards
Low wastages	Good shopping experience

low-cost retailing. This would include: local *kirana* shops, owner-manned general stores, convenience stores, hand cart and pavement vendors, etc. This is over-whelmingly popular in India.

The Indian retail sector is highly fragmented, with 97% of businesses being run by unorganized retailers. Organized retail is at a very nascent stage. The sector is the largest source of employment after agriculture, and has deep penetration in rural India, generating more than 10% of India's gross domestic product (GDP)" (NABARD, 2011).

Table 5.1 summarizes the advantages of organized and unorganized retail.

It may be also important to note there that there are certain disadvantages like poor hygiene at times, limited stock-keeping units and little or no shopping experience.

On the other hand, the organized retail in agricultural products, including fruits and vegetables, is picking up in India. Produce is sold through a number of registered outlets, comprising cooperative, private and sole-ownership stores. These establishments are becoming common in metros, urban centres and even in semi-urban centres with a population of 100,000 or more.

However, organized retail may have some disadvantages like higher wast-age, as responsiveness has to be high, with the challenges of managing stock and store, pilferage, incremental transaction costs, and in India, the high cost of real estate and poor parking facilities has proved to be spoil sport for buying experience.

5.2 Size of Indian Retail

It may be useful to discuss the pecuniary size of the retail industry in India and then relate it to the food and beverages that would be fall under agribusiness, so that we can then arrive at the unorganized component. A NABARD (2011)

Table 5.2 Size of Retail Industry in India

Year	Retail Industry Size in ₹ billion	Growth Year on Year (Nominal)	Growth Year on Year (Real)	GDP Growth Rate (Real)
2003–2004	10,559	–	–	–
2004–2005	11,295	7	0.5	7.5
2005–2006	12,661	12.09	7.66	9.5
2006–2007	14,096	11.34	5.92	9.7
2007–2008	15,623	10.83	6.22	9
2008–2009	17,497	12	10.93	6.7

Source: NABARD, *Organized Agri-Food Retailing in India*. Mumbai: NABARD, 2011.

report on organized agri-food retailing in India has brought out some estimates for the sector.

The study has used Central Statistical Organization (CSO) data on Private Final Consumption Expenditure (PFCE) at current prices (nominal) classified by types of goods to estimate the size of total retail. It considers the total PFCE less the PFCE contribution towards services to be the value of retail.

It may be observed from Table 5.2 that the size of the retail industry was around ₹17,497 billion as of March 2009, and has been growing between 7% and 12% between 2003–2004 and 2008–2009. The growth is directly linked to GDP growth, which determines disposable income and private spending abilities.

The NABARD (2011) study mentions that secondary data show that the food and beverage category forms a large proportion of total retailing. The study has used CSO data on PFCE classified in terms of current prices (nominal) to estimate the size of the food retail sector. The value of the food retail sector was expressed as the sum of all contributions of PFCE towards food and beverages. This includes expenditures on the following food categories:

- Cereals and bread
- Pulses
- Sugar and *gur*
- Oils and oilseeds
- Fruits and vegetables
- Potato and other tubers
- Milk and milk products
- Meat, eggs and fish

- Coffee, tea and cocoa
- Spices
- Other beverages, including processed food.

It may be observed from Table 5.3 that the share of food and beverages in total sales is about 61%, a marginal reduction over the years from 63.4%. This could be because of the increase in disposable income and increasing private spending on durables and other non-food items, which is a typical trend in an economy that is growing, and is especially typical of lower and middle income groups' aspirations to scale up their standards of living.

It may be observed from Table 5.4 that

1. Unorganized retail has a major share in purchases made in both urban and semi-urban/rural markets, as observed by this study—something we also agree with.
2. Compared to non-perishables, perishables are bought mainly from unorganized retail sources, to the extent of between 80% and 90% in the case of urban markets, and to an even higher extent, between 92% and 98%, in the case of semi-urban/rural markets.
3. Buyer behaviour tends more to unorganized sources because of the price factor, ease of buying and probably freshness of the stock, especially in case of non-vegetarian food items.
4. The presence of unorganized retail establishments in agribusiness could be a challenge to supply chains, though the ultimate buyers see a value in what these establishments deliver.

Table 5.3 Size of the Food and Beverage Retail Sector and Its Share in Total Retail

Retail Sales	Total Retail Sales (₹ billion)	Food and Beverage Retail Sales (₹ billion)	Share of F&B in Total Sales (%)	Growth in F&B Retail Sales (Nominal) (%)
2003–2004	10,559	6,697	63.44	4.1
2004–2005	11,295	6,954	61.57	3.8
2005–2006	12,661	7,702	60.83	10.8
2006–2007	14,096	8,563	60.75	11.2
2007–2008	15,623	9,585	61.35	11.9
2008–2009	17,497	10,673	61.00	11.4

Source: NABARD, *Organized Agri-Food Retailing in India.* Mumbai: NABARD, 2011.

Table 5.4 Purchase of Different Food Items by Percentage of Population

Particulars	Groceries	Fruits and Vegetables	Milk and Milk Products	Chicken	Other Meat	Fish	Egg	Other Food[a]
Urban Market								
Organized	32	20	18	10	14	10	11	31
Unorganized	68	80	82	90	86	90	89	69
Total	100	100	100	100	100	100	100	100
Semi-urban and Rural Markets								
Organized	21	6	2	5	4	2	8	19
Unorganized	79	94	98	95	96	98	92	81
Total	100	100	100	100	100	100	100	100

Source: NABARD, *Organized Agri-Food Retailing in India.* Mumbai: NABARD, 2011.

[a] Including processed food.

5.3 Unorganized Sector

Traditionally, "retailing in India has grown with the number of *kirana* shops (mom and pop format establishments). These shops are well distributed across all geographical areas, from metropolitan cities to small hamlets or villages. As the economy grew, reflected in changes in demography, and rise in income and wealth distribution, the number of such establishments grew and reached enormous levels. According to the economic census of 2005, the number of establishments in India engaged in retail trade was around 15 million. While 52% of these were in rural areas, the remaining 48% were in urban areas. During the earlier census of 1998, the total number of establishments engaged in retail trade was 10.6 million (with 49% in rural and 51% in urban areas). The period has seen an average growth of about 6% per year in the number of retail establishments" (NABARD, 2011).

5.3.1 Rationale

1. *Buying convenience*: Unorganized retail establishments are usually in close proximity to their customers. The NABARD (2011) study points out that organized retail outlets are located quite far away, that is, more than four times the distance (average 1180 m) of unorganized outlets (average 280 m) from consumers' residences in A class centres, while it was further away (1320 m) in case of B and C class centres. Though there is a high penetration of organized retail in A and A1 class cities, consumers have a clear preference for unorganized retail, especially for perishables. Consumers buy from the unorganized retail most especially when something needs to be bought quickly. The average amount of time spent per visit is significantly high in the organized sector compared to the average time spent at unorganized outlets. However, this additional time spent in organized retail organizations would span purchases of both food and non-food items. There are many reasons why this additional time is spent. They include the self-service systems, wherein customers go through different sections for choosing different items, more customers demanding attention and queues at billing/paying counters.

   In the same manner, the average amount of time spent per visit is significantly higher in case of the organized retail sector. Since customers spend more time, but are less likely to visit as often as in the case of unorganized retail outlets, the time and amount spent per visit could be higher. This makes it clear that unorganized retailers are likely to be highly preferred for buying perishables.

   This is clear from the structure of the marketplaces, where selling is done in small lots by persons using head loads, carts and other vehicles to provide door delivery as well. For example, one wonders how a hawker walking into a neighbourhood with his cart of vegetables and fruits knows the demand pattern of the families living in the area. It may be that one of his regular clients—an elderly lady—chats with him about the other households in the locality. Equipped with surprisingly precise knowledge about the demand pattern of each household, his cart would be stocked with the exact amount of perishables. If we observe his selling pattern a little more, we will find the hawker completes his round through the neighbourhood by 11 a.m.—a total work time of six hours—and makes a decent margin of profit with less variability in price, lesser necessity for stock-keeping items and smaller quantities. It's amazing. His customers are happy with his fresh stock and fair prices. The vendor is happy with his service, realization and the stability of his earnings. The cash-to-cash cycle is so perfect that his sourcing and distribution plans are also clear and well established. Synchronization of supplier relationship management and customer relationship management—the macro processes of supply chain management—is well in place. The hawker would have roughly around ₹3000 worth of stock, and made his sales at a 30% margin. Ideally, as per our enquiry, we observe that he sells about 80% of his stocks on his neighbourhood rounds, and sells the remaining 20% to small restaurants and hotels. There is also a likelihood of credit sales, and the vendor manages his cash flow by anticipating such credit sale requirements.

Apart from direct sales at homes, agricultural products are sold in marketplaces in each locality, either in the morning hours or in the evening hours. These marketplaces have assemblies of large numbers of sellers and buyers in locations either assigned by the local municipal authorities or captured by the traders by dint of long practice over the years. There would be no great convenience in buying in such marketplaces for high income groups, as these markets are unstructured and without much parking space; however, large middle- and low-income groups would find it attractive.

Thus, unorganized markets have come to stay because of convenience buying.

2. *Way of livelihood for vendors and retailers*: There are a large number of retailers and vendors in the unorganized market. For many of them, trading is their livelihood. They have taken up retailing of agricultural produce because the entry barriers are low. All that is required is possession of selling space in an advantageous location and limited working capital to invest on stock. When interviewed, many of the retailers gave us to understand that they are in the business by choice, and at times, because it was the family business. Further, they do not have any alternate means of livelihood, as they took up retailing with very little formal education, and started their careers as apprentices in retail stores. Hence, as they are engaged with their only livelihood, they stay focused, and try to create value for themselves by buying and selling goods and services. There is a common belief that such unorganized retailers thrive on low margins and low turnovers, benefiting from the corresponding low risks. As the population increases and new agglomerations of residential

dwellings arise, agricultural products for the ultimate consumers reach the markets through the proliferation of both long-time retailers and new generations of youth taking up a means of livelihood through trading of goods at the retail point.

3. *Competitive pricing*: Conceptually, organized retailing must provide strong price advantages, as they enjoy economies of scale; in reality, small and unorganized retailers are competitive in pricing. Our probing into the market brings out the fact that the small retailers have the advantages of low overheads and low interest charges on their stocks. This may seem to be contradictory to common belief, as large retailers do have overheads, but these overheads are spread over wide inventories of stock. Similarly, large players must be able to get working capital at a lesser cost of fund. The key advantage to small retailers is their limited stocks and lesser wastage, resulting in their ability to offer competitive prices. It is also important to note that small retailers provide the scope for buying any lot size, which is seen as an advantage to buyers. Hence, it is not just price, but also the ability to transact at low volume and outflow, that makes them attractive retail outlet choices for consumers.

Thus, it is our view that the small unorganized retailers are popular because of the convenience they offer, their long working hours, competitive pricing, good understanding of customers and relationship management, as also their focus on their retail activity as their necessary choice of livelihood.

5.3.2 Business Structure

1. *Petty, unorganized retailers who sell in carts, on pavements and in marketplaces*: These are businesses owned by operators who have typically organized their own capital and labour, and run their businesses for profits. The capital employed could be between ₹10,000 and ₹100,000. The focus is mainly on using the opportunity to maintain a steady livelihood with a regular customer base and the opportunity to serve repeat customers. The retailers earn a decent profit margin of 20%–30% and typically, do a cash turnover in a day or two. The more financially strong members of this group may have paid large sums in the range of ₹60,000–70,000 for the rights to set-up and run their petty shops on pavements or in marketplaces. This is true for perishables as well as for other agricultural products that are sold in rural and semi-urban markets. Normally, one would not come across non-perishables such as cereals and grains being sold in this format in urban centres.

2. *Petty shops*: Typically, these shops are formally structured in an area of 600 sq. ft or less. Again, they are operated by their owners, with two or

three helpers. The shop would sell many agricultural products and a few general format durable retail items such as personal care products, hair care products and home products. The total stock could be valued at around ₹300,000–500,000. Perishables will be replenished daily, and could be of value between ₹2,000 and ₹10,000. This is a key node in the agribusiness supply chain in India, as it caters to large sections of the population of India, especially low- and middle-income groups. This is especially so in rural markets. The node participants may not be acting as part of a supply chain, but well-honed market survival instincts compel them to keep extending their businesses. There is stiff competition and low entry and exit barriers, making it unattractive for players who want to ensure mature operations and long-term survival with normal profits.

3. *Single store:* The unorganized retail trade for agricultural products includes perishables. Typically, these stores have the following characteristics:

   a. Proximity, goodwill, credit sales, bargaining, loose items, convenient timings and home delivery, all of which are major drivers of these units.

   b. While *kirana* (mom and pop) stores sell on credit to retain customers, their own reliance on institutional finance remains very low.

   c. These stores depend on their customer bases and relationships with them, maintained by extending excellent service based on intimate knowledge of customer demand. They are responsive, and provide all support services such as home delivery, credit on purchases and information support on events and happenings around the locality.

   d. Quality is never an issue as the customers are well known. If a retailer falls short on quality, his customers would demand replacements. Hence, the challenge these retailers face are having the right inventory and the right quality, and offering the best prices and service.

   e. The viability of a single large store, retailing perishables and operating on a 3000 sq. ft area is given in Table 5.5.

It may be seen from Table 5.5 that a perishable retail outlet that could be selling fruits and vegetables along with a few more food items will probably have a stock valued at ₹1,800,000. The store may be employing about 27 staff, incurring a staff cost of ₹200,000 excluding their welfare expenses and bonus. We look at the other operating costs and determine the level at which the store must operate to earn profits.

It may also be observed from Table 5.5 that obsolescence cost, rental and interest on working capital are key line items that need discussion. Obsolescence cost in perishables is estimated to be 10% of the stock or 6.5% of turnover. It would be appropriate to see this cost as a percentage of turnover, as the success of business involving retail of perishables depends more on liquidation of stock and management of price and revenue using the discounting strategy. Rental is assumed to be

Table 5.5   Single Store Economics

	No	Salary/ month ₹	Total ₹
Estimated built-up area of the shop (sq. ft)	3000		
Estimated stocking space (cu ft)	6000		
Estimated value of stock in ₹ per cu ft of holding	300		
Value of stock at the store/display (₹/000s)	1800		
Number of employees required	40		
Store manager	1	20,000	20,000
Supervisors	6	10,000	60,000
Support staff	20	6000	120,000
Staff cost excluding welfare and bonus			200,000
Staff cost including welfare and bonus ₹/month			240,000
Electricity and utilities ₹/month			30,000
Rental ₹/month			75,000
Other overheads ₹/month			50,000
Interest on stock ₹/month			27,000
Obsolescence cost ₹/month			180,000
Total monthly cost ₹			602,000
Owners' revenue expectation ₹/month			100,000
Total cost along with return for the owner/month			702,000
Margin on sale			25%
Break even sales for a month			280,8000
Monthly sales (in 000s)			2808
Daily sale (in 000s)			93.6
Per customer basket ₹			250
Number of customers required/day			374
Customers' purchases in a year			134,784
Repeat purchases in a year			104
Number of basic family contacts			1296

₹25 per sq. foot, which is quite nominal in urban and semi-urban centres. If the retailer uses a designated market space operated by a large sellers' association, this could be lower. This requires long leases and a hold on such opportunities over a long period. There is a trend, especially in A class and A1 class urban areas, for some retailers to invest on space. For example, in our illustration, the retailer must invest around ₹3 million to own 3,000 sq. ft, and the capital cost per month, excluding repayment of capital, could be ₹50,000. In any case, the capital appreciates, and the retailer would make good on his investment. Another challenge for small retailers in India is the interest cost on stocks. In this case, we have assumed 18% per annum, or 1.5% on a monthly basis, which is very fair and optimistic. Often, small retailers do not have access to organized funding, and hence, the cost of their working capital is high. The retail trade, which is operated as a full-time engagement by the owner, provides ₹1,00,000 towards salary and return for risk-taking. This is abysmally low.

Let us understand the viability at this level of costs. The cost of operation is about ₹702,000, and at a 25% margin, the retailer has to achieve total sales of ₹2,808,000 per month. The daily sale must be about ₹93,600, which means at least 374 transactions at a basket value ₹250 each are required. This could be possible only if the retailer has about 1,296 customer accounts, with each customer buying at least twice in a week. This is based on actual observation and fair assumptions about customer loyalty. In reality, customer footprints to conversion and casual walk-ins would happen. But by and large, the economics is fairly supported on ground by a few experts. This case is not generalized, but is used to draw broad inferences about the sector.

The drivers or challenges for unorganized retail in India are:

1. the ability to build customer loyalty and ensure repeat purchases;
2. management of obsolescence and reduction of waste to a single-digit percent of turnover;
3. management of the interest cost on working capital;
4. most important, the cost of real estate, which is met through ownership or rentals. Since many of these categories of retailers do not have the support of institutions and are self-starters, capital is scarce. Hence, they go for rentals. Managing the space and holding on to the rental rate is a contemporary challenge, especially in class A and class A1 cities, where real estate prices have zoomed.

This leads us to the conclusion that given the clusters of dwellings and space constraints, agricultural retail trade in the unorganized format would continue to thrive in India, as it provides livelihood for a large community of small traders. There would be less recognition as parts of supply chain networks, but more effectiveness could be driven by instincts and node (retail) level business sense, and by providing value to small-quantity customers.

5.4   Organized Sector

Organized agricultural food retailing, as defined earlier, refers to food and beverage retailing by licensed retailers (registered for sales tax, income tax, etc.) with proper technical and accounting standardization. The organized retail trade in the food category also has a significant impact on supply chains related to the flow of material from the rural areas to the consumption centres and could directly affect the livelihoods of the farmers.

The hierarchy or structure of the agribusiness market has two components:

1. Agricultural inputs and services which are bought and consumed by farmers in their production activities as part of the backward linkages in agribusiness supply chains. However, we are not discussing this issue here, as it has been dealt with in the chapter 2 on agribusiness inputs. Moreover, it does not constitute ultimate individual consumption, but is seen as an input for an output to which value is being added. Though such products and services require retailing, they are not part of the discussion here for the reasons mentioned above.

2. Agricultural outputs move as products directly or through value additions to the retail points for ultimate consumption; this is a part of the forward linkages of agribusiness chains, and our discussion here is centred around those products and the outputs to the food, perishable and organized retail businesses where value-added agricultural produce is sold.

Table 5.6 provides the phases of evolution of the retail sector in India by bringing out the structures or hierarchies of retail formats.

Here, it is important to discuss the rural and urban definitions widely used in India, and how they need to be contextualized here. "As per the Census Report, the 'rural sector' means any place that meets the following criteria: a population of less than 5,000; a density of population less than 400 per sq. km, and more than 25% of the male working population engaged in agricultural pursuits." (Ministry of Home Affairs, 2011b)

Further, the census further mentions that the rural areas are the smallest areas of habitation, viz., the village generally falls within the limits of a revenue village as recognized by the normal district administration. The revenue village need not necessarily be a single agglomeration of habitations, but has a definite surveyed boundary. Each village is a separate administrative unit, with separate village accounts. It may have one or more hamlets. The entire revenue village is one unit.

"Rural primary markets are typically located in rural and interior areas, and serve as focal points for a majority of the farmers in the vicinity. These are mostly small and marginal farmers who come to market their farm produce and purchase their consumption needs. Typically, the markets are like *melas* (celebrations) on

Table 5.6   Phases in the Evolution of Retail Sector

Market	Description
Weekly markets, village and rural *melas*	Source of entertainment and commercial exchange
Convenience stores, mom and pop stores, *kirana* stores	Neighbourhood stores/convenience stores. Traditional and pervasive reach
Public distribution stores, cooperatives	Government supported availability/ low costs/distribution
Hypermarkets and supermarkets, department stores	Shopping experience/efficiency/ modern formats

Source: NABARD, *Organized Agri-Food Retailing in India.* Mumbai: NABARD, 2011.

a given day of the week in a common designated place, where traders from the villages and other nearby areas come to sell. *Melas* occur in a cluster on different days of the week, and traders keep moving to sell their produces. Further, these markets function as collection centres for adjoining secondary markets. The commodities collected in these markets find their way to the wholesale assembling markets in the process of further movement to consumers. It is estimated that at an all India level, the number of such markets are more than 47,000" (NABARD, 2011).

On the other hand, urban areas are still defined on the basis of the 1971 census, as described below.

1. all places with a municipality, corporation, cantonment or notified town area
2. all other places which satisfied the following criteria:
   a. a minimum population of 5000
   b. at least 75% of the male working population following non-agricultural activities
   c. a density of population of at least 400 per sq. km (i.e., 1000 per square mile).

A new concept that had been developed for the 1971 Census for the tabulation of certain urban data was the Standard Urban Area. The essentials of a Standard Urban Area are as follows:

1. It should have a core town of a minimum population size of 50,000.
2. The contiguous areas made up of other urban as well as rural administrative units should have close mutual socio-economic links with the core town.
3. The probabilities are that this entire area will get fully urbanized in a period of two to three decades (Ministry of Home Affairs, 2011a).

Further, "an urban agglomeration is a continuous urban spread constituting a town and its adjoining outgrowths or two or more physically contiguous town together with or without outgrowths of such towns" (Ministry of Home Affairs, 2011a).

The wholesale markets are located at *talukas* (a subdivision of a district; a group of several villages organized for revenue purposes), at the headquarters of a town or a municipality, or at important railway stations, and perform the assembling and distribution functions. The business is conducted according to market practices established by age-old customs, or as per the regulations of Agricultural Produce Market Committee (APMC), wherever these regulations are in force. "As per government estimates, there are about 6,300 such markets across India" (NABARD, 2011). Here, largely small retailers and processors come to buy their needs, and at times, even end customers come to buy. It is an important avenue for end customers to buy directly or at last mile, and hence is discussed here.

"Over the years, to achieve an efficient system of buying and selling of agricultural commodities, most of the state governments and union territories enacted legislations Agricultural Produce Marketing (Regulation) Acts (APMRActs) to provide for regulation of agricultural produce markets. Most of the wholesale markets and some of the rural primary markets have been brought under these regulations" (NABARD, 2011). Hence, they are discussed here under the organized retail category of agribusiness in India. Table 5.7 gives details of APMCs in different states in India.

APMCs are focal points for sellers, buyers and agents. The committees are responsible for providing the required infrastructure as well as for preventing unfair trade practices by functionaries. APMCs regulate the wholesale selling of agricultural produce in their defined areas of operation. The aim of regulating the market functioning was to establish a transparent trading market environment where sellers would bring their agricultural produce to sell in open auctions. The complete process of such sales, including the auction, weighing, grading, payment and temporary storage then would be controlled by a regulating body represented by democratically elected members. Moreover, even the behaviour of the functionaries operating in the market area is closely monitored. Updated information about the total inflow of various agricultural commodities, their prices, and month-wise demand/supply positions is easily available. It may be noted here that understanding of APMCs is important here, as small retailers and even large retail formats buy from here, and effective functioning of APMCs could lead to healthy retailing of agricultural produce.

One is not clear about the impact of the APMC, as only 17 out of 35 states and union territories have implemented the APMC Act and its amendments. According to Suparna (2011), "an important stakeholder group in the retail distribution chain are the multitude of small farmers and small agro-processors, who receive the most miniscule share of the revenue pie, and whose concerns are almost always overshadowed by the better lobbying power of the intermediaries in the agricultural supply chain." Only 17 states have amended their APMC Act to allow direct marketing,

Table 5.7   Number of APMCs in Major States (2008)

State	No. of Regulated Markets (Market Yards and Sub-Yards)
Andhra Pradesh	889
Maharashtra	871
West Bengal	684
Uttar Pradesh	584
Karnataka	492
Madhya Pradesh	488
Punjab	437
Rajasthan	416
Gujarat	405
Orissa	314
Others	1977
Total	7557

Source: NABARD, *Organized Agri-Food Retailing in India.* Mumbai: NABARD, 2011.

contract farming and markets in the private and cooperative sectors. Key grain producing states such as Haryana, Punjab and Madhya Pradesh have initiated only partial reforms. Further, seven states and Union Territories do not have any APMC Act to govern agricultural trade. The states that have APMC Acts allow retailers to procure outside the regulated local markets, known as *mandis*. According to the author, creating an effective multiple and competitive market channel choices for farmers by means of APMC reform will be the first step towards fighting the persistent inflationary conditions assailing the country. We also concur that given the proliferation of agricultural production and markets, setting up APMCs and e-tailing connecting APMCs, like Gujarat has done, may have a substantial impact in terms of better value creation in agribusiness retailing. There are analysts and advisors who have opined that APMCs have not delivered in accordance with the intended purposes.

The Tamil Nadu government came up with a different approach to marketing of agricultural produce. Since most of the farmers sell their produce through village level markets, fairs, *mandis*, cooperative societies and so on, the middlemen could exploit farmers as well as consumers. It introduced *"Uzhavar Santhais"* (loosely translated as farmer markets) in 1999 to enable farmers to reach consumers directly at consumption centres. There were about 103 *Uzhavar Santhais* functioning as of 2011. All *Uzhavar Santhais* are maintained and staffed by personnel from the Departments of Agricultural Marketing, Agriculture and Horticulture.

Though this is seen as an alternative to APMCs, one cannot say conclusively which approach is more ideal for improving supply chain efficiency.

Urban organized retail outlets are usually located quite far away compared to unorganized retail stores in the same location. Further, in terms of numbers, there are more unorganized stores for every organized store, as the size and economics of an organized store is five times and more in terms of size and stock. The differences also vary. For example, in large urban areas, one may find the organized stores better in terms of ratios than semi-urban and rural areas.

On an average, consumers spend more in an organized retail store on every purchase compared to that in an unorganized store. This is mainly because of two reasons: (a) the number of visits to organized stores is less compared to visits to unorganized stores, and (b) the range of items available, including items in the food and non-food categories, is higher in an organized store. We may have to keep in mind here that we cannot make an apple to apple comparison, as a customer would also buy general format items, which are not related to agribusiness, from an organized store. As footprints are driven by demand, and sales happen on bundles of many items, organized stores have an advantage. This is true mainly for processed agricultural products, especially in an urban centre, as customers feel organized stores offer better quality and grade, even with private labels. When an item that is available without a brand in the unorganized sector is available with an established brand, the push in organized stores is high because of promotions. Hence, customer spending is high.

However, organized retail outlets continue to be less preferred by consumers for all categories of food items such as milk and milk products; fruits and vegetables; non-vegetarian food items and eggs. These food items are purchased frequently mostly at unorganized outlets. Typically in India, fruits and vegetables required for one to four days are purchased, while milk and eggs are purchased on a daily basis. Other food items are purchased in quantities meant to last between 7 and 15 days.

Customer loyalty towards organized retail is less compared to loyalty towards unorganized retail. The reasons are as follows: (a) Organized retail is in the early stages of growth and is providing new experiences and discounts for attracting footprints. (b) Customers believe all organized retail is at a given level of quality, and swapping loyalty is acceptable, as they can gain from every purchase. In the case of unorganized retail, loyalty has to be higher because of familiarity, value-added service offered and so on.

One would expect organized retail to score over unorganized retail in cleanliness and freshness for all food categories. Of the two factors, freshness is more important for grocery and non-vegetarian food items. However, for non-vegetarian items and other perishables, customers in India still prefer the unorganized retail market.

On price competitiveness between organized and unorganized retailers, there is no perceptible difference. Though theoretically organized retail should offer better value for customers, current market structures and the cost of real estate, as well as working capital costs and losses due to wastage bring in competitive disadvantages.

It is also observed that customers are unhappy with organized retail because of long queues for billing, heavy customer rush during peak hours and lack of attendants who could facilitate the buying experience.

Thus, it may be observed that organized retail is at an evolutionary stage only, and initiatives taken by the government are expected to have less impact because of the costs of real estate and capital, footprints and value of basket bought, and customer choice. In India, it is difficult to imagine that there are stores where the value of customer experience can be higher, because there are no adequate parking spaces, and most stores are not well laid out, with lots of walking bays and huge displays that would bring the customer's experience in line with stores in developed markets.

5.5   Supply Chain Challenges

While discussing supply chains, an analyst would be interested in knowing how many nodes and flows are there between the origin of a produce, in the value addition process and in the final reach to ultimate customers. The more the nodes, the more the intermediaries, the lesser the efficiency and the higher the supply chain costs. This is what is typically happening in Indian agribusiness.

There are companies such as ITC, Pepsi and many others in processed food business, with beverages such as coffee, tea and milk, who organized direct procurement with the support of the government and were able to reap benefits. Since a number of examples are discussed in this book in different chapters, we avoid repeating them here.

It may be important to note that direct procurement of fruits and vegetables by organized food retailers from farmers not only results in the farmer realizing a better share of consumer spending, but also in reducing costs to the consumers. This supply chain design requires certain preconditions.

1. Government policy support is required to enable direct procurement from farmers. It is not forthcoming in all states in India. There are a number of guidelines and approvals required even in states where it is permitted.
2. Farmers need to be organized into clusters or groups to produce in contiguous areas so that there are economies of scale and feasibility in cost management. Non-contiguous planting could impose difficulty in harvesting, post-harvest management and so on. More important, crop management costs within the support system would be higher.

3. Adequate infrastructure is needed for grading, sorting and cleaning of produce and for moving produce to markets/stores.
4. Excellent communication and supplier relationship management with farmers are needed to ensure seamless flow of products, information and finance. Farmers cannot accept reverse flows, and it is the responsibility of direct procurement agents/retailers to ensure that produce is liquidated and there are no returns. Typically, our observation has shown that farmers are open to grading while procuring and to fixing prices according to grades, but they do not encourage returns. We also observed that farmers received better prices in cases of direct procurement, when compared to the prices available at nearby wholesale markets; over and above this advantage, they also save on transportation and handling costs. Retailers who procure directly are able to get the right produce of good quality and standards and realize better prices, as wastage is reduced.

Thus, supply chain efficiencies can be improved through direct procurement or by reducing intermediaries wherever possible in agribusiness, mainly because a number of role agents do not have adequate responsibilities compared to the reward they receive.

5.6 Role of Intermediaries

Retailing is the final stage of delivery of produce to the consumer. However, there is a long chain of intermediaries enabling the transfer of produce between the producer and the retailer. Further, most agricultural products require processing and value addition before they can be retailed. Agricultural products can be classified broadly into two groups on the basis of the degree of value addition.

Agricultural products requiring processing: These include grains, oils and pulses, as well as all processed and packaged foods (including packaged milk). They form the bulk of the merchandise of the retailers. Here, the role of the processor is important, as it brings value addition through grading and quality control, and often, changes in the form of the produce, like edible oil being extracted from oilseeds. Organized retailers directly procure from the processors or through wholesale markets. Processors/millers are an integral part of supply chains for this category, thus completing the channel; the length of the channel is determined by the processors' ability to structure procurements directly from farmers. If they have to go through intermediation like agents and markets, products become more expensive.

Agricultural products requiring minimum or no processing: These include fresh fruits and vegetables, and eggs, which can be retailed with no significant value addition. Here, the minimum value addition, such as grading and packaging, can be done by the retailer or by the supplier. Fruits and vegetables, with their need to be retailed fresh, provide ample opportunity for the retailer to come near to the farmers and associate with them. However, a majority of the purchases of the organized and the unorganized retailers are through private sector wholesalers or through the APMCs.

The other commodities, including chicken, meat and milk, would come under the "processed food" category, and are supplied directly through the processors, intermediaries like agents and wholesalers and distributors. There are some excellent brands in India such as Amul, Aavin and Mother Dairy in milk and milk products that could go direct to the sources for bulk items and through intermediation for items of longer shelf life. Similarly, in chicken products, brands such as Suguna and Venky's are successful, apart from others. Long life shelf products such as sugar, tea, coffee, edible oil and processed grains all have to go through certain levels of intermediation. Spices comprise one of the largest markets in India, and some brands such as Sakthi Masala, Aachi Masala and MTR have created a large presence by managing their supply chains efficiently and as also by better retailing through their networks, such that they have a high brand recall and loyalty.

Thus, organized retailing in India is a challenge, and supply chain efficiencies are critical. Supply chain efficiencies must exist through all drivers of supply chains, but most importantly in procurement and inventory management, along with facility management and transportation, all of which provide value to role players.

5.7 Foreign Direct Investment

"In January 2012, the government removed restrictions on FDI in the single brand retail sector, allowing 100% FDI. The government has put a condition in respect of proposals involving FDI beyond 51%, making it mandatory to source at least 30% of the value of products sold from Indian 'small industries/village and cottage industries, artisans and craftsmen'" (www.business-standard.com). More recently, in September 2012, the government has proposed allowing FDI of up to 51% in the multi-brand sector. This has created a lot of opposition from many interested groups, who claim that it would kill the domestic retail business and that the agricultural sector would also be exploited by the large MNCs. Contrary to this, the government and those who are supportive of FDI in retail are of the view that there is scope for improving supply chain efficiencies by eliminating intermediaries, and customers would benefit from low prices and wide choices. According to them, both the organized and the unorganized sectors would be able to co-exist.

We would like to submit the following points for consideration:

1. Organized retail in India is a miniscule part of the business, and different studies have given different projections. However, in a decade or more, its share cannot go up beyond the single-digit level, given the complexity of Indian operations. There would be advantages for large foreign players, but they cannot do away with the unorganized domestic players. In the short run, the domestic players may have a disadvantage, but would soon regain their competitive position.
2. FDI coming in would help with infrastructure investments and with better processes and practices in agriculture. These benefits would pass on commonly as farmers learning would not compel their selling decisions.
3. FDI would find it difficult to bring in big changes, as real estate costs are high and would not allow such operations to be viable. FDI may have advantages in working capital costs, but would have disadvantages in labour and employee costs.
4. Given the complexity of trade, it may be worth allowing FDI in retail with certain conditions relating to procurement guidelines and the need in invest on common facilities that could benefit trade as well.
5. At the outset, it is not very attractive for FDI to try to make a big bang effect. In India, we already have corporate entities such as Reliance, the Goenkas and the Birlas spreading out in the retail sector and taking a long time to make deep impacts. FDI can be no different from them.

5.8 Conclusion

Retail in agribusiness is challenging because of proliferated demand and supply market conditions. Speaking in ideal terms, the Indian sub-continent's culture and social set-up are such that they have promoted retailing as a livelihood for many, and intermediation processes have benefited many. Unorganized retail is deep-rooted and is supported by consumers as well. Without taking positions, structural changes are not possible through statute or intentions; they can be achieved only by setting economic values for the various players in the supply chain system, namely, the producers, ultimate consumers and a host of intermediaries. Considering the fact that real estate is expensive in India and customers largely prefer fresh produce, organized retail will slowly pick up in semi-urban and downtown areas of large cities. There cannot be two opinions on the need for agribusiness to benefit from a structured business approach, especially in retailing. A few sets of practices such as fair pricing, transparent systems and structured market operation supported infrastructure may usher in a lot of good for the sector.

References

Ministry of Home Affairs. Census terms. 2011a. http://censusindia.gov.in/Data_Products/ Library/Indian_perceptive_link/Census_Terms_link/censusterms.html

Ministry of Home Affairs. Provisional population totals: Urban agglomerations and cities. 2011b. http://censusindia.gov.in/2011-prov-results/paper2/data_files/India2/1.%20 Data%20Highlight.pdf

NABARD, *Organized Agri-Food Retailing in India*. Mumbai: NABARD, 2011.

Suparna, K. APMCs hold key to retail reform, *Business Line*, 2011. http://www. thehindubusinessline.com/opinion/article2660044.ece?homepage=true

INTRINSIC ISSUES

Chapter 6

Perishability

OBJECTIVE

The objective of this chapter is to understand the challenges faced in handling perishable commodities with respect to distribution of produce, short shelf life, lack of temperature control or appropriate facilities for storage and transportation, wastages and pricing pressures. The chapter also addresses channels of distribution of perishable commodities and issues of wastages in the channel.

6.1   Introduction

India is most often portrayed as a country of hungry people deprived of food and basic nutrition. Is this a problem of insufficient production or of inefficiencies in the supply chains dealing with distribution of agricultural products including food grains, fruits and vegetables, meat and dairy products? There are reports which mention that wastage in the post-harvest stages for fruits and vegetables is about 30%–40%. Mounts of food grains that go rotten in storage godowns have also been drawn to our attention. This highlights the scope for improving supply chain efficiency in the perishables segment.

It is important to understand that the dimension of perishability varies for different types of farm produce, and that as these produce are processed, either the freshness of the produce or other original qualities of the produce may probably change. This in turn impacts the supply chain, depending on whether the consumption at the customer's end is soon after harvest, without processing, or whether the converted product from the harvest is consumed after some form of processing.

This being the case, it is a question of how the efficiency and responsiveness of the supply chain adapt to whether unprocessed raw produce is consumed or processed produce is consumed. In other words, the issue is how long the supply chain should be to optimize the benefits for the various stakeholders.

When we talk about perishability, we basically need to look into the following:

■ Depending on the basic qualities of the farm produce, the processes to add value and the ultimate shelf life of the final product, how to map out the supply chain components and the necessary logistics
■ How to extend the shelf life of the produce, even if not be in its "harvested fresh" form
■ The fact that wherever extending the shelf life does not serve any purpose, the supply chain cycle should be shorter than the "perishable cycle" of the produce.

In this chapter, issues related to the management of the supply chain mechanism for perishable agro-based products are discussed. This could help to understand the challenges faced not only by the central government and its agencies like the Food Corporation of India, as well as state government agencies, but also by corporations or companies involved in sectors relating to milk and milk products, flowers, perishable fruits and vegetables, meat and seafood processing, pharmaceuticals and other related industries. This chapter also covers topics related to facility management, processing, distribution networks, retailing and price management.

Along with the 4P's of marketing, agricultural products and agri-related products demand that another "P" be given close attention: perishability. Transporting fresh produce from one point to another is certainly not an easy task. Many factors affect perishable product logistics in today's market, such as rising fuel costs, food safety and statutory issues. The quality and freshness requirements associated with the demand for these products put constraints on the time available for the supply chain steps to be completed. The demand for fresh produces virtually dictates a shorter supply chain. Quick decision making and faster operations only will help to meet targets and may or may not fulfil customer demand for fresh and quality products fully. The supply chain set-up has also to assimilate quick transport, maintenance of ideal temperature ranges and efficient marketing through wholesale markets and retailers to match customer expectation. Minimization of wastage and the costs involved in transportation and storage can help increase profits. This happens because revenue is improved through better realization and costs are reduced as wastage is reduced, and thereby, the supply chain profits. Both revenue increase and cost reduction help ensure better profitability. Energy usage reduction and quality packaging also go a long way in ensuring better supply systems, longer life and increased returns. These elements also pave the way for food grains and other perishable products to be served to larger customer and population bases, which improves another supply chain goal, namely, service level, which is a measure of demand being served on a number of occasions.

6.2 Facilities at Source: Case of Different Products

One of the integral and essential parts of managing a "perishables" supply chain is choosing a proper facility, as this plays a major role. These decisions are more influenced by industry structure and product characteristics, and hence, are strategic in nature. They are typically made over a longer time horizon because the industry/product management practices are being evolved. They are also closely connected to the corporate strategy in the case of supply chain focal firms.

The creation of a supply chain necessitates appropriate geographical placement of sourcing points, such as the farmers, production facilities and storage points. Resources would depend on a long-term basis on the facility selected. Therefore, selecting the facility is crucial to revenue, cost and service level. Consider a farmer whose farm production point is located close to the production unit. Obviously, this can facilitate the efficiency of the chain. However, with disaggregated farming and geographical spread, the challenges are in determination of size, number and location of facilities. Further, there must be synchronization with optimization of production and distribution costs. Production limitations contribute immensely to reducing the cost and quality issues in supply chain management.

The perishables' facilities near the farm could be as follows:

1. Washing, cleaning and grading centres
2. Preservation centres such as in the case of fish, icing and salt mixing
3. Chilling centres and temperature control units
4. Intermediate and full-scale processing units.

After harvest, agricultural produce such as fruits, vegetables, milk, meat and fish could be impacted by accelerated physiological, chemical and microbial processes that invariably lead to deterioration and loss of freshness, physical quality and volume. This mandates the creation of facilities with some measure of processing, such as reduction in moisture content, denaturation of endogenous enzymes and micro-organisms or suitable packaging to reduce or eliminate perishability. It is important to have such facilities; in the absence of such facilities, farmers could incur huge post-harvest losses right at the farm gate itself. It is the responsibility of the players in the supply chain network to work with technologists to understand the underlying processes contributing to food deterioration and spoilage, and to devise appropriate measures and methods of preservation in order to ensure availability, acceptability and safety of foods. We now highlight the practices and challenges in select agriculture and allied businesses in India.

6.2.1 Seafood

It may be interesting to look at aspects of supply chains in the Indian seafood industry, especially from the exports perspective, and to understand the importance of facilities at source. India exports about $2132.84 million worth of seafoods. Major destinations include the United States, Japan, the European Union and South-East Asian nations. Table 6.1 gives details of such exports.

Ports such as Pipavav, Mumbai, Kochi, Chennai, Vishakapatnam and Kolkata have a large share of exports, as shown in Table 6.2, in terms of port-wise exports of seafood.

Most of the exports comprise frozen fish from India. Processed fish export is reported to have less than a single-digit share in total exports as shown in Table 6.3. A large percentage of fish exported to South-East Asia gets re-exported to northern markets after value addition. The fish market has a demand–supply imbalance due to depleting supply at source. However, the unorganized businesses run by Indian exporters and inadequate facilities lead to low competitive advantages.

The supply chain of fish exports, in general, comprises of:

Fisherman →Commission agent →Supplier (Pre-processor) →Exporter

The distribution of income across the supply chain is shown in Table 6.4.

We consider the roles of supply chain actors in Table 6.5.

Table 6.1 Export of Seafood for April 2009 to March 2010—Country-Wise Total

Country	Exports (April 2009–March 2010)			Exports (April 2008–March 2009)		
	Quantity (tonnes)	Value (in ₹ crores)	$ (million)	Quantity (tons)	Value (in ₹ crores)	$ (million)
Japan	62690	1289.58	278.56	57271	1234.01	278.61
United States	33444	1012.52	213.52	36877	1021.55	227.29
European Union	164800	3013.33	637.40	155161	2854.07	635.34
China	144290	1790.89	379.70	147312	1296.39	281.90
South-East Asia	149353	1479.55	314.85	88953	873.09	191.08
Middle East	34907	553.55	117.05	27177	475.72	105.20
Others	88953	909.11	191.77	90083	853.11	189.22
Total	678436	10048.53	2132.84	602835	8607.94	1908.63

Source: SEAI, Statistics, 2012.

Table 6.2   Port-Wise Export of Seafood, April 2009 to March 2010—Port-Wise Total

Ports	Exports (April 2009–March 2010)			Exports (April 2008–March 2009)		
	Quantity (tonnes)	Value (in ₹ crores)	$ (million)	Quantity (tonnes)	Value (in ₹ crores)	$ (million)
Pipavav	182052	1673.74	361.21	163866	1408.35	307.69
Kochi	104281	1576.19	333.02	98537	1504.98	335.35
J N P	129318	1564.42	331.46	126853	1487.28	329.52
Chennai	45991	1314.10	277.80	39043	1078.44	240.80
Vishakapatnam	31863	943.29	199.24	32277	897.93	199.85
Calcutta	46901	892.48	188.10	33625	720.36	159.96
Tuticorin	27782	686.45	145.22	29354	693.76	153.59
Mumbai	2349	462.67	97.97	2319	176.56	38.60
Mangalore/ICD	59000	400.33	85.43	33083	238.44	52.81
Goa	29409	219.24	46.49	21146	185.16	42.04
Ahmedabad	1145	153.98	32.80	123	65.97	14.96
Trivandrum	2985	92.66	19.66	2209	66.16	14.69
Mid Sea	6289	36.82	7.86	5482	46.08	10.27
Hill Land Customs	8236	19.12	3.83	13960	25.16	5.48
Mundra	671	11.02	2.34	661	11.00	2.50
Calicut	28	1.16	0.24	33	1.03	0.21
Agartala	18	0.52	0.11	13	0.40	0.09
Karimganj	117	0.25	0.05	173	0.36	0.08
Delhi	0	0.07	0.02	1	0.07	0.02
Okha	1	0.02	0.00	1	0.01	0.00
Kandla	0	0.00	0.00	75	0.41	0.10
NSICT	0	0.00	0.00	1	0.02	0.00
Port Blair	0	0.00	0.00	0	0.01	0.00
Total	678436	10048.53	2132.84	602835	8607.94	1908.63

Source: SEAI, Statistics, 2012.

Table 6.3 Item-Wise Export of Seafood during 2009–2010

Items	Exports (April 2009–March 2010)			Exports (April 2008–March 2009)		
	Quantity (tons)	Value (in ₹ crores)	$ (million)	Quantity (tons)	Value (in ₹ crores)	$ (million)
Frozen shrimp	130553	4182.35	883.03	126039	3779.80	839.28
Frozen fish	260979	2032.33	430.94	238544	1722.34	375.24
FR cuttle fish	63504	923.83	195.69	50750	761.05	168.27
FR squid	61445	622.63	132.24	57125	632.35	142.87
Dried item	47053	981.11	208.72	31688	420.75	92.51
Live items	5492	139.14	29.52	3434	99.00	21.82
Chilled items	28817	264.49	55.87	21453	217.34	48.39
Others	80592	902.64	196.84	73801	975.33	220.24
Total	678436	10048.53	2132.84	602835	8607.94	1908.63

Source: SEAI, Statistics, 2012.

Table 6.4 Distribution of Income across Supply Chain in Seafood

Node→	Fishermen	Commission Agents	Suppliers (Pre-processor)	Exporters
% share on price	25–35	1.5–4	20	40–50

Source: Kulkarni, P., The marine seafood export supply chain in India: Current state and influence of import requirements, International Institute for Sustainable Development, Canada, 2005.

From the above, we can appreciate the role and importance of having handling facilities close to the fishermen's docks. Table 6.6 gives a summary of the facilities available in India. It is believed that the fishermen lack adequate knowledge of fish handling, especially for exports. Further, there are issues with respect to the quality of the ice on which the fish are stocked and of the salt used. Generally, the supplier will have small depots alongside docks or harbours where products are sorted and cleaned. Again, the facilities are to be evaluated from the viewpoints of sanitation, labour and

Table 6.5 Role and Responsibilities of Supply Chain Players in Seafood

Fishermen	Commission Agents	Suppliers	Exporters
Inputs: diesel, ice, food, nets, boats, 6–12 helpers	Receive fish from boat	Receive fish from agent	Receive fish as raw material
Undertake 4–8 days fishing trip	Weigh fish	Stock fish in crates filled with ice	Wash with portable water
Classify caught fish as per fish category	Grade fish as defective and non-defective	Sort fish in four grades as per quality standards of exporter	Process using Hazard Analysis and Critical Control Point (HACCP) procedures
Store fish in ice	Negotiate price with fishermen ad suppliers	Transfer fish to pre-processing unit	Pack processed fish
Unload fish on docks after preliminary wash		Clean fish	Perform export procedures and dispatch
Negotiate with agent and receive money		Negotiate price with exporter and agent	Negotiate price with importer and with supplier

Source: Kulkarni, P., The marine seafood export supply chain in India: Current state and influence of import requirements, International Institute for Sustainable Development, Canada, 2005.

management for export handling. The significance of having the facility close to the supply centre for an efficient supply chain is clearly understood by the discussion so far.

The socio-economic statuses of the role players in terms of education, culture and access to health, water and sanitation facilities could impact the quality and value of agricultural and allied industry produce, as depicted in the case of fisheries.

6.2.2 Oil Palm

"Oil palm is another interesting example of the importance of locating processing centres closer to sources for value creation. Palm oil or palmolein is the oil produced from the red oil palm tree. Palm oil is extracted from the pulpy portion (mesocarp) of the fruit of the oil palm. The crude palm oil is deep orange red in

Table 6.6   Facilities Overview of the Industry

Category	Registered as on March 31, 2010	Capacity (in metric tonnes)
Manufacturer exporter	403	NA
Merchant exporter	428	NA
Route through merchant exporter	27	NA
Ornamental fish exporter	50	NA
Fishing vessels	6809	NA
Processing plants	411	15113.03
Ice plants	65	1858.80
Peeling sheds	586	5387.47
Conveyance	154	1056.34
Storages	490	162859.60
Fresh/chilled fish	27	1372.40
Live fish handling centre	24	2371.63
Salted/dried fish handling centre	43	646.48

Source: SEAI, Statistics, 2012.

colour, and is semi-solid at a temperature of 20° centigrade. Palm oil contains an equal proportion of saturated and unsaturated fatty acid, containing about 1–40% oleic acid, 10% linoleic acid, 44% palmitic acid and 5% stearic acid" (oilpalmindia.com).

Oil Palm India Limited, the leading processor of oil palm, was established in the year 1977 with the objective of propagating oil palm cultivation in the country. This organization is located in Kerala, where agro-climatic conditions suit farm economics. "From 1983 onwards, the company started functioning as a joint venture of the Government of Kerala and the Government of India. Oil Palm India Limited has got a total planted area of 3,646 hectares of plantation spread over in three estates, viz., Yeroor, Chithara and Kulathupuzha in Kollam District, Kerala" (oilpalmindia.com).

It is important to note here that in a fresh ripe and un-bruised fruit, the free fatty acid (FFA) content of the oil is below 0.3%. Generally, the FFA in a damaged part of the fruit can increase rapidly to 60% in an hour. There is great variation in the composition and quality within a given bunch, depending on how much the individual pieces in the bunch have been bruised. Moreover, harvesting involves the cutting of the bunch from the tree and allowing it to fall to the ground. Fruits

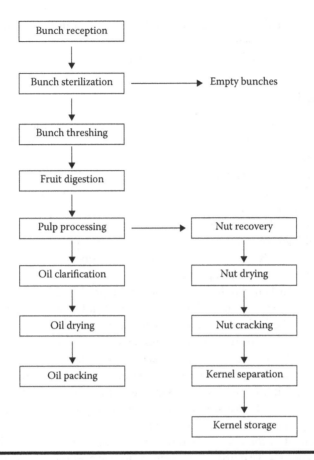

Figure 6.1 Palm oil processing unit operations. (From Poku, K. Palm oil process-ing. Ch. 3 in: *Small-Scale Palm Oil Processing in Africa*. Rome: FAO, 2002.)

may be damaged in the process of pruning palm fronds to expose the bunch base to facilitate bunch cutting. As the bunch hits the ground, the impact bruises the fruit. During loading and unloading of bunches into and out of transport containers, there is further scope for the fruit to be bruised. Hence, the practices adopted for harvesting, transportation and handling of bunches can damage the fruit. Given the above, it is advisable to process the fruit as early as possible after harvest, say within 48 hours. Ideally, this would require processing at the oil palm estate itself. The general oil palm process is shown in Figure 6.1.

Thus, oil palm, like fish, explains how important it is to correctly locate at least the immediate processing centres of agricultural produces that may undergo deterioration in quality if not handled appropriately. Another product which goes through a similar situation is milk. In India, oil palm is not produced in large quantities, and has a minor share in terms of local edible oil production. But dairies are better distributed in most parts of rural India (except in Eastern India), which may not otherwise be well connected. Let us see how the perishable nature of the product is better handled with facilities support.

6.2.3 Dairy

India is one of the top-ranking countries in milk production. The milk production of India is about 13% of the world's total milk production. The top five milk producing countries are India, the United States, Russia, Germany and France.

Milk production went up from 55.7 million tonnes in 1991–1992 to 121.8 million tonnes by 2010–2011, with a compound annual growth rate of about 4.2%. It may be interesting to note that the per capita consumption of milk has gone up to 281 g per day, which is still low compared to international standards, as shown in Table 6.7 and Figure 6.2. Further, with the growth in population, demand for milk is growing. Another issue is with respect to surplus milk, which is being marketed for consumption within the population.

As per popular belief, only 50%–60% of milk produced is available for distribution in the markets. Of this, a large portion of milk is sold through the unorganized sector. Of the milk distributed through the organized sector, more than 80% is sold as bulk milk. This really complicates handling, and we need to understand the process of milk movement to the markets from the farmers.

From Figure 6.3 on milk distribution, we can see farmers providing milk at the collection centres almost twice a day. As mentioned earlier, though India is the largest milk producer in the world, only 20% of its milk production is processed and moves through the organized sector. The remaining 80% of the total production is still distributed in the unorganized sector. In order to boost their milk procurement, many cooperative dairies and private sector have set up societies or collection centres in the rural areas.

"The traditional system of milk collection in the dairy cooperative societies mainly used volume to measure milk accepted, carried out fat testing using the Gerber method and entered the fat and SNF data manually in registers. Quite often, this approach did not receive adequate confidence from farmers. In order to avoid such issues, a unique system to automate milk procurement operations of village milk cooperative societies was developed. The Automatic Milk Collection Station is an integrated unit that combines different functions of a milk collection centre, such as measuring the weight and fat content, and providing the price output of the milk collected at the centre. The equipment also helps the milk cooperatives/milk collection centres to maintain operational data on milk supplied, amounts payable based on fat content, and so on. The system comprised of an electronic milk weighing unit, an electronic milk tester and a data processing unit. At this point, the local facilities of collection centres play a significant role" (NABARD, 2011).

From here, trucks on a milk run collect the milk and carry it either to chilling centres or to processing plants, depending upon the volume and coverage of the

Table 6.7   Dairy Production and per Capita Consumption in India

Year	Production (million tonnes)	Per Capita Availability (g/day)
1991–1992	55.7	178
1992–1993	58	182
1993–1994	60.6	187
1994–1995	63.8	194
1995–1996	66.2	197
1996–1997	69.1	202
1997–1998	72.1	207
1998–1999	75.4	213
1999–2000	78.3	217
2000–2001	80.6	220
2001–2002	84.4	225
2002–2003	86.2	230
2003–2004	88.1	231
2004–2005	92.5	233
2005–2006	97.1	241
2006–2007	102.6	251
2007–2008	107.9	260
2008–2009	112.2	266
2009–2010	116.4	273
2010–2011	121.8	281
2011–2012*	127.3	

Source: National Dairy Development Board. Milk production in India. 2013.

* Anticipated achievements.

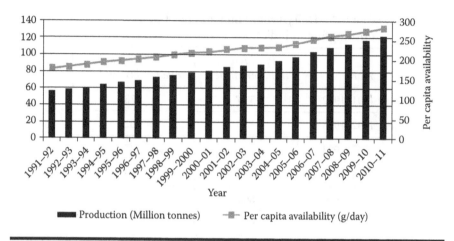

Figure 6.2 Milk production in India.

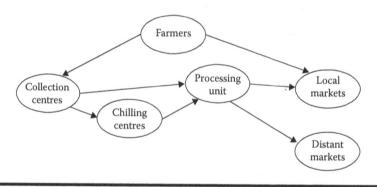

Figure 6.3 Milk distribution network in India.

processing plants. The task of procuring milk and getting it ready for processing is a complex problem involving ownership, pricing, collecting, grading, measuring, weighing, testing, bulking, transporting and chilling, packaging and heating. Chilling, packaging and transportation of milk constitute a component of dairy technology that deals with the processing of milk on an industrial scale. Milk must be cooled as soon as possible after it is collected from collection centres. A temperature of 4°C or less is recommended. It is necessary, because as long as this temperature is maintained, bacterial action in the milk is retarded. As soon as micro-organisms get into the milk, they start growing rapidly because milk contains all the nutrients required for their growth. If the growth of micro-organisms is not checked, then several biochemical changes will take place in milk. Due to these changes the quality of milk is adversely affected, and the milk becomes unfit for human consumption. Hence, in dairy products which are highly perishable in nature, collection centres and chilling centres play a significant role.

If there are more collection centres in a contiguous location, the transport vehicles would directly take the milk to a district level processing plant where the milk is processed using technology and then distributed. In India, some states such as Gujarat, Tamil Nadu, Karnataka and Andhra Pradesh are leaders in milk production and distribution, especially through cooperatives.

After discussing facilities close to farms/production centres for the fish, palm oil and dairy industries, we will highlight the situation prevailing in sugar manufacture.

6.2.4 *Plantation White Sugar*

Given the characteristics of sugarcane, such as perishability and bulkiness, there is a need for shipping the sugarcane within a shorter time span; the longer the duration, the less the sugar extract. The production of sugar depends on the quantity and quality of its raw material, sugarcane. Sugarcane is the raw material for jaggery and *khandsaria*s well. Unlike in many other countries, sugarcane is the only source of sugar in India and hence any imbalance between demand and supply of sugar impacts the economics of sugarcane cultivation to a great extent. Sugarcane is cultivated in fertile agricultural lands that are developed virtually as satellite areas of sugarcane mills by both private companies and cooperative societies. According to the Sugarcane Control Act, the nearest sugar mill with whom a farmer registers his crop issues the cutting order for the fully grown cane, which is then transported by lorry, tractor or cart to that crushing mill. Since cane is a stem crop, some percentage of water content will be lost during transit. Thus, in order to benefit the farmers in terms of price per ton, weight loss is minimized by locating facilities near resource centres. This system also improves the production of sugar and other by-products, and helps the factories to achieve their goals. Crucially, as cane dries up, its sucrose content diminishes. Hence, plants must process cane within 48 hours of harvesting. In India, apart from logistics issues, involving transportation and facilities, a larger concern is behavioural factors, as the farmers may resort to bunch planting, which impacts harvesting. This leads to unevenness in the day schedules of processing, unless they are coordinated very well by the focal firm, the sugar plant. Moreover, the issue of poor economics and the cyclical nature of the industry lead to problems wherein after years of abundant supply of cane, farmers tend to burn some portion of cane in their farms, leading to unhealthy agricultural economic practices. A more detailed discussion would be out of the scope of our discussion here, but may be useful to readers. In fact, sugarcane supply chains could be ideal examples of the woes of agribusiness in India.

Here, we reiterate that for the success of the sugar industry, facilities must be closer to the farm production centres, preferably in a centroid position, and operate efficiently, so that all players get good rewards.

6.2.5 Other Perishable Products

There are a number of agricultural produce types such as fruits, vegetables, flowers, meats and poultry products, which require facilities near production centres. Most of the fruits and vegetables that are perishable in nature require washing, cleaning, sorting and grading even for sale in local and urban markets. In India, we are in the process of the creating necessary infrastructure for such activities. Efficiency in perishable supply chain can be achieved only when an efficient infrastructure is realized. A mature industry like the tea industry is able to keep improving because of quick and quality processing immediately after plucking. It is important that such developments occur in a much more aggressive manner to reduce losses of perishables at source or supply centres.

An efficient supply chain can increase surplus by reducing inefficiencies in production, processing, storage and transportation. It ensures that farmers get better prices, prompting them to invest more capital on the essential inputs, so that productivity increases. It expands market opportunities for products, and thus, helps in maintaining an ever increasing demand for them. Well-integrated supply chains stimulate the growth of agro-based industries, helping farmers to plan their production in advance, based on demand forecasts. Local decisions have a direct implication on the global (across players and processes) efficiency of an operation, and consequently, on its performance, despite the fact that they are primarily based on strategic factors.

The challenge for managing perishables is that goods value depreciates significantly with time taken in the supply chains, as they are extremely dependent on temperature and humidity. Taking dairy products as an example, raw milk received from farmers should be processed and treated within a few hours; this is because time-sensitive products may deteriorate without proper cold storage facilities. Shortening the distance between the source and the production point contributes to reduction of waste. Second, certain fruits and vegetables which face seasonal customer demand can be stocked utilizing optimum technology to reduce waste, thus enhancing profits. For perishable products, a decision about supply chain strategy involves a choice between responsiveness and efficiency. Proper management of facilities results in reducing wastage of resources, minimization of cost and maximization of profit not only for the company, but also for the customer, who enjoys low prices on the final product.

6.3 Distribution and Value Addition

Effective management of a supply chain is all about getting the right product at the right time to the right place and in the right condition. This has never been a simple task, especially in case of supply chains dealing with perishable products. Regional distribution plays a significant role in the cost of goods, especially in the execution of processing of goods. To begin with, goods are received in bulk amounts and stored. They are retrieved and assembled into shipment lot as and when required. It is essential to possess a meticulous processing system at the distribution centres, so as to reduce the labour requirement and positively affect the final cost of the product. Hence, the regional distribution centres can adversely impact profitability if they do not understand their roles and responsibilities.

Let us look at the case of onions and potatoes in India. Storage becomes important for these vegetables, and lack of adequate storage facilities leads to inefficient supply chains. India is the second largest producer of onion after China. Onion is grown in many states in India, and is consumed in all parts of the country. Recent onion production data are shown in Table 6.8.

Onion production in India is about 15.48 million metric tonnes (MT) as of 2011–2012. The key onion-producing states are Maharashtra, Karnataka and Gujarat, which together constitute about 50% of the country's production. In fact, Maharashtra alone contributes 30% of India's onion production. The other significant onion-producing states are Madhya Pradesh, Bihar and Andhra Pradesh. However, the yield of onion in India is one of the lowest in the world, as seen in Table 6.9.

The low yield of onion in India is due to non-availability of storable high-quality seeds and lack of post-harvest facilities and policy support. Beyond these factors, the wide seasonality in onion production impacts production, as the crops are skewed among the three primary crop production seasons in India. The distribution of crop production is as follows: *kharif/early kharif*, 20%; late *kharif*, 20%; and *rabi*, 60%.

While production is seasonal, like with many other agricultural crops in India, consumption is more even, except during winter in northern India. According to

Table 6.8 Onion Production in India

Crop Year	2006–2007	2007–2008	2008–2009	2009–2010	2010–2011	2011–2012*
Production in million MT	8.885	9.138	13.588	12.191	15.117	15.748

Source: National Horticultural Research and Development Foundation.

* Estimated.

Table 6.9 Onion Yield in Different Countries

Country	Korea Republic	United States	Spain	the Netherlands	China	India
Yield/ha MT	66.67	56.56	53.53	48.81	22.21	15.16

Source: SFAC, Baseline data for onion and potato. Apr. 2012.

a report, total household consumption in 2011–2012 was about 9 million MT; indirect demand was about 3.50 million MT and exports accounted for about 2.00 million MT.

Cold storage of onions is not common in India. This is mainly because cold storage is prohibitively expensive, and the technical viability is also not clear. If we look at the high cost of cold storage, the ability to match demand and supply in the onion market becomes nearly impossible. Only low-cost and efficient storage can improve the economics of the onion market. *Rabi* crop onion needs to be stored between May and October for better economics for both farmers and consumers. Thus, the example of onions clearly establishes the case for intermediate storage, for managing of demand and supply mismatch.

Potato is another important crop in India, though per capita consumption is still low compared to many other countries in the world. Potato is cultivated in about 1.8 million hectares of land in India, and crop production is about 2 million MT. In India, Uttar Pradesh and West Bengal contribute to 50% of the production. Bihar, Punjab and Gujarat are the other states that contribute substantially to the output.

Again, potato is a *rabi* crop, with about 80% of output occurring during the period November to February. There is a huge seasonality in arrivals, and therefore in prices. During the post-harvest season, farmers stock up, as prices hit rock bottom as arrivals peak in February. In May, prices move up, to fall in August and September and move up again during early winter. Data on potato cold storage capacities and production in major states are shown in Table 6.10, which clearly shows a gap in the availability of facilities.

Due to this mismatch in production and storage, potato prices fall below the cost of production during the peak harvesting period but skyrocket during the lean season.

Apart from storage of goods, distribution centres perform value-added functions such as cleaning, grading, sorting, labelling and preserving the quality of the product against spoilage. In fact, distribution centres have evolved into essential hubs in supply chain networks, and have become a specialized professional area. Advancement of new processing technologies is essential for ensuring value addition, particularly for perishables. Depending upon their temperament, some types of agricultural produce must be processed without delay, as otherwise, they will undergo a significant loss in quality and value.

6.4 Processing

Processing refers to any activity that maintains or improves quality or modifies the physical characteristics of any material. A process can be as simple as cleaning vegetables or it can be as lengthy and complicated as making cheese. Various farmhouse yields must be processed before they can be used as intended. Perishables, such as, for instance, most fruits and vegetables, are cleaned, graded and stored or

Table 6.10 State-Wise Potato Production and Availability of Cold Storage

States	No. of Cold Storages	Total Capacity in MT as on 31/12/2009	Potato Production in 2010 (MT)	% Storage Capacity to Production
Bihar	228	1069841	5748000	18.61
Gujarat	213	967000	1881800	51.39
Orissa	39	139630	191400	72.95
Punjab	344	1097609	2088000	52.57
Rajasthan	19	65896	75700	87.05
Uttar Pradesh	1286	8719533	13576000	64.23
West Bengal	402	5460000	13391000	40.77

Source: SFAC, Baseline data for onion and potato. Apr. 2012.

processed before they are finally consumed. Many agricultural goods are perishable, and are only available for a particular period of time. Processing can extend and go beyond the seasons during which produce is available. Examples are the conversion of mango into pulp for juices and fruits into jam, so that consumers can consume these products throughout the year. However, because of its natural characteristics, the mango is a far easier product for handling through processing, and hence, gives better efficiency and value to all stakeholders.

Processing of fruits and vegetables can be very simple when it involves just picking, sun-drying and/or some basic process. Due to rapid development and shifting lifestyles and food habits of consumers, the demand for processed (ready to serve) food stuff such as beverages and pickles is increasing. The Indian beverage industry has made significant growth, and is contributing considerably to global trade. Instituting proper cold storage and added services at the farmer and trader levels enhance opportunities for the processing industry. At present, mango, grapes, tomatoes, lemon and cucumber are being processed on a large scale.

Generally, the main causes of spoilage of agricultural produce are microbiological (bacteria, yeasts, moulds), chemical (enzymatic discolouration, rancidity, oxidation) and physical (bruising) factors. Processing facilitates prevention of postharvest losses and eliminates waste. It is important to preserve quality and thereby preserve the nutritive value of the raw materials. It also helps to even out product availability to make seasonal horticultural produce available throughout the year, in forms convenient for users. Further, processed agricultural produce helps to handle emergencies. Thus, processing helps to develop new products and to increase the value of the product. This is again a critical aspect of supply chain efficiency in agribusiness.

6.4.1 Benefits

There are a number of benefits in processing agricultural produce. First, processing helps with storage of perishable and semi-perishable agricultural commodities. This facilitates the matching of demand and supply throughout the year. Second, apart from increasing availability of seasonal products in all seasons, it helps avoid gluts in the market that would otherwise lead to price volatility and wastages. Further, it helps in designing marketing and distribution channels so that products can penetrate markets effectively. More important, processing enables transportation of delicate perishable foods across long distances. Food processing can also help create nutritional and food security, especially in case of exigencies and disruptions in supply chain. Finally, in India, processing provides potential for exports, which fetch foreign exchange, as processed agricultural produce is exported in large volumes.

6.4.2 Products in India

Fruits and vegetables are processed as jam, jelly and marmalade and preserved by concentrating fruits to nearly 70% solids (Total Soluble Solids, TSS), with addition of sugar and heat treatment. The preparation of jam requires several unit operations, viz., selection of fruit, preparation of fruit, addition of sugar, permitted preservatives, mixing, cooking, filling, closing, cooling and storage. This is not very popular with a large segment in India, especially among the rural populations and poor and middle-income groups. A large volume of processing happens in cottage, micro and small-scale industries.

Another product, called chutney, is a mixture of fruits or vegetables with spices, salt and/or sugar, vinegar and other flavours or preservatives. Most popular chutneys are those from tomato and mango, among many others. Similarly, sauces are prepared from tomato, papaya and a few other fruits. Vinegar, salt, sugar and spices are the common preservatives used for the preservation of these products. A large volume of processing happens in cottage, micro and small-scale industries.

Fruit juices are preserved in different forms, such as pure juices and beverages. Fruit beverages can be classified into two groups: unfermented beverages and fermented fruit juices. Fruit juices that do not undergo alcoholic fermentation are termed as unfermented beverages. They include natural and sweetened juices, ready-to-serve beverage, nectar, cordial, squash, crush, syrup, fruit juice concentrate and fruit juice powder. These beverages can be distinguished on the basis of the differences in TSS content and minimum juice percentage. On the other hand, fermented beverages are fruit juices that have undergone alcoholic fermentation by yeasts and lactic fermentation by bacteria. They include wine, champagne, port, sherry and cider.

Other kinds of processing of fruits and vegetables include chemical processing, preparation of pickles and drying of products. Again, a large volume of processing happens in cottage, micro and small-scale industries. Except for a few dominant players, a large proportion of the market is being served by fragmented small players.

6.5 Direct Marketing through Wholesale Markets

The most important concern whenever a product is carried is to decide on the type of distribution networks that have to be used, and whether a direct channel (direct to customer) or an indirect channel (customer through a channel intermediary) is preferable. Choosing the right distribution channel for perishables in agribusiness is critical, as time plays a crucial role. "The type of intermediary channel is highly dependent on the target customers. Wholesaling comprise all the activities involved in selling goods to those who purchase for resale or for commercial purpose. The main function of the wholesale distributor is to balance supplies against retail requirements and to take the initiative of bringing produce from areas where it is plentiful and cheap to those where it is relatively scarce and expensive. Wholesalers usually have a good understanding of the market, with access to first-hand market information. Wholesalers generally go directly to the producers/farmers. In some cases, they get produce from central wholesale markets, agents, collectors and local markets. Many wholesalers have their own storage space facilities. Wholesale distributors may engage specialized transport agencies to conduct their collection and distribution activities" (www.ruralfinance.org).

To provide a quick understanding of wholesale market operations in perishables in India, an example of a wholesale market operating at Chennai is discussed in Exhibit 6.1.

Exhibit 6.1   Koyambedu Perishables Market in Chennai

The vegetables, fruits and flower market, called the Koyambedu market, is located at a prominent part of the Chennai city. Koyambedu market serves as the centralized distribution centre for perishable such as flowers, greens and vegetables, and for fruits of both inland and imported origin. Established by the State Government of Tamil Nadu under the jurisdiction of the Chennai Metropolitan Development Authority (CMDA), Koyambedu market is located within the Chennai city limits, with adequate road connections and proximity to the government-operated long distance (Chennai to other cities) bus terminal and private bus terminals. This makes transportation easy without causing disturbance to the city traffic, as the market is accessible through bypass roads.

The entire market premises are divided into three blocks:

1. The vegetable section, known as the "Periyar Vegetable Market"
2. The fruits section, known as the "Anna Fruits Market"
3. The flower section.

In each block, shops occupying space of various sizes were allotted and sold to wholesale merchants, dealers and commission agents by the CMDA of Chennai. A cold storage facility is provided by Tamil Nadu Cooperative Marketing Federation Limited (TANFED) (State Government of Tamil Nadu operated cold storage depot) within the market premises for merchants and dealers and also the general public. All common amenities are available for merchants/dealers and the public within the premises.

There is sufficient space provided in the market premises for movement of transportation vehicles carrying goods in and out of the market, with adequate parking facilities. Most of the perishable goods arrive at the market at night by truck, and are parked at one end, adjacent to the Periyar Vegetable Market.

As per the association's market guidelines, traders in the fruit market can trade only on fruits of any variety, and similarly, the vegetable traders and merchants can trade only in vegetables. Similar guidelines are applicable to the flower traders.

Supply/procurement: Vegetables are procured and also supplied to various places in Tamil Nadu, Kerala, Karnataka, Andhra Pradesh, Maharashtra, Gujarat and New Delhi. Fresh garlic is imported from China, while cinnamon is imported from Sri Lanka. Fresh vegetables are also exported to Singapore, Malaysia and Canada by some of exporters who operate at Koyambedu. Small and large retailers and members of the public who plan to purchase large volumes of vegetables and fruits for family functions or other occasions visit the market for procuring their needs.

Fruits are procured from and supplied to various places in Tamil Nadu, Karnataka, Andhra Pradesh, Maharashtra, Himachal Pradesh and New Delhi. Apples are imported from the United States, Austria and Australia. Mangoes are exported to Singapore, Malaysia, Sri Lanka, Mauritius and Dubai. All exports of fruits and vegetables are made through air freight.

Exporters of vegetable and fruits arrange to collect them at the market, get them packed at the market premises and then move the packed commodities to air cargo handlers for export. Commission agents and contract labourers provide the services of handling, packing, unloading and loading of bags, boxes and baskets of fruits and vegetables for transporting to various local markets as well as for export purposes.

Transportation and Handling: Fruits imported in 20- to 40-feet containers are brought to the fruit market after customs clearance and are stored in the cold storage at the first floor of the middle block in the fruit market. Fruits and vegetables procured from various states arrive at the market late at night by trucks, mini lorries and vans. Transport vehicles with the fresh vegetables that arrive by night halt in the parking area. The arrival of the transport vehicles is intimated to the contract merchant through phone by the driver of the vehicle, trader or any other seller's representative, who accompanied the goods in the vehicles. The vegetables and fruits are handled by the unorganized and contracted labourers available in and around the market premises. Handling, packing and marking of the consignments for exports are undertaken by organized labourers of commission agents at the market for exporters who do not belong to the market association. Exporters within the market association make their own arrangements. Retailers make their own arrangements collectively for procuring their needs and transporting them to their destination by pre-arranged mini-vans, vehicles, autos or any other convenient form of transport; usually, several retailers along one route share the transportation costs.

Storage and Warehousing: There are two cold storage facilities maintained at the premises of the Koyambedu Market. One is situated at the Anna Fruit Market and is meant exclusively for storage of fruits. The other, next to the Periyar Vegetable Market, is run by TANFED for vegetables, fruits or any other commodities, and even members of the public can make use of the facility as long as they pay the specified storage rates.

Distribution: The Koyambedu Market at Chennai functions as the largest distribution centre for vegetables, fruits and flowers in Tamil Nadu. Vegetables, fruits and flowers are supplied and brought to the market by wholesale dealers and commission agents from large farmers and growers at various places in several states. Wholesale traders and merchants, directly or indirectly through their representatives, as well as exporters of vegetables, fruits or flowers, small and large retailers and individual customers visit the market every day looking for better deals.

The distribution and sale transactions begin early in the morning, at about 3 a.m. The peak time, when the most wholesale transactions and distribution activities take place, is between 3 and 8 a.m. every day. Thereafter, retail sales continue till 6 p.m. All transactions are closed with cash on the spot. In a couple of cases, balance amounts and short payments are accepted by the Koyambedu merchants for retail customers who are regular and trusted by the traders.

Pricing: Prices per unit/unit weight for vegetables are fixed by group of merchants on competitive methods based on the freshness of goods and information collected by the representative of the traders and merchants of the Koyambedu Market from the transport drivers and representatives of the commission agents who arrived with the trucks during the preceding night. The higher the quantities that have been brought in, the lower the unit prices fixed. Conversely, the lesser the quantities that arrived, the higher are the unit prices fixed. For fresh vegetables, a higher unit price is fixed. Vegetables and fruits that arrived in trucks should be sold only to the traders and merchants of the Koyambedu Market, and not to any person who does not own a shop in the market. The entire wholesale transactions and distributions end by 8 a.m. every day. During the distribution, if demand for particular vegetables is higher and the available quantities are noticed to be inadequate, the sale prices per unit/unit weight of particular vegetables are raised, sometime even doubled, during the later hours of the distribution activities. Correspondingly, the prices per unit/unit weight of the vegetable sold to retailers also vary based on the quantities, that is, the larger the quantity, the lesser the unit price.

Competitive Markets in Chennai: Similar vegetable markets are located at T. Nagar, Mylapore and Kothawal Chawadi. The markets at T. Nagar and Mylapore receive their supplies of vegetables every day from places and rural areas immediately outside Chennai and in neighbouring districts. Fresh vegetable are available in the afternoon and evening vegetable markets of T. Nagar, Mylapore and Kothawal Chawadi. For the most part, the fresh vegetable from the T. Nagar market are supplied to Kothawal Chawadi and other places. The T. Nagar market is located close to the Mambalam Railway Station, and a large number of working-class employees, who commute by local trains, buy vegetables from the T. Nagar market. However, these markets are very small, and can be treated more as retail vegetable shops serving direct individual customers.

A substantial quantity of fruits and vegetables in India is wasted due to poor infrastructure, especially in terms of poor development of wholesale markets, improper handling and lack of sensitivity to pricing at sourcing points and at markets catering to final customers. Generally, there is high information inefficiency in terms of availability at source and at market, and in terms of demand patterns. There are too many intermediaries, leading to market inefficiencies and wastages. Government initiatives like the Koyambedu Market provide the necessary infrastructure and cooperative framework to improve supply chain operations.

Source: Chandrasekaran, N., *Supply Chain Management,* New Delhi: Oxford University Press, 2010. With permission.

The roles of wholesalers in the distribution of perishables are adequately explained by the above examples. There are a number of such initiatives operating in most of the cities, including metros, urban and semi-urban centres in India, supported by local governments. Even in rural areas, there is a practice of having a weekly day market wherein players sell both in wholesale and retail for the benefit of the players in the supply chain. Thus, wholesale distributors are significant links in perishables supply chains in India.

6.6 Intermediaries

There are several players and intermediaries, such as farmers, dealers, carriers and retailers, involved in satisfying the needs of consumers in the supply chains for perishables. Many agri producers do not have the economic resources and expertise required for direct distribution. Producers usually harvest a large quantity of a particular variety of goods. On the other hand, consumers generally need only a limited quantity of a wide variety of goods. Intermediaries act as important links between producers and consumers by bridging the gap between production and consumption and creating place, time and possession utilities.

Supply chain management in perishables has to be improved at all the stages of the supply chain by implementing worldwide best practices in storage, packaging, handling, transportation, value-added services, and so on, with a view to benefiting both farmers and intermediaries as well as ultimate consumers. Ease of access to markets is a significant factor. The transport facilities that are available to the market remain a key issue. The role of intermediaries has become important due to the ever-spreading markets and the increasing intricacies of distribution. A channel of distribution performs the work of moving goods and services from producers to consumers. It fills the place, time and possession gaps, and connects producers and consumers.

There are interventions at times to reduce aberrations in prices, which could be seasonal. In India, there are cooperatives and government bodies that facilitate these interventions. Exhibit 6.2 cites one such practice initiated in Kerala.

Exhibit 6.2   Market Intervention Was Strong at Kerala during the Festive Season

Market interventions and increased supplies dampened the price run of vegetables and bananas during the Onam festive season. Officials at the Vegetable and Fruit Promotion Council Keralam said that market intervention was strong ahead of Onam with Big Bazaars being opened in all the major cities early June onwards. The Kerala State Horticultural Products Development Corporation has two special outlets in each of the districts for the Onam season. These outlets have increased local procurement and have done away with middlemen to improve returns for the farmers and to ensure sales at reasonable rates to the buyers.

The major reason for the softening vegetable market during Onam season was intervention by government agencies. There has been a reported four-fold increase in procurement of vegetables and bananas from districts such as Wayanad, Palakkad, Idukki, Ernakulam and Thiruvananthapuram.

Horticorp procured 800 tonnes of vegetables directly from farmers and another 1,650 tonnes from the six wholesale markets across the state.

Source: The Hindu, Aug. 28, 2012.

6.7   Wastage

India is endowed with good agro-climatic conditions and a skilled labour force for the cultivation of a wide variety of fruits and vegetables. It has the distinction of being the second largest fruits and vegetables producer in the world. However, there is massive waste during the post-harvest storage, transportation and handling stages due to power interruptions, lack of temperature control mechanisms, inadequate infrastructure and lack of cold store facilities for preserving the produce. Wastage is one of the primary reasons for loss occurring in the supply chain of any perishable good. Waste occurring within the supply chain of a product results in increased costs and reduced efficiency for the business. For example, wastage on account of pest infestation is significant—if not more, definitely on par with loss during transportation. Similarly, wastages occur because of improper temperatures during storage and transportation, arising out of an unreliable power situation. It is not uncommon to see freezers switched off to save power; generally, the backups provided are insufficient to run freezers during periods of power blackout. Hence, businesses are continuously on the lookout for options and techniques to reduce wastage during the various activities in the supply chain, such as transportation and storage. Wastage at source is also an increasing concern.

6.7.1 Source

Minimizing wastage in an agri-based supply chain gains more importance, since the items produced are perishable and damaged easily. To get the appropriate and required quantity of output in terms of the number of units of weight or volume of vegetables, fruits or milk sold in the market, it becomes essential to reduce wastage and preserve as much as possible.

Many agricultural products such as vegetables, fruits and grains get damaged or destroyed at the source itself. This could happen due to causes arising while the items are yet to be readied for transportation, or during the time of loading on to trucks or vans for transportation to godowns, retailers or to the markets. Such loss of goods at the source causes loss of income to the cultivator as well as his buyers. When a huge loss occurs at the source, which in most cases is the first stage in the supply chain, it drastically reduces the quantity of goods that ultimately reaches the markets. This could lead to shortage of these goods and a subsequent hike in their prices at the end stages of the supply chain. This increase in price often affects the inflation levels in the country.

6.7.2 Storage

The piling up of inventory in godowns is yet another form of wastage, more so in the case of perishable goods. The resources lie unused and often rot away, or are destroyed by moisture, vermin, fungus or other factors. There are reports that huge piles of wheat and other grain stocks lie unused in government and private godowns. Exhibit 6.3 mentions risky storage practices by the government agencies. However, FCI maintains that wastes are less than 0.001% (Kumar, 2011). Lack of proper facilities for storage, such as moisture-free zones for grains, and cooling facilities for dairy and meat products, together with frequent power outages, result in the perishable goods getting damaged.

Exhibit 6.3   Wheat Under Risk of Damage

More than 6.6 million tonnes of wheat meant for the public distribution system are lying in the open, running the risk of damage from rain. The government has stocks to the tune of 82.3 million tonnes, against a storage capacity of 64 million tonnes. Of the 50 million tonnes of wheat, 27 million tonnes are kept in the open under the Covered and Plinth (CAP) method, of which 6.6 million tonnes are stored in an "unscientific way." As a contingent measure, the FCI has been directed to raise the stack size in case of both wheat and rice. "This way, the FCI would be able to accommodate 5–10 lakh tonnes more of stock."

Source: Parsai, G., *The Hindu*, June 22, 2012.

Inventory management is the most important issue to take care of to reduce storage wastes. Inventory piled up in storage houses without demand outflow increases holding costs, and also reduces the shelf life of the product. It also results in loss or wastage of valuable produce. New market options, like exporting to African, Latin American or other Asian countries, need to be explored so that the perishable products do not go waste.

Another option to prevent wastage at source could be to process the excess perishable goods into other forms that can be stored over a longer period. For example, excess mangoes from the summer season could be processed into mango syrups and squashes, and thus, preserved for longer times. Additionally, they would also fetch a premium in the markets. Similarly, cold storage items like fish can be turned into pickles in case of non-availability of refrigeration facilities. These items, too, could be considered for exports.

An example of boosting sales by providing vendors with improved storage facilities is that of Coco Cola and Pepsi. They provide their retailers with small refrigerators which can be used to keep the soft drinks cool and hence satisfy customers better. Cold storage facilities also help in preserving other perishable goods such as milk and meat and thereby play an important role in reducing spoilage and retaining the quality of the products. Processing of perishables into products with longer shelf life, such as pickles and syrups, also ensures reduced spoilage.

Despite India being a surplus producer of grains, many people in the country, especially those who live below the poverty line, still go without two square meals a day. The primary reason behind this condition and the rising prices is not lack of production, but inefficiencies in the supply chain. The surplus grains harvested get wasted, as they are left to rot in the rains and be eaten by rats and rodents. Removing such inefficiencies can help eliminate hunger from India.

6.7.3 Transportation

Transport of perishables starts early in the morning. A customer needs fresh vegetables to start the day. Moving of garden-fresh goods all the way from the point of origin to the end consumer is often a challenge. An uninterrupted transport chain is thus essential.

Large quantities of goods are damaged or destroyed in transit. The lack of good roads and rail freight systems is one main reason for loss or damage of materials during transportation. Often, vehicles used for transportation are not refrigerated properly, mostly with the aim of cost cutting for the company concerned. But much more revenue is lost due to the amount of items wasted by this negligence.

Perishable goods, having very limited shelf lives, should be transported as soon as possible to the markets from their sources. This calls in for an efficient farm-to-retailer supply chain with minimal transport delays. Choosing the appropriate mode of transport is also essential to ensure the freshness and quick delivery of such perishables. Short transport times reduce costs and spoilage risks.

Emerging technologies and innovative cold chain management have ensured that spoilage during transportation of perishable goods can be minimized. The government needs to step in to help farmers to ensure safe delivery of their perishable produce to the market, since transportation costs, especially with cold chains, can become almost out of reach.

By reducing wastage at source, storage and transportation of perishable goods can ensure better returns and customer satisfaction. The cost of cold chain operation is the single most important factor that increases wastage, and India has little scope to overcome this drawback, considering the high energy costs and the inability to pass them on across the supply chain. However, any conscious effort to handle perishables prudently by segmenting markets on quality requirements and the ability to pay would bring in efficiency and improve overall supply chain profits. For example, one of the authors observed crates of fruits and vegetables going rotten in a day in a retail chain and getting dumped. Had there been focus on better deployment of efforts such as liquidation, product offering at appropriate point as juices or movement of products to lower strata by selling at a lower cost would have brought better results than zero recovery of cost.

6.8 Quality Management

In today's business atmosphere, proper quality management is essential for perishables in agribusiness; everywhere, customers are not only demanding value for money but are also expecting the business community to demonstrate some concern for society at large. Consumers' increasing concerns about food quality and safety, government laws and trade regulations have increased the nature and degree of responsibilities for producers all over the world. In developing countries and emerging economies, however, companies face particular challenges in adapting to these changing requirements. Perishable goods demand the highest quality standards. A major share of the consumable goods market is comprised of food items: grains, bread, milk, meat, etc and so on. Hence, consumers demand the best quality and the freshest choice of products. There are also regulatory bodies that determine the market standards to be followed. "Agmark" is one of the premier quality standard indicators used in Indian markets. Further, in their roles as marketers and manufacturers, the companies also have the moral and ethical responsibility to sell only quality products. The retailers, too, have to ensure that goods past their expiry date are taken off the shelf. Since the rate of spoilage among perishable goods is the highest, proper ambient conditions for storage should be maintained, to ensure that products do not get damaged before their due dates.

McDonald's is a global fast food major that has perfected its supply chain system in India. Its supply chain involves procurement and transport of agri-based raw materials such as potatoes, tomatoes, lettuce leaves and milk and milk products. By procuring agri-products from farmers within India on a timely basis and using efficient transport systems that ensure temperature controls, McDonald's ensures uniformity, freshness and quality in all its products. Quality checks at the various procurement points, good storage and inventory facilities and strong reporting and documentation practices have helped McDonald's in building up a strong supply chain system. Clean and on-time delivery drives all members involved in the supply chain of this fast food major.

6.9   Price Management

Price management includes planning, implementation and governance of pricing of the various goods produced by a company. It may involve continuous monitoring of prices, development of pricing techniques and determination and redefining of pricing strategies and tactics. Pricing strategies of perishable goods are dynamic in nature, and have to be monitored constantly and altered whenever necessary. Perishable goods have certain distinct characteristics, such as short sale cycles, low value of unsold goods and uncertain demand, which makes their pricing even more sensitive. Perishable items can include agricultural and related produce, dairy products, fish, meat, flowers and others.

The use of dynamic pricing strategies helps to keep track of inventory and price changes. For example, the price of flowers is highest early in the day, then gradually decreases through the day; flowers are sold at a pittance towards the evening as the flowers start to dry up and new stock comes in. In the same way, the prices of fruits and vegetables depend on quality. Thus, quality and freshness play a major role in the pricing of perishable goods. There is scope for using more supply chain analytics for dynamic pricing, and this aspect needs more efforts and focus by all stakeholders.

6.10 Supply and Demand Side Risks

Meeting customer needs at the right time, at the right place and at the right quantity and quality has always been a hectic job for marketers. This is all the more so for perishable goods that cannot be displayed on shelves for long periods. Sales of this nature of products cannot be deferred for even a few days. The market for fresh products is relatively higher than for processed variations in India. During certain periods, the consumption of certain crops will go down, and during festivals, specific crops face great demand. The success of supply chain management will depend on the actions taken to match demand and supply. However, even with the best efforts of a marketing manager performing to the best of his or her capabilities, success in this function can be obtained only with the support of an efficiently operating supply chain mechanism that the company must have in place.

Supply chain management plays a crucial role in ensuring timely delivery of goods to the right markets. There is always the fear of excess supply not matching up to demand expectations, with inventory piling up and goods going waste. On the other hand lies the reverse situation, that is, demand exceeding supply, causing revenue loss to the company and affecting its top line and bottom line. Accurate forecasting of demand and supply becomes even more essential if products have short life cycles, with the consequent inability to meet future demand with current inventory.

6.10.1 Types

Many different combinations of situations can arise when there is a demand and supply risk. Examples are situations where

1. expected demand exceeds the normal supply;
2. demand is not met due to lower than expected supply;
3. expected supply exceeds the normal demand;
4. higher than usual supply results in goods going unsold; and
5. mismatch between the required supply of goods and the demand.

The various challenges arising out of the different situations given above are discussed below.

Inflation: Inflation is one of the serious consequences of demand and supply side risks. Often, the demand for perishable goods such as vegetables and fruits goes up, while the supply side is not able to keep pace with the rising demands. This could be due to various reasons, such as change in food preferences, and increased demand for a certain variety of produce. A delay in meeting the demand on time can lead to demand pull inflation, where lack of availability of adequate amount of goods pushes up the price of the existing stock/supply.

Exhibit 6.4   Rise in Food Prices in India

The inflationary pressure on food products remained high from January 2008 to July 2010. It again went up from April 2011 to March 2012. Rather, it continues, with minor reprieves in between. Food articles continue to face upside risk to prices with almost double digit inflation.

It is reported that the high food prices experienced over recent years are due to the rising demand for high-value agriculture products such as pulses, milk, livestock, fishery, vegetables and fruits. This situation arises from rising per capita income and the consequent diversification of Indian diets.

Supply has not been matching demand. Hence, prices continue to remain high, thereby providing a structural character to food inflation. This reasoning has gained wide acceptance in government and policy circles, including the Reserve Bank of India.

Source: Nair, S.R. and L.M. Eapen, *The Hindu Businessline,* Jul. 5, 2012.

Supply side issues can also crop up even when the demand side remains stable. Such situations can arise because of issues such as poor crop production due to bad monsoons, labour strikes by transport workers, government rules and norms and other factors. The existing supply is not able to meet the average demand in such cases, and prices shoot up. The frequent inflation in the prices of onions and other vegetables is the result of such situations.

6.10.2   Deflation

India also often experiences bumper crops, when the production levels exceed the predicted levels. This increases the supply of goods into the market. Due to excess supply, the value of the produce goes down. In the case of perishable goods, such a phenomenon affects the producer even harder because the life of his produce is limited to a couple of days. The farmer is, therefore, forced to sell his produce at a lower price and incur a loss.

Storage issues: When supply volumes exceed predictions, they can also lead to storage problems. The capacities of the godowns would have been designed for the normal production quantities only. Excess crops often get wasted due to inadequate facilities, or are sold for very cheap prices.

6.10.3 Resource Crisis

There are often situations, much out of the control of producers and distributors, that affect the supply chain. For example, there has been alarming news about how the number of fish in the sea is getting depleted due to the heavy extent of fishing done over the past decades and the other commercial activities carried out in the seas. The fishing industry could get badly affected by such a resource crunch. Further, environmental norms might be instituted to protect these marine creatures. Such scarcity of resources can result in limited supply of fish to the market, leading to higher prices for these items. Further, government rules that come into force during certain seasons, like those banning trawling (as in states like Kerala), also lead to reduced catch and higher prices during those seasons.

Mismatch of supply and demand: Certain vegetables and fruits tend to follow the product life cycle of a fad. They are in high demand for a few weeks and then fade off. Such changes in food fashion and food fads can upset the supply arrangements and the market as a whole. The short-term popularity of certain vegetables or meat can lead to increased demand, which cannot be met easily by quick supplies. Such food fads result in the decreased consumption of other items, while the demand for the fad items skyrockets and markets are in no situation to meet this growing demand.

Reading the markets and predicting demand and supply trends accurately is a tough task, especially in the case of perishable goods that are generally fast moving. A single calculation error or slight disturbance in the system due to internal or external factors can cause an imbalance that can even hurt the whole economy.

A supply chain plays an important role in the success of any company, all the more so if it is an agri-based industry. Efficiency at all the points in the supply chain is essential to meet the ultimate goal of satisfying customers. Following best practices in supply chains can help not just the corporate sector, but also government agencies and non-governmental organizations (NGOs), to meet the growing needs and expectations of millions of people around the world.

6.11   Conclusion

Supply chain challenges in perishables are aplenty in India. There are tough demand side issues and equally daunting supply chain issues. The government and its agencies are fighting with regulatory mechanisms and support systems to ensure that there is food security and that the human development indices are not adversely affected because of supply chain challenges. A lot of these issues are structural in nature, and they arise more out of economic issues rather than the lack of understanding or will. Taking this itself as a unique aspect, more supply chain professionals must use analytics and fill in the gaps by using appropriate levers such as product conversion and pricing and information technology, rather than by focusing on storage and inventory issues.

References

Chandrasekaran, N., *Supply Chain Management*, New Delhi: Oxford University Press, 2010.

The Hindu, Vegetable prices under control. Aug 28, 2012. http://www.thehindu.com/news/cities/Kochi/article3830941.ece (accessed on August 31, 2012)

Kulkarni, P. The marine seafood export supply chain in India: Current state and influence of import requirements, International Institute for Sustainable Development, Canada, 2005. http://www.iisd.org/tkn/pdf/tkn_marine_export_india.pdf

Kumar, C. *Tribune* rejoinder. Food Corporation of India, Aug. 4, 2011. http://fciweb.nic.in/app/webroot/upload/pressrelease/Tribune%20rejoinder.pdf

NABARD, *Organized Agri-Food Retailing in India*. Mumbai: NABARD, 2011.

Nair, S.R. and L.M. Eapen, Food inflation not due to income. *The Hindu Businessline*, Jul. 5, 2012. http://www.thehindubusinessline.com/opinion/article3606405.ece

National Dairy Development Board. Milk production in India. 2013. http://www.nddb.org/English/Statistics/Pages/Milk-Production.aspx

National Horticultural Research and Development Foundation. http://www.nhrdf.com/

Parsai, G. 6.6 m. tonnes of wheat under open sky faces rain fury, admits Centre, *The Hindu*, June 22, 2012. http://www.thehindu.com/news/national/article3555824.ece

Poku, K. Palm oil processing. Chap. 3 in: *Small-Scale Palm Oil Processing in Africa*. Rome: FAO, 2002. http://www.fao.org/docrep/005/Y4355E/y4355e04.htm

Seafood Exporters Association of India (SEAI), Statistics, 2012. http://seai.in/filecategory/statistics

Small Farmers' Agribusiness Consortium (SFAC) Baseline data for onion and potato. Apr. 2012, http://www.sfacindia.com/Docs/Onion%20&%20Potato%20Baseline%20Report.pdf

Chapter 7

Quality

OBJECTIVE

The objective of this chapter is to understand the importance of quality systems, including that of grading. Practices and procedures of quality management for supply chain efficiency are discussed here. The discussions include international standards with respect to foreign trade in the agribusiness sector and the role of government in adopting quality systems.

7.1 Introduction

The role of quality systems in the agriproduct sector has been gaining ever-increasing importance in recent years. During the early days, buyers and sellers met and bargained over their produce; there was not much requirement for standardization because they knew each other very well, as they all mostly lived within a small region. However, today products are handled in huge volumes and move beyond borders. Hence, a well-regulated system of grading and quality is of the utmost importance to provide a common language for defining product values. Today, customers in various parts of the world demand certifiable evidence of *traceability* and more and more refined levels of quality in food and related products. With rising income levels and shifts in lifestyles, an increasing proportion of the populace depends on ready-to-eat meals. This has a significant effect on the role that agriculture would play in the future. Not only will it have to meet increasingly stringent food safety requirements and the need for adequate returns on investment, but it should also ensure sustainable environmental norms in agriculture.

This change in focus from high production yield to better quality and sustainability has resulted in demand for the creation and development of a traceable supply chain. Companies are persistently searching for ways and means that allow the production of goods that meet the variety of characteristics demanded by the marketplace. The quality of agricultural products is evaluated by a customer based on the freshness of the product and the time since its removal from the farm. Effective supply chain practices will help ensure the freshness of agriproducts by helping to reduce the time taken in transportation and by using favourable storage conditions and better processes and systems. The quality of farm products brought to the market by cultivators varies over space and time. To ensure that farm produce remains fresh, safe, and nutritious and is capable of fetching good prices in the market, it is necessary to improve infrastructure in the agriculture supply chain, reduce the length of the supply chain, and improve essential practices such as grading and quality management. This chapter focuses on the quality systems and procedures for supply chain effectiveness.

7.2 Grading

Before discussing the importance of grading and quality management systems, it would be appropriate to define and understand the role of these concepts in the agriproduct sector.

Grade refers to lots having similar quality features, whereas grading refers to the classification of products into different homogeneous lots, that is, each lot has similar characteristics. Grading describes the quality of the package and simplifies the marketing of agricultural products. Unlike industrialized products, which are manufactured to precise engineered specifications and are basically consistent, agronomic produces are more diverse and varied in characteristics such as colour, fragrance, location, heaviness (bulk), thickness, shape, ripeness, length of staple, taste, stains, stickiness, and so on. For instance, the characteristics may be based on location of origin (Nagpur orange, Salem mango, Byadagi chilli), shape (small, round, big), length of staple (cotton), and so on. Grading helps the farmer to get suitable prices for his products, because based on the lot characteristics different prices may be fixed for different lots.

Furthermore, quality refers to how well the characteristics of a produce meet the requirements of the end user. The quality of agriproducts requires preservation or maintenance of the basic characteristics of the products. To preserve quality, it is imperative to devise a mechanism that controls bacteriological spoilage and chemical degradation.

Although quality connotes the level of requirement, standards are set up and established by statutory authorities as well as reputed authorities like quality standards organization as rules for the measure of quantity, weight, volume, extent, value, age, and quality. In the agriproduct sector, an established set of guidelines or standards facilitates determination of grades and provides a common language for describing product value. In the same manner, quality standards should have a common language.

Hence, quality standards are established by government agencies in order to provide a common language for farmers, traders, agents, and end users for buying and selling.

A quality management system is a method by which the agriproduct sector aims to reduce and eventually eliminate non-conformance to specifications, standards, and customer expectations in the most cost-effective and efficient manner. A quality management system guarantees that agricultural products move through the market faster and without any hindrances.

7.2.1 Process

In a typical agri-supply chain, grading occurs after drying or cooling. Figure 7.1 explains the basics of the grading process.

It can be noted that grading may happen before drying/cooling in the case of many fruits and vegetables and other perishables. Thus, there could be scope for multi-stage grading. Similarly, grading is done based on taste as well. Tea is graded for size and shape and also based on the taste quality. Similarly, ground coffee beans, which are part of the instant coffee market, are tasted after being processed.

By way of example, let us consider the processing of tea leaves. Once the tea leaves are picked, they are transported to factories for processing. They undergo a process called withering, where moisture content is extracted from them. Then the tea leaves are placed in a rolling machine. During this process, the leaves are broken open, which starts the process of oxidization. Next, the tea is dried using hot air at a temperature of approximately 85°C–88°C in order to stop the oxidation process. During this stage, the colour of the leaves will change. Finally, dried tea is sorted into grades before packing. This process includes grading and then packing them in various sizes and weights for inspection. The leaves must be conditioned. The fineness of grading is determined by the type of leaves and the local custom. Colour, position of the leaves on the plant, maturity, size, and other qualities are required for grading. The orthodox production method provides teas of all leaf grades: leaf, broken, fanning, and dust.

7.2.2 Importance

Assuring quality from farm to fork is one of the key elements of efficient agricultural supply chain management systems. Grading and quality management play an important role in maintaining and improving quality in agri-supply chains. The quality of agriproduce that arrives at the market differs considerably from lot

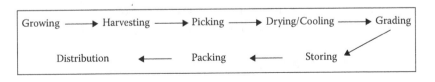

Figure 7.1 Process of grading in an agri-supply chain.

to lot. This is because of the wide range of varieties, farming practices, seasonality issues, and different agro-climatic conditions that prevail at the sources of the various lots. Thus, it is essential to grade the products brought to the market on scientific grounds, in order to get competitive prices.

For example, the following dimensions are taken into account while grading pomegranates: size, appearance, shape, seed content, total sugars, acids, colour, and weight of the pulp. The producers or sellers must be able to satisfy consumers on the quality they are offering. Tables 7.1 and 7.2 explain grade designation and sizing for pomegranates as per AGMARK standards.

Size is determined by the weight or maximum diameter of the equatorial section of the fruit, in accordance with the data shown in Table 7.2.

From Tables 7.1 and 7.2, it is clear that there are variances in terms of size and weight while meeting grading requirements.

This is true for all of the agricultural produce, as size, colour, and content elements such as fibre and juice can influence the quality of the produce and the premium it can demand for direct consumption and for processing.

7.2.3 Benefits

Grading and quality management are a must for the agro-based industry because of natural compulsions and the likely diversity of multiple players and geographical and climatic conditions. Adherence to quality standards is expected to improve value for all players across the supply chain, as the produce moves from farmer to consumer either fresh or processed. These are discussed in the following sections.

7.2.3.1 Uniformity

Customers can buy any kind of domestic and international products, as the quality is certified and verified by a reputed authority. Because of grading and quality management, there is uniformity and consistency.

Therefore, a customer can purchase an item for consumption without any risk. For example, when fruits such as apples, oranges, and grapes are sold, they are graded on the quality of uniformity, which enables customers to pick a sample and finalize a sale.

7.2.3.2 Facilitates Buying and Selling

It provides valuable market information that makes it easy for sellers to explore new markets and helps buyers in making useful quality or price assessments. It also allows different market mechanisms such as commodity exchanges and futures trading.

7.2.3.3 Competitive Advantage

Improved grading and quality management present an opportunity to attain competitive advantage in certain products in the worldwide markets. For example,

Table 7.1   Grade Designation of Pomegranate

Grade Designation	Grade Requirements	Grade Tolerances
Extra class	Pomegranates must be of superior quality. They must have the shape, development, and colouring that are typical of the variety and/or commercial type. They must be free of defects, with the exception of very slight superficial defects, provided these do not affect the general appearance of the produce, the quality, the keeping quality, and presentation in the package.	5% by number or weight of pomegranate not satisfying the requirements for the grade, but meeting those of Class I grade, or exceptionally, coming within the tolerances of that grade.
Class I	Pomegranates must be of good quality. They must be characteristic of the variety and/or commercial type. The following slight defects are acceptable, providing these do not affect the general appearance of the produce, the quality, the keeping quality, and presentation in the package: • Slight defects in shape • Slight defects in colouring • Slight skin defects (i.e., scratches, scars, scraps, and blemishes), provided these do not exceed 5% of the total surface area.	10% by number or weight of pomegranate not satisfying the requirements of the grade, but meeting those of Class II, or exceptionally, coming within the tolerances of that grade.
Class II	This grade includes pomegranates that do not qualify for inclusion in the higher grades, but satisfy the minimum requirements. The following defects may be there, provided the pomegranates retain their essential characteristics with regard to the quality, the keeping quality, and presentation: • Defects in shape • Defects in colouring • Skin defects (scratches, scars, scrapes, and blemishes), provided these do not exceed 10% of the total surface area.	10% by number or weight of pomegranates not satisfying the requirements of the grade, but meeting the minimum requirements.

Source: Ministry of Agriculture, Fruits and vegetables grading and marking rules, *Agmarknet*, 2004.

Table 7.2   Provisions Concerning Sizing of Pomegranate

Size Code	Weight in Grams (Minimum)	Diameter in mm (Minimum)
A	400	90
B	350	80
C	300	70
D	250	60
E	200	50

Source: Ministry of Agriculture, Fruits and vegetables grading and marking rules, *Agmarknet*, 2004.

Note: Size tolerance: (i) For all grades, 10% by number or weight of pomegranate corresponding to the size immediately above and/or below that indicated on the package. (ii) The maximum size range of 8 mm between fruit in each package is permitted.

Washington apples are available throughout India; new varieties are being introduced into the market due to the ever-increasing demand for this sweet and juicy fruit. Premium Washington apple varieties such as Honeycrisp and Cripps Pink have been sold by leading retailers in urban centres such as Chennai, Mumbai, Delhi, and Bangalore. This is because consumers demand the best quality apples and are willing to pay for quality and unique taste. This provides a competitive advantage for retailers in the supply chain. Similarly, in the processed segment, high-quality tea, coffee, confectionary products, and sugar have a clear competitive edge. Hence, firms pursue quality standards over and above the statutory/threshold requirement for any product. India is a pluralistic socio-economic society where multiple levels of rich and poor live. Firms must have the ability to ensure a threshold quality for all products and all consumers; the ability to differentiate premium/high-quality/other grades is of paramount importance in winning a competitive advantage in the market. We will have a firm selling a popular brand of edible oil in a given market and selling a differentiated premium brand of high-quality edible oil with premium margins for a niche segment.

7.2.3.4   Price Fixing

The prices for goods are determined by their types or categorizations. Prices may be fixed in the market according to the different grades of a particular product. There is an acknowledgement of quality perception; consumers are fairly aware of quality standards, and therefore relate to price differentiations.

7.2.3.5 Market Access

Grading and quality management not only create value for local trades but also deliver value in terms of improvement of quality for international trade. Thereby, these practices facilitate easy access to better quality products and prevent poor quality products from being traded. This is because the standards set are high, ensuring compliance with international norms.

7.2.3.6 Transparency

A well-established quality management system increases the efficiency and transparency of markets. Moreover, it also helps to differentiate among different geographical origins. Mandatory labelling and other procedures governing the sale of certain edible products and visually clear practices in the case of fresh products bring out transparency. Market practices are increasingly adopting methods to showcase objectivity and ensure transparency. For example, there was an issue about using calcium carbide to artificially ripen mango in Tamil Nadu. Local administrations tried to intervene and validate proactive measures to ensure that the fruits are naturally ripened. One wonders how transparency and objectivity can be achieved here, because the retailer may find it difficult to isolate artificially ripened lots from naturally ripened ones. The retailer is also not in a position to grade the fruit. Traders' associations got together to ensure that the fears were addressed and communicated the relevance of natural ripening to all traders. Thus, lack of transparency could become an issue in supply chains. There was another case with respect to illegal meat being sold in some retail markets. The low price of the unhygienic meat is a major reason for its popularity among many small meat shops and hotels. Mutton, beef, and chicken sold at these places cost 40%–60% less than stamped meat. Local municipal corporations exercise vigilance, and wherever there is lack of transparency with respect to the source of the meat and stamping, the product is summarily withdrawn from the supply chain. Thus, objectivity and transparency are critical for maintaining quality standards in the agro-based industry.

7.2.3.7 Sustainable Agriculture

Grading and quality management relating to sustainability are recently emerging market requirements. The objective of these standards is to safeguard the environment and reduce the use of agrochemicals. The importance of these standards will increase, and in due course play a very important role in future product entry to markets.

7.2.3.8 Incentives for Fair Trade

It assists farmers to get premium prices for quality products and safeguards customers from unfair trade practices. Grading and quality management help not only farmers but also all stakeholders in agriculture and related activities, namely, wholesalers, cooperatives, agents, dealers, and the final consumers.

7.3 Stages of Supply Chain

A quality product available to customers at the tail end of a supply chain is a function of quality perspectives and support mechanisms across the various players in the network. The players, from farm to fork, could be farmers; intermediaries; market operators; primary, secondary, and tertiary processors; storage, transport, and distribution players; and so on. Here, we provide some insights on the scope for quality management.

The factors that affect quality in various stages are described in the following sections.

7.3.1 Farm Level

Farming and harvesting conditions impact fresh and processed food stuffs. Important factors affecting quality include selection of the best plant varieties and the best cultivation methods. Weather and seasonal changes also affect quality. At the farm level, quality is impacted significantly by the absence of good agricultural practices (GAPs). Like good manufacturing practices (GMPs) for factories, GAP is for farming. This is discussed later in this chapter in Exhibit 7.1. Farming practices are rudimentary in certain areas, which affects quality. For example, India is a leading producer of mango. Harvesting (plucking) of mango (in large quantities) from trees is done without proper appliances and tools, leading to damage. It should be mentioned that a similar situation prevails for plantation crops as well. The other aspects of quality at the farm level include right application of manure and plant protection. All these would be a part of GAPs when adopted appropriately by farmers.

7.3.2 Transportation

In the course of transport and storage, damage may occur to plants; hence, the following should be taken care of:

■ Loading density
■ Loading and unloading facilities
■ Transport time

In India, goods are mostly moved by normal trucks, and wherever temperature control is required over short distances, freezers or ice stuffed insulated boxes are used. However, refrigerated trucks are needed in order to protect the quality of the produce during transportation. This is specifically applicable for milk and related products, fruits and vegetables, drugs, chemicals, genetic products, frozen rudiments, and other items that are very sensitive to temperature or to transportation time limitations, and are likely to deteriorate. Temperatures differ from product to product: in the case of milk, 32°F–40°F; eggs, at or below 45°F; poultry, 26°F–32°F; and meat, under 26°F.

Goods having limited shelf lives should be transported as soon as possible to the market from the source. For instance, milk transport from the farmhouses to the collection centres, processing centres, or factories should always be as quick as possible to prevent spoilage of milk. Availability of good quality raw milk at neighbouring localities is critical, as milk is highly perishable. In the case of frozen fish, a temperature of –0.4°F to –16.6°F (–18°C to –27°C) is essential during transport. This calls for an efficient farm-to-retailer supply chain with reduced transport delays.

Selection of a suitable mode of refrigerated transport is essential to ensure the freshness and natural qualities from the point of produce to warehouses, distributors, retailers, and finally, customers. Short transport times reduce costs and spoilage risks. This means moving perishables with the highest efficiency, thus retaining freshness and natural qualities from the point of produce to warehouses, distributors, and retail customers. Pre-cooling freezer trucks help in transporting time-sensitive goods. Throughout the distribution process, even while loading and unloading of the truck, the temperature of the products must be maintained. Failure to do so may affect the quality of the produce.

Products that require specific care from ecological influences may be transported by surface means, with protective measures. Proper employee training is required to take care of perishable products; they should not be left on loading yards for lengthy periods, and the least possible time should be spent on loading and unloading. This will enhance the quality of the produce and ensure that spoilage during transportation of perishable goods can be minimized.

7.3.3 Storage and Distribution

Storage settings may impact the quality of fresh produces and processed foods. When considering current preservation techniques, there are limits that pertain to different factors such as temperature, sourness, water content, use of flavours and preservatives, and alteration of combinations. For instance, elevated temperatures are applied to reduce the number of micro-organisms, to inactivate enzyme activity and to increase chemical reactions.

As a further example, extremely perishable meat containing water and proteins requires specific temperatures, moisture, and exposure to air conditions. If the meat remains untreated, its decomposition occurs in a matter of hours or days, and results in contamination and infectious conditions. Raw meat should be preserved at 40°F or below in order to decrease the growth rate of bacteria that may exist on its sides.

India has distinctive topographical areas that produce a variety of fruits and vegetables. Furthermore, the extensive coastal areas enable harvesting of marine foodstuff in large quantities. Every fruit and vegetable has a certain lifespan. To protect the goods from damage, they should be preserved at the appropriate temperatures. Table 7.3 summarizes the requirements for storage of vegetables. The optimum storage conditions are shown for vegetables with water activity of 85%–90%, except for beet root, cabbage, radish, turnips, and peas, which have water activity of 90%–95%.

The location is a very critical factor for a cold storage unit, as it ensures the quality of the produce being handled. It should be near produce growing areas as well as marketplaces. Uninterrupted power supply is yet another pre-requisite.

Table 7.3 **Vegetable Storage Characteristics**

Name of Vegetable	Storage Temperature (°F)	Cold Storage (Life in Weeks)
Beans	32–35	2–3
Beet root	32–36	6–8
Brinjal	47–50	3–4
Carrots	32–35	13–17
Cauliflower	34–35	4–6
Cabbage	32–35	9–13
Onions	32–35	17–26
Potatoes	37–38	26–35
Radish	32–35	6–8
Turnips	32–35	13–17
Tomatoes, ripe	40–45	1–1.5
Tomatoes, mature (greenish yellow)	55–60	
Peas	32–35	3–4

Source: Ministry of Micro, Small & Medium Enterprises, Cold storage, Chap. 7 in *Food Project Profiles*, 2004.

7.4 Need for a Structured Approach

Effective quality systems, which assure the health and safety of agriproducts, have become more and more important. Implementing proper quality systems has become a necessity in the agrifood sector. This is primarily due to changing consumer requirements, increased competition, environmental issues, and governmental interests. The customer perception of a commodity can be based on two approaches: intrinsic and extrinsic attributes. Intrinsic attributes for customers include considerations of whether the product is hygienic, non-toxic, and nutritive. These factors are directly related to the physical properties of the product. Extrinsic attributes refer to additional aspects such as environmental impact and sustainability. They do not directly influence physical properties, but may affect acceptance of products by customers. Examples are the use of chemicals and pesticides.

Nowadays, different farming methods are used in different parts of the world. Quality methods are changed rapidly and are becoming innovative through the deployment of technology. Although quality systems have been adopted in many countries, there is still a need for a structured approach to make our agricultural supply chain efficient so that neither are the customers burdened with high prices nor are the producers incurring losses. There are still lots of variations in terms of grading and quality practices. The agriproduct sector uses various quality assurance systems such as Hazard Analysis Critical Control Point (HACCP), International Organization for Standardization (ISO), British Retail Consortium (BRC), and other national and international standards.

In spite of all these approaches, human handling plays a crucial role in quality management, and is rather unpredictable and changeable. As a consequence, results in the agribusiness and food industry, as the combined action of individuals striving for quality, are much more uncertain than is often assumed. To face this challenge, continuous improvement in agri-quality management methods is required, where knowledge of modern technologies and management methods play a crucial role.

7.5 Overview of Policies

Agricultural policies are sets of regulations relating to national and international trade in agricultural products. Governments and other agencies devise these agricultural policies in order to accomplish product quality, guaranteed supply, and price stability. While devising these policies, countries must make sure that there is adequate production capability in order to meet the home country's needs in the event of external supply interruption. Dependency on other countries puts a nation at risk in case of natural disasters, war, or other disruptions. Sustaining sufficient domestic product capability allows for self-reliance, which reduces the risk due to supply disruptions. Agriculture is considered to be the strength of our economy.

Thus, there is an increasing need to confirm good quality in order to develop our agrarian economy. To ensure high quality of produce, several quality standards and systems are in place. This section provides an overview of different agriculture policies that may have an impact on the quality of goods in the agriproduct sector. Some of the Indian quality management systems and standards relating to this sector are discussed in Exhibit 7.1.

Exhibit 7.1   Indian Quality Standards for Agriproducts

GAP—INDIA

The GAPs standard covers the basic requirements for all farming practices in a sustainable manner for maintaining quality and food safety of agricultural produce. GAP, in addition to improving the yield and quality of the products, has environmental and social dimensions. Its social dimension would be to protect agricultural workers against risk to health arising from improper use of chemicals and pesticides. The main components of this standard are base modules and crop-based modules. Crop-based modules include fresh fruits and vegetables, combinable crops, tea, and green coffee. There are various standards available for control measures in value addition through processing of food meant for human consumption. Although grade standards on size, shape, colour, and local preferences are available for most of the fruits and vegetables marketed and consumed in India, their quality in terms of maturity standards, residues of pesticides and other contaminants, microbial loads, and so on, has not been adequately addressed. The Indian Good Agricultural Practices (INDGAP) guidelines take into account not only the quality and quantity of the produce obtained from a unit area, but also the care taken in integrating pre-harvest practices such as soil and water management, nutrient management, pest management, harvesting, post-harvest handling, and other logistics.

DIRECTORATE OF MARKETING AND INSPECTION—MINISTRY OF AGRICULTURE

The Directorate of Marketing and Inspection (DMI) implements the agricultural marketing policies and programmes of the Government of India. DMI performs the prime function of standardization, grading, and quality control of agricultural and allied produce. DMI has framed AGMARK standards for agricultural and related produces. Standards outlined under the provisions of the Agricultural Produce (Grading and Marking) Act, 1937 (http://agmarknet.nic.in/apgm1937.htm) are popularly known as AGMARK standards.

AGMARK

AGMARK is a quality certification mark provided by DMI, an agency of the Government of India for agricultural products in India. This certification guarantees that the product conforms to a set of prescribed quality standards. The legal enforcement of AGMARK in India is controlled by the Agricultural Produce Act of 1937. The AGMARK standard covers quality grading and certification of products for the domestic and export markets. Nearly 205 commodities have been brought under the umbrella of the AGMARK standard. AGMARK is used to certify different commodities covering a wide spectrum of fruits and vegetables, pulses, cereals, vegetable oils, vermicelli, essential oils, roasted Bengal gram, *vanaspati*, cotton, sugar, honey, jute, wool, and so on. Appendix 7.1 outlines the grading of cashew kernels. DMI has a set-up for quality certification of agricultural produce through a network of 22 regional AGMARK laboratories at different places in the country with the Central AGMARK Laboratory at Nagpur as the apex laboratory. The regional AGMARK laboratories are engaged in analysis of agricultural and food commodities for evaluation of quality. Testing is conducted on all types of agricultural and allied commodities and meat products across regional laboratories for chemical and microbiological analysis, pesticide residue, and aflatoxin.

FRUIT PRODUCTS ORDER

The Fruit Products Order (FPO) is a certification mark employed on all fruits and vegetables. FPO certification in India has been "effective since 1955, but became mandatory only in 2006, after passing of the Food Safety and Standards Act. It ensures that any food under consideration is suitable for purchase and is fit for consumption, and guarantees that it has been manufactured in a hygienic environment. FPO certification products include fruit crushes, beverages, syrups, sauces, ketchups, pickles, vinegars, canned juices, pulps, frozen vegetables and fruits, cereal flakes containing fruits and other similar products containing processed fruits and vegetable" (www.indiatrademarkregistration.com).

7.6 International Policies

Agricultural policies differ from country to country. For instance, in the United States, nutrition labelling is mandatory. However, in France, the source confirmation, that is, display of country of origin of fruits and vegetables is mandatory in all types of markets. Peru requires packing dates for all meat and poultry products. In the case of exports, fruits and vegetables shall comply with the residue levels of heavy metals, pesticides, and other food safety parameters as laid down by the Codex Alimentarius Commission for exports. The Japanese JAS Law and Food Sanitation Law require that the label on retail packages for imported food products should include the name of the product, country of origin, net weight, and ingredient information in Japanese.

Exhibit 7.2 outlines the different international quality policies and standards relating to the agriproduct sector.

Exhibit 7.2 International Quality Policies and Standards

AGRICULTURE POLICIES IN THE UNITED STATES

The United States has various "agricultural policies, with goals that range from the traditional objectives of stabilizing agricultural production and supporting farm income to those that have more recently increased in importance, such as assuring adequate nutrition, securing food safety, encouraging environmental protection, and facilitating rural development" (www.oecd.ilibrary.org). The US Department of Agriculture quality grade marks are employed on chicken, eggs, beef, and butter. There are five specialized acts relating to agriculture: the "Food Security Act of 1985; the Food, Agriculture, Conservation, and Trade Act of 1990; the Federal Agriculture Improvement and Reform Act of 1996; the Farm Security and Rural Investment Act of 2002 (2002 Farm Act); and the Food, Conservation and Energy Act of 2008" (www.oecd.ilibrary.org).

INTERNATIONAL STANDARDS IN AGRICULTURE

Agriculture and agricultural produce have international standards set by many organizations such as ISO, the United Nations Economic Commission for Europe (UNECE), the World Trade Organization (WTO), the National Agricultural Products Quality Management Service (NAQS), and others. UNECE, ISO, and Codex Alimentarius standards are globally recognized standards, and hence, they are commonly used as the basis for many grades and standards. A few of these standards are elucidated below.

ISO AGRICULTURAL STANDARDS

ISO international standards certify that products and services are safe, reliable, and of good quality. For agribusinesses, they are considered as strategic tools, as they decrease costs by minimizing waste and errors and increasing productivity. The ISO standards for agriculture are given in standards catalogue 65. They include standards relating to the fertilizers and pesticides that can be used, to animal feeding stuffs, to agricultural machinery, and to farm buildings. They also specify the standards for fishing, fish breeding, and beekeeping. They have a specific section for tobacco products and related equipment. ISO/TC 34 deals with food products. These standards are most widely accepted, and most of the time, are required for the products to be successful and accepted.

UNECE AGRICULTURE QUALITY STANDARDS

UNECE develops "global agricultural quality standards to facilitate international trade. The standards encourage high-quality production, improve profitability and protect consumer interests. UNECE standards are used internationally by governments, producers, traders, importers, exporters and international organizations. They cover a wide range of agricultural products: fresh fruit and vegetables (FFV), dry and dried produce (DDP), seed potatoes, meat, cut flowers, eggs and egg products" (www.unece.org). It collaborates with the WTO, the OECD Scheme for the Application of International Standards for Fruit and Vegetables, and also the FAO/WHO Codex Alimentarius Commission to prevent any duplication of standards. Together, these bodies have developed around 100 standards for agricultural products and international trade. They also conduct seminars and awareness programmes to make the standards well known to the agriculturists and also to the public who are the users of the products.

CODEX ALIMENTARIUS INTERNATIONAL FOOD STANDARDS

The Codex Alimentarius international food standard has established procedures and principles that enable the safety, quality, and fairness of the international food trade. The Food and Agricultural Organization of the United Nations and the World Health Organization have together come up with this safety standard for food and agricultural products. They also unify the standards set by all governmental and non-governmental organizations to ensure the well-being of food consumers and also to see to it that fair trade practices are followed for food and agricultural products. The Codex Alimentarius, or the food code, has become the global reference point for consumers, food producers and processors, national food control agencies, and the international food trade.

WTO-SANITARY AND PHYTOSANITARY (SPS) STANDARDS

The WTO has specified certain norms for food safety standards and standards for animal and plant health. The SPS measures are applicable to regionally produced or traded products, and lay down safety measures such as checking whether the agricultural produce has come from a disease-free area, carrying out general inspections of products, treatment or processing of products, and making sure that the products are safe for consumers. They also make sure that pests or diseases do not spread among plants and animals.

NATIONAL AGRICULTURAL PRODUCTS QUALITY MANAGEMENT SERVICE

NAQS, a specialized organization in agri-food quality management, promotes quality enhancement and fair trade through national inspections on agricultural products, and boosts the convenience of producers and buyers by improving inspection methods. NAQS is upgrading product quality through various certification systems, including good agricultural practices (GAP), and enhancing certification capability gradually by nurturing credible private certification agencies. NAQS also supports the exports of GAP products, the quality of which is thoroughly managed from production to consumption, and lays a foundation for producing high-quality and safe agricultural products through quality manager systems.

INTERNATIONAL FEDERATION OF ORGANIC AGRICULTURE MOVEMENTS (IFOAM)

IFOAM's organic guarantee system facilitates the development of organic standards and third-party certification worldwide, and also provides an international guarantee of these standards and organic certification. Organic agriculture is based on the factors discussed here.

THE PRINCIPLE OF HEALTH

It deals with the health of individuals and communities. Organic agriculture is envisioned to promote high-quality, nutritious food that contributes to preventive health care and well-being. For instance, it avoids the use of fertilizers, pesticides, animal drugs, and food additives that may have adverse health effects.

THE PRINCIPLE OF ECOLOGY

It is based on ecological processes and recycling. Organic agriculture should attain ecological balance through the design of farming systems, establishment of habitats, and maintenance of genetic and agricultural diversity.

THE PRINCIPLE OF FAIRNESS

This principle emphasizes that those involved in organic agriculture should conduct human relationships in a manner that ensures fairness at all levels and to all parties—farmers, workers, processors, distributors, traders, and consumers.

THE PRINCIPLE OF CARE

Organic agriculture is a production system that sustains the health of soils, ecosystems, and people. It relies on ecological processes, biodiversity, and cycles adapted to local conditions, rather than the use of inputs with adverse effects. Organic agriculture combines tradition, innovation, and science to benefit the shared environment, and promotes fair relationships and a good quality of life for all involved (www.naqs.go.kr).

7.7 Role of Government Agencies

Government and other allied agencies play a pivotal role in promoting and strengthening grading and quality infrastructure for agriculture. For instance, governmental support agencies such as the Food Safety and Standards Authority of India (FSSAI), the National Bank for Agriculture and Rural Development (NABARD), the Agricultural and Processed Food Products Export Development Authority (APEDA), and the commodities boards promote grading, standardization, and quality certification systems to give a major thrust to promotion of agricultural markets in order to stabilize market systems and increase farmers' income. Exhibit 7.3 discusses some of the agencies promoting quality systems in the agriproduct sector. The Department of Agriculture provides technical assistance and advice to the state in framing suitable quality systems. The Government of India launched a scheme to develop the agricultural marketing infrastructure by encouraging private and cooperative sector investments. In order to increase investment in the development of marketing infrastructure as envisaged above, the ministry has formulated a scheme for development/strengthening of agricultural marketing infrastructure and grading and standardization. Under this scheme, the government provides subsidies for strengthening infrastructure facilities such as grading. Table 7.4 shows the subsidies sanctioned by various government agencies. The following observations on Table 7.4 could be useful.

1. Maharashtra is the state that has the maximum of projects (475) involving a financial outlay of ₹741.92 crores in the year 2012, with a subsidy component of ₹80.37 crores. Most of these projects (421), with a financial outlay of ₹674.62 crores, have come from the NABARD nodal agency (bank for growth of agriculture and rural development) in India. It will not be out of place to mention here that Maharashtra leads in the export of certain fruits such as pomegranates, grapes, and oranges in India.
2. Gujarat ranks second in terms of the number of projects, with 410 projects and a financial outlay of ₹48.21 crores. It is interesting to note that most of the projects (389) are funded by the National Cooperative Development Corporation (NCDC), an agency for developing cooperatives in agriculture. This goes well with the culture of the state. From the data, it can be observed that the average project size is relatively small.
3. With 82 projects, Andhra Pradesh is the third largest in terms of the number of projects, with a financial outlay of ₹223.24 crores that is next only to Maharashtra. All of this support has come from NABARD.
4. This is followed by Madhya Pradesh, with 180 projects involving a financial outlay of ₹204.81 crores, all of which comes from NABARD.
5. Interestingly, some of the leading agro-based states such as Punjab, Tamil Nadu, Himachal Pradesh, and West Bengal have not drawn as much for projects aimed at strengthening infrastructure and building quality.

Exhibit 7.3   Agencies Promoting Quality Systems in the Agriproduct Sector

FOOD SAFETY AND STANDARDS AUTHORITY OF INDIA

FSSAI has been established under the Food Safety and Standards Act, 2006, which consolidates various acts and orders that have hitherto handled food-related issues in various ministries and departments. FSSAI (http://www.fssai.gov.in/)

has been created for laying down science-based standards for articles of food and to regulate their manufacture, storage, distribution, sale, and import to ensure availability of safe and wholesome food for human consumption.

The functions of FSSAI include:

- The framing of regulations to lay down the standards and guidelines in relation to articles of food and specifying appropriate systems of enforcing various standards thus notified
- The laying down of mechanisms and guidelines for accreditation of certification bodies engaged in certification of food safety management systems for food businesses
- The laying down of procedures and guidelines for accreditation of laboratories and notification of the accredited laboratories
- The provision of scientific advice and technical support to the central government and the state governments in the matters of framing policies and rules in areas that have a direct or indirect bearing on food safety and nutrition
- The collection and collation of data regarding food consumption, incidence and prevalence of biological risk, contaminants in food, residues of various contaminants in food products, identification of emerging risks, and introduction of a rapid alert system
- The creation of an information network across the country so that the public, consumers, *panchayats*, and others receive rapid, reliable, and objective information about food safety and issues of concern
- The provision of training programmes for persons who are involved or intend to get involved in food businesses
- Contributions to the development of international technical standards for food, including SPS standards
- Promotion of general awareness about food safety and food standards

NATIONAL BANK FOR AGRICULTURE AND RURAL DEVELOPMENT

NABARD is an apex development bank in India that facilitates credit flow for the promotion and development of agriculture and cottage and village industries. It promotes quality practices in agriculture. With a view to strengthen the marketing grading infrastructure, it encourages investors to invest

Table 7.4 Scheme for Development/Strengthening of Agricultural Marketing Infrastructure, Grading, and Standardization during 2011–2012

State/Union Territory	Project Sanctioned by NABARD				Project Sanctioned by NCDC				Project Sanctioned to State Agency				Total Projects State-wise			
	Number of Projects	Total Financial Outlay	Total Eligible Subsidy	Total Subsidy Released	Number of Projects	Total Financial Outlay	Total Eligible Subsidy	Total Subsidy Released	Number of Projects	Total Financial Outlay	Total Eligible Subsidy	Total Subsidy Released	Number of Projects	Total Financial Outlay	Total Eligible Subsidy	Total Subsidy Released
Andhra Pradesh	82	223.24	24.96	28.13	0	0.00	0.00	0.00	0	0.00	0.00	0.00	82	223.24	24.96	28.13
Madhya Pradesh	180	204.81	60.27	27.41	0	0.00	0.00	0.00	0	0.00	0.00	0.00	180	204.81	60.27	27.41
Punjab	60	47.52	11.34	3.68	0	0.00	0.00	0.50	0	0.00	0.00	0.00	60	47.52	11.34	4.18
Kerala	46	5.78	4.97	3.53	2	31.88	0.48	0.27	0	0.00	0.00	0.00	48	37.66	5.45	3.80
Tamil Nadu	23	5.38	3.70	2.24	0	0.00	0.00	0.00	0	0.00	0.00	0.00	23	5.38	3.70	2.24
Rajasthan	3	4.37	0.88	7.96	0	0.00	0.00	0.00	0	0.00	0.00	0.00	3	4.37	0.88	7.96
Chhattisgarh	12	11.37	2.54	2.44	0	0.00	0.00	0.00	9	11.78	3.11	0.00	21	23.15	5.65	2.44
Himachal Pradesh	14	5.85	0.17	0.67	0	0.00	0.00	0.00	0	0.00	0.00	0.00	14	5.85	0.17	0.67
Maharashtra	421	674.62	66.93	79.99	0	0.00	0.00	0.00	54	67.30	16.85	0.38	475	741.92	83.78	80.37
Odisha	0	0.00	0.00	0.03	0	0.00	0.00	0.00	0	0.00	0.00	0.00	0	0.00	0.00	0.03
Gujarat	13	2.94	0.70	8.82	389	39.92	10.24	9.32	8	5.35	1.34	0.00	410	48.21	12.28	18.14
Sikkim	0	0.00	0.00	0.00	0	0.00	0.00	0.00	0	0.00	0.00	0.00	0	0.00	0.00	0.00
Karnataka	21	7.21	1.76	0.12	84	85.77	12.64	0.00	1	1.78	0.44	0.00	106	94.76	14.84	0.12
Assam	2	3.01	0.60	1.79	0	0.00	0.00	0.00	0	0.00	0.00	0.00	2	3.01	0.60	1.79
Nagaland	0	0.00	0.00	0.00	0	0.00	0.00	0.00	0	0.00	0.00	0.00	0	0.00	0.00	0.00
West Bengal	0	0.00	0.00	0.00	0	0.00	0.00	0.00	0	0.00	0.00	0.00	0	0.00	0.00	0.00

Source: Ministry of Agriculture, DMI statistics 2011–12, *Agmarknet*, 2012.

in agricultural infrastructure facilities. From Table 7.4 we can see that NABARD sanctioned the highest subsidies for grading infrastructure.

AGRICULTURAL AND PROCESSED FOOD PRODUCTS EXPORT DEVELOPMENT AUTHORITY

APEDA was established by the Government of India to provide standards and specifications for scheduled products for the purpose of exports. It is responsible for export promotion and development of scheduled products such as fruits, vegetables, meat, poultry, dairy, confectioneries, biscuits and bakery products, honey, jaggery and sugar, cocoa, chocolates, alcoholic and non-alcoholic beverages, cereal, groundnuts, peanuts and walnuts, pickles, and guar gum. In addition to this, APEDA has been entrusted with the responsibility of monitoring imports of sugar. APEDA has also been taking various steps for promotion and control of quality. APEDA provides subsidies for the creation of laboratories to analyse quality parameters such as pesticide residue, metal content, and so on.

COCONUT DEVELOPMENT BOARD

"Coconut Development Board (CDB) is a statutory body established under the Ministry of Agriculture, Government of India, for the integrated development of coconut cultivation and associated industries in the country, with focus on product quality, productivity increase and product diversification" (coconutboard.nic.in).

SPICES BOARD OF INDIA

"The Spices Board (Ministry of Commerce, Government of India) is the flagship organization for the development and worldwide promotion of Indian spices. The Board is an international link between Indian exporters and the importers abroad. The Board has been spearheading activities for excellence of Indian spices, involving every segment of the industry. The Board has made quality and hygiene the cornerstones of its development and promotional strategies" (http://www.indianspices.com).

THE CASHEW EXPORT PROMOTION COUNCIL OF INDIA

"The Cashew Export Promotion Council of India (CEPC) was established by the Government of India in the year 1955 with the active cooperation of the cashew industry, with the object of promoting exports of cashew kernels and cashew nut shell liquid from India. The Council has been set up such that it provides the necessary institutional frame-work for performing the different functions that serve to intensify and promote exports of cashew kernels and cashew nut shell liquid" (cashewindia.org).

7.8 Foreign Trade

Quality plays a very important role in the agriproduct sector, especially in foreign trade. Over the decades, international trade of agricultural products has rapidly increased between countries that have different environmental and development standards. This situation also increases the relevance of quality management in the international trade of agriproducts.

7.8.1 Emergence

Agricultural trade is not a modern phenomenon, but has been in existence since the times of the Indus Valley civilization. Historians have found evidence of trade between the Indian sub-continent and the Middle East and Europe. It is believed that agricultural products such as pepper, cinnamon, opium, indigo, and cotton were exported to Europe, the Middle East, and South East Asia in return for gold and silver. This trade grew further and was transformed by the coming of the Europeans to India.

Because of the comparative advantage theory, there is always bound to be trade in agricultural items, because no land is able to produce all types and varieties of products. Some areas of the world are blessed in the production of rubber, whereas other areas produce more wheat. As food is necessary for daily subsistence, agriculturally poor countries will be forced to barter, borrow, or buy from agriculturally rich countries, thus resulting in the international trade of agricultural products.

7.8.2 World Agricultural Trade

Food is one of the basic necessities of life, so every country would like to be self-sufficient and therefore provide support and assistance to its agricultural community. This assistance can be in the form of subsidies, higher tariff on imported agriproducts, or imposition of stringent quality requirements on imported agricultural products.

The developed countries have a mechanized system of agriculture, with better seeds, fertilizers, and large integrated agricultural units, whereas the developing countries still generally follow traditional methods, and have smaller land hold-ings. In combination with the lower capacity of the developing countries to provide agricultural subsidies, this results in imbalances in the foreign trade of agricultural products. World bodies such as WTO have brought in guidelines to reduce this divide and to ensure fair trade practices in agricultural products.

The Uruguay Round of the WTO aimed to establish a fair and market-ori-ented agricultural trading system, and stressed that its long-term objective would be to provide substantial reductions in agricultural support and protection over an agreed period of time, thus helping in correcting and preventing restrictions and distortions in world agricultural markets.

7.8.3   Exports

To compete in the international market, it is necessary that exports must meet the specific standards set by the governments of the importing countries. The export of agricultural products requires the exporter to look carefully into the issues of quality of the produce before and during shipment. International trade in agriproducts is strictly governed by different rules in different countries. These rules lay down specifications for the quality of the product, such as the tempera-ture at which it is to be kept, its moisture content, packaging types, and so on. Export guidelines often also ask for documentation regarding the activities car-ried out on the farm during cultivation of the crops. For example, for export of mangoes to the United States, the guidelines require that the fruits be harvested at half-maturity stage if shipment is to be done by sea and at full maturity stage if shipment is by air. The guidelines also ask for storing of harvested fruits in clean and disinfected plastic crates with clean polyurethane foam. Quality measures are stringent, especially in the case of agricultural products, and vary from country to country. Hence, ensuring good control over quality of produce is essential, par-ticularly in international trade.

India's share in the international agricultural trade has been very low. But it is promising to note that over the years since independence, our agricultural imports have decreased, whereas exports have increased. Our main export items include tea, coffee, wheat, cashew nut, spices, and other products. Indian agricultural exports have also come under the scanner in different countries such as the United States and the European Union for quality parameters.

7.8.3.1   Global Tea Trade

Tea as a beverage is widely used in almost all parts of the globe. India is the fourth largest exporter of tea along with other leading tea exporters such as Sri Lanka, Vietnam, China, and Kenya. Although tea traded internationally needs to comply with mandatory standards such as the European Union's food safety and labelling regulations, quality standards imposed by major private companies and certain countries have increasingly important implications for exporters of tea products from developing countries. The quality standards dictated in different end markets are affecting market accessibility for developing countries' tea products.

The EU rules and regulations place an obligation on food business operators to ensure that all stages of production, processing, and distribution of tea under their control satisfy the relevant hygiene requirements laid down in the regulations. They also call for onsite inspection of 10% of the goods imported and sampling tests for checking pesticide residues and other items. The exporters are also required to have the necessary entry documents to enter designated ports, as well as documents for testing.

7.8.3.2 Fruit and Vegetable Trade to the European Union

Similar to its restrictions on the import of tea, the European Union has a specific set of rules and regulations to monitor the import of vegetables and fruits Appendix 7.2. Some of the guidelines include the requirements that products packed together must be of the same origin, variety, commercial class, or quality; they must be intact, sound, free from pests and damage from pests, and clean. Imported items are also classified as Class I and Class II, based on their quality standards. The European Union also has stringent norms on packaging and labelling, with stipulations for visible information such as variety, origin, quality class, dispatcher name and address, and so on marked prominently. Each variety of fruit or vegetable should be packed separately. From the processed grain and agriproducts perspective, ISO 22000 is a very popular standard today. Most agencies, such as the Spices Board, APEDA, and the Ministry of Food Processing provide subsidies in the form of reimbursements to the companies opting for the certifications.

Although agricultural trade is a necessity, regulating it so as to ensure adherence to quality, safety, and hygiene norms is also important. However, it is also essential to ensure that all countries are given a level playing field in the foreign trade of agricultural products. Some of the major quality issues that affect the produce relate to:

1. Macrobiological factors such as insects
2. Foreign bodies such as stones
3. Pesticide residue, which is emerging as a major concern
4. Microbiological factors

7.9 Secondary and Tertiary Processing

The demand for ensuring quality standards in processing plants is high. There are a number of stringent procedures and statutory rules with respect to the setting up of facilities, and accurate record maintenance is mandatory. Companies have established routine quality checks, laboratories, and systems such as training of employees, reviewing of manuals, and so on. The quality system also requires high adherence to hygienic process handling, apart from temperature control and grading. For every product level industry such as beverages, dairy, sugar, edible oil, and so on, there are defined quality systems that must be transparent and inspected by statutory authorities and standards committees. The systems are well evolved and require lots of discipline for adherence. Appendix 7.3 to 7.7 provides a few useful references.

7.10 Conclusion

It is important that agribusiness must improve on its productivity through input efficiency and output management through a well-developed network of agricultural supply chain infrastructure so that wastage is reduced and the maximum produce can be utilized. Adoption of good quality management and standardized operating procedures with the right infrastructure are critical for meeting future challenges of the demand from this sector. Even though grading and quality management play a pivotal role in the agriculture sector, there is very limited recognition of the functions it carries out. There is a need to popularize quality standards, especially in the national and international markets, as awareness levels are very low, and the government can initiate methods to implement quality management systems in agriculture. There is also going to be growing emphasis on the international trade of agricultural products. In order to remain competitive in this market, it is essential that we adhere to the highest quality standards and efficient and hygienic practices. The world is definitely going to see a sea change in the way agricultural products are produced and sold globally.

Appendix 7.1

Grading of Cashew Kernels (Whole)

Grade Designation	Number of Kernels Per 454 g/kg		General Characteristics
1	2	3	4
W 210	200/210	440/465	Cashew kernels should have been obtained through shelling and peeling cashew nuts, *Anacardium occidentale*. They should have the characteristic shape; should be white, pale ivory, or light ash; reasonably dry and free from insect damage, damaged kernels and black or brown spots. The kernels should be completely free from testa.
W 240	220/240	485/530	
W 280	260/280	575/620	
W 320	300/320	660/705	
W 400	350/400	770/880	
W 450	400450	880/990	
W 500	450/500	990/1100	

Source: Ministry of Agriculture, Cashew kernels grading and marking rules, *Agmarknet*, 2005.

Note: Tolerance—broken kernels should not exceed 5% at the time of packing.

Appendix 7.2

Grading of Fruits and Vegetables—Minimum Requirements for Export to EU Countries

1. Premises should be clean and in hygienic condition.
2. Surroundings of the premises should be clean.
3. The premises should not be situated near tanneries, chemical plants, fertilizer plants, etc.
4. The walls of the premises should be properly plastered and free from crevices, holes, dampness, etc. Thatched roofs are not advisable.
5. The premises should be pest, insect, and rodent proof.
6. The premises should be free from cobwebs and spiders.
7. The premises should have proper drainage systems.
8. The premises should have arrangements for disposal of rejected, rotten, and otherwise wasted horticulture produce.

Source: Ministry of Agriculture, Instructions for grant of C.A. and certificate of AGMARK grading for exports of pomegranates, *Agmarknet*, 2010.

Appendix 7.3

American Spice Trade Association (ASTA) Cleanliness Specifications for Turmeric

Whole Insects Dead (by count)	Excreta Mammalian (mg/lb)	Excreta Other (mg/lb)	Mould % (wt.)	Insect Defiled Infested (% wt.)	Extraneous Foreign Matter (%/wt.)
3	5.0	5.0	3.0	2.50	0.50

Source: Ministry of Agriculture and Cooperation, Spices, *Market Intelligence,* 2009.

Note: Turmeric exported to the United States should conform to the cleanliness specification stipulated by the ASTA. (a) For exports, packaging is normally done in clean polythene laminated gunny bags. (b) For domestic markets, turmeric is packed in gunny bags and jute sacks.

Appendix 7.4

Codex Standard for Wheat and Durum Wheat: Quality and Safety Factors

Parameter	Wheat	Durum Wheat
Moisture, maximum	14.5% m/m	14.5% m/m
Organic extraneous matter, maximum	1.5% m/m	1.5% m/m
Inorganic extraneous matter, maximum	0.5% m/m	0.5% m/m
Test weight (weight of a hundred litre volume expressed in kg/hL), minimum	68	70
Shrunken and broken kernels, maximum	5.0% m/m	6.0% m/m
Edible grains other than wheat and durum wheat, maximum	2.0% m/m	3.0% m/m
Damaged kernels, maximum	6.0% m/m	4.0% m/m
Insect-bored kernels, maximum	1.5% m/m	2.5% m/m
Filth (impurities of animal origin, including dead insects), maximum	0.1% m/m	0.1% m/m
Ergot (sclerotium of the fungus *Claviceps purpurea*), maximum	0.05% m/m	0.05% m/m

Source: Ministry of Agriculture, Manual on standards of wheat, *Agmarknet,* 2005.

Appendix 7.5

Quality Information Requirements for Import of Food Products into India

In India, import of food products entails adherence to certain quality requirements. Import of frozen or chilled foodstuff must be approved by the Ministry of Agriculture, Government of India. The following information is essential for imports.
i) Full details of the number of packages
ii) Gross and net weights (kg)
iii) Method of transportation
iv) Estimated time of arrival (New Delhi)
v) Type of meat
vi) Storage temperature (°C)
vii) Country of origin
viii) Airport of departure
ix) Cost, insurance, and freight value

Source: India-ASEAN, Food/meat products, 2010.

Note: In addition, all shipments should be covered by certificates of origin and health certificates.

Appendix 7.6

Commodity Grading and Marketing Rules

1. Tobacco Grading and Marking Rules, 1937 (amended in 1982)
2. Hides Grading and Marking Rules, 1937
3. Skins Grading and Marking Rules, 1937
4. Ghee Grading and Marking Rules, 1938 (amended in 1991)
5. Rice Grading and Marking Rules, 1939
6. Creamery Butter Grading and Marking Rules, 1941 (amended in 1995)
7. Sannhemp Grading and Marking Rules, 1942
8. Canned Bottled Fruits and Fruit Products Grading and Marking Rules, 1942

(Continued)

Appendix 7.6 (*Continued*)
Commodity Grading and Marketing Rules

9. Lac Grading and Marking Rules, 1950
10. Arecanuts Grading and Marking Rules, 1952
11. William Pears Grading and Marking Rules, 1953
12. Vegetable Oils Grading and Marking Rules, 1955 (amended in 2009)
13. Curry Powder Grading and Marking Rules, 1956
14. Cashew Kernels Grading and Marking Rules, 1960
15. Goat Hair Grading and Marking Rules, 1960
16. Myrobalans Grading and Marking Rules, 1962
17. Seed Potato Grading and Marking Rules, 1963
18. Tendu (Bidi Wrapper) Grading and Marking Rules, 1963
19. Water Chestnuts Grading and Marking Rules, 1964
20. Ambadi Seed Grading and Marking Rules, 1964
21. Palmyra Fibre Grading and Marking Rules, 1964
22. Taramira Seed Grading and Marking Rules, 1964
23. Animal Casings Grading and Marking Rules, 1964 (amended in 1997)
24. Senna Leaves and Pods Grading and Marking Rules, 1964
25. Groundnut Grading and Marking Rules, 1965
26. Tapioca Products (Animal Feed) Grading and Marking Rules, 1965
27. Cereals Grading Rules, 1966
28. Table Eggs Grading and Marking Rules, 1968
29. Bristles Grading and Marking Rules, 1969 (amended in 1996)
30. Cotton Grading and Marking Rules, 1971 (amended in 1980)
31. Seedless Tamarind Grading and Marking Rules, 1971 (amended in 1981)
32. Dried Edible Mushrooms Grading and Marking Rules, 1972
33. Saffron Grading and Marking Rules, 1973
34. Wool Grading and Marking Rules, 1975 (amended in 1984)

(Continued)

Appendix 7.6 (*Continued*)

Commodity Grading and Marketing Rules

35. Besan (Gram Flour) Grading and Marking Rules, 1975 (amended in 1996)
36. Aloe Fibre Grading and Marking Rules, 1975
37. Kangra Tea Grading and Marking Rules, 1976
38. Coconut Grading and Marking Rules, 1976
39. Raw Meat (Chilled or Frozen) Grading and Marking Rules, 1991
40. Jute Grading and Marking Rules, 1978
41. Basmati Rice (Export) Grading and Marking Rules, 1979 (amended in 1987)
42. Suji & Maida Grading and Marking Rules, 1998
43. Vegetable Oil Cakes (Expressed or solvent extracted) Grading and Marking Rules, 1979 (amended in 1983)
44. Papain Grading and Marking Rules, 1979
45. Agar-Agar Grading and Marking Rules, 1979
46. Raw Cashew Nut Grading and Marking Rules, 1980
47. Bread Wheat Flour Grading and Marking Rules, 1981
48. Hand Picked Selected Groundnuts Grading and Marking Rules, 1982
49. Isubgol Husk Grading and Marking Rules, 1982
50. Mahua Seeds Grading and Marking Rules, 1982
51. Sal Seed Grading and Marking Rules, 1982
52. Gum Karaya Grading and Marking Rules, 1982
53. Linseed Grading and Marking Rules, 1982
54. Catechu Grading and Marking Rules, 1982
55. Niger Seeds Grading and Marking Rules, 1982
56. Sesame Seeds Grading and Marking Rules, 1982
57. Safflower Seeds Grading and Marking Rules, 1982
58. Sunflower Seeds Grading and Marking Rules, 1982
59. Cotton Seeds Grading and Marking Rules, 1982

(*Continued*)

Appendix 7.6 *(Continued)*
Commodity Grading and Marketing Rules

60. Tamarind Seed and Powder Grading and Marking Rules, 1983
61. Compounded Asafoetida Grading and Marking Rules, 1984
62. Tobacco Grading (commercial) Rules, 1987
63. General Grading and Marking Rules, 1988 (amended in 2009)
64. Vanaspati Grading and Marking Rules, 1990
65. Blended Edible Vegetable Oils Grading and Marking Rules, 1991 (amended in 2002)
66. Essential Oils Grading and Marking Rules, 1993
67. Wheat Atta Grading and Marking Rules, 1993 (amended in 2003)
68. Desiccated Coconut Grading and Marking Rules, 1994
69. Fat Spread Grading and Marking Rules, 1994
70. Nutmeg Grading and Marking Rules, 1995
71. Tejpat Grading and Marking Rules, 1996
72. Cocoa Powder Grading and Marking Rules, 1996
73. Cocoa Beans Grading and Marking Rules, 1996
74. Bura Grading and Marking Rules, 1997
75. Clove Grading and Marking Rules, 1997
76. Poppy Seed Grading and Marking Rules, 1997
77. Ajowain Seeds (Whole and Powdered) Grading and Marking Rules, 1997
78. Mace Grading and Marking Rules, 1997
79. Mixed Masala Powder Grading and Marking Rules, 2000
80. Wheat Porridge Grading and Marking Rules, 2000
81. Caraway & Black Caraway Grading & Marking Rules, 2000
82. Cereals Grading and Marking Rules, 2001
83. Roasted Bengal Gram (split) Grading & Marking Rules, 2002
84. Sundried Raw Mango Slices and Powder Grading & Marking Rules, 2003

(Continued)

Appendix 7.6 *(Continued)*

Commodity Grading and Marketing Rules

85. Vermicelli, Macaroni and Spaghetti Grading & Marking Rules, 2003
86. Pulses Grading and Marking Rules, 2004
87. Fruits & Vegetables Grading & Marking Rules, 2004[a] (amended in 2007, 2010,[b] and 2012[c])
88. Makhana Grading and Marking Rules, 2004
89. Spices Grading and Marking Rules, 2005[d]
90. Tapioca Sago Grading and Marking Rules, 2007
91. Sattu Grading and Marking Rules, 2007
92. Sheekakai Grading and Marking Rules, 2008[e]
93. Mahua Flower Grading and Marking Rules, 2008
94. Tamarind (with Seed) Grading and Marking Rules, 2008
95. Amla Grading and Marking Rules, 2008
96. Jaggery Grading and Marking Rules, 2008 (amended in 2011)
97. Honey Grading and Marking Rules, 2008
98. Organic Agricultural Produce Grading and Marking Rules, 2009 (amended in 2011)
99. Puwad Seeds Grading and Marking Rules, 2009
100. Karanj Seeds Grading and Marking Rules, 2009
101. Guar Grading and Marking Rules, 2011
102. Jatropha Seeds Grading and Marking Rules, 2011
103. Castor Seeds Grading and Marking Rules, 2011
104. Mustard and Rapeseeds Grading and Marking Rules, 2012
105. Soybeans Grading and Marking Rules, 2012

Source: Ministry of Agriculture, List of the grading and marking rules notified under the Agricultural Produce (Grading and Marking) Act, 1937 (as on 16-04-2012), *Agmarknet*, 2012.

(Continued)

Appendix 7.6 (*Continued*)

Commodity Grading and Marketing Rules

[a] In exercise of the powers conferred by Section 3 of the Agricultural Produce (Grading and Marking) Act, 1937 (1 of 1937), and in supersession of (1) the Grapes Grading and Marking Rules, 1937; (2) the Plums Grading and Marking Rules, 1938; (3) the Onion Grading and Marking Rules, 1964; (4) the Banana Grading and Marking Rules, 1980; (5) the Mangoes Grading and Marking Rules, 1981; (6) the Pineapple Grading and Marking Rules, 1982; (7) the Guavas Grading and Marking Rules, 1996; and (8) the Garlic Grading and Marking Rules, 2002. Fruits and Vegetables Grading and Marking Rules, 2004 was published vide GSR 220 dated 14 June 2004, on pages 1209–1285 dated 26 June 2004 in the Gazette of India Part II Section 3, sub-section (i).

[b] In exercise of the powers conferred by Section 3 of the Agricultural Produce (Grading and Marking) Act, 1937 (1 of 1937) and in supersession of (1) Citrus Fruits Grading & Marking Rules, 1949; (2) Table Potato Grading & Marking Rules, 1950; (3) Table Potato (for export) Grading & Marking Rules, 1964; (4) Walnut Grading & Marking Rules, 1960; (5) Oranges (for export) Grading & Marking Rules, v1983, the Fruits and Vegetables G&M (amendment) Rules, 2010, is published in the Gazette of India vide GSR no. 186(E), dated 12 March 2010.

[c] In exercise of the powers conferred by Section 3 of the Agricultural Produce (Grading and Marking) Act, 1937 (1 of 1937) and in supersession of the Apples Grading and Marking Rules, 1984, the Fruits and Vegetables G&M(amendment) Rules, 2012, is published in the Gazette of India.

[d] In exercise of the powers conferred by Section 3 of the Agricultural Produce (Grading and Marking) Act, 1937 (1 of 1937) and in supersession of (1) the Chillies Grading and Marking Rules, 1962; (2) the Cardamom Grading and Marking Rules, 1962; (3) the Turmeric Grading and Marking Rules, 1964; (4) the Chillies Powder Grading and Marking Rules, 1964; (5) the Ginger Grading and Marking Rules, 1964; (6) the Fennel (whole and ground), Fenugreek (whole and ground), and Celery seeds Grading and Marking Rules, 1967; (7) the Pepper Grading and Marking Rules, 1969; (8) the Coriander Grading and Marking Rules, 1997; (9) the Cumin Seeds Grading and Marking Rules, 1997; and (10) the Large Cardamom Grading and Marking Rules, 1998. The Spices Grading and Marking Rules, 2005 was published vide GSR 257 dated 21 July 2005 appearing on pages 943–991 dated 30 July 2005, in the Gazette of India, Part II, Section 3, sub-section (i).

[e] In exercise of the powers conferred by Section 3 of the Agricultural Produce (Grading and Marking) Act, 1937 (1 of 1937) and in supersession of Sheekakai Powder Grading and Marking Rules, 1973; the Sheekakai Grading and Marking Rules, 2008 was published vide GSR 66(E) dated 31 January 2008, in the Gazette of India, Part II, Section 3, sub-section (i) dated 02 February 2008.

Appendix 7.7

National and International Standards—Websites List

1. http://agmarknet.nic.in
2. www.iso.org
3. http://www.unece.org
4. http://www.wto.org/english/tratop_e/sps_e/spsagr_e.htm
5. http://www.naqs.go.kr/english/vision/vision.jsp
6. http://www.codexalimentarius.org
7. http://www.apeda.gov.in/apedawebsite/Latest_Notification/Latest_Notice.htm
8. http://www.mofpi.nic.in/
9. http://coconutboard.nic.in/
10. http://www.indianspices.com/php/trade_notifc.php
11. http://www.cashewindia.org/
12. http://www.indiatrademarkregistration.com/fpo-certification/
13. http://standardsdata.in/
14. www.ifoam.org/

References

India-ASEAN, Food/meat products, 2010. http://www.india-aseanbusinessfair.com/e/guidelines-food%20-products.pdf

Ministry of Agriculture, Fruits and vegetables grading and marking rules, *Agmarknet*, 2004. http://agmarknet.nic.in/fveggmrules04.htm#pomegranate

Ministry of Agriculture, Manual on standards of wheat, *Agmarknet*, 2005. http://agmarknet.nic.in/Wheat_manual.htm

Ministry of Agriculture, Cashew kernels grading and marking rules, *Agmarknet*, 2005. http://agmarknet.nic.in/cashewkernelsgmr.pdf

Ministry of Agriculture, Instructions for grant of C.A. and certificate of AGMARK grading for exports of pomegranates, *Agmarknet*, 2010. http://agmarknet.nic.in/Pomecert.pdf

Ministry of Agriculture, DMI statistics 2011–12, *Agmarknet*, 2012. http://agmarknet.nic.in/agmstat2012.pdf

Ministry of Agriculture, List of the grading and marking rules notified under the Agricultural Produce (Grading and Marking) Act, 1937 (as on 16-04-2012), *Agmarknet*, 2012. http://agmarknet.nic.in/listgradinglatest.pdf

Ministry of Agriculture, Notification (Fruits and vegetables grading and marking rules), *Agmarknet*, 2012. http://agmarknet.nic.in/FruitVegGmRules2012.pdf

Ministry of Agriculture and Cooperation, Spices, *Market Intelligence*, 2009. http://mi.agri.net.in/Default.aspx?mypath=Domestic/Spices

Ministry of Micro, Small & Medium Enterprises, Cold storage, Chap. 7 in *Food Project Profiles*, 2004. http://www.dcmsme.gov.in/publications/pmryprof/food/ch7.pdf

Chapter 8

Risk

OBJECTIVE

The objective of this chapter is to understand the different sources and classification of risks that affect agribusiness supply chains. Further, discussions include risk mitigation strategies, and detail price and market risks and support system for handling these issues. Insurance support is also discussed.

8.1   Introduction

Indian agriculture is subject to various risks and uncertainties. It is dependent on the monsoon, as perennial river resources are inadequate. Indian agriculture is hampered periodically as well as sporadically by droughts due to insufficient rains, and at times even by the total failure of monsoon. This dependence has stayed at such a level that Indian agriculture is described as "a gamble against monsoon."

One would expect that the technology explosion in modern society can help in managing the risks in agribusiness. Modern management science facilitates understanding of and controlling the macro environment's impact on business. In contemporary business, technological advancements are judged by their effectiveness in managing "uncertainties." It may be noted that risks manifest themselves in two forms: normal and radical uncertainties.

On a different note, situations relating to risk can be discussed based on the probabilities of occurrence and the possibilities of damages. This refers to situations where the possible damages and their probabilities are known, and also to situations where the probabilities are unknown or the decision takers are confronted with ignorance ("unknown unknowns" or "don't know what we don't know" and how relevant the issue is for our decisions), indeterminacy (issue conditions and causal chains open, outcomes dependent on how intermediate actors behave), complexity (open behavioural systems and multiplex, often non-linear processes) and so on. Typically one of the ways of differentiating risk and uncertainties is that the former can be understood and captured through probability of occurrences, whereas the latter is too unsure to be assigned with probability of occurrences.

As in any economic system, agriculture and agribusiness carry a number of risks and uncertainties. There are a number of controllable and uncontrollable factors that could affect agribusiness economic activities. These include weather climatic conditions, economic cycle, crop diseases, market factors like excess supply and low demand, and so on. There are a number of role agents in agribusiness supply chains. Risk impact and the ability to handle risks vary widely among them. For example, farmers are not generally well informed, and lack training in forecasting risk possibilities and pursuing mitigation strategies. On the other hand, corporates in agribusiness may be better off.

This risk would affect farm production decisions and role agents. How a role agent like a farmer manages such risks is an interesting issue. Like any decision maker, a farmer has to decide on various issues: farming, using portfolio techniques of his holdings and investment and using market instruments and other tools traditionally available for managing risks. The government, as a facilitator and direct interventionist in this sector for the larger good of society, provides a number of support programmes to manage risks. Financial schemes like crop insurance and support financing, including subsidies and other forms of assistance that are routed through banking and insurance companies, facilitate managing risks. One may note that government policies and actions, the schemes of financial agencies and the overall farmer decision scenario in terms of on-farm and off-farm activities form a unique interdependent ecosystem.

There is also a need to study these interdependencies and develop an integral system. The integral system will help in better understanding of the actors, their actions and the implications of those actions. The risks can then be understood in terms of:

- sources of risks;
- risk management strategies and tools;
- government policies.

These three actors form an axis, and the interrelations among them and across the axis add to the complexity of risk management.

Agriculture by definition is a risky business. It is governed by the fact that due to a complex physical and economic environment, the produce from agricultural activities is subjected to great variation. Further, the possible outcome (produce) may vary depending upon the interactions of different actors. The distinctions between danger and risk are not clearly drawn. Hence, it becomes necessary that the dangers and risks are segregated in the case of agriculture. Thus, in order to have clearly set goals, we would define "risk as exposure to uncertain and unfavourable economic consequences. Further, the term 'vulnerability' is often used to define the likelihood that a risk will result in a significant decline in well-being, that is, it refers to the resilience or lack of resilience against a given adversity. Vulnerability does not depend only on the characteristics of the risk, but also on the asset endowment of the entity at risk and the availability of insurance mechanisms" (www.oecd.org).

We will now try to look at risk at a broader level and analyse it to form a framework for risk mitigation in supply chains. Risks basically include three elements:

- Risk assessment and evaluation
- Risk management
- Risk communication.

Risk assessment is generally performed by systematically processing all available information. Such a process would help to identify events and triggers that may be potential risks for business. On the other hand, the risk evaluation process would assign priorities to risk events and then evaluate the impact. On completion of risk evaluation, it is important that the results are communicated to the different stakeholders who may need such information. This is referred to as risk communication.

All these processes require a comprehensive risk management system. The system should address the different sources of risk that affect farming, and facilitate strategies to manage them. It should have appropriate tools earmarked for combating various kinds of risk. A typical approach to risk management would include (a) assessing and measuring the risks and their impact on income and assets; (b) collecting and analysing information to understand the optimal risk management tool for a farmer, while accounting for his endowments and risk preferences, and (c) deciding on appropriate policies to improve this risk management strategy.

These three elements are not related in a linear fashion. They interact in a complex manner. It is often not possible to isolate and identify individual risks and strategies for individual farmers against the backdrop of government policies, and thus, a holistic approach is needed for the analysis of the system.

8.2 Sources of Risks

"The risks and sources of risks that are relevant in agriculture have different characteristics, and they can be classified in very different ways. These can be classified in varying terms according to the context of the study. Some technical characteristics of risks apply across different classes, and these characteristics can be very significant in terms of determining the appropriate and available strategies to deal with each risk" (OECD, 2010).

We will further discuss the possible classifications of the sources of risk, the correlations among them, and the links between agricultural risk and climate change.

The risks in agriculture may be defined as systematic and non-systematic risks. Systematic risks are related to events that repeat over time. A clear pattern can be drawn on the basis of past trends, and the pattern can be relied upon. The non-systematic risks are characterized by very short and/or imperfect records of their past manifestations, and hence, it is not possible to predict their future occurrences.

It may be noted here that it is downside risk that matters most. Further, the probability of downside risk occurring is high as and when risky variables are interdependent and non-linear in their relationships. This is particularly relevant in agriculture. For instance, yields depend on several factors, such as rainfall and temperature, but large deviations from the mean values of these variables in either direction have an adverse effect on all agribusiness elements, including farm output, primary, secondary and tertiary sector processing and marketing.

In agriculture, a normal crop season could be one where all defined variables are likely to realize their expected outcomes. Hopes for a normal season are usually far-fetched, as it is more unlikely to happen in India, considering the past performances of the sector. In the event of abnormal, or rather, dismal performance of the variables, the downside risk could be higher. Though theoretically the downside must be supported by a similar upside in returns, it seldom happens in such a fashion. It is always up to analysts and experts to estimate distribution based on risk and arrange for cover with suitable protection.

The frequency of occurrence and the intensity of occurrence, that is, the magnitude of loss, are the important attributes of risks. "The links among the distributions of different risks are very important for any risk evaluation. An individual risk that is independent or uncorrelated with any other risk, including market risk, is called idiosyncratic risk. However, in reality, a risk has some degree of correlation with other risks. If there is a high degree of correlation among individuals in the same region or country, the risk is called systemic risk" (OECD, 2010). There could be correlation with other risks, and the nature of relationship could be positive or negative.

Another risk type is the catastrophic risk, associated with a risk that may be of low frequency, but results in high losses. It relates to the extreme of the negative tail of the distribution of outcomes. This is more linked to the possibilities of huge overall losses, which could occur to a region or a country from causes such as swine flu or a tsunami. Such incidences are simultaneously catastrophic and systemic.

8.3 Classifications of Agricultural Risks

The risks in agriculture have been segregated into a number of different categories. There are a number of ways to classify these risks. A broad and general way of risk classification in agriculture is detailed in Table 8.1.

Table 8.1 Types of Risks

Type of Risk	Micro Risk	Meso Risks	Macro Risks
Impact level	Individuals or households	Communities	Nation or regions
Market/price		Changes in price of land, new requirements from food and agribusiness industry	Changes in input/output prices due to trade policy, new markets, endogenous variability
Production	Hail, frost, non-contagious diseases, personal hazards (illness, death), asset risks	Rainfall, landslides, pollution	Floods, droughts, pests, contagious diseases, technology
Financial	Changes in income from other sources (non-farm)		Changes in interest rates/value of financial assets/access to credit
Legal/ Institutional	Liability risk	Changes in local policy or regulations	Changes in regional or national policy and regulations, environmental laws, agricultural payments, etc.

Source: OECD. *Managing Risk in Agriculture: A Holistic Approach (Extracts).* Paris: Organisation for Economic Co-operation and Development, 2010.

- "The risks that are common to all businesses (family situation, health, personal accidents, macroeconomic risks)
- The risks that affect agriculture more specifically: production risk (weather conditions, pests, diseases and technological change)
- Ecological risks (production, climate change, management of natural resources such as water)
- Market risks (output and input price variability, relationships with the food chain with respect to quality, safety, new products)
- Regulatory or institutional risk (agriculture policies, food safety and environmental regulations)" (OECD, 2010).

However, a more precise and acceptable way to distinguish agriculture risk is described here.

The first category is "business risk, which includes production, markets, institutional and personal risks. Production risk is linked to natural phenomena and their relationships with agriculture. Production risk is due to unpredictable weather and performance of crops and livestock. It is concerned with variations in crop yields and in livestock production due to weather conditions, diseases and pests. Market risk is related to uncertainty about the prices of outputs (and sometimes inputs, too) at the time production decisions are taken. Marketing risk is related to variations in commodity prices and quantities that can be marketed in the post-production situation. Institutional risk is due to government actions and regulations such as the laws governing disposal of animal manure, the use of pesticides, tax provisions and payments. Legal and environmental risks are concerned with the possibility of lawsuits initiated by other businesses or individuals and changes in government regulations related to the environment and farming practices. Personal risks are due to uncertain life events, such as death, divorce or illnesses. These are associated with the personal matters of farm households. The human resources risk is concerned with the possibility that family or employees will not be available to provide labour or management" (OECD, 2010).

The second category includes "financial risks arising from different methods of financing the farm business. These risks are related to the nature and source of borrowings. The use of borrowed funds means that interest charges have to be met before equity is rewarded, which may create risk due to leverage" (OECD, 2010). Additionally, there is financial risk when there is an increase in interest rates or a decrease in the quota of loans that are available to agribusiness. A financial risk relates to the ability to pay for expenses and dues when they are liable to be paid so that farmers and processing units have money to continue farming and to avoid bankruptcy.

"Price and production risk are two important components or types of farming risk. However, they have different origins.

Production risk is determined by weather conditions and animal or plant diseases. Price risk originates in the markets for inputs and outputs" (OECD, 2010). This type of risk could also be generated in an economic environment due to the dynamics of the markets. Price and production risks are different with respect to all the important characteristics mentioned above, like information inefficiencies on systemic factors and market. Price and production risks differ, though the common perception would be that they must be highly correlated, as failure in production impacts with an upswing in price.

Price risk can be weakened by arbitrage or borrowing. "Normally, this makes prices for all farmers move in parallel, with very high correlation across farmers and regions whose markets are linked by trade" (OECD, 2010). Price risk is also influenced by factors like transportation and storage costs across locations and time.

Production risk could afflict farmers in a location because of large events like drought, floods and diseases that impact at a regional or a farm level due to inefficient management. A situation can arise where a farmer suffers a bad production year, while his neighbours have an average year. However, it is very unlikely that a farmer will receive a low price while his neighbours sell at much higher prices. It is difficult to use available information and collect more information to measure the magnitude of risks and their likely impact on earnings and asset positions. However, one may conclude that in general, it is far easier to predict and assess production risks compared to price risks.

8.3.1   Production

There are a number of facets of production risks that are enumerated here in this section.

8.3.1.1   Land Use Pattern

Land use patterns and usages have been changing in India. The increasing pressure of population, economic forces, livestock pressure and weakening of land laws and implementing institutions have further marred production. Added to this are the changes caused by natural factors. These factors are aggravated by the absence of any thinking and planning for the future.

India has a large land area of 328.73 million hectares (ha). Out of this, the land area in use is nearly 306 million ha (Appendix 8.1). The total land area under agriculture and related use is nearly 141 million ha. In addition to this, India has 17 million ha of barren and uncultivable land, 13 million ha of cultivable waste land, 10 million ha under permanent pasture and other grazing and 3 million ha under miscellaneous tree crops and groves. Over the years, there has been an increase in the land used for non-agricultural uses. Over the last 40 years, the net sown area has remained constant. The area under non-agricultural uses has increased from 16 to 26 million ha, while the land area that is barren and uncultivable has come down from 28 million ha in 1970–1971 to 17 million ha in 2008–2009.

This diversion of land from agriculture to other usages has resulted in an abrupt increase in the area under non-agriculture use land. However, the government has played a critical role in mitigating this negative impact of the land use trend. It has worked to improve the quality of barren and cultivable land and make it fit for agriculture use. Cropping intensity has increased from 118% to 138% during the same period.

The above trend shows a shift in land use from agriculture to urban use. This pose a serious threat to agricultural output, as the rate of urbanization is 2.4%, compared to a population growth rate of 1.132%. This rate of urbanization is the highest in the world. The increasing pressure of population and urbanization and increasing family income and literacy rates coupled with growth in the service and manufacturing sectors will negatively affect agriculture farm production. It is projected in studies that the population explosion will result in a decrease in the per capita land availability from 0.32 ha in 2001 (against the world average of 2.19 ha) to 0.23 ha in 2025 and 0.19 ha in 2050. Further, it is reported that about 120 million ha of land is degraded in India, and about 5334 million tonnes of soil are lost annually through erosion. Out of 120 million ha of degraded area, a large portion is degraded because of water erosion (68%). The next worst factor is chemical erosion, which accounts for 21%, and then, wind erosion, which accounts for 10%. It is high time that policymakers in India address the management of these issues, which could be caused by ecological degradation, improper management of water tables and discharge of effluents by manufacturers and processors. Apart from these factors, India also suffers from small size and fragmented holdings, tenancy, ceiling limits, improper land records, acquisition and diversion of productive land and climate change.

The above description suggests that there is very limited scope to expand the area under cultivation. This is primarily due to increase in industrialization, urbanization, housing and infrastructure. With the increasing population pressure and the decreasing per capita availability of cultivable land, there is a need to enhance cropping intensity without compromising land productivity. The green revolution has improved productivity, but its effects are limited, and now it is also having adverse effects. This situation has forced authorities to seek alternate sustainable solutions for land use and improve the efficiency of resources and inputs.

The above trend of land use suggests an increasing shortage of land in use, compounded by the need to achieve production increases. The shortage of cultivable land will translate into a decrease in production. Other factors, like the population, will further hamper productivity by reducing per capita land availability. Some suggestions to assuage these effects are discussed here.

- All resources, including land, need to be planned and managed in an integrated manner for maximization of environmental, social and economic factors, apart from ensuring livelihood and equity.
- Comprehensive land inventories and detailed policies have to be prepared.
- A perspective plan for treating degraded lands has to be prepared.
- Infrastructural facilities have to be developed, and the technology support system needs to be strengthened.
- Investments have to be made in capacity building, education and awareness on the importance of conservation and optimum use of all natural resources.
- Continuous monitoring and assessment of the impact of climate change on land use has to be done, and appropriate remedial measures suggested and undertaken.

8.3.1.2 Soil

Soil is a highly complex, multi-component system of water, air, rock strata and living materials. The interactions among these elements determine the properties of the soil. Soil is classified into types, orders, groups and sub-groups. It is reported that 90% of soils are low to medium in nitrogen (N), 80% in phosphorus (P) and 50% in potassium (K). Incidences of micronutrient deficiencies are also increasing.

Soil health from the agricultural point of view refers to the capacity of the soil to ensure proper physical, chemical and biological activities. Effective management practices are required for sustaining higher crop productivity. Good soil would ensure proper retention and release of water and nutrients. It would also aid in promoting and sustaining root growth. It also resists degradation.

The major threats to soil quality come from

- loss of organic carbon,
- erosion,
- nutrient imbalance,
- compaction,
- salinization,
- water-logging,
- decline in soil bio-diversity, urbanization,
- contamination with heavy metals and pesticides, and
- the adverse impacts of climate changes.

A number of programmes have been initiated at the national level to counter the ill effects on soil. All of these programmes have been started by the government at the central and local levels. Primarily, these programmes are aimed at imparting knowledge to the farmers. Some of these initiatives are

- the use of chemical fertilizers along with organic manures based on soil test outcomes, and fertility maps based on crops and other practices;
- providing soil health cards, which are farm level repositories of information;
- encouraging integrated farming practices balancing varied factors;
- corrective actions in locations where soil is affected by salinity or leaching of salts;
- deploying technology like GPS to collect data, analysis and decision making. The use of this technology should also be increased by the government.

8.3.1.3 Water and Irrigation

Water shortage is a serious issue in India, and would be compounded in the decades to come. There are spatial variations because of the wide spread of the country. There is a temporal variation because large tracts of land are dependent on the monsoon, which fluctuates between a good one and a bad one year on year. Water position and usage is also affected by high evaporation levels and climate changes. On top of all these factors is diversion, or competing demands for water, which affects agribusiness.

To handle such precarious water availability position, extensive knowledge and information practically applied to solve specific area- and factor-wise problems are required. Specifically, for example,

- More rigorous understanding and integration of the biophysical and socio-economic processes involved in water resources generation and agricultural production is required. These aspects include climatic, hydrologic, hydraulic and soil-water processes, ground recharge patterns, crop growth patterns, and economic processes at the farm, regional and global levels.
- Extensive data, information and knowledge systems are required. These have to be created across regions and times and periods. They have to be analysed by policy and decision makers in various networks.
- Improved understanding of system behaviour needs to be accompanied by efficient methods of water control and execution in the conveyance and field application areas.
- Importance should be given to development of recharge measures and prudent use of efficient methods like micro irrigation.

Irrigation efficiency is at present 35% in the case of surface water systems and 65% in the case of ground water systems. In principle, there is scope to increase the former to 60% and the latter to 75%. Even a rise by 5% can impact about 10–15 million hectares. Though new irrigations systems like drip, trickle, micro jets and sprayers have been attempted in India, overall adoption has been very minimal.

8.3.1.4 Weather and Climate

Weather and climate variability, covering the quantity and temporal and spatial distribution of rainfall during June to September, are very important determinants of agricultural performance during the *kharif* season. The key aspects comprise quantitative forecasts for seven weather parameters, namely, rainfall, maximum temperature, minimum temperature, wind speed, wind direction, relative humidity and cloudiness. The risks to the Indian agriculture sector due to climatic variability and extreme events would be felt at the basic levels of crops, livestock, farms, cropping systems and food systems. Needless to say, failure to adopt risk mitigation strategies would lead to issues like hyper food inflation and engulf food security issues.

Therefore, it becomes imperative that Indian agriculture is transformed into a climate-resilient production system through suitable adaptation and mitigation measures in the domains of both crops and animal husbandry. The measures would include research and development in crop management, development of technology for agribusiness activities, adoption of best practices in farm and processing, creation of physical infrastructure, allocation of financial support through appropriate institutional framework and so on. The focus should be on developing integrated agriculture development in rain-fed areas as well as on developing dry-land agriculture. The ultimate goal would be to ensure sustainable agricultural development.

This would mean that the focus of agricultural investments must be on irrigation development, research and commercialization of crop varieties that could resist drought, development of information technology using satellite and communication systems and development of a pool of scientists, researchers, statisticians and economists in agribusiness.

Some of the risk mitigation strategies are discussed below.

- Arrangements should be made to appropriately replace the *rabi* and *kharif* crops with each other in case of low yield or disasters. This strategy would enable loss to be minimized and impact nullified. This may include irrigation facilities as a primary requirement.
- The climate risk information system should be strengthened. This should be done not only at the information gathering and analysis levels, but also at the information dispensing level. This would require building up the capacity of climate assessment agencies.
- Better water management approaches like water budgeting, water conservation, artificial recharging of water, development of community ponds, and so on.
- To achieve targets in agricultural production, priority investments would be required in building up the climate resilience of vulnerable rural communities with a portfolio of adaptation options that can address climate risks.

8.3.1.5   Farm Inputs and Management

The key inputs that change the complexion of agriculture include high-yielding variety (HYV) seeds, chemical fertilizers, irrigation, pesticides, farm machinery and equipment, credit and labour.

8.3.1.5.1   Fertilizers

"Fertilizer accounts for a significant portion of total operating costs, averaging around 30% for wheat production, 20% for sorghum and up to 40% for intensive corn production" (Kenkel, 2010). If fertilizer price volatility could be completely eliminated, the results would eliminate yield variability, as farming would consistently use fertilizers. In such a situation, this leads to overall risk reduction by managing input costs and output productivity. In the case of developed nations like the United States, "fertilizer prices have tended to move somewhat in tandem with crop prices. The scenario of low crop prices coupled with high fertilizer prices would stress farm profits.

Farms which have higher than average fertilizer usage or a limited ability to rotate to less fertilizer-intensive crops would be particularly vulnerable" (Kenkel, 2010).

The impact of a slowdown in the usage of fertilizer on the growth of agricultural output can be seen from the data presented in Table 8.2.

"After 1991–92, the growth rate in fertilizer consumption turned out to be only a little more than one-third of what it was between 1966–67 and 1991–92, and the growth rate in the crop sector declined to less than half in the corresponding periods. The growth rate in fertilizer consumption declined further after 1998–99, while the growth rate in output of the crop sector has declined, but this may not be significant" (Chand and Pandey, 2008).

Subsidies given by the central government on various fertilizers increased from ₹891 crore during the early 1980s to ₹22,452 crore during 2006–2007. Though these numbers are on current prices, a large portion of the increase could be due to inflation.

At present, the greatest risk in subsidy allocation is the uneven distribution of subsidies. The present standard prescribes distribution of fertilizer subsidy according to the area under cultivation in any state. This does not take into account variation in productivity. Hence, this causes imbalance in fertilizer distribution and creates a void, further hampering productivity in states that require subsidies the most.

Fertilizer pricing policy has been a potential risk element, and has been adversely impacting the fertilizer industry, which is a key agribusiness input sector. According to A. Velliyan, President, Fertilizer Association of India, prices of urea in India were cheaper than those in other major agrarian countries. In the United States, urea is priced at around $526 per tonne; in the Philippines, it is $577 a tonne; in China, it costs $295 and in Pakistan, $266 a tonne. In India, however, it is priced at $96 a tonne. FAI claimed that nearly 5–6 million tonnes of urea was being illegally diverted for non-farm use and also being smuggled to Nepal and Bangladesh due to the lack of a rational pricing policy between the prices of

Table 8.2   Growth Rate of Fertilizer versus Growth Rate of Crop

Period	Rate Fertilizer (% per year)	Growth Rate Crop (% per year)	NPK/Ha
1950–1951 to 1966–1967	17.7	2.4	0.5–7.0
1966–1967 to 1991–1992	9.2	2.8	7–70
1991–1992 to 2006–2007	3.4	1.3	70–113
1998–1999 to 2006–2007	2.6	1.1	86.9–113

Source: Chand, R. and L.M. Pandey, Fertiliser growth, imbalances and subsidies: Trends and implications. New Delhi: NCAP, Apr. 2008.

urea and those of phosphatic and potassic (P&K) fertilizers like di-ammonium phosphate (DAP) and muriate of potash (MoP) (*The Hindu,* 2012). This increases the production risk and the subsidy burden, and thereby, the financial risk.

Fertilizer price movements have not been consistent with the prices of crops. This means that the fertilizer business has not seen any increase in profitability, at least, not at par with the agricultural sector. This shows that the business is not lucrative, and has suffered a serious production bottleneck, as can be seen from Table 8.3.

Hence, it is important that prices of fertilizers increase in suitable proportions with increases in food grain prices to maintain parity. This will ensure that the real prices of fertilizer continue to decline; it is the real prices of fertilizer that determine fertilizer use. The prices of inputs that go into the production of fertilizer have increased sharply, and international prices of all kinds of fertilizers have also witnessed sharp increases in recent years. In this kind of situation, if the increase in cost of production is not passed on to the consumer prices, it is bound to increase the subsidies. If subsidies do not compensate for the entire increase in cost of production, it would adversely affect the incentives to fertilizer producers to expand production. India's fertilizer industry seems to be facing a vicious cycle of low profitability/high subsidy; low economic drive for domestic fertilizer production; and the poor economic position of farmers unable to encourage market economics in the fertilizer business.

Unless this situation is resolved by taking the fertilizer sector closer to market economy, development of indigenous production facilities could be difficult. This would result in a movement away from self-reliance and also from isolation from international fluctuations.

Table 8.3 Production, Imports and Consumption of Fertilizer in India (Lakh Tonnes)

Year	Production	Imports	Consumption
1981–1982	41	20	61
1985–1986	58	34	85
1990–1991	90	28	125
1995–1996	113	40	137
2000–2001	147	21	167
2005–2006	156	53	203
2009–2010	162	91	265

Source: Ministry of Agriculture, *State of Indian Agriculture 2011–12.* Department of Agriculture and Cooperation, 2012.

8.3.1.5.2 Farm Machinery and Equipment

The productivity of any farm depends on the use of modern technology. Farm mechanization saves time and labour, cuts down crop production costs in the long run, reduces post-harvest losses and boosts crop output and farm income. There is a likelihood of a strong correlation between farm mechanization and agricultural productivity, as has been observed in modern farming across the globe. In India, there has been a popular belief that states like Punjab, which adopted modern farm mechanization measures, have reaped higher productivity.

Thus, a low level of mechanization poses a threat to productivity. This matter gains importance as the net sown area in India has remained constant over the past four decades. Hence, any improvement in productivity will come from improvement in inputs, processes and outputs. The inputs, namely raw materials, are given sufficient importance. The output factors like market development and prices mechanisms are also important. The process part is looked at only from the perspective of the key ingredients. The physical methods are rarely looked into. The importance of physical equipment is limited by the financial implications.

Some of the key benefits of using mechanization are discussed now.

■ The "threat to natural resources, notably land and water, has further necessitated a switching over to machine-assisted resource-conservation techniques, such as zero tillage, raised-bed planting, precision farming, drip or sprinkler irrigation, etc.
■ Mechanization allows some mitigating the effects of climate change by allowing the readjusting of crop sowing schedules.
■ A greater degree of farm mechanization can also address the issues of scarcity of farm labour during the peak agricultural seasons of sowing and harvesting, even with the implementation of the Mahatma Gandhi National Rural Employment Guarantee Act (MGNREGA)" (Ministry of Agriculture, 2012).

8.3.1.5.3 Mechanization Process

Farm mechanization has become a dire necessity in view of the significant reduction in the supply of labour for agricultural operations in recent years. With continued shrinkage in average farm size, more and more farms will fall into the adversely affected category, thereby making individual ownership of agricultural machinery progressively more uneconomical. The higher risk due to "uncertain demand" and "immature markets" has barred even seasoned business entities from entering this segment. Non-availability of the much needed technical and institutional support, as well as of after sales infrastructures, has further prevented this segment from developing and maturing.

Risk mitigation strategies include

- increasing the reach of farm mechanization to small and marginal farmers and to regions where availability of farm power is low;
- offsetting adverse "economies of scale" and "higher cost of ownership" of high value farm equipment by promoting "custom hiring centres" for agricultural machinery;
- passing on the benefit of hi-tech, high value and highly productive agricultural machinery to farmers through creation of hubs for such farm equipment;
- promoting farm mechanization by creating awareness among stakeholders through demonstration and capacity building activities;
- ensuring quality control of newly developed agricultural machinery through performance evaluation and certification at designated testing centres located all over the country.

8.3.2 Prices and Markets

Food and agricultural commodity prices in India are primarily determined by domestic demand and supply factors influenced by domestic price policies. The nature of markets catering to agricultural commodities and the imperfections in these markets also influence price transmissions and final consumer prices. India meets a bulk of its large food demand through domestic production, barring a few commodities like edible oils and pulses.

In India, domestic demand and supply are the key determinants of agricultural commodity prices, as trade in agricultural products is especially unlikely to dampen prices or stimulate demand. India imports significant quantities of pulses and edible oils. They are sensitive, and prices are volatile. It is not just these two products, but also all other agro-based commodities that suffer market imperfections, which do not benefit the growers, the processors or the customers. However, intermediaries are able to benefit from these imperfections in the commodity markets.

The rise in per capita income and its implications on food commodity prices need to be understood from the perspective of their impact on agribusiness. Per capita GDP has increased at an average rate of 6% per annum over the past five years. This means that the demand for products and commodities, especially for agro-based products, would go up. Production in the agricultural and allied sectors has to catch up with demand, or else prices cannot be controlled. A small decline in supplies can sharply affect the prices of these commodities.

In comparison to wholesale prices, retail prices in different consumption centres display divergent trends in different markets.

Such variations are due to differences in consumption patterns across regions in India. There are also supply and movement constraints which further aggravate such aberrations in prices. Added to these are infrastructural constraints, transportation costs and government levies and taxes, which could further limit or widen market imperfections.

8.3.2.1 Market and Price Risks Drivers

The agriculture market is highly affected by local changes. This sector is highly fragmented, and hence the market forces and their effects are not so widespread. Retail prices in different consumption centres display divergent trends. Some of the key aspects driving the uncertainties in the markets are discussed in the following paragraphs.

The regional differences in consumption patterns and supply condition the prices, and their movements vary across the major markets. Imperfect market conditions, supply chain issues such as restrictions on the movement of agricultural commodities due to infrastructural constraints, transport costs and local taxes, and so on, influence retail price trends across the major markets and consumption centres.

Differences in tastes and varieties consumed across the centres also pose problems for comparison of retail prices. In the case of fruits and vegetables, milk, egg, meat and fish, prices have gone up despite an increase in per capita availability. This is due to a changing pattern in increasing aggregate household demand for high value items—a result of increasing income levels.

Apart from demand side factors, one supply side factor, namely, constraints in gearing production and marketable surplus, also makes for an important factor in influencing price rise, both nationally and globally. One may note that supply side constraints have both long-term and short-term features. At times, due to weather or related factors, production may fall, which could have an impact for a season or longer—say, a year. Alternatively, the gap may be due to changes in consumer behaviour and demands, or changes in crop economics, leading to shifts that may require issues to be addressed on a long term. This leads to time lags between supply and demand, which in turn leads to a spiralling effect on prices.

8.4 Management

In agribusiness, risks can be across the supply chain at various nodes. Such risks could be impacting the actors like farmers, intermediaries, processors, marketers and consumers. Most part of the production risks are up to the level when the produce is grown at the farm or processed and moved down the supply chain. Here, production risk is more in terms of volume of output and other cost side economics with respect to production. Once the product moves towards the market, one would experience market related risks which predominantly exposes to price risks. Apart from this, there is another perspective of looking at ex-ante and ex-post risks which are based on trigger of events and responses.

In agricultural risk management, a distinction is necessary between measures that aim to create and foster the management of risk by markets (particularly insurance, savings, and formal lending) on an ex ante basis and the management of risks by the government (particularly emergency humanitarian relief, compensation for catastrophic events, and reconstruction of public goods) normally on an ex-post basis.

Facilitating the use of market-based approaches can reduce the needs and scope for government interventions, and thereby, decrease the costs incurred by government in ex- post coping activities. For this reason, many governments are active in the promotion of market-based risk management and insurance (although they are normally operated through public–private partnerships). Examples of such activity are government subsidies, information and extension, and legal and regulatory measures, including those for insurance.

However, when disasters strike or when there are losses that were not managed by the agricultural sector's stakeholders, government intervention will be necessary, partly because not all risks are insurable and partly because not all farmers or stakeholders can or want to access commercial insurance.

8.4.1 *Production*

Production risks affect volume and quality of output in agribusiness. These impacts financially at the micro level economic units namely farms, farmers, processors and even consumers. There are a number of initiatives taken by policy makers and system to manage such risks. Some of the important initiatives which are in vogue are discussed here.

8.4.1.1 *Financial Inputs and Calamity Relief: Insurance*

Given the sensitivity of yields and production to weather conditions and other hazards, there is a potential demand for crop insurance. Crop insurance has been successfully used to protect farmers from risk in many countries. There are instances where government support for insurance being higher compared to fair play of insurance market, which is a service business to trade and commerce. Since the cost of managing insurance is high, it is also difficult to expect private operators to successfully provide support through insurance. In developing nations like India, funding costs are high, as is the incidence of failures, and further, avenues for managing insurance investment portfolios are less. Obviously, all these factors make insurance unattractive for fund managers to manage. Overshadowing these factors, information imperfections make the insurance operation position weak, demanding high expertise in operation.

There are single peril insurance contract and multiple peril insurance contract options available to the farmer. However, owing to the high information asymmetry, the premium for such contracts is very high. The premium amount increases with the level of indemnity, that is, single peril contracts are cheaper than multiple peril contracts. The "high costs of offering insurance contracts are associated, at least in part, with information asymmetries. Moral hazard in this context occurs when it is impossible or excessively costly to write a contract based upon everything a farmer might do that would affect his yields. Adverse selection occurs when contracts based on all the relevant environmental parameters are unfeasible" (OECD, 2010). These two drawbacks of the insurance contract severely affect the feasibility and even availability of insurance contracts. The insurance contracts types most used, and the implications of each, are discussed below.

8.4.1.1.1 Area Yield–Based Crop Insurance

Area yield insurance provides support based on the average yield of a wide area. It also considers potentially reducing adverse selection. It may be noted that area yield–based insurance is done at the cost of adding basis risk to be borne by the farmer.

Another fundamental issue is that the level survey and preliminary study are done on a global scale. Such a study focuses on the macro level and does not take the micro aspects into consideration. It is not possible for such a mass level study to include minor risks. This hampers the risks coverage borne by such contracts. Thus, such an arrangement does not indemnify the farmer against all risks.

Further, the information available about the risks associated with farm products is minimal. This issue is not paid much attention. The lack of study about the perils and their probabilities makes it impossible to calculate the financial costs related to such events.

8.4.1.1.2 Strategies to Improve the Insurance Contracts

A primary classing of risks into different layers is imperative in order to understand the various risks and devise suitable insurance contracts. These layers segregate the risks according to the likelihood of their occurrence (probability) and the severity or magnitude of loss caused by such occurrence.

Table 8.4 shows the magnitude of damage with respect to expected income vs. probability of occurrence.

The retention layer shows that the expected losses or gains are limited in this region. They have a high occurrence rate, but are less severe. Hence, farmers can manage these risks with strategies available at the farm/household level.

The second layer is of "risks that are more significant and less frequent. Both frequency and magnitude are in the middle ranges. In this layer, there is scope for

Table 8.4 Magnitude of Damage and Impact on Expected Income

Layer	Probability of Occurrence	Realizable Amount of Expected Income	Remarks
Retention	0.1–3	58% to –20%	High occurrence rate but less severe
Market insurance	0.025–0.1	–20% to –45%	Farmers use market specific instruments like insurance
Catastrophic	0.01–0.025	–45% to 53%	Very large losses

Source: Antón, J. and C. Moreddu, *Non-Distorting Farm Support to Enhance Global Food Production.* Rome: FAO, 2009, pp. 187–207.

farmers to use additional specific market instruments, such as insurance or options designed to deal with farming risk. This is the market insurance layer.

The third layer is of risks that are catastrophic in nature because they generate very large losses, even if their frequency is low. This type of risk is more difficult to share or pool through any kind of market mechanism, particularly if it is systemic. This is the catastrophic risk or the market failure layer" (OECD, 2010).

The insurance or market layer may have different types of instruments for different types of risks. The insurance could be for more independent hail or frost risk, which covers production loss to futures and options for covering price risks. In between the spectrum, there are hybrid options. These could be to cover crop yields or revenue from operations. Apart from these, cooperatives provide risk pooling by joint action for protecting against risk. One may note that in India like in many other countries, cooperatives exist not only at the production end, but also at the consumers' end, supported by consumer cooperatives.

Finally, the normal risk layer, to be retained by the farmer, can typically be managed through normally available instruments in the public finance system, such as taxes, duties and subsidies that may have general or agriculture-specific income smoothing properties. Farmers must be encouraged to use savings and investments opportunities available through such mechanisms. Unfortunately, in India large agribusiness players are vulnerable, just like small and medium farmers or small and medium traders in the processing sector.

There are other strategies that could be considered:

■ One strategy is to bring the insurance unit down to the village level so as to minimize the base risk.

- Another is to use a longer time yield series when fixing the guaranteed yield, to ensure more stable coverage.
- A third strategy is to increase the levels of indemnity (coverage).
- Yet another strategy is to cover horticultural crops such as vegetables and fruits.

8.4.1.1.3 Weather-Based Crop Insurance

"Monsoon-linked insurance indices can provide adequate collateral in terms of encouraging banks to lend more liberally to the agriculture sector. Weather indices can be used to make early payouts in areas covered by yield crop insurance. A macro weather index can be an ideal tool for protecting a large portfolio at the district/regional/state levels against drought or floods. The success of a weather insurance programme in India would depend on the product design, the steps taken to minimize the base risk, the adoption of reliable and sustainable pricing mechanisms (including government subsidies) and the resolution of the issues of product servicing and timely payouts" (Kolli and Bockel, 2008).

8.4.1.2 Post-Harvest Management and Value Addition

The post-harvest handling and processing of agriculture products ensure high-quality products. This, along with increase in productivity, is essential to the agriculture markets. Too often, even when the yields are high, producers lose income due to poor post-harvest practices.

Food processing converts agricultural produces into more palatable and nutritious food stuff. Such activities also extend the shelf life of the products, and add value in various other ways. Food processing assumes further importance in India due to seasonal variations, which lead to high levels of wastage or shortages. One of the reasons is the lack of adequate measures to preserve and store the foods. The important aspect in food processing is that it covers all the steps, right from the farm to the fork. These steps include basic cleaning, grading and packaging, as in the case of fruits and vegetables, and also alteration of the raw material to a stage just before the final preparation. All the ready-to-eat food segments like bakery products, instant foods, flavoured and health drinks, etc., form part of the food processing sector. Such proliferated economic agents in the food processing industry can offer livelihood for many workers. Incidentally, they offer wide strata of communities several options at reasonable costs, as they can achieve low costs with the proper focus. In addition, these entities can play a primary role in the development of the rural community. Food processing has come a long way in the last few decades. The ever-changing lifestyles, food habits and tastes of customers globally have altered the dynamics of the industry. The world's food production and consumption patterns are evolving with changes in the needs of the customers. Recent corporate diversifications into the food industry and the new marketing strategies adopted by companies have opened new opportunities for this sector.

Food processing can benefit the agriculture sector in the following ways:

■ Farmers get a higher yield, better revenues and drastically lowered risks.
■ Consumers have access to new products, a greater variety and better prices.
■ The economy gets benefitted with new business opportunities for entrepreneurs, which means employment for a bigger workforce.

The most important challenge in this sector is the problem of wastage. This is a widespread problem and is aggravated at each level. The quantum of wastage multiplies at each level in the value chain. According to one estimate, the average level of wastage in the value chain alone amounts to six percentage points. Some of the important points of wastage are mentioned in Table 8.5.

It may be noted that avoiding wastages at every level and improving value addition is the key challenge. We now discuss some measures to improve the condition of the post-harvest processing industry and to alleviate the risk levels.

■ Stress should be laid on the production of and value addition to high-value commodities such as fruits and vegetables.
■ There should be strong marketing support in deciding upon the value addition process. This is due to the fact that marketing is more remunerative than the value addition process itself.

Table 8.5   Wastages at Different Levels

Farmer/village level	Crop damage Improper harvesting techniques Poor handling, storing and packaging Poor transportation Moisture loss
Wholesaler level	Poor handling Poor transportation Multiple handling Storage Grading and sorting Moisture loss
Retailer	Poor handling Poor transportation Handling by customers Moisture loss

Source: Based on interaction with agriculture experts.

- There is also a lack of appropriate and cost-effective packaging technologies. This also adds to the wastage of raw material during processing.
- The focus of processing activities should be pull driven, that is, demand driven, rather than push driven, that is, production driven, as has been traditional.
- Intermediaries in the food chain lock value and add to the cost of the raw materials, sometimes by as much as 80%–100%. Effective linkages need to be built between Indian agriculture farmers and processors on a mutually beneficial contractual agreement.
- A comprehensive analysis of the present situation is needed. Such a study can delineate effective strategies for network plans, warehouse locations, and so on. A complete study of supply chains at the regional level will produce a detailed plan of action.
- The processing of agricultural products results in by-products that are generally termed waste. However, with the evolution of new technology and methods, this so-called waste can be processed and utilized in an eco-friendly manner.

8.4.2   Prices and Markets

Demand and supply side risks impact agricultural products prices. This could further impact the markets in terms level of maturity and conditions towards fairness. There are a number of initiatives taken by policy makers to manage such risks. Some of the important price related policy and market changes are discussed here.

8.4.2.1   Government Policy Instruments

8.4.2.1.1   Minimum Support Prices (MSP)

"The government has pursued a policy of ensuring a remunerative and stable price environment for so long as a result of many political, social and economic compulsions. Specific instruments include the minimum support prices (MSP) scheme and the market intervention schemes (MIS). MSP is available for as many as 25 different agricultural commodities, accounting for 80% of the production of the gross farmed area and approximately 75% of the gross value of crop production.

MSP has been largely successful in keeping prices under control from the consumer point of view. On the other hand, the utility to producers has long been questioned, as only farmers from a few states (for only a few commodities) have been able to take advantage of MSP. Causes of worry include the extent of government funds involved in the MSP-based procurement operations; storage wastage; and inefficiencies in handling operations. While the Government wants to make MSP universal, many economists feel that the Government's procurement of food grains should be limited to its requirement for buffer stocks and welfare schemes like public distribution systems" (Kolli and Bockel, 2008).

The elements of rationale behind the determination of MSP include:

1. the need to provide incentives to producers and farmers for adopting improved technology and for developing a production pattern broadly in the light of national requirements;
2. the need to ensure rational utilization of land, water and other production resources;
3. the likely effect of the price policy on the rest of the economy, particularly in terms of the cost of living, level of wages, and so on; and
4. the terms of trade between the agricultural sector and the non-agricultural sector.

However, the challenge in MSP lies in reducing the wastage of products and in efficient storage and transportation systems.

8.4.2.1.2 Market Intervention Scheme (MIS)

Though the MSP is useful and effective for certain food grains, it cannot be applied across all types of crops. The Market Intervention Scheme (MIS) is a support programme for procurement of horticultural commodities that are perishable in nature and applies to commodities not covered under MSP. The objective is similar to that of MSP except that MIS is triggered when there is a bumper crop, and the focus is on avoiding distress sales by farmers.

The criteria for implementation of MIS are discussed below.

1. The MIS is implemented when there is at least a "10% increase in production or a 10% decrease in the ruling market prices over the previous normal year.
2. The MIS is implemented at the request of a state/UT government which is ready to bear 50% of the loss (25% in the case of the north-eastern states), if any, incurred on its implementation" (Ministry of Agriculture, 2012).

8.4.2.1.3 Price Supports Scheme (PSS)

The Department of Agriculture and Cooperation implements PSS for procurement of oil seeds, pulses and cotton through National Agricultural Cooperative Marketing Federation of India Limited (NAFED), which is the central nodal agency, at the MSP declared by the government. NAFED undertakes procurement of oil seeds, pulses and cotton under the PSS as and when prices fall below the MSP. This is typically done to support growers who reach the markets in a short time window in the post-harvest period. It continues to procure under PSS till prices stabilize at or above the MSP, so that farmers are ensured of a certain assured return.

However, these are government schemes that are supportive in nature. The aim is to prevent huge losses to farmers. This has been adopted to help upliftment of marginalized farmers. They in principle violate the free market economy principle.

8.4.2.2 Electronic Spot Exchanges

"The present marketing system, as governed by the Agricultural Produce Marketing Committee (APMC) Act, is replete with problems, such as a non-transparent auction system; poor incentives for quality-consciousness; multiple layers of intermediaries; poor infrastructure; informal credit linkages; significantly less buyers vis-à-vis sellers, etc. All these factors contribute to creating disadvantaged transaction terms for primary producers. As a result, the need for a near-perfect physical market network in India is becoming increasingly acute with the emergence of national-level futures exchanges. Over the last few years, the former is being a prerequisite for the effective functioning of the latter. The objectives of an electronic spot market include transparency of physical markets; better links to quality parameters; improved information availability across players' groups; reduced wastage through the creation of better infrastructure; value-added a long the supply chain; and better price leads for futures markets.

The electronic spot exchange market is highly scalable because of its standardized operation, high level of technology orientation and the potential availability of private investments. India is piloting this system though SAFAL National Exchange (SNX) in Bangalore. The model presents an electronic platform that can be directly accessed by authorized brokers. Both buyers and sellers transact through these brokers. The presence of an electronic exchange absorbs counter-party risks and ensures the open dissemination of prevailing price levels. In addition, the auction system is technology-driven, where the parties involved enjoy complete anonymity. The scalability of this model is relatively easier because it attracts authorized brokers with investments to extend the technical infrastructure. Electronic spot exchanges are here to stay and provide huge marketing support to farmers" (Kolli and Bockel, 2008).

8.4.2.3 Commodity Markets

Across supply chains in agribusiness, farmers, producers, processors, traders or even consumers are concerned with commodity prices that disproportionately benefit and/or hamper only a few partners in their networks. For example, farmers are the most affected especially when they trade in regulated environments or sell through private traders. One of the ways to resolve imperfections of market affecting commodity prices is enabling fair operations of open markets, especially of commodity markets. With future markets in place, partners in supply chain are able to shift risk based on perceptions of participating members and on better reach to market information on demand and supply and prices. Further, futures prices, as well as influencing spot prices, are influenced by information availability. Such economic activity is a rather welcome development in commodities markets.

The price risk is the risk associated with sharp fluctuations in prices. These fluctuations could be highly uneven and adverse. Agricultural products are seasonal in nature. Due to this seasonality, the supply–demand balance is highly disturbed during harvest seasons. Forward and futures contracts are efficient risk management tools that insulate buyers and sellers from unexpected changes in future price movements. The futures market provides the lead in economic activity. Futures prices give necessary indications to producers and consumers about the likely direction of future spot prices. They also give an idea of physical volumes in terms of the demand and supply conditions of the commodity traded. However, one should also note that the cash market, or the spot delivery market, on the other hand, is a time-tested market system, especially in agriculture, as farmers and processors, who have low holding capacities, have the tendency to deliver and cash in early.

There are 15 futures markets operational in India. For a variety of reasons, however, these markets are not operating efficiently. The reasons include operational and management issues to infrastructure and trader issues. Future options are still in the developmental stage in India, and need some time to mature.

8.4.2.4 Contract Farming

"Contract farming is a structural issue in terms of agriculture production and marketing. Its chief aim is to bring agricultural management in line with the best agricultural production practices while ensuring a competitive and pre-arranged price for farmers. Thus, contract farming can play an important role as an alternative risk management instrument. While India is catching up quickly with the contract farming model, the requisite developmental and regulatory framework is not yet in place" (Kolli and Bockel, 2008). At present, most contract farming contracts are loose and heavily loaded against the farmers. With the creation of an environment suitable for private–public partnerships, clear defining of roles, and given the limitations of corporate farming, one would expect contract farming to be a useful risk mitigation tool for small and marginal farmers.

From the processing side, risks can be business or operating risks and financial risks. Operating risk is measured by the variability in earnings before interest, depreciation, amortization and taxes. In agribusiness, variability is high, as prices are volatile and input costs, like the costs of procuring farm products, are also volatile. This is where contract farming helps. Financial risk is measured by the debt-to-equity ratio, which means that the higher the debt, the higher the financial risk. In the agribusiness sector, financial risk is also high because of the need to borrow, especially for working capital. Contract farming helps handle risks in agribusiness.

8.5 Conclusion

In this chapter, we present a holistic taxonomy and the implications of risks in agriculture supply chains. It takes into account the design, aggregate planning, policy framework, long-term strategy and the latent facts of agriculture supply chains. Supply chain management at the global level is of critical importance. Today, the issues are tackled at tactical or regional levels. Modern-day studies, and therefore, the models they work on, look at only simple tools to take relevant decisions at this level. They fail to take complex issues into consideration, due to their strong assumptions about isolation and *ceteris peribus*. Existing problems are not considered at a complex global level.

The brief classification of risks is the first step towards comprehensive risk management. Appendix 8.2 provides indicative lists of potential risks and their management which are discussed in this chapter. It can present a holistic and integrated picture of the complex issues. The model for future agriculture supply chain management should also capture the stochastic and shelf life features of agricultural products. The future of any supply chain in the present context depends on energy management as well. The agriculture sector is energy intensive, and thus, has to take into account its energy usage while working on the policy framework.

Logistics and SCM have emerged as areas of critical importance for the energetic utilization of waste biomass and organic substrates. Unfortunately, the existing models address only a minor subset of the decisions needed to be taken at a strategic, tactical and operational level; moreover, they fail to capture the existing complex and stochastic issues due to their severely limiting steady-state assumptions. Moreover, the problem becomes even more challenging when considering the numerous variables, parameters and constraints that could be taken into account in the formulation of such a decision support model.

The use of modern technology has to be increased. Given the static nature of the net sown area over the last four decades, it is clear that any improvement in productivity can be brought about only by the better usage of available resources. Available satellite data relating to weather news, long-term and short-term weather forecasts, production information, market prices and policy developments pertaining to agriculture, apart from the number of advisory services in the public or the private domain that disseminate information should be utilized adequately. New initiatives are required to develop data and knowledge bases, as well as knowledge delivery mechanisms in different sectors of agriculture, such as water management, soil management, plant protection, animal and fish health cover, protected cultivation in high value agriculture, genome analysis for genetic transformation in plants and animals, and so on.

Evidence of a large quantum of waste is abundant in the agriculture supply chain. This waste presents an easy and quick opportunity to improve the performance of the supply chain. Adequate government expenditure on building up infrastructure, improving delivery networks, extending the reach of technology, enabling closely knit supply chains and putting into place better information systems can help in reducing the wastage of products.

Improved farm management and practices have also enabled a year-wide sowing season. In this context, the *zaid-kharif* and *zaid-rabi* gain prominence (in between the *rabi* and the *kharif* seasons, there is a short season during the summer months known as the *zaid* season). These crops can be developed so as to improve productivity. Mapping of crop rotations and cropping patterns and geospatial modelling to assess performance have to be initiated at the national level. The output of this activity is linked with crop diversification and intensification planning under the National Food Security programme.

Before the consumer gets the food from the farmer, it goes through many hands. Traders buy and ship produce and commission agents arrange the transactions between farmers and traders. Since a typical farmer owns and sows only a small portion of land, the middleman has the advantage in terms of information and power. This leads to multiple marginalization of the supply chain and reduces its performance.

Appropriate studies and local adaptability are also critical. It is clear that food policy design and approaches to managing food sector risks will vary widely, depending on each country's context; thus, different strategies have to be adopted in different regions. The reform agenda in India could also play an important role. This refers to the entry and use of the private sector in the agricultural sector.

Discrete risk management strategies at the farm and global levels will have only limited effects. "A single source of risk and a single instrument considered in isolation from other relevant sources of risk may induce unintended results in terms of revenue variability and welfare" (OECD, 2010). The produces and output from the farm should be correlated to other factors such as production variables and others to get a holistic model. At present, the cost of risk avoidance is very high. Therefore, finding ways to extend the reach of the risk market is important.

Hence, it is far more important to look at agribusiness and risks in a holistic sense. This can improve performance at the farm level and ensure food security. The policies should take multiple factors into account before devising any instrument.

Appendix 8.1
Agricultural Land Use in India

Sr. No.	Classification	1950–1951	1960–1961	1970–1971	1980–1981	1990–1991	1999–2000	2007–2008 (P)	2008–2009 (P)	2009–2010 (P)
I.	Geographical area	328.73	328.73	328.73	328.73	328.73	328.73	328.73	328.73	328.73
II.	Reporting area for land utilization statistics (1–5)	284.32	298.46	303.75	304.16	304.86	305.02	305.61	305.59	305.61
	1. Forest	40.48	54.05	63.83	67.46	67.81	69.16	70.02	70.03	70.04
	%	14.24	18.11	21.01	22.18	22.24	22.68	22.91	22.92	22.92
	2. Not available for cultivation (A+B)	47.52	50.75	44.61	39.55	40.48	41.13	42.7	42.86	42.95
	(A) Area under non-agricultural uses	9.36	14.84	16.48	19.6	21.09	23.6	25.71	26.06	26.17
	%	3.29	4.97	5.42	6.44	6.92	7.74	8.41	8.53	8.56
	(B) Barren & un-culturable Land	38.16	35.91	28.13	19.96	19.39	17.54	16.99	16.8	16.78
	%	13.42	12.03	9.26	6.56	6.36	5.75	5.56	5.5	5.49
	3. Other uncultivated land excluding fallow	49.45	37.64	35.13	32.31	30.22	28.31	26.67	26.28	26.36

(A) Permanent pasture & other grazing land	6.68	13.97	13.26	11.99	11.4	10.85	10.2	10.18	10.15
%	2.35	4.68	4.37	3.94	3.74	3.56	3.34	3.33	3.32
(B) Land under miscellaneous tree crops & groves not included in net area sown	19.83	4.46	4.37	3.58	3.82	3.73	3.41	3.36	3.35
%	6.97	1.49	1.44	1.18	1.25	1.22	1.12	1.1	1.1
(C) Culturable waste land	22.94	19.21	17.5	16.74	15	13.74	13.06	12.75	12.86
%	8.07	6.44	5.76	5.51	4.92	4.51	4.27	4.17	4.21
4. Fallow lands (A+B)	28.12	22.82	19.33	24.55	23.37	25.34	24.84	24.48	26.24
(A) Fallow lands other than current fallows	17.45	11.18	8.73	9.72	9.66	10.29	10.33	10.29	10.48
%	6.14	3.75	2.87	3.2	3.17	3.37	3.38	3.37	3.43

(Continued)

Appendix 8.1 (Continued)
Agricultural Land Use in India

Sr. No.	Classification	1950–1951	1960–1961	1970–1971	1980–1981	1990–1991	1999–2000	2007–2008 (P)	2008–2009 (P)	2009–2010 (P)
	(B) Current fallows	10.68	11.64	10.6	14.83	13.7	15.05	14.51	14.19	15.75
	%	3.76	3.9	3.49	4.88	4.49	4.94	4.75	4.64	5.15
	5. Net area sown (6–7)	118.75	133.2	140.86	140.29	143	141.06	141.38	141.93	140.02
	%	41.77	44.63	46.37	46.12	46.91	46.25	46.26	46.44	45.82
	6. Total cropped area (gross cropped area)	131.89	152.77	165.79	172.63	185.74	188.4	195.14	195.36	192.2
	7. Area sown more than once	13.15	19.57	24.93	34.63	42.74	47.33	53.76	53.43	52.18
	8. Cropping intensity[a]	111.07	114.69	117.7	123.05	129.89	133.55	138.03	137.64	137.26
III.	Net irrigated area	20.85	24.66	31.1	38.72	48.02	57.53	63.29	63.74	63.26
IV.	Gross irrigated area	22.56	27.98	38.2	49.78	63.2	79.22	87.98	88.87	86.42

Source: Ministry of Agriculture, *State of Indian Agriculture 2011–12.* Department of Agriculture and Cooperation, 2012.

Note: 1. Figures in parentheses indicate percentage to the § Reported Area. 2. The decline in net area sown in 2002–2003 reflects the impact of the severe drought of 2002–2003 on agriculture operations.

Abbreviation: P, Provisional.

[a] Cropping Intensity is percentage of the gross cropped area to the net area sown.

Appendix 8.2
Potential Risks and Their Management (Indicative)

Agribusiness Risks	Identify and Prioritize	Minimize, Monitor and Control Impact of Unexpected Events	
	Assessment and Evaluation	Management (Generic Strategies: Avoidance, Prevention, Reduction, Sharing, Retention)	Communication (Reach Out)
Market/Price	Vagaries of demand and supply. Price risk due to demand and supply or intermediary	Ability to capture demand and production forecast more accurately and reassess at frequent intervals. Government to facilitate with market operations through procurement and releases to manage price	Communication is mainly to all stakeholders with near real time data on arrivals, local/ regional and national trends of price movements
Production	Volume of production affected by weather conditions impacting through monsoon, excessive heat, cold waves, and floods and so on. There could be local and regional level factors like crop diseases, labour conditions and lack of capital	Weather-related risk cannot be avoided but prediction and post-damage revival are critical aspect. Government support in managing contingencies is critical and hence government departments need to be agile. On micro level risks, better field level management practices are important. These are more handled through a collective approach. These risks for primary, secondary and tertiary processors are far more higher, and the extent of impact could be severe	Communication on weather-related issues is through multiple media like radio, TV, websites, direct broadcasting through departments and university centres and so on. Government agencies play a critical role.

(Continued)

Appendix 8.2 (Continued)

Potential Risks and Their Management (Indicative)

Agribusiness Risks	Identify and Prioritize — Assessment and Evaluation	Minimize, Monitor and Control Impact of Unexpected Events — Management (Generic Strategies: Avoidance, Prevention, Reduction, Sharing, Retention)	Communication (Reach Out)
Financial	Inadequate flow of capital, interest rates and inflationary pressures, making cost management difficult; risks due to non-receipts or failure to honour commitment of procurement and so on	Estimation of fund requirement by financial institutions and ensuring adequate velocity of money is important. Interest costs are defined by credit policy and approach towards agricultural finance. Government of India leads the support efforts. Inflation is also handled, but some concerns are still there. With respect to payment risks, all participating entities are expected to do risk cover by designing individual and collective approaches.	Approaching experts for advice on managing at the micro level; decision makers are important, as many of these risks stem from macro environment forces. There are traditional approaches, and a number of covers are available. If costs are uncontrollable elements, then economic agents will have to see if it could be shared among network partners!
Institutional	A number of institutions which have a major stakes in agribusiness supply chain effectiveness. Risks emanate from the failure or inadequate performance of any such agent. Often, it is lack of timely action and/or inadequate support which triggers risk	Farmers and network partners to map the potential risks and ensure they are represented in appropriate forums. Supply chain partners must be in synchronization over such risk and share responsibility of handling the same.	Communication is through different media and forums; it is positive and with orientation to resolve rather than enhance risk triggers

Acknowledgement

The authors thank Shailesh Tiwari, PGPEx VLM, IIM Calcutta, 2013 who worked on an internship with N. Chandrasekaran on Agriculture Supply Chain-Ensuring Food Safety.

References

Antón, J. and C. Moreddu, Risk management in OECD agriculture: From a holistic approach to the reality of support measures. Ch 5 in *Non-Distorting Farm Support to Enhance Global Food Production*. Rome: Food and Agriculture Organization of the United Nations, 2009, pp. 187–207. ftp://ftp.fao.org/docrep/fao/012/i1098e/i1098e11.pdf

Chand, R. and L.M. Pandey, Fertiliser growth, imbalances and subsidies: Trends and implications. New Delhi: National Centre for Agricultural Economics and Policy Research, April 2008. http://www.ncap.res.in/upload_files/others/oth_13.pdf

The Hindu, FAI favours rational pricing policy, September 21, 2012. http://www.thehindu.com/business/companies/fai-favours-rational-pricing-policy/article3918936.ece

Kenkel, P. Managing fertilizer price risk. Oklahoma Cooperative Extension Fact Sheet AGEC-262. Stillwater, OK: Oklahoma State University, 2010.

Kolli, N. and L. Bockel, Risk management as a pillar in agriculture and food security policies: India case study. Rome: Food and Agriculture Organization of the United Nations, 2008. http://www.fao.org/docs/up/easypol/764/risk_mngmnt_pllrfs_policy_indiacs_209en.pdf

Ministry of Agriculture, *State of Indian Agriculture 2011–12*. Department of Agriculture and Cooperation, 2012. http://agricoop.nic.in/sia111213312.pdf

OECD. *Managing Risk in Agriculture: A Holistic Approach (Extracts)*. Paris: Organisation for Economic Co-operation and Development, 2010. http://www.oecd.org/tad/agricultural-policies/45558582.pdf

SUPPORT SYSTEMS

Chapter 9

Infrastructure

OBJECTIVE

The objective of this chapter is to understand the importance of infrastructure such as distribution centres, organized market, cold storages and communication networks in managing efficiency in agribusiness supply chains. Issues covered include transportation networks; quality of roads; cold storages; organized markets with storages, communication network and training of personnel.

9.1 Introduction

Infrastructure plays a significant role in the development of agriculture and industry, and in boosting overall economic growth. In a globalized economy, production facilities tend to converge on one location because of resource advantages, while the products are consumed at different locations, which are the demand points. It is the infrastructural facilities that play a major role in determining the success of firms, industries and the nation in managing production, distribution and exchange in markets, so that consumers and suppliers are rewarded for their roles and responsibilities in the economic system.

A country's progress is often measured by how fast and effectively it can develop its infrastructure to help industries and businesses thrive, and also to ease the lives of the common people. Infrastructural support facilities needed for greater competence include efficient roads, railways, airports and shipping systems (transportation); buildings, water pipelines, sewers, electric power lines (physical infrastructure); and high-class storage facilities such as cold storage, moisture proof storage and so on (support infrastructure). The need for such facilities assumes greater importance in the case of perishable goods such as agricultural produce and dairy products. Timely shipment of these goods under the mandated temperature and moisture conditions is necessary to protect their quality and usability. The limited shelf life of agri produce and its demand and price variability make agri supply chains more intricate and tougher to manage than most other supply chains. The strengthening of rural infrastructure in terms of quality roads, proper irrigation and adequate storage facilities, including post-harvesting facilities, results in higher production and lower cost. Furthermore, better infrastructure paves the way to market expansion and economies of scale, and promotes employment and income for the rural people. In this chapter, we discuss the importance of infrastructure in managing agribusiness supply chains. The focus would be on transportation networks, quality of roads and storage infrastructures.

9.2 Definition

The theory of infrastructure derives mainly from the public goods theory in economics. Infrastructure in a broad sense refers to basic physical and organizational structures and facilities (e.g., buildings, roads, water and power supplies) required for the function of the general public or enterprises. Infrastructure is mainly comprised of physical structures, material or tangible structures that aid the competitiveness of the business world and the functioning of economic activities. This understanding would extend to productive agriculture and agribusiness sectors. In this sense, agribusiness has a number of related organizational systems that support managerial functions such as planning, organizing, administering and monitoring of economic activities, which may include buying, producing, distributing and consumption of goods and services. According to a Food and Agriculture Organization of the United Nations (FAO) report, "there are five infrastructure categories: (i) farm to market roads, (ii) water for irrigation, (iii) wholesale markets and trading centres, (iv) agro-processing facilities and (v) information and communications technology" (FAO and ODI, 2008). According to the Reserve Bank of India (RBI), "the infrastructure sector is defined as (i) power, (ii) telecommunication, (iii) railways, (iv) roads, including bridges, (v) sea ports and airports, (vi) industrial parks, (vii) urban infrastructure (water supply, sanitation and sewage projects) and (viii) mining, exploration and refining and (ix) cold storage or cold room facilities, including those for farm level pre-cooling and those for preservation or storage of agricultural and allied produce, marine products and meat" (Gangadharan, 2010). Thus, infrastructure is the set of interrelated structural elements that facilitate the production of goods and services, and also the distribution of raw material and finished products. For instance, roads and other transportation systems enable the movement of raw materials to production sites and facilitate distribution of finished products to marketplaces.

9.3 Role

Among the categories of infrastructure discussed above, agricultural infrastructure plays an important role, especially in a developing country like India, where a greater part of the society depends on this sector for existence. Infrastructure plays a significant role in poverty reduction, economic growth and empowerment of the rural society. In spite of agriculture occupying an important role in India's economy, it has been suffering from major barriers which have slowed down its development. Problems have been existent at each stage of the agri supply chain. Issues in relation to agri infrastructure are one of the most prominent challenges that need to be addressed with immediate effect. As India is one of the fastest developing countries today, there is a continuous call for good quality infrastructure. Investments in rural infrastructure, primarily in transportation, quality roads, irrigation, electric power, agricultural markets, storage facilities and so on, will be required to accelerate the growth in agricultural production and increase effectiveness. Table 9.1 shows various infrastructure elements connected to agribusiness supply chains, each of which is discussed subsequently in detail.

9.4 Transportation Network

A transportation network is usually a network of roads or other structures that allow uninterrupted movement of vehicles, products and services. Such a network must also facilitate multimodal movement of goods. The transportation system is a key player in sustaining the economic development of the country.

Table 9.1 Role of Infrastructure in Managing Agri-Supply Chain

Role Agents	Input Providers	Farmers	Processors/ Wholesalers	Channel Partners Including Retail	Consumers
Activity and systems	Input processing plants and facilities	Input production and output storage	Transportation at different stages from inputs to production	Markets– Facilities to physically sell agricultural goods	Communication links such as television, radio and support systems through other modes
Basic infrastructure	Roads	Power	Irrigation through dams and canals	Storage, land and building and so on	Communication system such as radio, television and computer linkages

The transportation network includes different modes of transportation, such as roadways, railways, shipping, airways and pipelines. Many agricultural products are either bulky or perishable (sometimes both), and the costs involved in transporting inputs can account for a high proportion of the value of the final products; inadequate transport facilities will be a hindrance when farmers try to sell their produces.

Each agriculture produce has its own distinctive characteristic. Understanding of the harvest and its characteristics are essential in managing transportation networks. The selection of the appropriate mode of transportation network is also vital. These decisions are more strategic in nature, and should be interconnected with inventory decisions. In several cases, the trade-off is between the cost of using the specific mode of transportation and the indirect cost of inventory associated with that mode. Further, quality problems and other regulatory issues also play a role in the decision-making process.

For example, wheat, which is produced in the northern states of India, is of a premium quality and is moved across the country. The farm produce of wheat, either procured directly or through agents, is transported to different processing mills by road, and at times by train. There are different locations where stocks are carried over. Based on the urgency of needs, the stocks are moved either by road or rail. Further, these decisions are dependent on volumes and the ability to aggregate loads for movement in case of movement by dedicated rail rake. There are certain produces like fruits and vegetables that need to be moved swiftly, and they depend upon road transportation. The travel is usually overnight and lasts less than a day. Of course, the traders also export and import agricultural goods wherein they use air and sea transport as modes for movement. It is important to note that since transportation is a major contributor to logistic costs, focusing on appropriate transportation networks makes good economic sense.

The configuration of an agricultural transportation network is determined by the nature of services required, and the costs, distances and volumes involved. Distances from the producers and the competence of the transportation network are important factors in managing agri supply chains. Batch sizes, routing and scheduling are crucial in effective management of the transport strategy. Farm roads and other road networks also affect the quality of the products. For example, if sugarcane has to be moved from farm to factory for processing cane sugar, the rural village roads determine the effectiveness of movement of goods. Normally, these roads are not well laid or maintained, and could be badly affected during monsoon. Moreover, these roads are not very wide. The goods are often moved through village roads maintained by local governments and state roads maintained by state highway authorities. The local and state governments may not have funds adequately available for upkeep of these roads. However, in the organized sector, there are a number of opportunities to overcome this drawback, using some special grants and soft loans through various schemes operated by nodal agencies.

It may be noted here that lack of suitable storage infrastructure, poor quality roads or slow and intermittent means of transport result in losses, as perishable items such as milk, fish, fresh vegetables, and tea deteriorate within a short span of time. India suffers an estimated food grain and agriculture produce loss of over ₹5 billion due to the lack of an adequate post-harvest infrastructure and inefficient supply chain management. It has been estimated that loss of primary produce before reaching the markets due to lack of proper handling, cleaning, sorting, grading and packaging facilities at the village level is about 7% of total production of food grains. Better transportation networks reduce the inputs cost to the farmer. Similarly, proper means of transport and connectivity from farm to marketplace is critical to efficient marketing structures. With increasing petrol prices, the influence of transportation cost will affect the cost-effectiveness and profitability of agribusiness supply chains.

Road infrastructure is the most important element in the agri supply chain, and it has a direct impact on the bottom line, as it affects both revenue and costs. Roads play a major role in the everyday food marketing system, which emphasizes the importance of the quality maintenance of road networks. The upkeep of efficient rural transportation networks is important for agricultural marketing, particularly in the case of seasonal and perishable products such as fruit and vegetables, dairy products, and fish. A bad network, that is, one with poor roads, and unsuitable and irregular means of transportation can result in low margin for farmers and high prices for consumers.

In India, the primary responsibilities for road construction, repairs and maintenance are vested with the governments (local, state and central). The state roads in India are administered by the state public works departments. The management of the national highway system is assigned to the Ministry of Surface Transport. The other minor roads in India are preserved by the relevant districts, municipalities and villages.

India has the third largest road network in the world. However, in terms of density and quality, India still lags behind many countries of the world. "The Indian road network is the one of the largest in the world, aggregating 3.32 million km, consisting of 65,569 km of national highways, 1,28,000 km of state highways, 4,70,000 km of major district roads and 26,50,000 km of other district and rural roads. Out of the total length of national highways, about 22% have single lane/intermediate lanes, about 53% have the standard two lanes and the remaining 25% is standard four lanes or more. Though national highways comprise only about 2% of the total length of roads, they account for about 40% of the total traffic across the length and breadth of the country" (Planning Commission, 2011b). Roads are the primary factor that decides the magnitude of damage to agricultural produce that is being conveyed from source to market. Building good quality roads will help in transportation of agricultural products without loss of time, thus protecting their quality, which is a fundamental factor in the agri product sector. Increasing rural connectivity can ease market access for farmers and bring about an overall improvement of supply chains, in addition to having other constructive effects on rural households.

9.5 Challenges

One of the main objectives of supply chains is to ensure that goods reach on time and at a reasonable cost. But due to poor road conditions, delivering "on time" becomes a daunting task. Some of the challenges faced are mentioned below.

1. One of the major reasons for excessive damage to roads in India is overloading of commodities. It is quite common for trucks to be overloaded. There are efforts now underway to regularize this. When a truck is overloaded, it affects the road, and further, in case of failure on the road, there is a major impact on vehicle movement. The ability to regain normalcy takes longer. The social cost of overloading had never been considered by the players.

2. India's roads are congested, and getting more so. India is home to quite a large populace. Traffic congestion is a common problem in most of the cities because of the excessive population. Congestions are common, as there are a number of bottlenecks on the roads, by way of bridges, narrow connecting roads, rail crossings and so on. Congestions are also due to poor driving sense and lack of driver discipline. Projects undertaken by the National Highways Authority of India (NHAI) are improving the situation, but these efforts are still far from the level of intervention required.

3. In India, most of the rural parts do not have access to all weather roads, and therefore, have a hard time during the rainy seasons. Large sections of Indian rural roads, even during normal times, pose tough driving conditions, as they are not fully topped with bitumen or well laid with other suitable substances. This reduces the speed of movement drastically. Further, the quality of vehicles used for transportation is often low, and thereby, efficiency is badly hit. Transportation operating costs rise due to breakdowns, especially on roads that get worse during the monsoon seasons. Thus, both supply chain objectives, namely, cost and responsiveness, are affected.

4. The sluggish road conveyance system has also affected the movement of goods, particularly in terms of transportation of fresh goods such as milk, vegetables and flowers. These perishable goods become rotten or get destroyed before they reach the local or export markets. Due to delays in reaching the marketplaces, a large proportion of merchandise is wasted in transit.

5. India is a vast country experiencing widely varying weather at different parts of the country. There are places that experience frequent floods, bringing about major damage to road infrastructure. There are terrains that are dry,

humid and poorly connected with markets. Again, weather in those places is not supportive of well-maintained local roads, which are a last mile issue in transportation.

6. Poor road connectivity hinders the efficient transfer of goods from the source to the destination. Poor connectivity must be understood in terms of the sectors involved: from farms to villages, from villages to markets and from markets and processors again to multiple consumption points. The roads are local roads, state roads and highways. They are not seamlessly connected. Rapid road connectivity to the main ports of the country and to important cities is very essential to increase trade and ensure enhanced prices for farm goods. It is a challenge at this point in most parts of India.

It may be noted here that the magnitude of the expenses incurred in repairing roads is alarming, and hence, the government lays emphasis on building quality roads. According to a Working Group report on the central roads sector, "several states have taken up improvement programmes for their states roads with funding assistance from the World Bank, ADB, JBIC, etc. Works completed are generally of good quality and high standards, with safety features" (Planning Commission, 2011a).

To conclude this section, road transportation, which is the main mode of transportation of agricultural produces and processed goods, continues to be a challenge for achieving supply chain efficiency. Availability of capital and the ability to service the same could be a limiting factor for decades to come.

9.6   Cold Storages

A cold chain is a temperature-controlled supply chain. The cold chain infrastructure normally comprises pre-cooling services, icy storing, refrigerated transporters, warehousing, integrated information systems and so on. The cold storage infrastructure is very essential for temperature-specific goods. It is used to lengthen the shelf life of fresh agricultural produce such as milk and milk-related products, fresh fruits and vegetables, meat, seafood, frozen food, chemicals and pharmaceutical drugs.

There are two logistic layers in cold chains:

Surface Storage: refrigerated warehouse for storage of time-sensitive goods

Refrigerated Transport: refrigerated trucks, containers and so on.

A cold chain logistics industry consists of public and private players who could be either cold storage owners or the owners of refrigerated trucks. Further, there are third party logistics firms that facilitate movement from sourcing to the final destination of temperature-sensitive products. Therefore, the success of cold chain companies depends on how proficiently they can transport temperature-sensitive

Table 9.2 Storage Recommendations on Temperature and Humidity for Fruits and Vegetables

Commodity	Temperature (°C)	Relative Humidity (%)
Apple	–1 to 3	90–98
Apricots	–0.5 to 0	90–95
Blackberry	–0.5 to 0	95–97
Cherries	0.5–0	90–95
Grapes	–1 to 1	85–90
Mango	11–18	85–90
Watermelons	2–4	85–90
Orange	0–10	85–90
Beetroot	0–2	95–97
Brinjal	0–2	90–95
Broccoli	0–2	90–95
Cabbage	0–2	90–95
Carrots	0–2	90–95
Cauliflower	0–2	90–95
Cucumber	7–10	90–95
Green beans	4–7	90–95
Potato	1.5–4	90–94

products from the places of origin to their destinations without any loss of produce quantity or quality. Different products require different temperatures (Table 9.2) to impede the activities of micro-organisms.

The provision of sufficient storage, under systematically controlled temperature conditions, makes certain that a produce reaped over a period of one or two months is maintained to meet a year round demand. For instance, production of potato peaks between December and March, but the consumption needs last throughout the year, and have to be met out of stocks available in storage. The commodity-wise distribution of cold storage is presented in Table 9.3.

Table 9.3 Commodity-Wise Distribution of Cold Storages (as on December 31, 2009)

Commodity	Numbers	Capacity (in tons)
Potato	2,862	18,426,316
Multipurpose	1,584	5,644,659
Fruits and vegetables	160	96,427
Meat and fish	497	188,496
Milk and milk products	191	68,230
Others	87	26,524
Total	5,381	24,450,652

Source: Agmarknet. Commoditywise distribution of cold storages as on 31/12/2009, 2009.

9.6.1 Status

Three layers of structures exist in the storage infrastructure in India, viz., "at the national/state level, the district level and the village level in the country. The Central Warehousing Corporation (CWC) has been providing warehousing services at centres of national importance and the State Warehousing Corporations (SWCs) and the state governments at centres of state and district level importance. Cooperatives provide storage facilities, generally located in villages, at the primary and marketing society level" (NABARD, 2011).

Public warehouses are commonly used in India, especially by government and government corporations, as a public policy initiative. Cash crops and produces of the same are commonly stored.

Table 9.4 shows the capacity of different warehouses in India. The capacity available with FCI is used mainly for storage of food grains, while that with the CWC and SWCs is used for storage of food grains as well as certain other items. Besides the above classifications, the storage industry in India includes large private players such as Snowman, Voltas Ltd and Blue Star, which provide refrigeration and storage services. If we observe Table 9.5 the share of the private sector is dominant, as compared to that of the public sector.

9.6.2 Challenges

There are a number of challenges faced by the cold chain infrastructure system in India. There has been a general belief that the cold chain network is inadequate in India and is not very popular. This is mainly because of the following challenges.

Table 9.4   Capacity of Different Types of Warehouses in India

Sr. No.	Type of Warehouse	Capacity in Million Metric Tons
1	CWC	8.5
2	SWC	20
3	FCI	16
4	Private warehouses available for hire (estimated)	12
5	Private warehouses used for private needs (estimated)	20

Source: NABARD, *Organized Agri-Food Retailing in India.* Mumbai: NABARD, 2011.

Table 9.5   Status of Various Cold Storages in India

Sector	Number of Cold Storages	Total Capacity in Tons	Share by Numbers (%)	Share by Capacity (%)
Private	4,875	22,571,475	90.5	95.4
Cooperative	377	990,595	7	4.2
Public sector	134	100,742	2.5	0.4
Total	5,386	23,662,812	100	100

Source: NABARD, *Organized Agri-Food Retailing in India.* Mumbai: NABARD, 2011.

9.6.3   Inadequate and Poor Infrastructure

The cold chain industry in India is highly fragmented. Because of inefficient supply chain management and inadequate post-harvest infrastructure, India incurs huge agriculture produce loss. A poor cold chain infrastructure not only affects the quality of goods, but also leads to extensive wastage. In fact, India wastes more fruits and vegetables than it consumes. Reports describe the overall losses on account of poor post-harvest technology in fruits and vegetables as follows: "around 30% of fruits and vegetables grown in India get wasted annually due to gaps in cold chains, infrastructure, insufficient cold storage capacity, unavailability of cold storages in close proximity to farms, poor transportation infrastructure, etc. Operating costs for Indian cold storage units are over $60 per cubic metre per year, compared to less than $30 in the West" (Maheshwar and Chanakwa, 2006). According to an Assocham-KPMG report on food processing and agribusiness, the country requires over 31 million tonnes (mt), but has facilities for only 21.7 mt of cold storage. Further, the export-related infrastructure for agricultural products is grossly inadequate at ports and airports.

A restrictive land use policy in terms of floor space indices limits the extent to which buildings can be constructed, and this has resulted in a tendency to increase the per square foot investment cost of cold storages.

9.6.4 Rising Energy Cost

"Energy expenses make up about 28% of the total expenses for Indian cold storages, compared to 10% in the West" (Maheshwar and Chanakwa, 2006). "This is the main constraint to setting up cold chains in India. India's peak power deficit is around 17%–18%" (NABARD, 2011). As a result, cold storage units have to provision for adequate backups, which increases capital investment costs. The intermittent power supply adversely impacts the operations of cold storages, and almost all units are using alternate power supply systems. This involves a high operating cost and requires the capability to manage and maintain alternate systems so that backup is available at all times. Though this is common across the globe, in India the dependency is very high, and hence, makes cold storage highly uneconomical for large parts of the agribusiness sector.

9.6.5 Poor Utilization

Different types of produce need different temperature conditions, and this leads to poor capacity utilization and higher fixed costs, reducing profitability. Further, many storage houses concentrate on a single commodity. This also leads to poor utilization. Coupled with this is the problem of high operational costs in India for cold storages, examples being as the high costs of power and insurance against spoilage. There is a hidden cost arising from in efficiency and failure, which adds to the operating costs.

Though there are numerous challenges, the growth in the processed food sector, the bio-pharma sector and the organized retail sector presents huge opportunities in the area of cold chain infrastructure. Furthermore, the government also plans to invest in partnerships with private players. Identifying and filling gaps in the infrastructure and adopting best practices, such as partnering with players in the logistic chain and adopting energy efficient technology, will benefit the entire agri supply chains.

9.7 Organized Markets

Agricultural produce moves through the marketing chain using different channels of distribution. In order to eliminate the middlemen and facilitate farmers to sell their produce directly to the consumers at reasonable rates, organized markets are established. These markets can help to reduce costs by bringing the growers and retail merchants/dealers closer, leading to lower prices for the consumers and higher realization for the farmers. This way, both the farmers and the consumers are benefitted. The development of organized markets in recent years has boosted the demand for quality food stuffs. Organized markets can enable agrarians with technology and provide them better market access. The regulated markets are managed by Agricultural Produce Market Committees (APMCs). The main purpose of regulation of agricultural markets was to safeguard farmers from the exploitation of intermediaries and traders, and also to guarantee better prices and timely payments for their goods. Under the APMC Act, state governments are permitted to set up markets. These organized markets follow certain rules and procedures while carrying out their transactions. Some examples of these channels are discussed now.

9.7.1   Rythu Bazaars

Rythu bazaars were started by the Government of Andhra Pradesh on January 26, 1999. "The following are the objectives of *rythu* bazaars:

1. To ensure remunerative prices to the farmers and provide fresh vegetables to the consumers at reasonable rates fixed every day
2. Facilitate prompt realization of sale proceeds by the farmers, without any deductions
3. Curb malpractice in weighment
4. Provide direct interface between farmers and consumers, eliminating intermediaries in the trade process" (Ministry of Agriculture, n.d.).

Transport facilities for picking up and dropping off registered farmers at the *rythu* bazaars are to be arranged by the market functionaries through tie-ups with the state transport department. Stalls are to be allotted to the farmers on a first come first serve basis daily. The farmers are not allowed to occupy the stalls permanently. No fee is to be collected from the farmers. The market committee will every day communicate moderate wholesale prices of vegetables through fax between 7.30 and 8.00 a.m. Based on the these inputs, the prices of vegetable are fixed in consultation with the farmers' committee, and may be 20%–25% higher than the wholesale price and lower than the local retail market prices in the area. The maintenance expenditure of *rythu* bazaars is being met from the financial sources of Agricultural Produce Market Committees.

9.7.2   Uzhavar Santhai

The *Uzhavar Santhai* system was introduced by the Government of Tamil Nadu in order to provide more benefits to farmers as well as consumers. The location of the *uzhavar santhai* is of utmost importance. The *uzhavar santhais* are located in consumer areas, allowing easy access to consumers. "The first *uzhavar santhai* of the state was inaugurated in 1999. At present, 103 *uzhavar santhais* are functioning in Tamil Nadu. All *uzhavar santhais* are maintained by the Agricultural Marketing Department and manned by the staff of the Department of Agricultural Marketing, Agriculture, and Horticulture. The prices of the vegetables and fruits are fixed daily at an average of 20% higher than the wholesale prices and 15% less than the retail prices by the committee, consisting of the representatives of farmers and officials" (Ministry of Agriculture, n.d.).

9.7.3   Facilities

Indicative list of infrastructure required in a wholesale market are referred to in Appendix 9.1.

Regulated markets deliver services electronic weigh bridges and weighing balances that facilitate accurate weighing, storage facilities, bank facilities, daily price information, rest sheds, drinking water facilities, cattle sheds, free medical aid to farmers, input shops, phone and fax facilities, etc. Under the "AGMARKNET" centrally sponsored scheme, 93 regulated markets have been provided with computers and Internet connectivity for effective price dissemination among farmers through the AGMARKNET website. Information on commodity prices prevailing in various markets is made available, enabling farmers to get better prices by moving their produce to the markets that pay higher. During the year 2007–2008, 100 regulated markets were computerized through AGMARKNET" (TNAU, 2013). NABARD sanctioned 25,682 godowns for farmers (Exhibit 9.1).

Exhibit 9.1   Rural Godown Scheme for Creation of Scientific Storage

The Ministry of Agriculture is implementing the Rural Godown Scheme for creation of scientific storage capacity with allied facilities in rural areas to meet the various requirements of farmers for storing and processing farm produce and agricultural inputs, including onion. As on 31 October 2011, 25,682 godowns, having a capacity of 294.83 lakh metric tonnes (MT) have been sanctioned by the National Bank for Agriculture and Rural Development (NABARD) and the National Cooperative Development Corporation (NCDC) all over the country. The scheme is demand driven, and back-ended subsidies are made available for construction of godowns to all categories of farmers, agricultural graduates, cooperatives, individual companies and corporations and women farmers.

Source: Ministry of Agriculture. Rural godown scheme for creation of scientific storage. Press release, Dec 20, 2011.

9.8 Communication Network

A communication network is the infrastructure that allows two or more people to communicate with each other. It consists of interconnected individuals who are linked by the patterned flows of information, and its analysis identifies the communication structure in the system (Rogers and Kincaid, 1981). In agriculture, the need for such a network should be obvious: it allows farmers from different parts of the country, with different operating characteristics, to know the market conditions prevailing at any given time.

Agricultural information networks interact with and impact agricultural efficiency in several ways. At present, there are certain communications networks existing in the country through which farmers gather information from various sources.

- *Radio*: The radio is a very popular means of information dissemination among farmers. Various educational and informative programmes are being broadcast on radio to help the farmers at large. Agricultural universities and state agricultural departments use this media extensively to spread knowledge about agricultural practices and also information on farming, best practices, extension services, markets, financial assistance opportunities and so on.
- *Television*: There are educational and informational programmes on television targeted at creating awareness for farmers. An instance is the "*Vayalum Vaazhvum*" on the DD Podhigai channel. Radio and television are also widely used in all developed states for outreach to farmers.
- *Toll-free call centres*: The toll-free call centres such as ANGRAU and *Kisan*, started by the government help farmers to interact and get access to information whenever they need it. This call centre process is highly effective, as it involves close interaction with farmers.
- *Workshops*: Through agro-based workshops, farmers get useful information. These workshops are organized by local, state and central government departments, agricultural universities, NGOs, farm input companies and research organizations.

Besides the above, farmers also get information through direct interaction with other farmers and dealers and from display boards in market yards. Such market information is gathered by APMCs, the Department of Agricultural Marketing, the Department of Food and field staff of the Directorate of Economics and Statistics. The Agricultural Marketing Information System Network (AGMARKNET), a public portal integrated with information technology, displays information on about 400 commodities every day. This AGMARKNET project benefits farmers by providing them with new global market access opportunities, and also strengthens the internal agricultural marketing system in India. It connects almost 2900 markets all over the country and is the most successful intervention in this area. Many agricultural marketing boards (state and central) and agribusiness organizations have developed web portals on market information. Their websites' URL are given in Appendix 9.5.

The communication networks for price discovery and the commodity markets are discussed in further detail in a later chapter.

The Government of India has been spending on different programmes, under the five-year plans and the annual budgets, for the creation of information systems. There should be more coordination between public and private information sources in order to encourage the farmers to learn about and adopt reforms in agricultural practices so that they become proficient and yield quality products. Proper information networks are the essential infrastructure elements that help farmers to produce quality goods and lead a better life.

A popular case of an effective communication network in the private sector is that of ITC's *e-choupal*. "ITC's *e-choupal* business model is mainly based on communications infrastructure. ITC has made significant investments to create and maintain its own IT network in rural India and to identify and train a local farmer to manage each *e-choupal*. The computer, typically housed in the farmer's house, is linked to the internet via phone lines, or increasingly, through a VSAT connection, and serves at least 10 surrounding villages within a radius of about 5 km.

The host farmer, called a *sanchalak*, incurs some operating costs, and is obligated by a public oath to serve the entire community. He needs to be a well-respected local leader to run this service. The *sanchalak* benefits in the form of commissions paid to him for all *e-choupal* transactions. The farmers can use the computer to access daily closing prices at local *mandis*, as well as to track global price trends or find information about new farming techniques—either directly or, because many farmers are illiterate, via the *sanchalak*.

They also use the *e-choupal* for buying farm inputs directly through ITC. Farmers order seed, fertilizer and other products like consumer goods from ITC or its partners at prices lower than those available from village traders. The *sanchalak* typically aggregates the village demand for these products and transmits the order to an ITC representative.

At harvest time for farm produce output, ITC offers to buy the crop directly from any farmer at the previous day's closing price; the farmer then transports his crop to an ITC processing centre, where the crop is weighed electronically and assessed for quality. The farmer is then paid for the crop, and also receives a transport fee. 'Bonus points', which are exchangeable for products that ITC sells, are given for crops with quality above the norm. Thus, the ITC communication infrastructure benefits agribusiness supply chains by reducing intermediaries and providing near real-time information to different stakeholders, mainly farming communities, and improves value creation in the system" (ITC, n.d.).

On similar lines is the establishment of the commodity exchanges in India. There are three national exchanges, and in total, 24 approved commodity exchanges are operating in India. The three national exchanges approved by the Government of India are the Multi-commodity Exchange of India Limited (MCX), Mumbai; (ii) the National Commodity and Derivatives Exchange Limited (NCDEX), Mumbai; and (iii) the National Multi-commodity Exchange of India Limited

(NMCE), Ahmedabad. These on-line national commodity exchanges have been organized for conducting forward/futures trading activities in all commodities to which Section 15 of the Forward Contracts (Regulation) Act, 1952 is applicable. Trading in the other commodities that are neither listed under Section 17 nor under Section 15 of the said Act is subject to the approval of the Forward Markets Commission, Government of India. These national commodity exchanges have in place the best international practices in trading, clearing, settlement, and governance (FMC, 2013). Technology for information handling with respect to commodity arrivals and volumes offered, prices and financial flows are vital aspects of the communication network that makes the initiative successful.

Thus, establishment of appropriate communication infrastructures brings the benefits of a reduced number of intermediaries, and thereby, of improved margins and better responsiveness of the supply chains.

9.9   Training

Training of personnel in different disciplines of agriculture is a pre-requisite to improve effectiveness. This is mentioned here as a part of human resource capability and the effectiveness of personnel in using the infrastructure currently available.

Educating farmers in agricultural technology will increase yield and land productivity. This is core to agricultural production systems, especially in a country like India, where farmers do not have adequate scope for learning and absorption of tools and techniques through customary channels of the informal awareness networks.

More specifically, farmers and other operators in agribusiness require training in logistical drivers, namely, operations of warehouses and regional distribution centres, transport decision making and inventory decisions. Training is provided to personnel involved in grading, handling, processing, storage and so on. According to a report on Indian warehouse marketing, there are huge skill gaps in the warehouse sector. Table 9.6 presents the profiles of the people employed in the warehouse sector in India.

It is estimated that by 2015, India will need approximately 35,000–40,000 warehousing managers. There are a few training institutes like the CII Institute of Logistics and other specialized institutions that can train people in the managerial skills needed to cater to relevant needs as Level 3 managers in the warehousing sector. A number of private players use their own programmes, delivered by experts in the system, to improve capabilities at Levels 1 and 2. The demand for Levels 1 and 2 would be in large numbers, an issue that needs to be addressed to achieve growth in the long run.

The Ministry of Agriculture introduced special programmes for empowering farm women, designed to train them to contribute equally towards agricultural productivity. Farmers' training classes are conducted periodically on various topics at the local level through *Krishi Vigyan Kendras* and field demonstrations. The farmer training programmes usually consist of training centres, vocational agricultural schools, agricultural market level training for farmers, information and

Table 9.6   Manpower Skill Sets for Warehouse Sector in India

Skill Level	Position Level	Qualification and Years of Experience
Level 1	Operators	10th and 12th class with/without any prior experience
Level 2	Supervisors	Diploma holders/graduates with four to five years of experience
Level 3	Managers	Graduates with 8–10 years of warehouse industry experience
Level 4	Trainers, researchers and policy analysts	Post-graduates with research degrees with conceptual and analytical skills to train, research and impact people and policies in the system

guidance centres, training of women in agriculture and farmer exchange programmes. The central government runs a scheme to promote agricultural development by providing extension functionaries and farmers with information, training and other extension support on a continuous basis to Union Territories and the states of Goa and Sikkim (National Portal of India, 2011). The training programme makes the farmers' daily farming activities much easier. This also enhances productivity, which results in better profits. Besides these government programmes, the public–private partnership (PPP) model can also be developed to foster research and development in agriculture and allied sectors.

9.10   Investment

Most of the infrastructure connected to agri supply chains has been set up by the government, as it involves huge capital outlay, lengthy conception phases, high risk and low rate of return on investment. But at present, a lot of private venture is increasingly engaged in building up agri supply infrastructures. "The eleventh Five Year Plan emphasized the need for removing infrastructure bottlenecks for sustained growth. Accordingly, it proposed an investment of US $500 billion through a mix of public and private actors to reduce deficiencies in identified infrastructure sectors. The private sector is expected to contribute nearly 36% of this investment. During the Twelfth Plan period, investment in infrastructure is to go up to 50 lakh crore, with half of this expected from the private sector" (Union Budget & Economic Survey, Appendices 9.2 and 9.3). The report on "Agricultural marketing infrastructure and policy required for internal and external trade for the XI five year plan 2007–2012" outlines the investments proposed for the 11th five-year plan; the details are given in Appendix 9.4. The government is implementing various schemes and programmes to increase investment in the infrastructure development in the agricultural sector.

9.11   Conclusion

To conclude, India's agricultural sector functions now in a new and competitive business and social setting. In the present-day context, there is a necessity for reorienting the views of the numerous players involved in providing infrastructure. The situation presents enormous opportunities for investments if various interested parties work in partnership and develop innovative models that not only address the problems in India's agri supply chain infrastructure, but bring about best agricultural practices as well. Public and private players in collaboration, with the right policies, need to play a crucial role in promoting investments in proper agri infrastructural facilities. This increases the comparative advantages of the states and regions, thereby touching the potential for overall improvement of agri supply chains. Furthermore, this transformation will bring enormous benefits to all the stakeholders from farmers to end-users, including governments and other allied agencies. The agriculture sector being the primary employer for India's population, developments in infrastructure could go a long way in improving the standard of living of the poor.

Appendix 9.1
Infrastructure Required in the Wholesale Market

Core Facilities	Platforms for automatic weighing Auction platforms Packaging and labelling equipment Drying yards Loading, unloading and dispatch facilities Grading facilities Standardization facilities Price display mechanism Information centres Storage/cold rooms, offices Ripening chambers Public address system Extension and training to farmers
Support Infrastructure	Water supply Power Veterinary services Sanitary facilities Posts and telephones Banking Input supply and daily necessity outlets POL Repair/maintenance service Office Computerized systems Rain proofing
Service Infrastructure	Rest rooms Parking Sheds for animals Market education Soil testing facilities Drainage
Maintenance Infrastructure	Cleaning and sanitation Garbage collection and disposal Waste utilization Vermi-composting Bio-gas Production/Power

Source: Planning Commission. *Report of the Working Group on Agricultural Marketing Infrastructure and Policies Required for Internal and External Trade for the 11th Five Year Plan 2007–2012,* 2007.

Appendix 9.2
NHDP Projects as on December 2011

Sr. No.	NHDP Components	Total Length (Km)	Completed 4/6 Lane (km)	Under Implementation		Balance for Award of Civil Work (km)
				Length (km)	No. of Contracts	
1	GQ	5846	5831	15	8	–
2	NS-EW	7142	5914	803	76	420
3	Port connectivity	380	341	39	4	–
4	Other NHs	1390	946	424	5	20
5	SARDP-NE	388	5	107	2	276
6	NHDP phase III	12,109	3024	6514	90	2572
7	NHDP phase IV	20,000	–	2549	18	17,451
8	NHDP phase V	6500	709	2768	22	3023
9	NHDP phase VI	1000	–	–	–	1000
10	NHDP phase VII	700	7	41	2	659
	Total	55,455	16777	13265	227	25,421

Source: Ministry of Road Transport and Highways, Achievements of national highways development projects, 2011.

Abbreviations: GQ, Golden Quadrilateral connecting Delhi, Mumbai, Chennai, and Kolkata; NS-EW, North-South and East-West corridor; SARDP-NE, Special Accelerated Road Development Programme in the North-Eastern Region.

Appendix 9.3
Expenditure Budget

Five-Year Plan/Annual Plan	Agriculture and Allied Activities Plan Outlays		Total Plan Outlay Actual Expenditure		Percentage Share of Agriculture and Allied Activities to Total (crore) Plan Outlays	
1	2	3	4	5	6	7
Ninth Plan (1997–2002)	**37546**	**37239^d**	**859200**	**941041^d**	**4.9**	**4.0**
Annual Plan (1997–1998)	6974	5929	155905	129757	4.5	4.6
Annual Plan (1998–1999)	8687	7698	185907	151581	4.7	5.1
Annual Plan (1999–2000)	8796	7365	192263	160608	4.6	4.6
Annual Plan (2000–2001)	8281	7577	203359	164479	4.1	4.6
Annual Plan (2001–2002)	9097	8248	228893	186315	4.0	4.4
Tenth Plan (2002–2007)	**58933**	**60702^d**	**1525639**	**1618460^d**	**3.9**	**3.8**
Annual Plan (2002–2003)	9977	7655	247897	210203	4.0	3.8
Annual Plan (2003–2004)	9940	8776	256042	224827	3.9	3.9
Annual Plan (2004–2005)	11109	10963	287843	263665	3.9	4.2
Annual Plan (2005–2006)	13840	12554	361239	247177	3.8	5.1

(Continued)

Appendix 9.3 (*Continued*)

Expenditure Budget

Five-Year Plan/Annual Plan	Agriculture and Allied Activities		Total Plan Outlay		Percentage Share of Agriculture and Allied Activities to Total (crore)	
	Plan Outlays		Actual Expenditure		Plan Outlays	
1	2	3	4	5	6	7
Annual Plan (2006–2007)	16163	16573	441285	309912	3.7	5.3
Eleventh Plan (2007–2012)	136381	Na	3644718	Na	3.7	Na
Annual Plan (2007–2008)	19370	18770	559314	361255	3.5	5.2
Annual Plan (2008–2009)	27274	26598	867828	477430	3.1	3.5
Annual Plan (2009–2010)[c]	10629	11014	447921	406912	2.4	2.7
Annual Plan (2010–2011)[c]	12308	14362[a]	524484	502250[a]	2.3	2.9
Annual Plan (2011–2012)[c]	14744[b]	Na	592457[b]	Na	2.5	Na

Source: Statement 13, Expenditure Budget Vol. I, State Plan Division, Planning Commission.

[a] Revised estimate for central plan.
[b] Budget estimate for central plan.
[c] For centre, excluding states & UTs.
[d] Indicates 10th and 9th plan realization.

Appendix 9.4

Projections for Agri Supply Chain Infrastructure in the Public Sector

Sr. No.	Infrastructure	Number	Unit Cost (₹ lakh)	Total (₹ crores)	Appropriate PSP Option	Private Sector Outlay
1	Development of wholesale markets					
	(a) Principle markets	2,428	300	7,284	BOT	3,000
	(b) Sub markets	5,129	100	5,129	BOT	1.000
2	Rural primary markets	5,000	25	1,250	–	–
3	Primary value addition centers and soil health management infrastructure	50,000	30	15,000	Concession	5,625
4	New wholesale markets	75	1,000	750	–	750
5	Live stock markets	1,000	20	200	–	–
6	Terminal markets	35	5,000	1,750	Concession	1,300
7	Apni mandis/Direct markets	1,152	50	576	–	–
8	Markets for spices crops	50	50	25	–	–
9	Storage capacity (Million tons)	6.67	0.03	2,000	–	–
10	Cold storage (Lac tons)	45	0.045	15,708	Concession	11.500
11	Specialized commodity markets (F&V)	241	2,000	4,820	Concession	3,600

(Continued)

Appendix 9.4 (Continued)

Projections for Agri Supply Chain Infrastructure in the Public Sector

Sr. No.	Infrastructure	Number	Unit Cost (₹ lakh)	Total (₹ crores)	Appropriate PSP Option	Private Sector Outlay
12	Flower markets	10	1,500	150	Concession	100
13	Medical and aromatic and forest produce markets	500	100	500	–	–
14	Modern abattoirs	50	1,000	500	BOT	500
15	Retail market infrastructure for poultry	1,000	500	5,000	BOT	2.500
16	Centre for perishable cargo	15	2,000	300	–	–
17	Farm road infrastructure/green corridors	100	500	500	–	–
18	Quality and food safety infrastructure	500	100	500	BOT	250
19	Specialized quality and safety Infrastructure	50	500	250	–	–
20	GAP and certification infrastructure	1,00,000	1	1,000	–	–
21	Model farms for India GAP certification	1,000	1	10	–	–
22	Farmers organizations support infrastructure	5,000	5	250	–	–
23	Production risk management infrastructure setting up of automatic weather stations	50,000	–	860	BOO	500
	Total	–	–	**64,312**	–	**30,625**

Source: NABARD, *Organized Agri-Food Retailing in India.* Mumbai: NABARD, 2011.

Abbreviations: BOT, Build, Operate and Transfer; BOO, Build, Own and Operate.

Appendix 9.5

Web Portals on Agri Market Information

• www.rakmb.com
• www.msamb.com
• www.maratavahini.kar.nic.in
• www.agri.rajasthan.gov.in
• www.market.ap.nic.in
• www.mandiboardpunjab.com
• www.mpmandiboard.com
• www.gov.ua.nic.in/uamandi2
• www.upmandiparishad.in
• www.bsamb.com
• www.megamb.nic.in
• www.osamboard.org
• www.nafed.com
• www.assamagribusiness.nic.in
• www.hortibizindia.org
Private players
• www.agriwatch.com
• www.kisan.com
• www.indiagriline.com
• www.echoupal.com

Source: Planning Commission. *Report of the Working Group on Agricultural Marketing Infrastructure and Policies Required for Internal and External Trade for the 11th Five Year Plan 2007–2012,* 2007.

References

Agmarknet. Commoditywise distribution of cold storages as on 31/12/2009, 2009. http://www.agmarknet.nic.in/commcold3112009.htm

FAO and ODI. Market-oriented agricultural infrastructure: Appraisal of public-private partnerships. London: Food and Agriculture Organization of the United Nations and Overseas Development Institute, January 2008. http://www.odi.org.uk/resources/docs/2131.pdf

Forward Markets Commission (FMC). IV. Exchanges and their Role. 2013. http://www.fmc.gov.in/index3.aspx?sslid=141&subsublinkid=59&langid=2

Gangadharan, S. External commercial borrowings (ECB) Policy. Reserve Bank of India, March 2, 2010. http://rbi.org.in/scripts/NotificationUser.aspx?Id=5519&Mode=0

ITC. e-Choupal. n.d. http://www.itcportal.com/businesses/agri-business/e-choupal.aspx

Maheshwar, C. and Chanakwa, T.S. Post harvest losses due to gaps in cold Chain in India: A solution, *Acta Hort.* (ISHS) 712:777–784, 2006. http://www.actahort.org/books/712/712_100.htm

Ministry of Agriculture. Model project on development of direct marketing infrastructure: Rythu Bazar / farmers' market. *Agmarknet,* n.d. http://agmarknet.nic.in/amrscheme/rythu_bazarmodel.htm

Ministry of Agriculture. Rural godown scheme for creation of scientific storage. Press release, Dec 20, 2011. http://pib.nic.in/newsite/erelease.aspx?relid=79068

Ministry of Road Transport and Highways (MoRT&H). Achievements of national highways development projects, 2011. http://mospi.nic.in/Mospi_New/upload/SYB2013/CH-21-ROADS/Table%2021.7.xls

NABARD, *Organized Agri-Food Retailing in India.* Mumbai: NABARD, 2011.

National Portal of India, Farmers' training, 2011. http://www.archive.india.gov.in/citizen/agriculture/index.php?id=51

Planning Commission. *Report of the Working Group on Agricultural Marketing Infrastructure and Policies Required for Internal and External Trade for the 11th Five Year Plan 2007–2012,* 2007. http://planningcommission.nic.in/aboutus/committee/wrkgrp11/wg11_agrpm.pdf

Planning Commission. *Report of the Working Group on Central Roads Sector,* 2011a. http://planningcommission.gov.in/aboutus/committee/wrkgrp12/transport/report/wg_cen_roads.pdf

Planning Commission. *Report of the Working Group on Logistics.* New Delhi, 2011b. http://planningcommission.gov.in/reports/genrep/rep_logis.pdf

Rogers, E.M., & D.L. Kincaid. *Communication Networks: Toward a New Paradigm for Research.* New York: Free Press, 1981.

TNAU. Agricultural marketing in Tamilnadu, TNAU Agritech Portal, Tamil Nadu Agriculture University, 2013. http://agritech.tnau.ac.in/agricultural_marketing/agrimark_Tamilnadu%201.html

Chapter 10

Information Technology

OBJECTIVE

The objective is to understand the scope for deploying technology in agribusiness at the farm level and at the processing stage. Discussions are on various information technology systems, adoption of new technologies like Internet, mobile computing and cloud computing in this sector. A number of application examples are discussed here.

10.1 Introduction

Information technology (IT) plays a key role in achieving the supply chain goals of efficiency and responsiveness. Over the last three decades, we have seen the emergence of the role of information technology in decision-making processes in agribusiness as well. The progression has been from data capture and use of office tools for analytics and decision making, to the modern-day phenomenon of real-time data capture and transfer, and application of sophisticated models for decision making. This progression has been seen not only at the level of firms and organizations, but also at the level of individual farms. In this chapter, we highlight some of the impacts the India agribusiness sector has experienced over the years, and how the sector has improved supply chain efficiency and responsiveness.

273

10.2 Legacy Systems

IT is one of the cross-functional supply chain drivers that are critical for realizing the full potential of role agents in terms of meeting their various responsibilities, and for the equitable sharing of rewards. To appreciate this, we need to map out supply chain network constituents and information needs. Table 10.1 explains this aspect.

It may be observed in Table 10.1 that agribusiness requires a large quantity of data and information for operations as well as for planning periods. There are certain requirements like capacity creation and system level decisions like setting up of procurement centres, technology-related information and so on, which are strategic in nature. These require evaluation over a long-term horizon. Often, government sources provide a lot of data and information for this sector.

When we discuss legacy systems, we mean the traditional business focus on the approach to information system management. In the early 1980s, we had firms focusing mainly on the management information aspect of IT systems, which captured ex post data of events and transactions for analysing decision situations and

Table 10.1 IT and Supply Chain Decision Role Agents

Sr. No.	Role Agent	Remarks
1	Farmer	Farm production: Information on climate, irrigation, input management and support services; on harvest; on production trends, prices, arrivals at market and other related information; financial information like credit and support for production and capital investments
2	Production input supplier related	Data and information required are with respect to sowing plans and actual harvests, estimates of requirement, direct and channel sales forecasts and related information for cost management and responsiveness
3	Agro-based focal firm	Planting-related data, coordination if there are satellite/contract farming arrangements, input management, support services, harvest coordination, arrival management, market arrivals and supplement procurement if required; prices and management of financial flows, business process management, e.g., plant operations; inventory and resource management
4	Channel partners	Supply management, procurement, market data and demand management for secondary and tertiary markets
5	Government and its agencies	Macro data such as production estimates, irrigation and water management, support services, price and subsidies; credit flow management through nodal agencies like agricultural refinance and financing; propagation of best practices and so on.

impacts. For example, a large dairy firm would consolidate data on daily procurement from villages, distribution of payment for procurement and process analysis. If there was a funding agency like a bank that had lent fixed and working capital, it would require operational efficiency data for fund flows. IT systems were mainly focused on this and other similar issues.

A number of sugar mills, edible oil processors, dairy plants, plantation crop processors and channel partners used such legacy systems effectively then.

10.3 Enterprise Resource Planning

During the late 1980s and the early 1990s, businesses started implementing enterprise resource planning (ERP) systems from large vendors for intra-firm resource utilization efficiency. The central theme of ERP systems is that better resources planning would enable enterprises to execute and manage internal business processes efficiently. The basic principles of management used for achieving internal efficiencies are the following:

1. Accounting principles
2. Production management principles
3. Value chain principles
4. Management accounting principles for control

If one looks at these principles, agro-based businesses require application of these systems more effectively. For example, let us look at the case of a sugar plant which receives input from about 25,000 farmers. Each of their accounts needs to be managed all through the planting time, the input of services and the receipt of produce for crushing. The cycle is completed with the payment of procurement price for the produce. Further, sugar plants will have a multitude of vendors who provide process material and other goods and services.

Apart from the procurement aspect, the plant has to manage its production systems, such as juice extraction, clarification and vaporization, condensation and granulization of sugar. There are a number of efficiency parameters and operational aspects in managing these process centres, as well as auxiliary equipment stations, operating boilers and power generation equipment, which, respectively, provide steam and power for processing. Effective operations require appropriate production planning systems that cover receipt of material, resource allocation and scheduling of operations. ERP facilitate the operations of a sugar factory to achieve the same.

ERP facilitates not only the production and operations management of the firm, but also its value chain activities. Table 10.2 relates the value chain activities and their links to information system in a sugar plant.

Thus, value chain activities show the need for ERP-based systems, especially for organized agro-based units, to have appropriate information set-ups for achieving internal efficiency and intra-firm-focused financial perspective-driven IT systems.

Table 10.2 Value Chain Activities of a Sugar Plant and Relevance of IT

Activity	Details	Remarks (based on Porter's framework)
Primary activities		
Inbound	Receipt of material for processing	Requires material resource planning, coordination, harvest plan and actual receipt of material by production scheduling
Processing/manufacture	Conversion of input material into output	Managing of different process sections, e.g., capacity balancing, and manpower deployment
Outbound	Movement of goods into primary and then into secondary distribution. Primary would include mother warehouse and then from there to CFAs if they are part of the distribution network. Secondary distribution would include moving further from the channel to wholesalers, retailers and customers.	In case of a sugar plant, outbound is a crucial function in India. Sugar movement is regulated through releases by the government. Since it is a seasonal production with demand throughout the year (with spikes during festive times), outbound management is important. Primary distribution means holding stock in the company warehouses; based on which funding is done. A good and effective IT system is of paramount importance.
Marketing	In agro-based business where products are commoditized as in the case of sugar, the focus of marketing is more on sales and distribution. ERP is to take care of distribution resource planning and distribution management.	In the case of sugar, after the releases, the plants sell to local traders and wholesalers through intermediaries, namely, commodity brokers. The deciding criterion is the short-term hourly, daily and weekly price movements, which determine the sale transactions. ERP facilitates data analytics and also management of relationships with brokers. More importantly, on completion of a sale transaction, it facilitates physical movement from the factory's primary warehouse.

Post-sale service	Reverse flow for warranty or replacements and so on is generally referred here.	In sugar, there no such instances happen. The firms may internally have their equipment and/or the components requiring reverse flow management, like in case of rollers, crushers, pumps and motors and so on. IT facilitates planning and managing of maintenance, repairs and overhaul.
Support activities (indicative)		
Procurement	Procurement planning and coordination is an intense activity.	In the sugar industry, it is one of the important activities that require data analysis and information sharing. This could be different for industries where procurement is done from the markets.
Finance and accounting	Involves organizing capital for operations and capital expenditures and financial reporting as per statutory requirement	The sugar industry as well as other agro-based and allied industries like dairies requires effective financial information flow to ensure that operational flow is timely. Only when farmers are paid on time can loyalty be ensured. IT facilitates data capture, analysis, reporting and control.
HR systems	People resource planning and deployment is important. It needs to be in synchronization with business plans, mainly production plans	Information systems play a crucial role in human resource management, especially in industries like sugar and those where extensive field works are involved and production runs can vary based on crop availability and other operating parameters. Apart from crop-related manpower, operations use regular, seasonal and temporary resources, and an effective information system is essential.
Information systems	Providing necessary hardware and software for information capture and analysis	In agro-based industries like sugar, appropriate IT systems are essential. Many units make ERP investments that are either home grown and custom built or off-the-shelf market products that are then customized to specific requirements. There are a number of mandatory filings of information systems at government offices and other different stakeholders.

In general, ERP revolves around managing different types of resources, namely, machines, material, men and money. Resources are used to manage functions and execute processes to achieve business objectives.

Finally, ERPs are used for management accounting and controls such as breakeven analyses for decisions, budgeting, investment appraisals, managing cost centres and arriving at decisions like apportionment of cost based on activity-based costing, and consequently, pricing decisions. In agro-based businesses, these become crucial, and hence, ERP plays a major role.

However, given the current business supply chain orientations, where one may have to move from the intra-firm perspective to network-based information systems, ERP may bring in certain limitations because of the design and rigidity it requires in processes.

10.4 Supply-Chain-Centric Applications

There are three layers of supply chain management tools, which are supply chain planning tools, supply chain collaboration tools and supply chain execution tools. However, from the enterprise information system dimension, these automated tools could be classified based on their purpose, as we now consider.

1. *Supply chain planning*: Planning is a key decision area with focus on resource allocation, demand and supply planning, analysis and reporting. The planning process provides key inputs for upstream and downstream linkages and for firm level (internal) input–output operation. All those companies that operate under an organized structure with IT in place would require supply chain planning tools.

2. *Supply chain collaboration*: Supply chain collaboration involves team and work group collaboration of upstream and downstream partners. It facilitates seamless operation of the firm's supply chain. Collaboration could include sharing of documents, data and plans for effective management of supply chain costs and responsiveness. In agriculture input businesses like fertilizer and capital equipment like tractors, the need for collaboration is widely felt and used, based on availability of tools and techniques and the sophistication afforded by network partners. In modern times, even the farming sector uses collaboration tools such as mobile phones and the Internet.

3. *Supply chain execution*: The supply chain execution decision area relates to operations and execution management decisions on buying, storing, processing and moving, and so on. It would also include management of financial flows to various stakeholders. Execution information management involves using work support engines and/or frameworks. This would facilitate data capture, instruction triggers for action and documents such as delivery reports, stock reports and so on. Typically, all types of agro-based firms require capturing execution data, and in certain cases like regulated business, statutory filling may be mandatory. Capture of execution data and reports are vital for the firm.

There are many information system vendors who offer a wide range of supply chain management solutions using different information technology frameworks. However, the purpose of the frameworks would be to enable the tools across three supply chain management layers: planning, collaboration and execution. Each of the supply chain management activities relies on supply chain processes that span across various functions inside and outside the enterprise boundaries.

It may be useful to discuss the supply chain council reference model processes such as plan, source, make, distribute and return. This could be related with agricultural business activity and information system requirements. Table 10.3 highlights this issue.

Thus, we see how a supply chain IT system improves the firm's supply chain network and its partners' business effectiveness and responsiveness to ultimate customers. In agro-based business, the benefits to the ultimate customer have been unclear, as they are large groups at the farthest end of the supply chain.

Here, we would like to highlight an example of IT application for enhancing the supplier relationship management process of a dairy company. Farmers deliver milk at collection centres twice a day. "Based on the quality and the number of litres supplied by a farmer, his or her total amount is calculated. A sample from the supply is taken and tested for fat and solid not fat (SNF) percentage. Once all the tests are done, each farmer's data (quantity, fat & SNF percentages along with the farmer's unique number) is entered in a scannable data sheet. This sheet is sent to the company's computer centre, where it is scanned. The farmer is paid on a fixed day of every week for a week's collection. The entire data base is managed through a state-of-the-art computer software system. It would not be out of place to mention that the company has been paying the farmers on time every week, without any exception" (www.hatsun.com).

It is important to note one of the areas where IT is used for distribution management, and hence, supply chain networks: the government and its agencies like the Food Corporation of India (FCI), which supports the public distribution system (PDS) in India. This involves procurement, storage and distribution of essential items for economically less privileged people in India, and thus, comprised a large supply chain. An IT system is crucial, as it has to link up various government departments, FCI offices and storage centres, movement plans and distribution effectiveness at the level of every outlet in every district. Government departments, supported by the National Informatics Centre and other state level departments and agencies, have embraced technology for the benefit of the common people, and the agribusiness supply chain, too, has improved its effectiveness.

It may be useful to note here that the Internet and mobile and cloud technologies have dramatically impacted the supply chain partners' decision areas. Here are a few instances of success based on technology deployment.

Table 10.3 Supply Chain Processes, Activities and Layers in an Agro-Based Industry

Process	Enterprise Supply Chain Activity	Relevant Supply Chain Management Layer	Industry–Information System Relevance
Plan	Sourcing, materials, process, distribution (warehouse, transportation and channel), returns	Supply chain planning	Planning is a key function of agro-based industries because of seasonality, bulkiness (order quantity and movement) and perishability.
Source	Identify and evaluate potential farmers and markets to procure, organize procurement, procure materials, monitor supplier relationships	Supply chain collaboration for identifying, evaluating, organizing and monitoring suppliers Supply chain execution for procurement	Since individual farmers may have small quantities, procurement needs to be from a large set of farmers and could be spread over villages. Further, procurement often happens from the markets. For all these activities, a robust IT system could be important.
Make (process)	Critical function for primary, secondary and tertiary processing of goods for market	Supply chain collaboration and execution	Use of machinery and system appropriately requires synchronizing production, procurement and distribution plans. IT would play a crucial role in value capture.
Distribute	Organize storage locations, manage storage locations, organize transportation mode, manage transportation mode, organize distribution channel, manage distribution channel	Supply chain collaboration for organizing and managing distribution channel Supply chain execution for organizing and managing storage and transportation	IT system would focus on optimizing decisions of storage and movement
Return	Organize sales returns, organize purchase returns	Supply chain collaboration for organizing sales and purchase returns	Wastages could be significant in India. But there is still a need for IT and process improvement in this area.

10.5 Internet Technologies

E-business enables execution of business transactions over the Internet, which improves on efficiency and impacts responsiveness: the two primary supply chain objectives. There are two different important models of e-business: business-to-business (B2B) and business-to-customer (B2C). B2B depicts transactions between focal firms and other partner firms in a supply chain through cyberspace. One can think of e-procurement of goods, wherein the focal firm submits a tender bid electronically, prospective vendors participate and negotiation takes place. Finally, a successful vendor is finalized through these interactions. Thus, e-procurement is an Internet-based value-added application of e-commerce to facilitate, integrate and streamline the entire procurement process, from buyer to supplier and back. There is no substitute of digital certificate/signature in e-commerce. This aspect is approved by the IT Act, and has legal sanctity.

Cyber transactions bring the advantage of transparency to the procurement process, reduce the opportunities for abuse of power, improve competition without geographical limitation, reduce procurement cycle times and provide cost efficiency. These are just some of the great advantages that are enabled through e-procurement.

Exhibit 10.1 discusses the e-procurement process of Indian Farmers Fertilizer Cooperative Limited (IFFCO), a leading fertilizer manufacturer in India.

Exhibit 10.1   IFFCO e-Procurement System

■ Vendor submits their bids online using digital signatures
■ Vendors can upload/update/delete uploaded files
■ Vendors can take printouts of quotations for their reference and records
■ Vendors can give advance intimations about dispatches
■ Vendors can see the history of purchase orders placed and can raise queries about them
■ Opening and evaluation of bids are done in a secured and legal environment using public key infrastructure (PKI), encryption and digital certificates
■ A summary of technical and priced bids quoted by different vendors has been provided on the e-procurement website for access limited to the vendors who participated in the bidding process for that particular enquiry.

Source: Chandrasekaran, N. *Supply Chain Management.* New Delhi: Oxford University Press India, 2010. With permission.

In e-procurement, it is possible that vendors and buyers have the scope of having access to more information about the transactions than they would have in a normal supply chain management structure with disparate information flows. Vendors participating in bids may not be serious, and could fail to execute contracts awarded to them. Hence, focal firms need to do pre-bidding evaluation more rigorously and get scrutinized prospects for the bids.

Apart from e-procurement, which is information intensive, there are also other B2B transactions such as collaborative initiatives for product development, production information and product and market research, testing and analysis of results, literature development and so on. One may note that providing information across the supply chain, negotiating prices and contracts, allowing customers to place orders and tracking the same and managing financial flows are primary aspects of B2B e-business models. From the supply chain perspective, the system enables effective information flow, flawless financial flow and streamlined physical flow. B2B is more to do with sourcing, pricing and information, and logistical drivers such as warehousing, inventory and transportation enable the same. New business models evolve around these drivers and management of flows.

In agro-based business, ITC's *e-choupal* business is a classic example of use of technology for business growth and for improvement of the overall effectiveness of the farm to process and process to market networks.

10.5.1 ITC E-Choupal

ITC, one of India's largest private sector companies, is into diversified businesses, and has a long history. It has a turnover of US$ 7 billion and a market capitalization of US$ 35 billion. The company's shares are listed with three stock exchanges: Kolkata, Mumbai and the National Stock Exchange. ITC's businesses include fast moving consumer goods (FMCG), hotels, paperboards and specialty papers, packaging, agri business and information technology.

ITC is a market leader in its traditional businesses of cigarettes, hotels, paperboards, packaging and agri exports. It is rapidly gaining market share in its new businesses of packaged foods and confectioneries, branded apparels, personal care products and stationeries. It may be interesting to note here that its non-cigarette sector revenue, which was around 27% of the total revenue in 2002–2003, went up to 44% of the total revenue of the company by 2011–2012, which was a significant change in the composition of the revenue structure. Though ITC's businesses are domestic market centric, they also have exports and thus, earn foreign exchanges. The earnings in foreign exchange over the last 10 years amounted to nearly US$ 4.9 billion. ITC's agribusiness operation is one of India's largest exporters of agricultural products and contributed to the extent of 56% of the last 10 years of ITC's foreign exchange earnings.

ITC's *e-choupal* initiative is a path-breaking move to improve the competitiveness of Indian agriculture, benefiting farmers as well as the company by bringing transparency in procurement prices, reducing intermediation processes and empowering local villagers. All these are possible through use of the Internet at the village level through ITC kiosks.

Traditionally, farmers sell their produce to traders either at the village level or at local markets, and traders operate in government-regulated wholesale agricultural markets (called *mandis*). The food processing companies used to buy from the *mandis*. The drawback in this marketing system is that the farmers rarely have any information on market conditions, such as data on prices and arrival patterns, when they decide on a sale. The traders, on the other hand, are well informed about crop prices prevailing in different markets and the prices offered by processing companies. This helps intermediaries to exploit the farmers and earn super normal profits for their roles, while all other network partners, such as the farmers, the processing companies and the consumers are at a disadvantage.

We now take a look at ITC's *e-choupal* model. It is a unique hub-and-spoke model where villages near are linked to a village procurement centre with internet connectivity. *Choupal* means marketplace, and the prefix e stands for electronic, representing the Internet thus, the *e-choupal* is an electronically linked market where farmers can sell produce to ITC after making informed decisions (TechSangam, 2011).

1. It has two dimensions, namely, transparency and dynamic information on prices and operational procurement perspectives. The first dimension is the Internet kiosk (*e-choupal*) set up in villages to enable farmers to access daily wholesale prices for produce such as soybean, wheat, tobacco and coffee in the local *mandis*. The kiosks also display the price offered by ITC ABD. The prices are updated every day in the evening by ITC. The second dimension is the hub, which includes operational aspects such as procurement, storage of goods and farmer training. There are warehouses and farmer training centres to support every 40–60 *e-choupals*.

2. The kiosk is installed at the home of the *sanchalak,* who is the manager. Typically, he is a well-respected person in the local community and is acceptable to them. He plays the role of disseminating information about crop prices in major markets and weather information. Each *sanchalak* manages three to six villages. There are *upa-sanchalaks,* who are deputy managers, operating at the village level.

3. *Sanchalaks* also facilitate the purchase of farming inputs, crop sales, and non-ITC products such as insurance policies. The information dissemination is a free service. In all other cases where the *sanchalak* enables transactions, he gets paid a commission ranging from 0.25% (fertilizers) to 15% (insurance policies). *Upa-sanchalaks* share the commission with the *sanchalak* for transactions enabled by them.

4. IT provides scope for connecting with farmers directly and giving them information on market prices and ITC's procurement prices, and matching or improving upon market price. Farmers get value-added services like inputs

and support extension services, which impact favourably on the crop economics. Other value-added services also include about 25 warehouse hubs, which are full-service *Choupal Sagars* housing retail stores, fuel stations, soil testing labs and food courts.

5. ITC is operating in about 10 states in India with about 6500 *e-choupals*, 110 warehouses and is impacting the lives of about 4 million farmers in about 40,000 villages. The system contributes nearly two-thirds of ITC's agribusiness division revenues, a percentage that is likely to increase over the years.

6. Aparajita Goyal (2009), of the Development Research Group, World Bank, shares the following inferences based on research on ITC intervention through kiosks:

   a. Significant increase in the average price of soybean (1%–3%) in local markets after introduction of kiosks

   b. Heterogeneous effects of kiosks by distance. The further away the kiosks are from the markets, the lower is the effect on price

   c. Dispersion in prices across markets decreasing over time

   d. Increase in the area under soya cultivation by 19%. Evidence of substitution moves away from rice cultivation

   e. Effect of warehouses on average prices in local markets is small and insignificant, indicating the sorting effect offsets the competition effect.

7. ITC also reduced its procurement cost, improved loyalty and increased value across the supply chain.

10.6 Mobile Technology

Mobile phones represent one form of information and communication technology. Personal computers, laptops, the Internet, television, radio, and traditional newspapers are all used to promote improvement of agriculture and rural development. Hence, if mobile phones can use the Internet and other applications for information and transaction, then it is going to be important to understand how they improve supply chain effectiveness.

Mobile phones are used to deliver content and services that can help foster inclusive growth in India by digitally empowering citizens across all cross-sections of society, both urban and rural, in India. These services are called utility mobile value added services (MVAS). In urban areas, mobile penetration is believed to be 100%. However, in rural areas it is around 23% as of 2011.

With the increase in smart phones availability at a low cost, and with the prevalence of content and applications that are made popular and economical by telecommunications companies and content providers, the scope for adopting use of mobile for agro-based businesses is high.

Rural segment services include selling and procurement information, support for farm commodities, education for farming communities on best practices and delivery of health care and education to remote villages via the mobile broadband network (World Bank, 2011). Thus, the proliferation of mobile phones is being used to help raise farmers' incomes, make agricultural marketing more efficient, lower information costs, reduce transport costs and provide a platform to deliver services and innovate.

There are already a number of applications in India where mobile networks are used for agribusiness effectiveness. Some popular instances are discussed now.

1. Mobile phones can serve as the backbone for early warning systems to mitigate agricultural risks and safeguard agricultural incomes. The M.S. Swaminathan Centre, as well as a number of other interest groups, disseminates information on the weather through mobile phones, which helps farmers to plan their activities carefully under difficult agronomic and climatic conditions to manage their input–process–output management. Fishermen communities have deployed systems to avoid going into rough seas during high tides. Similarly, farmers could plan fertilizer applications and deployment of labour much better using mobile networks.

2. Mobile phones make markets efficient and help farmers to sell goods better. They also facilitate intermediaries in understanding the demand, the arrival pattern and prices. It is understood that ITC has also scaled up its information system with mobile applications. Just the proliferation of usage itself becomes a boom.

   Jensen (2007), in his study on information (technology), market performance and welfare of the South Indian fisheries sector, has observed that as mobile phone coverage increased in Kerala, fishermen bought phones and started using them to look for beach auctions along the coast, where supplies were lower and prices higher than at their home beaches. Fishermen rapidly learned to calculate whether the additional fuel costs of sailing to the high-priced auctions were justified.

   Their group actions reduced price volatility and wastages. Price dispersion was dramatically reduced, declining from 60%–70% to 15% or less. There was no net change in fishermen's average catch, but more of the catch was sold because wastage, which previously averaged 5%–8% of the daily catch, was effectively eliminated. The rapid adoption of mobile phones improved fishermen's profits by 8%, and was coupled with a 4% decline in consumer prices.

3. There are a number of companies that use mobile networks to enable price information and link it to procurement processes from primary and secondary processors. Marico, a leading company in the consumer goods industry,

especially in edible oil and personal and hair care products, is believed to be encouraging the use of mobile networks.

4. Small and medium enterprises in the processing of agro-based products have their own networks through mobile service providers. There is a coconut oil producers' association that operates as a registered society at Kangeyam, providing necessary price and transaction information to its members. Large buyers for secondary and tertiary processing have promoted information and price efficiency by encouraging community networks. They are bidding to purchase quantity at certain price points at regular intervals during the day through short messaging services using mobile telephone networks. This gives confidence to millers and improves information efficiency for market trading activities. Appendix 10.1 discusses a supply chain network for such a business.

Thus, mobile usage and evolution of telecommunications along with smart phone usage have improved supply chain efficiency in agro-based businesses in India.

10.7   Cloud Computing

It may be useful here to highlight cloud computing and applications deployment in agribusiness. Appendix 10.1 and 10.2 throw light on a few perspectives of cloud computing.

A cloud-based agro system is a useful monitor for the agricultural sector and its constituents, such as farmers, markets, governments and intermediaries. One can provide online service facilities that are available to all the users, from any part of the country and at any time. In order to render the services, the agri-sector-oriented system may have the following services:

1. *Communication*: There is a need for communicating information on soil, testing, weather, crop-related issues, credit and so on to farmers and various other stakeholders.
2. *Research and knowledge sharing*: Agriculture requires extensive field research and analysis of data. The opportunity to include large sampling from different locations and use technology for research analysis and synthesis would help all participants in the supply chain.
3. Further, there is a wealth of best practices generated by agricultural experts, farmers and researchers that are to be propagated through the sharing of experiences. The explicit knowledge quotient is high, and a cloud-based system can help in sharing. Further, implicit knowledge could also be shared through forums and hierarchy-defined interactions.

The scope and advantages of cloud computing applications in the agro-based sector would include

1. Firms in agro-based businesses can join up to widen the network of firms that can be brought into the information technology processes, as even small players in the network can adopt cloud computing at a low cost.
2. Firms can move to cloud-based computing and thereby improve their business processes with upstream and downstream network players. Some of the potential upstream benefits are
   a. reduction in procurement cycle;
   b. conversion of request for quotations into Purchase Order and issue in a short time, especially during peak season process time at secondary and tertiary processing units;
   c. reduction of process time for accounts payable invoices and quick release of payments, which would make farmers and intermediaries, including small-time primary processors, happy;
   d. elimination of redundant processes, data entry and thereby improved process efficiency of processing firms.
3. Further, associations and institutions in support services can adopt more IT processes and greatly improve the quality of research, laboratory functions and information sharing.

Thus, cloud computing and its applications can play a significant role in the impact of IT on Indian agro-based businesses and effectiveness of their supply chain.

10.8   Conclusion

There are some unique advantages of IT usage in agribusiness supply chains for farmers as well as for processors.

1. Farmers are empowered with cell phones, and hence, they are networked and better informed about weather, superior seeds, more attractive prices, and so on.
2. Farmers are now getting immediate payments for most of their produce because the buying companies and traders are empowered by IT.
3. IT replaces some of the intermediaries and hence results in better efficiency.

IT systems and e-business models help achieve effective supply chains. Services supply chains, including that of governments, governmental bodies and research institutions have moved up in maturity level in using Internet-enabled customer support in business transactions. Manufacturing and processing units in agro-based industries and distribution-driven networks are in process of evolving in e-business adaptation, as there are still conflicts between traditional and contemporary business models. However, technology developments and adoption of equipment like smart phones and other handhelds would certainly improve the role of information technology as one of the supply chain drivers.

Appendix 10.1

SME Cluster: Processing of Coconut Kernel and Extraction of Oil—A Case Study at Kangeyam, Tirupur District*

Introduction

Karthekeyan, son of Periyaswamy Gowndar, returns to his family business after completing his MBA at one of the premier institutions in Western India and working for some time in the sourcing operations of an MNC which is in the FMCG business. As the only son of his parents, Karthekeyan inherits farmland of 50 acres of coconut groves in Pollachi near Coimbatore, and the family business of merchandising coconut kernels and processing and selling coconut oil. When he was a student, Karthekeyan had the opportunity to study at Ooty Public School, after which he did his mechanical engineering at the Premier Technology Institute. Immediately thereafter, he enrolled as a fresher for a master's course in business administration. He has nurtured dreams of making it big in life. The sudden ill health of Periyaswamy Goundar compels him to take over the trusteeship of the family wealth and settle at home.

Periyaswamy Goundar, understanding the aspirations of Karthekeyan, briefs the latter on the nuances of coconut cluster operations. He sets the tone for Karthekeyan's future moves, impressing on him the need to address the challenges of industry structures and supply chains to ensure effective operations.

Sourcing of Coconut and Kernels

Farmers in and around Coimbatore district take pride in calling Pollachi the coconut city. Pollachi enjoys favourable agro climatic conditions because of its strategic location along the Palghat pass, where it benefits from both the southwest and the

---

* *Source*: Chandrasekaran, N. *Supply Chain Management*. New Delhi: Oxford University Press India, 2010. With permission.

northeast monsoons. Further, the temperature is always pleasant throughout the year, as it is neither too hot nor too cold, which makes it conducive to nurturing coconut that give the best of yield. The farms at Pollachi are well-irrigated and fertile. The disciplined labourers and landlords and an inclusive culture are conducive for nurturing coconut groves. Apart from coconuts, farmers also cultivate betel palm trees as interregnum crops to improve their earnings.

The best period for coconut yield falls between March and October, contributing almost 75% of the annual yield. Coconuts from Pollachi are sold as tender coconuts for consumption, which are sold by and large at premiums across Tamil Nadu and Kerala. Approximately 30% of the yield during the peak season and 10% during the off season are sold as tender coconuts. The larger part of the coconuts goes in for processing.

Processed coconuts can be sold directly as kernels and used as edible products. This usage accounts for about 25% of the coconuts available for processing. The remaining 75% of coconuts is predominantly used for extraction of oil. This process occurs at two levels. First, premier FMCG companies such as Marico Industries, CavinKare, Hindustan Uni Lever (HUL) and others procure kernels from vendors and process them either on their own or through contract manufacturers for selling as branded oil. Second, a large portion of material is being crushed by small processers and sold in bulk through commodity markets; their produce goes primarily to rural market and semi-urban markets for consumption, and also as intermediary to the personal care product segment.

Direct Consumption Trade

Produce that moves as tender coconuts from Pollachi is traded by middlemen/brokers at different consumption centres. They fix a farm gate price and move produce through road transport, mainly in trucks of 10/16 tonne capacity, to urban locations such as Chennai, Coimbatore, Trichy, Madurai, Cochin, Bangalore and so on. Typically, more tender coconut is sold during the March to September period every year, when the rest of Tamil Nadu goes through summer climatic conditions. The volume of sale is generally observed to be stable, and the influencing factor has been the availability of trucks at reasonable rates. A truck approximately carries 10,000 units of coconuts; the cost of transportation to Chennai, including secondary distribution to tender coconut sale points, is about ₹1.50 per piece. The market price for a tender coconut is ₹15. The farm gate price is about ₹10.

Process Cluster at Kangeyam

There are about 500 units that process the kernels brought from Pollachi. There are about 70 units that have crushing and oil extraction capacities. The average size of the oil extraction plants is about 10 tonnes per day. These units buy coconuts from Pollachi and also from other locations such as Theni, Uttamar Palam of Kambam district, Rajapalayam, Thenkasi, Dindigul, Vedasandur, Pudukottai and Thanjavur districts. Pollachi is about 80 km by road from Kangeyam, and it is possible transport coconuts at a transportation cost of ₹2000 per 10-tonne load. The other locations from where about 30% of the requirement is brought would average about ₹6000 per 10-tonne load.

One may like to understand two aspects of this cluster. First, the climatic conditions at Kangeyam support kernel processing, as the area receives less rainfall but has good climatic conditions. Second, entrepreneurial attitudes, commitment of labour and the natural reinforcement of family orientation towards this business have helped develop this business in this region. Apart from processing of kernels, rice mills are also key agro businesses established in this location.

It may be noted that out of the 500 units, only 100 process kernel throughout the year. The remaining 400 units are predominantly seasonal in operation. Further, from the statistics on the distribution of process activity, only 70 out of the 500 units, which is about 14%, are in the business of extraction of oil. One must understand the structure of this business. In sun drying, an operation of large size, grading is manual. These two activities are not easily amenable to automation at viable costs. Another important characteristic is that there is a limitation to sun drying and grading, as it required about 10–15 days of labour, and can be done only by those who have acquired the skill for fast handling through years of experience. One of the major risks is the kernels getting wet by exposure to rains. In such an event, there will be huge losses, as the material mix is as follows:

Oil content 63%–64%; oil cakes about 31% or 32%; vapour 4%–6%. The cluster gets justified because of organizational efficiency and environmental factors.

One would like to know some more details regarding the dispersion of process capacities. About 86% of the units that process kernels typically have two tonnes a day capacities. They procure coconuts and make kernels and trade them. They would not be able to process kernels to edible oil because of economic viability, especially due to the material mix of the grades of kernels. Typically, only 20%–25% of kernels would undergo direct consumption. The remaining 75%, a mix of full graded kernels and broken kernels, determines the ability to process additionally. Further, the economic viability of oil processing depends on the ability to reduce wastages. Another important factor in oil processing is ensuring utilization during the off season as well.

The typical economics of coconut, kernel and processing are shown here.

Price of per kernel grade wise (₹)[a]	28,000	27,000	25,000
Average grade distribution of kernel	30%	30%	40%
Yield—edible oil per tonne	64%	62%	60%
Yield—oil cake per tonne	31%	32%	34%
Market price of edible oil per tonne (₹)	48,000		
Market price of edible oil cake per tonne (₹)	4,000		
Processing cost per tonne of oil (₹)	1,000		
Coconut landed cost (₹)	10–11 per unit		
No. of coconuts for 1 tonne of kernel	2,500 units	2,600 units	2,700 units

[a] Adjusted for wastage.

One may note that the economic viability of edible coconut oil processing depends on the oil recovery percentage and the quality of kernels purchased. In terms of price recovery, edible kernels are sold to intermediaries who take the product to Andhra Pradesh and North India for sale there at a premium of ₹30,000 per tonne. The next grade of kernels is purchased by branded oil processers at a price of ₹29,000 per tonne. There will be wastage of 3%–5% while selling to these two markets, and shape and grade are important. The balance would be disposed of as low value kernels. Details of the utilization of coconut in processing are given below.

Processing Activity

1. Coconuts still in their shells are received at the yard and stocked in heaps covered with coconut fibre.
2. Manual separation of kernels and shells is done, and the shells are stacked in heaps that go for usage in boilers or for direct sales.
3. Kernels go for sun drying in open yards at slotted locations.
4. After six to ten days of sun drying, kernels are graded manually into three lots, namely, one for consumption, two for sale to processors and three for in-house processing. While grading, the labourers use two bins for grading, with two bins available for usage, similar to the *kanban* practice.
5. Apart from their own purchases of coconuts, units also buy sun-dried kernels from process units to supplement their business volumes. This input reaches the grading stage.

6. The edible grade lot is aggregated and sold through intermediaries. Kernels for processing outside are again sold through processors, and delivery is made within a mutually agreed time window from the unit.

7. The left over kernels are moved through hoppers to crushers or through an intermediate steam drying process before being transferred to crushers, depending on the moisture content. Two levels of crushing take place, and oil cakes with left over oil are sold to solvent extraction process units.

8. The finished oil is taken to storage units by tanker trucks of 10–15 tonne capacity. A sizeable portion of the material is also packed in 187 kg barrels; on specific demand, 15 kg tin packing is also undertaken.

9. The unit employs families who provide labour and live inside the campus. During this season, mills operate three shifts; one production manager and a promoter's representative take care of the managerial demands.

Information and Price Efficiency

There is a coconut oil producers' association which operates as a registered society at Kangeyam, providing necessary price and transaction information to its members. Operators like Marico Industries have promoted information and price efficiency by encouraging a community network, and there is bidding for purchase quantities at certain price points at regular intervals during the day through short messaging services using mobile telephone networks. This gives confidence to millers, and improves information efficiency for trading in the market.

Karthekeyan inherits a mill which is involved in trade of all three grades of kernels. Pattern of selling kernels is as follows: directly (30% of total volume); selling another 30% to millers; and the remaining 40% is used for processes in his 10-tonne plant. He wants to critically examine his supply chain drivers, namely, facility planning and utilization, labour management, storage, inventory, sourcing, pricing and product mix. He would like to evaluate and arrive at strategic decisions in terms of going bigger, setting up a regional brand and establishing himself as a commodity trader. You may evaluate his options and prepare a report accordingly for submission to Mr. Periyaswamy Goundar.

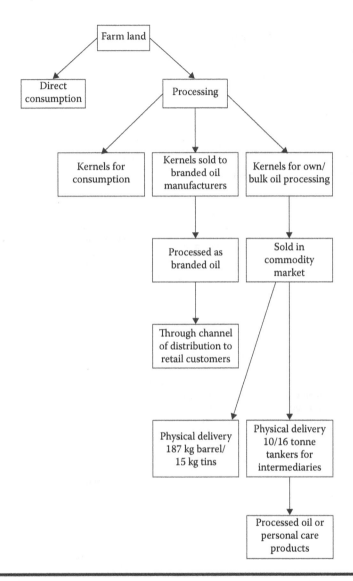

Figure A10.1   Options for coconut utilization across value system.

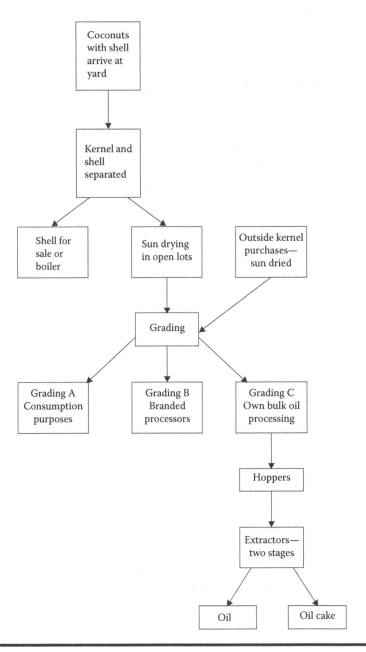

Figure A10.2   Kernel processing at a unit level.

Appendix 10.2

*Cloud Computing and Its Perspectives**

Cloud computing has caught on for IT applications in business, government and individual segments. The following points may be noted on cloud computing and its perspectives.

1. An end-to-end cloud service offering can be designed using a combination of any of the four layers comprising of software as a service (SaaS), platform as a service (PaaS), infrastructure as a service (IaaS) and desktop as a service (DaaS) which is the cloud computing network clients. Table A10.2 gives an idea of cloud structures.
2. A cloud could be deployed as:
   a. A public cloud, where storage and other resources are made available to the general public by a service provider. These services are free or offered on a pay-per-use model. Generally, public cloud service providers such as Amazon AWS, Microsoft and Google own and operate the infrastructure and offer access only via the Internet.
   b. A community cloud, which shares infrastructures between several organizations from a specific community with common concerns (security, compliance, jurisdiction, etc.). This can be managed internally or by a third party and hosted internally or externally. The costs are spread over fewer users than a public cloud.

Table A10.2   Cloud Computing Structure

Structure Node	*Description*
Cloud clients	Web browser, mobile apps, thin client, terminal emulator and so on
SaaS–Application	CRM, email, virtual desktop, communication, games
PaaS–Platform	Execution runtime, database, web server, development tools
IaaS–Infrastructure	Virtual machines, servers, storage, load balancers, network

---

* *Source*: Nath, B. and Chaudhuri, S. Application of cloud computing in agricultural sectors for economic development. International Conference on the Interplay of Economics, Politics, and Society for Inclusive Growth, Oct 15, 2010.

    c. A hybrid cloud is a composition of two or more clouds (private, community or public) that remain unique entities but are bound together, offering the benefits of multiple deployment models. Hybrid clouds lack the security and certainty of in-house applications. Hybrid clouds provide the flexibility of in-house applications with the fault tolerance and scalability of cloud-based services.

    d. A private cloud, which is a cloud infrastructure operated solely for a single organization, whether managed internally or by a third-party and hosted internally or externally.Undertaking a private cloud project requires a significant level and degree of engagement to virtualize the business environment. They have attracted criticism because users "still have to buy, build, and manage them" and thus, do not benefit from a lesser hands-on management style.

3. An International Data Corporation (IDC) study commissioned by Amazon Web Services (AWS) found that the five-year total cost of ownership (TCO) of developing, deploying and managing critical applications on AWS represented a 70% savings compared with deploying the same resources on-premises or in hosted environments. As stated earlier, the IDC findings show that the average five-year ROI of using AWS is 626%. Moreover, over a five-year period, each company saw cumulative savings of $2.5 million per application, IDC said. And TCO savings included savings in development and deployment costs, which were reduced by 80%. In addition, application-management costs were reduced by 52% and infrastructure support costs cut by 56%, while organizations were able to replace $1.6 million in infrastructure costs with $302,000 in AWS costs. The IDC study also showed that benefits also increased over time. The study found a definite correlation between the length of time customers used AWS services and the returns they gained. At 36 months, these organizations are realizing $3.50 in benefits for every $1 invested in AWS; at 60 months, they are realizing $8.40 for every $1 invested, according to the report.

4. While calculating return on investment (ROI) in cloud offering, one must note that the IT capacity and IT utilization would have a significant impact on infrastructure. IT capacity is measured by storage, CPU cycles and network bandwidth or workload memory capacity as indicators of performance. IT utilization is measured by uptime availability and volume of usage as indicators of activity and usability.

There are other benefits that determine ROI on cloud offerings. They are

    a. *The speed and rate of change*: Cost reduction and cost of adoption/de-adoption is faster in the cloud. Cloud computing creates cost benefits by reducing delays in decision costs by adopting pre-built services and a faster rate of transition to new capabilities. This is a common goal for business improvement programmes that lack resources and skills, and are time-sensitive.

b. *Total cost of ownership optimization*: Users can select, design, configure and run infrastructure and applications that are best suited for business needs.

c. *Rapid provisioning*: Resources are scaled up and down to follow business activity as it expands and grows. They are redirected.

d. *Increased margin and cost control*: Revenue growth and cost control opportunities allow companies to pursue new customers and markets for business growth and service improvement.

e. *Dynamic usage*: Elastic provisioning and service management targets real end users and real business needs for functionality, as the scope of users and services evolve seeking new solutions. This would be helpful in the case of clusters and business community networks where business process solutions are to be deployed. For example, ERP for SMEs through cloud in a cluster using a cloud would help with dynamic usage.

f. *Enhanced capacity utilization*: IT avoids over- and under-provisioning of IT services to improve smarter business services. Capacity additions can be staggered, as economies of lumpiness in capacity can be spread among users instead of set-up functions in case of individual deployment.

g. Access to business skills and capability improvement: Cloud computing enables access to new skills and solutions through cloud sourcing on demand solutions.

These measures define a new set of business indicators that can be used to create a "score card" of an organization's current and future operational business and IT service needs relating to cloud computing potential.

References

Chandrasekaran, N. *Supply Chain Management*. New Delhi: Oxford University Press India, 2010.

Chandrasekaran, N., Supply network management of Chennai Engineering Ltd; A case authored for a national level management school competition organized by NITIE, 2012.

Goyal, A. Information technology and rural market performance in central India. Mobile Innovations for Social and Economic Transformation Workshop, September 16, 2009. http://siteresources.worldbank.org/EXTEDEVELOPMENT/Resources/Aparajita09.pdf

The Hindu. Rural India to drive mobile phone market growth: Report. January 23, 2011. http://www.thehindu.com/business/Industry/article1118628.ece?homepage=true

Jensen, R. The digital provide: Information (technology), market performance, and welfare in the South Indian fisheries sector. *Quarterly Journal of Economics*, 122(3), 879–924, 2007.

Nath, B. and Chaudhuri, S. Application of cloud computing in agricultural sectors for economic development. International Conference on the Interplay of Economics, Politics, and Society for Inclusive Growth, October 15, 2010. http://www.rtc.bt/ Conference/2012_10_15/4-BhagwanNath-CloudComputingAgriculture.pdf

Public Distribution System issue, *Informatics*, 19(3), 2011.

TechSangam. ITC's e-Choupal model to increase farmer revenue and its own bottom-line, September 20, 2011. http://www.techsangam.com/2011/09/20/ itc-e-choupal-model-to-increase-farmer-revenue-and-its-own-bottomline

World Bank. Mobile devices and their impact, in *ICT in Agriculture Sourcebook*, 2011. http:// www.ictinagriculture.org/sourcebook/module-3-mobile-devices-and-their-impact

Chapter 11

Financial Systems

OBJECTIVE

The objective of this chapter is to highlight the importance of financial flow and then enumerate the various financial challenges impacting supply chain. Discussions are on enabling technology including that of commodity markets, on role of corporate in improving financial flows and on credit and institutional structure for financing.

11.1 Introduction

Financial flow is one of the key aspects for achieving supply chain efficiency. It also ensures completion of supply chain activities by all role agents, who then get rewards matching their responsibilities. If there is any strain on financial flow, it would affect the overall working of the supply chain. Typically, such shortfalls lead to inefficiency by way of delays and low volumes, and as a result, drops in efficiency levels in business. Further, they lead to a collapse of the supply chain such that, at times, the firm can be thrown into financial distress. There are a number of instances in India where unorganized markets have exploited gaps in the financial systems in agribusiness, leading to a lot of media attention and public debate, all calling for stern action. This has been most common in respect of fair prices for produce and timely payment of the procurement prices. Farmers have been known to launch intense agitations, accusing firms and the government of mismanaging financial flows. This is happening not only in cooperative sector farming and agribusiness, but also in private sector agro-based industries. In this chapter, we discuss financial flows and how they impact agribusiness.

11.2 Role and Relevance

The financial flow of any enterprise is vital for its business combined with its product/service (physical) flow and information flow. It is important that supply chain flows are synchronized to enable coordination among role agents and achievement of supply chain objectives such as cost-effectiveness and responsiveness. Though obstacles in supply chains occur due to disruptions in these three flows, it is important to understand some of the critical aspects of financial flow.

Typically, a process plant performs the following activities.

1. It plans for procurement of agricultural produce from farmers, either directly through intermediaries or at the markets. Direct procurement involves the establishment of loyalty. Hence, the firm must financially support farmers during the crop production stage by way of credit advances, and adjust these advances against the final payment for procurement. In case of procurement through intermediaries, financial support may still be required, but maybe not to the same extent as with direct procurement. When the firm goes for market procurement, transactions would be more on a cash and carry basis, where financial settlement is immediate.

2. In these three formats, there are possible impacts on balance sheet items and income and expense statement entries. For example, if advances are paid towards crop production, they may involve cost of capital by way of interest on working capital. But when a firm has loyalty built with its supplier farmers, that would help it to procure their produce at comparatively reasonable prices, with low acquisition cost overall. This is because the financial benefit of disintermediation could be shared by farmers and the firm. Further, this would reduce risk by way of variability in revenue compared to market procurement and direct procurement.

3. To enjoy such financial benefits, two things are important: timely payment and fair procurement prices, which are managed through relationships.

4. Typically, in agro-based industries, material cost would be around 70%, whereas processing costs could be very low. Other significant costs could be distribution and financial costs. What is more important from the financial perspective is productivity, which is measured by material balance, yield or conversion ratio. This would affect profitability, as costs could go up and the next cycle of procurement may be affected, as in some cases prices are linked to certain basic yield parameters.

5. During the post-production stage, many agro-based products go through a long period of storage, mainly because of seasonality in input availability. For example, milk, edible oil and sugar are all seasonal, where demand is far more evenly spread out. Dairy produce is processed and converted into

butter oil and milk powder, as well as many other value-added products like ghee and butter. Similarly, though edible oil has been regulated in terms of the number of days' stock that can be carried over, the tendency would be to build stock across the network. In the case of sugar, it is normal for the output produced over six months to be sold over about 14 months. Thus, working capital flow is critical to manage stocks. Banks and institutions would fund up to 75%, and hence, a firm has to provide the balance through equity. These are the real challenges of financial flow management in agro-based businesses.

6. An example of a coffee export unit is discussed here, with the name being withheld. India is a leading manufacturer of coffee. There has been a huge demand for instant coffee powder in eastern Europe. Coffee is exported in bulk to those countries. The demand for exports would be from January to September every year, as the sales are high during the winter months. The physical flow process would be as follows:
   a. The order to buy is received.
   b. The coffee beans are procured.
   c. The coffee beans are processed into powder.
   d. Export packaging and dispatch arrangements are made.
   e. Goods are received by the customers.
7. This process goes through a financial process that is linked to the physical flow, and is described in Figure 11.1.
8. A normal operating cycle is about 90 days, and inventory turnover would be four times in a year. On receipt of an order, a company estimates its production schedules and plans its cash flow requirements. It is commonly found that 50% of the total order value may be required to start with for buying of material and planned division of capacity utilization between domestic and export demands. Coffee production needs to be committed on order-wise receipts for improving realization and consistency of quality. The promoter is expected to finance up to 25% of the value of orders/inventory, and the

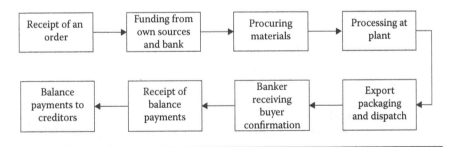

Figure 11.1 Typical financial flow cycle.

remaining balance of 75% is funded by banks. The order processing thus starts with the exporter's own funds and funds released by its banks. On completion of the order and on receipt of the letter of credit (LC), the company dispatches the goods. Depending on the type and nature of the LC and acknowledgement by the buyer or buyer's bank, the focal firm's banker releases the balance money. It is observed again that sundry debtors stay on the books for up to 120 days in this export business. The Export Credit Guarantee Corporation (ECGC) advises companies and their bankers about the creditworthiness of buyers, based on its information on defaulters and other market intelligence. There have been, however, cases when the company assumes a greater risk and takes on more contracts. Such companies get into financial trouble with a few deals, which again affects their ability to fund other orders. One observes here that smooth realization of funds against orders that are delivered on time, as per committed quality and quantity, is critical for streamlined financial operations (Chandrasekaran, 2010).

9. The importance of financial flows can be understood from this discussion: processing of agro-based goods is vertically and backwardly dependent on the orders. Any obstacle arising in financial transfers will jeopardize the entire supply chain. If an order is rejected on quality grounds, it is not just the primary vendor who gets affected, but also those who are part of the supply chain indirectly are also impacted negatively. It is important that the financial credibility of buyers and partnering organizations in supply chains be understood and shared for healthy operations. To some extent, the ECGC and banks facilitate this information sharing. Processors' associations, like in the case of tea and other plantation crops and APEDA in the case of small-scale fruit exporters themselves, provide their networks with reliable information. Their associations get valuable information on the creditworthiness of buyers, on risk perceptions and on relevant aspects, which they share among their communities. However, notwithstanding such information bridges, failure in terms of financial performance by buyers or by their bankers is common in agro-based businesses, and often, the lack of consistent quality and standards is reported as the cause of failure.

Thus, the key learning for managing financial flows is that, for processing of an order, units in the downward stream require timely release of funds through the purchaser's honouring of the invoice after delivery. The focal firm, which is the exporter, pays upstream and makes commitments for the business. If there is a financial flow disruption, then the business fails. It should be clear that the integration of financial flow with information flow and physical flow is critical. Unless the information system captures product movements and confirm them to the funders, fund release cannot be made on time.

11.3 Problems in Synchronization

1. Financial bankruptcy or failure of the organization occurs when supplies are not being paid for, especially when they are agro-based products. The focal firm in the supply chain in such cases goes in for a recovery mechanism, which can be arduous and time-consuming. It is possible that a farmer who is a producer and markets his goods is also the focal entity. Any problem with respect to disruption in financial flow could be due to an insufficient amount of money recovered through sales, or because of differences over the quantity or price of the produce marketed. Appendix 11.1 gives the details of the prevalence of farmers' debt scenarios in various parts of India.

2. It is theoretically possible for farmers to hold stock for want of the right price or market response. But in reality, this does not happen, as farmers' average land holding is low, and the pressure to realize money is high. Farmers borrow capital for raising farm output. Debt is based on estimated yield, expected price and additional collateral, as the case may be. A sudden reduction in yield or the failure of market agents or processing channel partners to buy adequate quantities puts pressure on the supply chain. Such situations lead to huge inventory holdings, involving cost, and poor realization on farm output.

3. There could also be occasions of quality variance observed at the customers' end, with resulting lot rejection. Though this is a physical flow management issue, it has got a serious impact on financial flow management. This is mainly because when the goods are rejected, the network firm needs to perform a reverse logistic operation and liquidate stock at huge discounts. Thus, cash flow and credibility are affected, and future funding becomes difficult for the farming community involved.

4. From the manufacturing and sourcing nodes of supply chains, the following could lead to financial obstacles:
   a. Financial solvency of the firm is affected by poor management of funding, and by diversion of funds from operation to capital accounts or personal accounts
   b. There could be over-exposure to risk and aggressive order booking without adequate cover. There have been times when a focal firm has booked orders on trust and faith that are misplaced, and gets into financial distress.
   c. Similarly, there could be over-exposure to purchase of material, or the booking of capacity of outsourced partners may not match orders, and thereby, the outflow of cash leads to financial imbalance. This is true for secondary and tertiary processors who could hedge on stocking and price variances.
   d. An over inventory management policy leads to the blockage of funds. This is a contraposition of the above example, where the firm or the individual farmers, depending upon the position in the supply chain, try to hold stock, blocking huge funds. At times, this becomes a speculative move, rather than a calculated risk.

e. Poor fund management and improper funding sources lead to high cost of funds. This leads to a loss of financial equilibrium. This is true for Indian agro-based businesses, as unstructured and informal sources of funding are often used.

f. Further, there could be poor checks and controls for receivables and payable management, and thereby, a lack of trust among the partners, leading to strain on physical and financial flows.

g. Instances of inability to provide documentation for banks to fund on time, and thereby, getting into financial stress are common. This is because of inadequate knowledge and discipline with respect to record keeping, which may be encouraged by the informal financial sector. This leads to a creeping inability to raise capital from the organized sector.

One of the financial aspects that influence operating decisions and thereby could prop up as obstacles is fiscal factors like taxes, duties and direct controls. Proper document handling is fiscal major requirement for purposes of clearance and movement of goods; documents must be properly presented and validated. There are third party service providers who enable this function. Second, government levies, especially of sales tax in India, are different for movement within states and outside states. To take advantage of sales tax, companies earlier used to locate warehouses in major states with substantial demand. This could create an obstacle especially arising from obsolescence of inventory, the behavioural patterns of warehouse operators, and so on. With implementation of a general sales tax that may lead to a uniform sales tax structure, the location of warehouses could be more rational. Another important fiscal element is direct physical checking at state borders, which leads to delays. Implementation of technological solutions like whole truck scanning might improve this situation.

11.4 Role of Technology

Technology plays an important role in managing financial flows. Financial institutions, banks, large corporates and intermediaries have implemented technological solutions. This provides visibility across departments and enables partnering organizations to monitor critical financial chocking points and expedite decisions to streamline the flow. Creditors and debtors no longer need to do excessive follow-up. This is because required flags are set to provide alerts for prioritizing payment; an easy flow of financial resources and physical goods is enabled through IT systems.

The level of openness and transparency with which an organization operates determines its information and financial flow efficiency. For example, when a vendor completes delivery against an order, the vendor becomes eligible for payment on the 21st day after delivery of goods and receipt of goods received note. A trigger is automatically generated for cash flow planning and any gap is highlighted, so that the concerned parties can take necessary steps to adjust the financial flow and make the supply chain more effective.

However, there are a number of cases that one comes across in real life, where such synchronization does not take place. Now this can be baffling to analysts: how and why do process and technology fail to streamline the financial flow? This is more common in cases of payments against supplies to farmers and small firms who have limited bargaining power in the supply chain.

IT systems must be linked to backward and forward linkages and to all partners in the system. The system must be transparent and provide access for partners to check their status, as relevant to their part of the network, using Internet connections. Any aberration in the schedules must be openly discussed among the partners, instead of efforts of back-door follow-up and intelligence sourcing by partners, which can lead to strains or even break up in network relationships. It is a common occurrence that a focal firm has some cash flow crunch. In such an event, the payment schedules may have to be prioritized, and technology cannot capture such intervention and scheduling. Even if there are valid reasons for a delay, some of them can be shared only orally, through trust-based relationships, rather than through technological processes. Hence, the supply chain's financial flows must be sensitive to such occurrences and must take cognizance of the long-term supply chain goals of establishing equitably shared normal profits.

Thus, we have discussed the importance of financial flows, areas where there could be bottlenecks and how they can be approached. It may have been observed that most of the discussions were firm-centric, whether primary, secondary or tertiary processing firms. Wherever we have mentioned farmers, it has been more from the perspective of the upward linkages in supply chains, rather than from the individual economic unit. In fact, as individual economic units, farmers can themselves become focal points and form their own supply chains. In such cases, for understanding financial flows, it becomes important to understand credit structures in India. These discussions throw light on the credit structure for farmers as well as for others who are funded in the agribusiness sector.

11.5 Credit Structure in India

As may have been noted from the discussions in earlier chapters, Indian agriculture has grown phenomenally. This was possible with the growth and evolution of credit structures for agro-based businesses. Our objective here is to highlight these credit structures; we suggest readers must go to open sources to understand the extent, scope and limitations of the agricultural credit structures in India. This is mainly because of the huge volume of material available at open sources and the span of issues covered, which could digress from the focus of this book.

The Government of India (GOI) plays a large role in shaping up agriculture. The Ministry of Agriculture consists of three departments through which it directly acts, regulates and coordinates with other entities in the ecosystem for growth. These departments are

1. the Department of Agriculture and Co-operation (DAC);
2. the Department of Agricultural Research and Education (DARE); and
3. the Department of Animal Husbandry and Dairying (DAH&D).

    These departments are vested with numerous responsibilities for shaping Indian agriculture. This link provides details: http://agricoop.nic.in/Orghistory.pdf. For this discussion on financial flows and the government, the following aspects are relevant.

    a. Agriculture-based industries with limits to formulation of demands and fixation of targets; price control of agricultural commodities except food grains, sugar, vanaspati, oilseeds, cakes and fats, jute, cotton and tea; cooperation in agricultural sector;

    b. agricultural credit and debt of farmers; and

    c. general policies relating to the marketing of agricultural produce, including pricing and exports.

4. The central government, directly or through state governments and various bodies like the Reserve Bank of India (RBI), the National Bank for Agriculture and Rural Development (NABARD), department-managed institutions like public sector undertakings, autonomous bodies and national level cooperative organizations (refer to Appendix 11.2);

5. While there are many schemes, operated directly or through its agencies by the GOI, which impact financial flows, the following two schemes would be most relevant.

    a. Kisan Credit Cards—The Kisan Credit Card is a credit delivery mechanism for providing credit to farmers under a single window, with flexible and simplified procedures and adoption of whole farm approaches. This focuses on providing adequate and timely credit to farmers. Further, it meets farmers' short-term, medium-term and long-term credit needs for agriculture and allied activities, with a reasonable component for consumption needs. Appendix 11.3 explains the extent of Kisan Credit Cards.

    b. Mahatma Gandhi National Rural Employment Guarantee Scheme (MGNREGS)—The Mahatma Gandhi National Rural Employment Guarantee Act (MGNREGA) was enacted by legislation on August 25, 2005, to provide job guarantees. The scheme provides a legal guarantee for 100 days of employment in every financial year to adult members of any rural household willing to do public work–related unskilled manual work at the statutory minimum wage of ₹120 per day at 2009 prices. The central government's outlay for this scheme was ₹40,000 crore in FY 2010–2011. This act was introduced with the aim of improving the purchasing power of rural people by providing primarily semi-skilled or unskilled work opportunities to people living in rural India, whether or not they are below the poverty line. Around one-third of the stipulated work force is women. In 2011, the programme was widely criticized for being ineffective, or at the least, for being less effective than other poverty reduction programmes in India. MGNREGA is also inflicted with controversy about corrupt officials, deficit financing as the source of funds, the poor quality of infrastructures built under this programme, and its unintended destructive effect on poverty.

Thus, GOI schemes like MGNREGA can impact financial flows, as easy options for labour deployment in the exchange markets make labour costly and difficult to access. This impacts adversely the financial viability of farming and allied activities. However, one must pay some accolades to the number of programmes organized by the government to ease financial flows in the agro-based sector.

11.5.1 Reserve Bank of India (RBI)

RBI is the central bank of India and is responsible for managing monetary policy; thereby, it regulates credit policy and demand and supply for money. The objectives of monetary policy management by RBI include maintaining price stability and facilitating growth, especially through employment generation. The broad areas of economic sectors include agriculture and allied sectors, mining, manufacturing and services. In terms of GDP share, agriculture and allied sectors like forestry, logging and fishing accounted for 15.7% of the GDP in 2009–2010, and employed 52.1% of the total workforce. Industry accounts for 28% of the GDP and employs 14% of the total workforce. The balance is accounted for by the service sector.

Since agriculture engages a large workforce, monetary policy must be aimed at stimulating the agricultural economy. Apart from this, agricultural commodities have significant weight in price indices. Hence, RBI's role is significant.

With respect to financial flows to the agriculture-based business sector, RBI mainly regulates credit through the commercial banking system. Primarily, RBI enables financial flows through priority sector lending, which goes for agriculture, allied sectors and small and medium processors. The objective here is to enhance credit flow to agriculture by removing the bottlenecks in credit delivery. RBI has played a lot of constructive roles in streamlining the Indian cooperative sector, especially in areas of financial credit management. RBI is constantly increasing its efforts to set up a healthy rural cooperative credit system. This is initiated through the strengthening of regional rural banks (RRBs), providing incentives to commercial banks for investment in rural economies and ensuring an adequate and timely delivery of credit at a reasonable price. RBI also works with banks and several state governments for financial inclusion of farmers and agro-based economic agents like labourers and small-scale processors, by adopting modern technology. RBI closely works with NABARD on improving credit and other financial services to the sector.

11.5.2 *NABARD*

"NABARD was set up by the GOI as a development bank with the mandate of facilitating credit flow for promotion and development of agriculture and integrated rural development. The objective also covers supporting all other allied economic activities in rural areas, promoting sustainable rural development and ushering in prosperity in rural areas. It was started with a capital base of ₹2,000 crore provided by the GOI and the RBI. It is an apex institution handling matters concerning policy, planning and operations in the field of credit for agriculture and for other economic and developmental activities in rural areas. Essentially, it is a refinancing agency for financial institutions offering production credit and investment credit for promoting agriculture and developmental activities in rural areas." (NABARD, 2007a)

NABARD's main functions include managing credit cover planning, dispensation and monitoring of credit. These activities involve framing policies and guidelines for rural financial institutions; providing credit facilities to issuing organizations; and preparation of potential-linked credit plans annually for all districts for identification of credit potential and monitoring of the flow of ground-level rural credit. These are critical for managing financial flows to the agriculture and allied sector, as delay and inadequacy lead the players to seek the support of the informal sector, which is expensive and makes agribusiness unviable.

NABARD is also responsible for development and promotional functions of all participating entities in its charter for agriculture and rural development. Credit is deployed for capital formation, technology adoption, and production and trade. Strengthening rural financial institutions is one of the thrust areas handled by NABARD. These include facilitation of operational aspects, legal issues, engaging in relationships with state governments and their departments, training of human resources and adoption of MIS and suitable technologies for the sector.

"Another important function of NABARD which is statutory in nature is that of supervision. Since it provides refinancing credit to many institutions (as it rarely lends directly), NABARD has been sharing with RBI certain supervisory functions in respect of cooperative banks and RRBs. NABARD has been entrusted with the statutory responsibility of conducting inspections of state cooperative banks (SCBs), district central cooperative banks (DCCBs) and RRBs under the provisions of Section 35(6) of the Banking Regulation Act (BR Act), 1949. In addition, NABARD has also been conducting periodic inspections of state-level cooperative institutions such as state cooperative agriculture and rural development banks (SCARDBs), apex weavers' societies, marketing federations, etc., on a voluntary basis." (NABARD, 2007a)

Apart from these, NABARD works closely on institutional and capacity building with different players in the network, and through its offices and institutions, offers training for human resource development in the sector. Figure 11.2 shows the institutional financial structure of agriculture and allied business in India, where the

GOI, RBI and NABARD play critical roles as nodal agencies responsible for policies, credit deployment, monitoring, statutory regulations, skill development and systems development all the way down to the lowest level in the hierarchy, which could be a cooperative society or a RRB or commercial bank branch.

Table 11.1 on the flow of institutional credit to the agricultural sector leads to the following inferences:

Institutional credit is provided for short-term, medium-term and long-term needs. Cooperative banks, RRBs and commercial banks are the major funding institutions.

Cooperative banks play a significant role in lending credit. They had a share of 43% of total institutional credit in the year 1998–1999, with an absolute value of ₹15,870 crore. In the year 2009–2010, after recording a 12% compounded annualized growth rate, the share of cooperative banks came down to 17%. Cooperative banks were the major lenders of short-term credit with 53% share of the money lent in 1998–1999. The total amount disbursed was ₹12,514 crore. Cooperative banks could achieve a CAGR of 13%, but their share came down to 21% of the portfolio, with total disbursement aggregating to ₹56,946 crore. In the medium- and

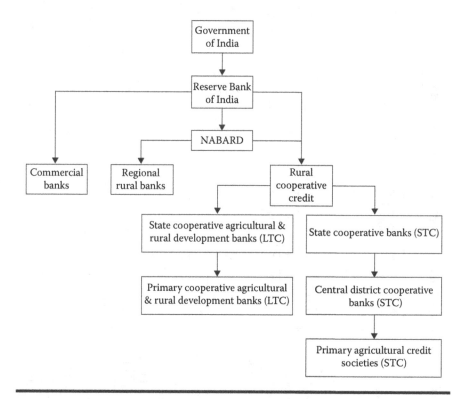

Figure 11.2   Institutional financial structure of agriculture sector in India. (From Mohan, R., *Reserve Bank of India Bulletin,* **November 2004.)**

Table 11.1 Flow of Institutional Credit to Agriculture Sector

(In Crore) ₹

Particulars/Agency	1998–1999	1999–2000	2000–2001	2001–2002	2002–2003	2003–2004	2004–2005	2005–2006	2006–2007	2007–2008	2008–2009	2009–2010	CAGR
I. Production (ST) Credit													
Cooperative banks	12514	14771	16528	18787	19668	22640	27157	34930	38622	40515	40230	56946	13%
RRBs	1710	2423	3245	3777	4775	6088	10010	12712	16631	20715	22413	29802	27%
Commercial banks	9622	11697	13486	17904	21104	26192	36793	57640	83202	122289	147818	189908	28%
Other agencies	59	74	55	41	39	57	104	68	0	0	0	0	
Sub-total (A)	23905	28965	33314	40509	45586	54977	74064	105350	138455	183519	210461	276656	23%
II. MT/LT Total													
Cooperative banks	3356	3489	4190	4737	3968	4235	4074	4474	3858	3169	5962	6551	6%
RRBs	750	749	974	1077	1295	1493	2394	2511	3804	4099	4352	5415	18%
Commercial banks	8821	13036	14321	15683	18670	26249	44688	67837	83283	58798	81133	95892	22%
Other agencies	30	29	28	39	41	27	89	314	0	0	0	0	
Sub-total (B)	12957	17303	19513	21536	23974	32004	51245	75136	90945	66066	91447	107858	19%
ST+MT/LT Credit													
Cooperative banks	15870	18260	20718	23524	23636	26875	31231	39403	42480	48258	46192	63497	12%
RRBs	2460	3172	4219	4854	6070	7581	12404	15223	20435	25312	26765	35217	25%
Commercial banks	18443	24733	27807	33587	39774	52441	81481	125477	166485	181088	228951	285800	26%
Other agencies	87	103	83	80	80	84	193	382	0	0	0	0	
Grand total (A+B)	36860	46268	52827	62045	69560	86981	125309	180485	229400	254658	301908	384514	22%

Source: Ministry of Agriculture, *State of Indian Agriculture 2011–12*. Department of Agriculture and Cooperation, 2012.

long-term segments of the portfolio, cooperative banks' share has again dwindled, and has achieved a CAGR of 6%.

RRBs play a significant role in institutional credit disbursement. RRBs had lent ₹2,460 crore in 1998–1999, with a share of 7%, and had achieved a CAGR of 25%. An amount of ₹35,217 crore of total credit was lent by all these institutions. RRBs have achieved a significant growth of 27% CAGR in the case of short-term capital for the period 1998–1999 to 2009–2010 and 18% CAGR for the same period in the medium- and long-term capital segments.

Commercial banks are the leaders in institutional credit for the agriculture and allied sectors. Commercial banks had deployed ₹18,443 crore in 1998–1999, with a share of 50%. And by 2009–2010, they had lent ₹285,800 crore, with a share of 74% of total money lent. The CAGR for this period was 26%. Commercial banks have the lead in the short-term capital portfolio with a CAGR of 28% for the same period; their share of the portfolio was at 60% in 2009–2010. Similarly, the growth rate has been even higher in the medium- and long-term portfolios, where commercial banks achieved a CAGR of 22% and their share went up from 68% in 1998–1999 to 89% in 2009–2010.

Thus, institutional credit to the sector has been growing at 22% (CAGR). This is higher than the growth rate in agricultural production in value terms. This leads to the inference that the informal sector is increasingly being replaced by institutionalized credit. Further, the more organized sector, comprising commercial banks and RRBs that are well regulated by apex bodies, are increasing their share. This proves that the sector is becoming increasingly formalized. However, critics express doubt as to whether the disbursements are actually reaching the real beneficiaries and helping the real sector to grow.

We consider such discussions out of the scope of this book, and are reasonably confident that the proliferation of commercial banks and RRBs in terms of geographical disbursement, together with the effective supervision and active role of apex institutions, has improved financial flow in agribusiness in India.

11.6   Commodity Markets

Generally, spot trading takes place mostly in regional markets (*mandis*) and unorganized markets, where farmers are paid in cash based on the prevailing price. With markets being isolated and the weak information available to farmers, spot markets are not truly reflective of demand and supply. Often, these spot markets do not consider stocks and future trading. Even if they do so, their considerations are based more on inadequate information. Such things lead to

1. price inefficiency, resulting in poor realization, and thereby, lowered profits;
2. improper allocation of supply chain profits, as a few intermediaries make disproportionate returns for their effort;

3. to allocating resources inefficiently, without realizing the future market, as may be reflected in crop patterns, sowing patterns and cropping season outputs.

The commodities market is expected to manage such financial flow more efficiently, leading to improved supply chain profits and responsiveness. Realizing such potential, the first commodity derivative market in the world was started in Chicago in 1865. After the Board of Trade of Chicago, other trading centres such as Kansas, Minneapolis and New York started functioning. Futures trading in commodities like rubber, soya bean, black pepper, etc., were started in the United States after 1920. India is one country that has also launched commodities futures markets. Early products were cotton (Mumbai), oilseeds (Gujarat and Punjab), wheat (Hapur, 1913) and raw jute (Kolkata, then called Calcutta, 1912). During World War II, futures trading was banned.

In the post-independent era, the Forward Contracts (Regulation) Act (FCRA, 1952) was passed in 1952. This led to the formation of the Forward Markets Commission (FMC) in September 1953. The FMC, headquartered at Mumbai, is a regulatory authority which is overseen by the Ministry of Consumer Affairs and Public Distribution, GOI. The structure of commodity markets in India is given in Appendix 11.4. Currently, "five national exchanges, viz., the Multi Commodity Exchange (MCX), Mumbai; the National Commodity and Derivatives Exchange (NCDEX), Mumbai; the National Multi-Commodity Exchange (NMCE), Ahmedabad; the Indian Commodity Exchange Ltd., Mumbai (ICEX); and ACE Derivatives and Commodity Exchange, regulate forward trading in 113 commodities. Besides, there are 16 recognized commodity-specific exchanges regulating trading in various commodities, approved by the Commission under the Forward Contracts (Regulation) Act, 1952.

The commodities traded at these exchanges comprise the following:

■ Edible oilseeds complexes like groundnut, mustard seed, cottonseed, sunflower, rice bran oil, soy oil, etc.
■ Food grains—wheat, gram, dal, *bajra*, maize, etc.
■ Metals—gold, silver, copper, zinc, etc.
■ Spices—turmeric, pepper, cumin seeds, etc.
■ Fibres—cotton, jute, etc.
■ Others—gur, rubber, natural gas, crude oil, etc.

Out of 21 recognized exchanges, the MCX, Mumbai; the NCDEX, Mumbai; the National Board of Trade (NBOT), Indore; the NMCE, Ahmedabad; and the ACE Derivatives & Commodity Exchange Ltd. accounted for 99% of the total value of the commodities traded during the year 2011–2012." (Ministry of Finance, n.d.)

Mukherjee, in his 2011 study on the impact of futures trading on the Indian agricultural commodity market, makes the following inferences:

The growth in agriculture commodities in India is found to be in 28–30 out of 113 commodities listed.

The national level exchanges are endowed with modern state-of-the-art technology, with electronic online trading systems. They provide their facilities at consumers' doorsteps. However, the commodity market functionaries and potential users, both hedgers and speculators in agricultural commodities and their related products, seem to be reluctant to avail of the exchanges' services and facilities. The commodity future market in India is still at a nascent stage compared to the equity market.

This unsatisfactory growth of futures contracts in agricultural commodities in India has raised questions about the benefits and feasibility of futures trading. There are critics who label them as the main factor behind rising inflation in the Indian economy.

The FMC, in collaboration with the exchanges and other related bodies, has been disseminating information on agricultural commodity prices across the country on a real-time basis by installing electronic price ticker boards in all *mandis*/APMCs that are networked under the AGMARKNET project. This is expected to propagate the adoption of futures agriculture commodity markets in India.

The Indian commodity market development can be seen from two extreme perspectives:

1. The first is protection of the essential commodity market through government intervention.
2. The other is opening up of the sector and getting the necessary protection through market-based instruments like commodity futures contracts.

There are a number of research studies on the impact of futures markets on spot prices and inflationary trends. The results are inconclusive, even though inflationary pressure on the prices of commodities, especially agricultural commodities, has gone up sharply after the introduction of commodity futures contracts. The destabilizing effect of the futures contracts is casual in nature for inflationary trends, and tends to vary over a long period of time.

Futures markets could provide a comparative advantage in disseminating information, leading to significant price discovery and risk management. Such outcomes can help to successfully develop the underlying commodity market in India.

There is a need to strengthen the market structure, using measures such as integration of widespread spot markets, wider participation in futures trading (e.g., by bankers, farmers, investors, etc.), making available necessary transport and storage infrastructure facility, enactment of necessary reforms and synchronization of acts of different regulatory bodies (such as FMC, RBI, SEBI), etc. A successful history of futures trading, with a proper regulatory framework, is essential to develop the underlying commodity market of a fast-growing economy.

It may be concluded here that the futures commodity market plays a significant role in agribusiness supply chains, especially in terms of better price returns and improved information flow, aspects that would facilitate the streamlining of supply chains.

11.7   Corporates in Agribusiness

The corporate sector entities in India who are into agribusiness have been able to showcase a few success stories in terms of managing financial flows, and thereby, achieving success. A number of such examples have been mentioned in different chapters. Here, a few of them are recapped quickly to reiterate the fact that effective management of financial flow is a prerequisite for successful participation in agribusiness in India, given the complexity of landholdings, diversity in agronomic conditions, market spreads, cultural patterns and the (lack of) maturity of the financial systems.

Amul, a member of GCMMF, established leadership in dairy production and marketing. This was possible with a cooperative structure because the suppliers could be paid the right prices on time. Daily milk collections are paid for on the following day based on the quality, which helps to win over the loyalty of farmers, and in turn, helps the growth of the milk producing unit. This also improved the overall effectiveness of the supply chain, whereby farmers as well as consumers benefited. A number of private dairy firms like Hatsun Foods, Heritage Food and CavinKare were able to establish similar models for dairy businesses and succeeded.

ITC's *e-choupal* helped farmers access real-time price information, thus enabling them to make informed decisions on selling their output. This helped reduce the number of intermediaries in the system, thus improving supply chain efficiencies. ITC ensured that farmers received a bigger piece of the financial flow pie, by ensuring transparency in all aspects. This built up farmers' loyalty and thus, a new business model was created.

There are a number of processors of items like edible oil, sugar and plantation crops like tea, coffee and rubber, where again streamlined financial flows help achieve effective supply chain management. Some of these industries feature the presence of private parties that are both listed and widely held companies and closely held companies. Further, there are public sector and cooperative sector units also present. This variety of ownership structures can be an impediment to smooth financial flows. However, widely held professionally managed companies have performed well through effective management of finance. Some of these are EID Parry, Shree Renuka Sugars, the K.K. Birla group of sugar companies, DCM Shriram and Bajaj Sugar Mills, to name a few. Exhibit 11.1, from Pepsi, discusses the improvement in supply chains brought about by corporate involvement.

Exhibit 11.1 Pepsi Contract Farming and Farmers Benefit in India

Since 1989, PepsiCo India has worked closely with farmers to help improve both their livelihoods and agricultural yield. They have introduced a high-yielding variety of tomato, and then helped paddy farmers increase their crop. Currently, they are into crop diversification and the farming of high-quality potatoes and other edibles.

The company's vision is to create a cost-effective, localized agri-supply chain for its business by building trust and relationships, leading to crop diversification and yield management. PepsiCo India's potato farming programme reaches out to more than 12,000 farmer families across six states. PepsiCo India has an assured buy-back mechanism at a prefixed rate with farmers. This insulates them from market price fluctuations. Through their tie-up with the State Bank of India, PepsiCo helps farmers get credit at a lower rate of interest. The company has arranged weather insurance for farmers through its tie-up with ICICI Lombard. According to company sources, in 2010, contract farmers in West Bengal registered a phenomenal 100% growth in crop output, creating a huge increase in farm income. Farmers recorded a profit between ₹20,000 and 40,000 per acre, as compared to ₹10,000–20,000 per acre in 2009.

It is also reported that there has been a yield improvement by over 300%, and the length of the tomato season more than doubled, resulting in a substantial increase of farmer incomes. The success of contract farming has spread, and PepsiCo engages with over 22,000 farmers across the country to grow a variety of crops.

Source: PepsiCo. Partnership with farmers. n.d.

It is important for the corporate sector to provide the right prices and timely payments, and give visibility to financial flows involving fair practices for agribusinesses to grow and prosper in India. Supply chain partners must work in unison to create a congenial climate, and stay together with trust and cooperation to sail through the vagaries of nature and cyclic trends in business. If any of the constituents is to apply pressure on the focal firm to perform during downtrends, it might stress and strain the entire supply chain.

11.8 Conclusion

For long the Indian agribusiness sector has been plagued by informal credit structures where farmers and small and medium processors are the losers. In addition, there is a need for a regulatory framework under the Essential Commodities Act and various welfare schemes to regulate agribusiness for achieving food security, scope for work and income and so on. With the growth of structured financial institutions led by the Ministry of Agriculture and Finance and apex institutions like RBI and NABARD, commercial banks, RRBs and cooperative societies have been able to infuse financial support for improving the sector. The advent of commodity futures market and the liberalization of policies allowing greater role for corporate, though short of expectations, are impacting the agribusiness sector positively, as financial flows are streamlined and even more important, constituents of supply chains are able to see visibility and objectivity in flow. This is typical of a growth economy which may have challenges but evolves in maturity to handle larger economic responsibilities, necessary to cater to the world's second largest population, a population that features huge diversity.

Appendix 11.1

Prevalence of Indebtedness in India

It is reported in various sources that about 17,500 farmers killed themselves every year between 2002 and 2006. The most suicides occurred in the states of Andhra Pradesh, Maharashtra, Karnataka, Kerala and Punjab. It may be observed from Table A11.1 that the level of indebtedness among farm households are high in these states. In 2006, the state of Maharashtra, with 4,453 farmers' suicides, accounted for over a quarter of the all-India total of 17,060, according to the National Crime Records Bureau (NCRB). NCRB also stated that there were at least 16,196 farmers' suicides in India in 2008, deaths between 1997 and 2008 is 199,132.

Our interest is not to discuss about suicides, but to highlight how obstacles in financial flows can hamper supply chains. It is important to note that the right prices and timely payments for role agents are critical for the synchronized flow of goods and services in agribusiness supply chains.

Table A11.1   India's Levels of Indebtedness Across States among
Farmer Households

Farmer Households' Level of Indebtedness	States
61%–82%	Andhra Pradesh, Karnataka, Kerala, Tamil Nadu and Punjab
41%–60%	Gujarat, Haryana, Maharashtra, Madhya Pradesh, Odisha, Rajasthan and West Bengal
21%–40%	Jammu and Kashmir, Himachal Pradesh, Uttar Pradesh, Bihar, Chhattisgarh, Jharkhand, Sikkim, Tripura, Mizoram, Nagaland and Manipur
4%–20%	Uttarakhand, Meghalaya, Assam, Arunachal Pradesh

Source: Sajjad, H. and Chauhan, C., *Journal of Geography and Regional Planning* 5, 397–408, 2012.

Appendix 11.2

Kisan Credit Cards

It may be observed from Table A11.2 that 94,864,425 Kisan Credit Cards were issued as of June 2010. Cooperative banks have a 41% share, RRBs have issued 15% of the total cards and the remaining 45% is issued by commercial banks. Though commercial banks share nearly 80% of the total credit disbursed, their share of cards issued is much lower (NABARD, 2007b).

Table A11.2   State-Wise Number of Kisan Credit Cards
Issued up to June 30, 2010

Sr. No.	State/UT	Cooperative Banks (CB)	Regional Rural Banks (RRB)	Commercial Banks	Total
1	Uttar Pradesh	6292418	3991882	6916545	17200845
2	Andhra Pradesh	4007436	2153853	9919395	16080684
3	Maharashtra	5407593	329834	2971921	8709348
4	Tamil Nadu	1649340	297334	4223972	6170646
5	Madhya Pradesh	3443912	592372	1813064	5849348

(Continued)

Table A11.2 (*Continued*)   State-Wise Number of Kisan Credit Cards Issued up to June 30, 2010

Sr. No.	State/UT	Cooperative Banks (CB)	Regional Rural Banks (RRB)	Commercial Banks	Total
6	Karnataka	1877896	1291404	2542616	5711916
7	Odisha	3636573	714575	1256635	5607783
8	Rajasthan	3032639	513681	1761734	5308054
9	Bihar	833013	1130949	1861816	3825778
10	West Bengal	1546898	425855	1534556	3507309
11	Kerala	1510837	477537	1511575	3499949
12	Gujarat	1245306	250138	1543120	3038564
13	Haryana	1262789	381698	860922	2505409
14	Punjab	910827	144123	1353314	2408264
15	Chhattisgarh	1132054	312422	316645	1761121
16	Jharkhand	278892	391270	502776	1172938
17	Uttarakhand	320366	50537	304349	675252
18	Assam	14523	160176	399122	573821
19	Himachal Pradesh	195518	57551	256109	509178
20	Tripura	3993	54186	64461	122640
21	Jammu and Kashmir	53742	32153	15411	101306
22	Meghalaya	10844	22036	46125	79005
23	Pondicherry	7211	133	59903	67247
24	Manipur	13532	2073	28230	43835
25	Nagaland	3076	1833	23300	28209
26	Mizoram	2116	9480	14852	26448
27	Delhi	2157		21741	23898

(*Continued*)

Table A11.2 (*Continued*)   State-Wise Number of Kisan Credit Cards Issued up to June 30, 2010

Sr. No.	State/UT	Cooperative Banks (CB)	Regional Rural Banks (RRB)	Commercial Banks	Total
28	Arunachal Pradesh	980	3333	19178	23491
29	Goa	4577		12465	17042
30	Sikkim	3411		8182	11593
31	Andaman and Nicobar Islands	3691		2863	6554
32	Dadra, Nagar and Haveli			3243	3243
33	Chandigarh			3210	3210
34	Daman & Diu			1765	1765
35	Lakshadweep Islands			680	680
36	Other states			47	47
36	Breakup not available			188005	188005
	For CBs (1998–1999)				0
	Total	**38708160**	**13792418**	**42363847**	**94864425**

Source: Ministry of Agriculture. Agricultural statistics at a glance 2011.

Appendix 11.3

List of Organizations under Department of Agriculture and Cooperation

Sr. No.	Organization
	Public Sector Undertakings
1	National Seeds Corporation
2	State Farms Corporation of India
	Autonomous Bodies
1	Coconut Development Board
2	National Horticulture Board
3	National Cooperative Development Corporation
4	National Oilseeds and Vegetable Oils Development Board
5	Small Farmers' Agri Business Consortium
6	National Institute for Agricultural Extension Management
7	National Institute of Agricultural Marketing
	National Level Cooperative Organizations
1	National Cooperative Union of India
2	National Agricultural Cooperative Marketing Federation of India
3	National Federation of Cooperative Sugar Factories Ltd.
4	National Heavy Engineering Cooperative Ltd.
5	National Federation of Urban Cooperative Banks and Credit Societies Ltd.
6	The All India Federation of Cooperative Spinning Mills Ltd.
7	National Cooperative Agriculture and Rural Development Banks Federation Ltd.
8	National Federation of State Cooperative Banks Ltd.
9	National Federation of Fisherman's Cooperative Ltd.
10	National Federation of Labour Cooperatives Ltd.
11	National Cooperative Tobacco Growers' Federation Ltd.

Source: Ministry of Agriculture. Organizational history of the Department of Agriculture & Cooperation.

Appendix 11.4

Structure of Commodity Markets in India

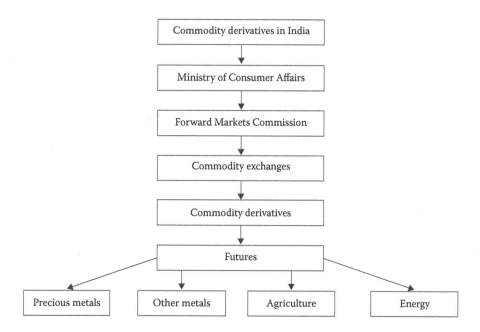

References

Chandrasekaran, N. *Supply Chain Management,* New Delhi: Oxford University Press India, 2010.

Ministry of Agriculture. Agricultural statistics at a glance 2011. n.d. http://eands.dacnet.nic.in/latest_20011.htm

Ministry of Agriculture. Organizational history of the Department of Agriculture & Cooperation. n.d. http://agricoop.nic.in/Orghistory.pdf

Ministry of Agriculture. *State of Indian Agriculture 2011–12.* Department of Agriculture and Cooperation, 2012. http://agricoop.nic.in/sia111213312.pdf

Ministry of Finance. Commission introduction. Forward Markets Commission. n.d. http://www.fmc.gov.in/index3.aspx?sslid=27&subsublinkid=13&langid=2

Mohan, R. Agricultural credit in India: Status, issues and future agenda, *Reserve Bank of India Bulletin,* November 2004.

Mukherjee, K.N. Impact of futures trading on Indian agricultural commodity market, 2011. http://mpra.ub.uni-muenchen.de/29290/1/Impact-of-Commodity-Futures-in-India_Full-Paper_Dr.KNM.pdf

NABARD. Role and functions, 2007a. http://oldsite.nabard.org/nabardrolefunct/nabardrole&functions.asp

NABARD. Development and promotional functions: Kisan credit card, 2007b. http://old-site.nabard.org/development&promotional/kisancreditcardmore.asp

PepsiCo. Partnership with farmers. n.d. http://pepsicoindia.co.in/purpose/environmental-sustainability/partnership-with-farmers.html

Sajjad, H. and Chauhan, C. Agrarian distress and indebtedness in rural India: Emerging perspectives and challenges ahead, *Journal of Geography and Regional Planning* 5, 397–408, 2012. DOI: 10.5897/JGRP11.107.

Chapter 12

Role of Government

OBJECTIVE

The objective of this chapter is to understand various initiatives taken by the government towards efficiency and effectiveness of agribusiness supply chains. Discussions include trade negotiations and subsidy; domestic support systems; price and distribution control; farm credit system and role of government in buffer stock creation.

12.1 Introduction

Agriculture plays a prominent role in the life of an economy. It is the strength of our economic system. The agricultural business system, which includes the gamut of activities from farm to fork, is the foremost creator of employment in the country. Apart from those who are directly involved in the agribusiness sector, a large number of the populace are also involved in agro-based and allied activities. The prominence of the agribusiness sector has developed significantly, as agricultural growth strategies lifted focus from production to a comprehensive system. This shift laid emphasis on agri supply chain synchronization and value creation. It is in fact a pre-condition for the country's improvement.

Over the past few decades, India has made significant advancements on the agricultural front. Agriculture growth strategies and policies would be a key factor influencing the business environment. In a country like India, the essential extent of outlay for the expansion of agribusiness sector is massive. Agriculture is contributing remarkably, not only to the growth of the economy, but also to the vitality of industry. It has been considered to be an indicator of both social and economic development, as the development of most of the industries depends on it. Perceived as a driver for growth, the agricultural business sector and its allied industries are gaining increasing visibility in government policies and strategies that aim to stimulate investments in agro-based value chains. Recognizing the significance of agriculture, the Indian government implemented several schemes to promote agriculture in the country. The state and central governments and the related bodies involved in the task of formulation of policies have been constantly trying to nurture the agribusiness environment that enables the sector to accomplish its role as the main driver of growth. This has created the need for a deeper understanding of the governmental role in strengthening the policy-making process for agribusiness supply chains. This chapter discusses various initiatives taken by governments towards efficiency in agribusiness supply chains.

12.2 Agencies

"Being the most important contributor to economic development and employment creation, the agribusiness supply chain plays a dominant role in reducing poverty and in achieving self-sufficiency. It ensures food security for the nation. Over the past five decades agriculture production improved significantly. Food grains production rose from 52 million tonnes in 1951–1952 to 244.78 million tonnes in 2010–2011" (Ministry of Finance, 2013). In spite of an exploding population, we have been able to meet the food requirements of the people. In India, agri supply chains have grown in size and complexity, handling great volumes of goods for domestic and international consumption. Since the agri supply chain ensures food security, it tends to be regarded as strategically important for the nation, and this necessitates more government intervention in this sector. This section discusses the role of governments in enabling agribusiness supply chains.

It has already been observed in previous chapters that agri supply chain improvement does not occur by itself, but requires a lot of commitment and capabilities from the various stakeholders involved. It calls for firm determination from all chain partners. The government plays an important role in enabling agribusiness supply chains, which involve different players.

The government gets directly involved in agriculture in order to ensure food security, fair prices and public distribution. It provides infrastructure facilities, market information and credit facilities for needy farmers, and evens out prices with proper price controls, buffer stocks and so on. It also regulates market competitiveness by establishing procedures or guidelines. During the recent past, we have seen a sturdy surge in the number of certification systems for sustainable and quality produce. Quality systems have been developed that set standards and procedures, including environmental and sustainable issues. This improvement is in accordance with government policies on supply chains to reduce the harmful aspects of agri production and supply. Thus, the government supports and promotes agriculture directly and through the roles of government agencies as members of agri supply chains. It also regulates and monitors agribusiness activities.

In short, the government takes on the following roles to improve efficiency in the agribusiness sector.

1. The government serves as a direct player.
2. It serves as a regulator through statutory bodies, agencies, etc.
3. It acts as a facilitator through research institutes, universities and nodal agencies.

The following paragraphs explain each of the above roles in detail.

12.2.1 As a Direct Player

Direct governmental intervention in agriculture has a long history, and it has passed through several phases. The most robust involvement began in the mid-1960s, a period that was very closely connected with vagaries in agricultural patterns. Due to numerous factors like bad harvests, high dependence on imports, inadequate food supplies and enormous food shortages in various parts of the country, the government decided to intervene by formulating an agriculture policy of self-reliance in food. After numerous attempts, the sharp rise in agriculture production during the Green Revolution of the 1970s saw the dream become a reality. The Green Revolution renovated farming practice in many districts for major food crops like rice, wheat and maize by adopting high-yielding new seeds, proper irrigation structures, upgraded cultivation techniques, etc. Subsequently, agriculture policy reforms were introduced.

This section outlines the government's direct interventionist measures to strengthen agri supply chains. Government measures like buffer stock creation, releases, price and distribution control, the Essential Commodities Act and its impact on supply chains are discussed. The role of the government in risk mitigation, infrastructure, farmer's credit and investment and commodity markets is not covered here, as it is elaborated in the previous chapters.

12.2.2 Food Corporation of India (FCI)

The Food Corporation of India (FCI) in India was set up under the Food Corporation Act, 1964, with a view to protect the interests of the farmers and to ensure smooth public distribution.

The following are the objectives of the FCI:

■ To provide fair remunerative prices for farmers
■ To make food grains available at reasonable prices, particularly to vulnerable sections of the society
■ To create buffer stocks as a measure of food security
■ To maintain a smooth flow of price and distribution.
  (FCI, n.d.)

Table 12.1 Stocks of Foodgrains and Sugar in Central Pool as on February 1, 2013 (Figures in Lakh MT)

	In Storage	In Transit	Total
Rice	352.13	1.71	353.84
Wheat	306.63	1.46	308.09
Wheat lying in *mandis*	0.00	0.00	0.00
Total	658.76	3.17	661.93
Coarse grains	0.90	0.00	0.90
Sugar	0.25	0.11	0.36
Grand total	659.91	3.28	663.19

Source: FCI, Stock position of foodgrains with Food Corporation of India as on 01.05.2013, 2013.

The primary responsibility of the FCI is to ensure food adequacy, security, timely and efficient procurement and distribution of food grains from producer to end consumer. This includes procurement of various food grains from different zones, monitoring of production, maintaining of stocks, their storage, movement and allocation to the distributing agencies. The FCI is the largest in terms of throughput and the value of produces. Table 12.1 summarizes the stocks of food grains and sugar in the central pool.

12.2.3 Public Distribution System

The Public Distribution System (PDS) is an Indian food security system constituted by the Government of India (GOI) under the Ministry of Consumer Affairs, Food, and Public Distribution. It is regulated by both the central and the state governments. The central government is in charge of procuring goods from various places and storing, transporting, and distributing food grains to the designated warehouses of the FCI. State governments are responsible for allocation, identification of families below the poverty line, issue of ration cards and supply of food grains to ration card holding families through fair price shops. The state government also supervises functioning of these shops.

The primary focus of the public distribution system is to support farmers in getting a fair value or the minimum support price for their produce, and to supply food grains to poor families and those that are below the poverty line (BPL). The government has also launched grain banks in scarce regions and has involved *Panchayat Raj* institutions in the public distribution system (PDS). The Department of Food and Public Distribution is responsible for regulating the sugar industry. It frames polices on the fair and remunerative price (FRP) of sugarcane payable by sugar factories, fixes the levy price for sugar and arranges its supply through the PDS, while also regulating the supply of free sale sugar. Further, it formulates policies on the export and import of food grains, sugar and edible oil.

Exhibit 12.1   Reforms in the Public Distribution System

According to the National Food Security Bill 2011, the reforms shall include

(a) doorstep delivery of food grains to the targeted PDS outlets;
(b) application of information and communication technology tools, including end-to-end computerization in order to ensure transparent recording of transactions at all levels, and to prevent diversion;
(c) leveraging the "Aadhaar" identification system for unique identification, with biometric information of entitled beneficiaries for proper targeting of benefits under this Act;
(d) maintenance of full transparency of records;
(e) preference to public institutions or public bodies such as *panchayats*, self-help groups and cooperatives, in licensing of fair price shops and management of fair price shops by women or their collectives;
(f) diversification of commodities distributed under the PDS over a period of time;
(g) support to local public distribution models and grains banks;
(h) introduction of schemes such as cash transfer, food coupons, or others to the targeted beneficiaries in lieu of their food grain entitlements prescribed by the central government.

Source: Ministry of Consumer Affairs, Food and Public Distribution. National Food Security Bill, 2011. Bill No. 132, 2011.

12.2.4   Buffer Stock Creation

Buffer stocks were created to ensure proper flow of food grains in the course of any uncertainties like bad harvests, adversarial weather conditions or other calamities. These stocks also even out prices during period of production shortage. The whole

Table 12.2   Buffer Norms (w.e.f. April 20, 2005) (Figures in Lakh MT)

	Buffer Norms			Strategic Reserve		
As on	Rice	Wheat	Total	Rice	Wheat	Grand Total
1st April	122	40	162	20	30	212
1st July	98	171	269	20	30	319
1st October	52	110	162	20	30	212
1st January	118	82	200	20	30	250

Source: FCI, Stock position of foodgrains with Food Corporation of India as on 01.05.2013, 2013.

stock of food grains is spread across four different quarters (January, April, July and October) of the year based on seasonality of production and procurement. Buffer stocking norms are calculated according to seasonal production and procurement off-take patterns. The government periodically appoints focal groups for buffer stock creation under the aegis of the FCI. Table 12.2 outlines the buffer norms with effect from 2005.

If we observe Table 12.2, we see that buffer norms vary according to season and procurement patterns. Besides buffer norms, the government also set strategic reserves of food grains to ensure continuous distribution.

12.2.5   Price Control Mechanism

In order to create a favourable price environment for the farmers, the Agricultural Prices Commission (APC) was established. The Commission was later renamed the Commission for Agricultural Costs and Prices (CACP). In order to maintain a remunerative and stable price environment, minimum support prices for principal agricultural crops are fixed by the government after taking into consideration the recommendations of the Commission for Agricultural Costs and Prices. The government's direct interventions are very important for agri supply chains in terms of distribution of the limited resources to various activities. Apart from government intervention in the form of minimum support prices, commodity exchanges have been an effective tool in maintaining price stabilization of such commodities.

12.2.6   Essential Commodities Act, 1955

The Essential Commodities Act was legislated in order to safeguard customers from unscrupulous sellers. The Act lays emphasis on fair prices. The Act provides a regulatory framework for the pricing and distribution of those commodities that are classified as essential. The commodities that are declared to be essential commodities are reviewed periodically in consultation with ministries governing the essential commodities. The following commodities are declared under the Essential Commodities Act, 1955:

1. Cattle fodder, including oilcakes and other concentrates
2. Coal, including coke and other derivatives
3. Component parts and accessories of automobiles
4. Cotton and woollen textiles
5. Drugs
6. Foodstuffs, including edible oilseeds and oils
7. Iron and steel, including products manufactured from iron and steel
8. Paper, including newsprint, paperboard and strawboard
9. Petroleum and petroleum products
10. Raw cotton, either ginned or unginned, and cotton seed
11. Raw jute
12. Jute textiles
13. Fertilizer, whether inorganic, organic or mixed
14. Yarn made wholly from cotton
15. (i) seeds of food crops and seeds of fruits and vegetables
    (ii) seeds of cattle fodder and
    (iii) jute seeds.
        (GOI, 1955)

12.2.7 Food Subsidy

The central government provides price support for paddy, coarse grains and wheat through the FCI and state agencies. All the food grains, conforming to the prescribed specifications, which are offered for sale at specified centres, are bought by the public procurement agencies at the minimum support price (MSP). The food grains are allotted for supply under the targeted public distribution system (TPDS) and other welfare schemes at subsidized prices. Besides this, the central government is also responsible for meeting the buffer stock requirements to ensure food security of the country. Hence, a portion of the total food subsidy also goes towards meeting the carrying cost of the buffer stock, in the form of buffer subsidy. (Dept. of Food & Public Distribution, n.d.)

Tables 12.3 and 12.4 show food subsidies handled by the FCI and the government.

Table 12.3a   Food Subsidy Release by FCI

Food Subsidy Incurred by FCI (₹ in crores)		
Year	Subsidy Incurred during the Year	% of Subsidy Released in the Year Incurred
2001–2002	18005.43	90.39
2002–2003	25321.90	89.54
2003–2004	21587.28	87.68
2004–2005	20773.60	92.37
2005–2006	21343.99	90.88
2006–2007	24027.84	80.64
2007–2008	30051.50	75.01
2008–2009	34787.47	83.62
2009–2010	42873.11	90.63
2010–2011	56394.42	80.36
2011–2012	68697.06[a]	82.55
2012–2013	81798[b]	

Source: FCI. Finance details section: Food subsidy due & released. 2013.

[a] Unaudited.
[b] Revised estimate.

Table 12.3b   Scheme-wise Subsidy Expected to Be Incurred and Claimed by FCI (₹ in crores)

Sr. No.	Scheme	2012–2013 (RE)		2013–2014 (BE)	
A	**Consumer Subsidy**				
1.	APL	16680		20725	
	BPL	27224		33037	
	174/150 Poorest Districts (BPL)	1223		1383	
	AAY	15032		17990	
	174/150 Poorest Districts (AAY)	304		354	
	MDM	4249		5184	

(Continued)

Table 12.3b (*Continued*)   Scheme-wise Subsidy Expected to Be Incurred and Claimed by FCI (₹ in crores)

Sr. No.	Scheme	2012–2013 (RE)		2013–2014 (BE)	
	Export	747		510	
	Open Sale	1417		1195	
	Others	1880		2190	
	Total Consumer Subsidy		68565		82578
B	**Buffer Subsidy**		13142		16136
	Food Subsidy Incurred during the Year		81798		98714
	Less: Deductions Pending Audit Certificate		4900		5736
	Net Food Subsidy		76898		92978
C	**Arrears for earlier year (on furnishing audit certification)**		19387		8466
	Total food subsidy to be claimed		96285		101444

Source: FCI. RTI manuals schemes. 2013.
Abbreviations: AAY, Antyodaya Anna Yojana; APL, above poverty line; BPL, below poverty line; MDM, Mid-day Meal.

Table 12.4   Food Subsidies Released by Government

Year	Food Subsidy (₹ in crores)	Annual Growth (%)
2001–2002	17,494.00	45.66
2002–2003	24,176.45	38.20
2003–2004	25,160.00	4.07
2004–2005	25,746.45	2.33
2005–2006	23,071.00	−10.39
2006–2007	23,827.59	3.28
2007–2008	31,259.68	31.19
2008–2009	43,668.08	39.69
2009–2010	58,242.45	33.38
2010–2011	62,929.56	8.05

Source: Ministry of Finance, Chap. 8 in *Economic Survey, 2011–2012.* 2013.

12.3 Policy Reforms

Over the past few decades, government policies for agriculture in India have evolved significantly. One strategy that has been incorporated into national agendas and five-year plans includes the strengthening of agricultural supply chains through government initiatives. In order to create a competitive economy, India has developed policies that lay emphasis on the improvement of supply chains at the national level. The underlying principle behind the agri supply chain policy would be to link partners at various levels of the supply chains to increase synchronization, invest funds for research and development and negotiate international trade policies and subsidies. As has been indicated, different channel partners play variously important roles in building sustainable and mutually supporting relationships. India's supply chain policy has focused on establishing the link between macro-level policy initiatives and their implementation, to make sure that whole supply chain partners obtain a justifiable share of the revenues, and thus, strengthen the sector's competitive edge.

Government policies can impact agri produce prices directly or indirectly due to social and other political reasons. The challenges of managing agricultural supply chains are evident in India, though the government has undertaken a series of policy reforms. The following are the challenges for the government in handling agri supply chains.

- Farmers in India hold small or marginal holdings of land, and are widely scattered.
- Support infrastructure facilities like roads are inadequate.
- Basic agriculture infrastructure facilities are in poor shape.
- There are too many intermediary levels.
- There is a major lack of standardization and quality certification systems.
- There is a lack of grading and cold storage facilities.
- There is limited access to information about agriculture markets.

The Ministry of Agriculture's policy lays emphasis on agriculture for improving supply chains. However, the success of those policies on agricultural supply chains depends on how partners within a supply chain cooperates. Lack of coordination in any stage will lead to major obstacles to the creation of a harmonized supply chain. Any policy reforms or transformations in the agriculture sector need constant support from all stakeholders, including farmers, state governments, and the central government, consumers and private players, to strengthen them further.

12.4 Direct Cash Subsidy

In the Union Budget for 2011–2012, the government announced a direct transfer of subsidies to below poverty line (BPL) households. This was a drastic departure from the earlier indirect or price subsidy system. In the earlier system, subsidies were routed through manufacturers who were required to sell goods at prices below the market rate. Under the direct transfer system, the difference between the market price and the subsidized price is directly transferred to the beneficiary in the form of cash in proportion to the quantity bought from the market. This would be based on estimated demand, and adjudged as being reasonable enough to deserve support from the government.

The earlier subsidy policy adopted by the government is believed to have the following shortcomings:

- *Dual-pricing*: This policy allows for one controlled price and another open market price at which a large portion of produce is allowed to be sold under market operations. There is a strong likelihood of traders diverting food grains from the regulated markets into the open markets to benefit from the premium. This both distorts the markets and creates inefficiency.
- *Unresponsiveness*: Subsidized goods are distributed through manufacturers, who lack the ability to respond to customers' needs in terms of categories of product, timing and quantity.
- *Poor target*: Though subsidies are meant to help the BPL population, under the current scheme they benefit people who are outside the target as well like even for people who are not poor or below poverty line.
- *Diversion and leakages*: This is one of the biggest problems as those who are in genuine need may still find it difficult to access products, while subsidized produce may get diverted into open markets either illegally, by administrating staff in the sales chains, or through the buyers themselves.

Direct cash distribution would overcome some of the problems, as beneficiaries would receive cash in their bank accounts and buy directly from the open markets. On the other hand, the challenges would be

- identification of right beneficiaries;
- disbursement through banking systems and cash withdrawal facilities, including access to ATMs;
- the ability of families, a lot of whom are still BPL, to use such systems properly and benefit from them;

- maintaining consumption baskets at the levels desirable for a healthy society; and
- the need to calculate the quantum of subsidy on the basis of the number of individuals per household. Normally, the government is likely to estimate average family size and distribute subsidies accordingly.

Though the scheme is well intentioned, and has been working in other countries, in India it has to evolve over the years, and the government must sternly implement the scheme, ensuring that the needy get support and that over the years, a reduction in the subsidy bill is possible.

12.5 As a Regulator

In olden days, farmers faced no difficulty in disposing of their produce, as they exchanged them with consumers under a barter system. In the present-day context, agro products have to go through chains of transfers from one entity to another before they reach the end user. All the produces may not be sold directly in the market; some may requires processing, storage, etc. Further, demand and supply may not be the same on all days. In an imperfect market, selling one produce depends on factors like seasonality, storage availability, demand, etc. Government agencies act as a facilitator in matching demand and supply across markets and supporting farmers in gaining access to global opportunities.

There are numerous central and state-owned enterprises under the Ministry of Agriculture that are involved in agribusiness supply chains, like RBI, NABARD, state-specific markets, etc. Besides these, various associations are grouped in the form of cooperative unions and export promotion councils to enhance cultivation, production, consumption, farm credit, marketing and export of agri products. There are also specialized marketing bodies for silk, sugar, tea, coffee, rubber, cereals and vegetables.

India is an agricultural nation, as most of the population is influenced by the agricultural sector directly or indirectly. With food being the basic requisite of humankind, importance has been given to agricultural production to a great extent. Adequate production and "even supply" of food are essential to satisfy the nation's populace. The growing population, increasing global demand, adverse weather conditions and rising energy prices have brought new opportunities and challenges. In order to meet the opportunities and challenges and to enable farmers to benefit from the recent global market opportunities, the agriculture sector in India needs to be reformed and strengthened. The government plays a prominent role in regulating agribusiness supply chains.

12.6   As a Facilitating Institution

Over the years, the government has actively engaged in promoting agribusinesses through a variety of facilitation roles. These include

1. Setting up of education, research and training institutes, universities and bodies across India and on different products, commodities and services in the sector;
2. Setting up promotion centres, mainly for marketing and business/trade development;
3. Setting up financial institutions and bodies for primarily promoting credit and also financial flows;
4. Setting up trade exhibitions and other means of promotion of and familiarization with Indian products across the globe;
5. Setting up auxiliary support services across the regions of India; creating products and services for promoting agribusiness.

12.7   Trade Negotiations and Subsidy

Over the last several years, agriculture has been the central theme of trade negotiations, as a majority of developing countries derive their livelihood from agribusiness. As food is one of the basic necessities of life, every country would like to be self-reliant in food production, and thus, would provide support and assistance to its agricultural sector. This support can be in the form of subsidies, special grants and rebates, tariffs, domestic subsidies and export subsidies, etc. While the developed countries have an industrialized system of agriculture, developing countries still face obstacles in providing subsidies because of poor capacity. This results in imbalances in international trade negotiations.

The government has a critical role to play in trade negotiations, and facilitates international trade. It creates market access, strengthens agri infrastructure, provides enabling policy interventions, and develops international collaborations. It endeavours to obtain support from worldwide organizations such as the World Bank, the Food and Agricultural Organization, the World Trade Organization and agriculture research institutes, and provides roadmaps for improvements.

These multilateral trade negotiations in the agriculture sector have significant effects for developing countries like India, which have agriculture as a crucial component of their food security and poverty reduction planning, and of their trade and industry growth. Though these multilateral trade agreements deal with more subjects, this section covers only issues pertaining to agriculture.

In order to get proper understanding of trade negotiations and subsidy, it is important to look at the historical developments relating to international agriculture agreements. The General Agreement on Trade and Tariffs (GATT) would be the first agreement one must look at.

12.7.1 General Agreement on Tariffs and Trade (GATT), 1947

The General Agreement on Tariffs and Trade (GATT), 1947, was a multilateral agreement constituted in order to remove barriers affecting international trade. GATT lasted till 1994, and was replaced by the World Trade Organization in 1995. Agriculture stayed outside the purview of GATT before the Uruguay Round. The Uruguay Round, signed in 1994 by 120 countries, was a significant step in the liberalization of world trade in agriculture. The Uruguay Round dealt with import restrictions aimed at reducing the distortions in trade in agricultural products, primarily in the form of farm subsidies offered by developed countries. India was one of the initial signatories to GATT 1947.

12.7.2 Uruguay Round Agreement on Agriculture (URAA), 1986–1994

The Uruguay Round of WTO aimed at establishing a fair and market-oriented agricultural trading system. At present, multilateral trade in agricultural products is governed by the Uruguay Round agreement on agriculture. This agreement imposes policies and directions on agricultural export subsidies, domestic subsidies and sanitary and phytosanitary (SPS) measures. Multilateral negotiations created four main parts of the agreement:

1. Agreement on agriculture
2. Market access, domestic support and export subsidies
3. Sanitary and phytosanitary measures
4. A ministerial decision concerning least developed and net food importing developing countries.

12.7.3 World Trade Organization (WTO), 1995

The agriculture sector is distorted by trade barriers in domestic and export subsidies. The WTO supports the GOI, stimulating significant reductions in trade distorting subsidies, tariffs and tariff quotas, export subsidies and protection measures used by other countries. During 1986–1994 in the Uruguay Round's WTO Agreement on Agriculture, the major phase of fair competition and a less distorted sector were constituted. WTO member countries agreed to reduce trade-distorting subsidies and increase market access to agriculture.

12.7.4 Doha Negotiations, 2001

The Doha Development Round is the most recent trade-negotiation round of the WTO. It took place in Doha (Qatar) in 2001. This round of trade negotiations was much extensive than earlier global trade negotiations, and was specifically designed for developing countries. The Doha negotiations reformed the international trading system through the introduction of lower trade barriers and revised trade rules. Developing countries oppose the fact that their own producers cannot compete against the surplus agricultural goods that the developed countries, primarily the EU and the United States, produce. India has made one of the most comprehensive proposals on agriculture to the Doha negotiations. The most vital aspect of India's negotiations would be to address implementation problems in the areas of market access, domestic support and export subsidy, with the aim of safeguarding its food security and improving market access opportunities that would encourage agricultural exports.

12.7.5 Domestic Subsidy

The WTO aims to achieve a significant reduction in domestic subsidies in developing countries. This has come as an after-effect of the Doha Development Round, which concluded that agriculture and the subsidies related to it are among the main reasons for the distortion of trade globally and the adverse influencing of prices. This, combined with import barriers in a few developed countries, poses a serious threat. It would require a lot of effort to bring about even small reductions in subsidies. As a part of the G20 initiative, it was mentioned that India has a responsibility to reduce these subsidies to a great extent as well. The Uruguay Round Agreement stated that "box shifting" (between the amber [production levels], blue [production limiting programmes] and green [causing minimal distortion of trade boxes]) should be minimized.

One of the proposed solutions for this is using a tiered approach to aggregate measure of support (AMS) in order to harmonize the effects of increased trade prices due to subsidies. But the effectiveness of this approach can only be determined once it is put into practice. On the whole, it is also suggested that the overall trade distorting support (OTDS), which is the sum of the AMS, be cut. There is no one-size-fits-all solution for the subsidies issue, as various crops such as sugarcane and tobacco need to be dealt with individually.

In short, with food security concerns topping the agenda of various global discussions, meaningful and conclusive trade negotiations and subsidies have become highly relevant. While protecting the interests of farmers is absolutely essential, it should not be at the cost of millions of people living in the third world countries. There is an urgent need to arrive at long-term agricultural agreements that benefit all countries and help in forming strong markets, in protecting geographical indicators and in encouraging active and robust international trade arrangements. International bodies like the WTO should see to active participation, and pay heed to the needs of all its member nations. Settling agricultural trade issues and protecting those involved in agriculture is absolutely essential to tackle food security concerns.

12.8   Conclusion

India has evolved into a service sector-based economy, but our policies and schemes still lay considerable emphasis on agriculture and food security. The government tries to support the farmers through various means like minimum support prices, fertilizer and diesel subsidies and restrictions on import and export of agricultural goods. Additionally, on the local government scale, too, like at the *panchayat* level, the government tries to support farmers by providing free seeds and advice on new farming techniques through its agricultural departments and officers. While the government's efforts in plugging the gaps in the agri supply chain and agri sector infrastructure have not been sufficient, it has in the recent past tried to invite public–private partnerships to work towards achieve these objectives. Governmental support in agricultural trade becomes necessary not only because of the many financial hardships faced by individual small-scale farmers, but also to protect India's interests in the international trade of agricultural produce.

Appendix 12.1

Important Agri Business Websites

1	http://www.agricoop.nic.in
2	http://www.dacnet.nic.in
3	http://www.nafed-india.com
4	http://www.iffco.nic.in
5	http://www.kribhco.net

(Continued)

Appendix 12.1 (*Continued*)
Important Agri Business Websites

6	http://www.ncui.net
7	http://www.agroecommerce.com
8	http://www.hafed.nic.in
9	http://www.nccf-india.com
10	http://www.ncdc.nic.in
11	http://www.nddb.org
12	http://www.itcibd.com
13	http://www.fieo.org
14	http://www.digitalmandi.net
15	http://www.nmce.com
16	http://www.ncdex.com
17	http://www.mcxindia.com
18	http://www.fert.nic.in
19	http://www.agmarknet.nic.in
20	http://www.faidelhi.org
21	http://www.krishiworld.com
22	http://www.ikisan.com
23	http://www.indiaagronet.com
24	http://www.mahindrakisanmitra.com
25	http://www.agriwatch.com
26	http://www.indiancommodity.com
27	http://www.uttam krishi.com
28	http://www.kisanayog.org

Source: NAFED. Useful websites. 2013.

Appendix 12.2
Minimum Support Price[a]

Sr. No.	Commodity	Variety	2008–2009	2009–2010	2010–2011	2011–2012	Increase in[b] MSP 2011–2012 / Over 2010–2011	2012–2013	Increase in[b] MSP 2012–2013 / Over 2011–2012
Kharif Crops									
1	Paddy	Common	850[c]	950[c]	1000	1080	80 (8.0)	1250	170 (15.7)
		Grade 'A'	880[c]	980[c]	1030	1110	80 (7.8)	1280	170 (15.3)
2	Jowar	Hybrid	840	840	880	980	100 (11.4)	1500	520 (53.1)
		Maldandi	860	860	900	1000	100 (11.1)	1520	520 (52.0)
3	Bajra		840	840	880	980	100 (11.4)	1175	195 (19.9)
4	Maize		840	840	880	980	100 (11.4)	1175	195 (19.9)
5	Ragi		915	915	965	1050	85 (8.8)	1500	450 (42.8)
6	Arhar (Tur)		2000	2300	3000[d]	3200[d]	200 (6.7)	3850	650 (20.3)
7	Moong		2520	2760	3170[d]	3500[d]	330 (10.4)	4400	900 (25.7)
8	Urad		2520	2520	2900[d]	3300[d]	400 (13.8)	4300	1000 (30.3)
9	Cotton	Medium Staple	2500[e]	2500[e]	2500[e]	2800[e]	300 (12.0)	3600	800 (28.6)
		Long Staple	3000[f]	3000[f]	3000[f]	3300[f]	300 (10.0)	3900	600 (18.2)
10	Groundnut in shell		2100	2100	2300	2700	400 (17.4)	3700	1000 (37.0)
11	Sunflower seed		2215	2215	2350	2800	450 (19.1)	3700	900 (32.1)
12	Soya bean	Black	1350	1350	1400	1650	250 (17.8)	2200	550 (33.3)
		Yellow	1390	1390	1440	1690	250 (17.4)	2240	550 (32.5)
13	Sesame		2750	2850	2900	3400	500 (17.2)	4200	800 (23.5)

#	Crop								
14	Nigerseed		2405	2405	2450	2900	450 (18.4)	3500	600 (20.7)
	Rabi Crops								
15	Wheat		1080	1100	1120§	1285	165 (14.7)		
16	Barley		680	750	780	980	200 (25.6)		
17	Gram		1730	1760	2100	2800	700 (33.3)		
18	*Masur* (Lentil)		1870	1870	2250	2800	550 (24.4)		
19	Rapeseed/mustard		1830	1830	1850	2500	650 (35.1)		
20	Safflower		1650	1680	1800	2500	700 (38.9)		
21	*Toria*		1735	1735	1780	2425	645 (36.2)		
	Other Crops								
22	Copra	Milling	3660	4450	4450	4525	75 (1.7)	5100	575 (12.7)
	(Calendar Year)	Ball	3910	4700	4700	4775	75 (1.6)	5350	575 (12.0)
23	De-husked coconut (Calendar Year)		988	1200	1200	1200	0 (0.0)	1400	200 (16.7)
24	Jute		1250	1375	1575	1675	100 (6.3)	2200	525 (31.3)
25	Sugarcane		81.18	129.84§	139.12§	145.00§	5.88 (4.2)	170.00§	25 (17.2)

Source: Ministry of Agriculture. Minimum support prices. Directorate of Economics and Statistics. Aug. 3, 2012.

a Refers to crop year—the time period from one year's harvest to the next for an agri produce. The crop year varies for each produce.
b Values in brackets indicate percentage increase.
c An additional incentive bonus of ₹50 per quintal was payable over the Minimum Support Price (MSP).
d Additional incentive at the rate of ₹500 per quintal of *tur, urad* and *moong* sold to procurement agencies is payable during the harvest/arrival period of two months.
e Staple length (mm) of 24.5–25.5 and micronaire value of 4.3–5.1.
f Staple length (mm) of 29.5–30.5 and micronaire value of 3.5–4.3.
g Fair and remunerative price.

Appendix 12.3

Minimum Support Price (₹/Quintal) for Nonfoodgrains According to Crop Year (Fair Average Quality)

Year	Sugarcane	Cotton	Jute	Groundnut (in shell)	Soya bean black	Soya bean yellow	Sunflower seed	Rapeseed/ mustard	Safflower
1980–1981	13.00	–	160.00	206.00	183.00	198.00	183.00	–	–
1990–1991	22.00	750.00	320.00	580.00	350.00	400.00	600.00	600.00	575.00
2000–2001	59.50	1,825.00	785.00	1,220.00	775.00	865.00	1,170.00	1,200.00	1,200.00
2001–2002	62.05	1,875.00	810.00	1,340.00	795.00	885.00	1,185.00	1,300.00	1,300.00
2002–2003	69.50	1,875.00	850.00	1,355.00	795.00	885.00	1,195.00	1,330.00	1,300.00
2003–2004	73.00	1,925.00	860.00	1,400.00	840.00	930.00	1,250.00	1,600.00	1,500.00
2004–2005	74.50	1,960.00	890.00	1,500.00	900.00	1,000.00	1,340.00	1,700.00	1,550.00
2005–2006	79.50	1,980.00	910.00	1,520.00	900.00	1,010.00	1,500.00	1,715.00	1,565.00

Year									
2006–2007	80.25	1,990.00	1,000.00	1,520.00	900.00	1,020.00	1,500.00	1,715.00	1,565.00
2007–2008	81.18	2,030.00	1,055.00	1,550.00	910.00	1,050.00	1,510.00	1,800.00	1,650.00
2008–2009	81.18	3,000.00	1,250.00	2,100.00	1,350.00	1,390.00	2,215.00	1830.00	1650.00
2009–2010	129.84	3,000.00	1,375.00	2,100.00	1,350.00	1,390.00	2,215.00	1,830.00	1,680.00
2010–2011	139.12	3,000.00	1,575.00	2,300.00	1,400.00	1,440.00	2,350.00	1,850.00	1,800.00
2011–2012	145.00	3,300.00	1,675.00	2,700.00	1,650.00	1,690.00	2,800.00	2,500.00	2,500.00
2012–2013	–	3,900.00	2,200.00	3,700.00	2,200.00	–	3,700.00	–	–

Source: Reserve Bank of India. Minimum support price for non-foodgrains according to crop year, Sept. 16, 2013.

Note: 1. For sugarcane, up to 2004–2005, the statutory minimum price (SMP) linked to a basic recovery of 8.5% of sugar with proportionate premium for every 0.1% increase in recovery above that level. The SMP for 2002–2003 includes the one-time drought relief. 2. Minimum support price (MSP) of cotton for H-4. 3. Cotton for 2006–2007, 2007–2008, 2008–2009 and 2009–2010 are of the long staple variety. 4. Crop year refers to the time period from one year's harvest to the next for an agri produce. The crop year varies for each produce.

Appendix 12.4

Public Distribution System—Procurement, Off-Take and Stocks (Million Tonnes)

Year	Procurement			Off-Take			Stocks		
	Rice	Wheat	Total	Rice	Wheat	Total	Rice	Wheat	Total
1980–1981	5.34	5.86	11.20	5.88	7.51	13.39	6.69	3.07	9.87
1990–1991	12.92	11.07	23.99	7.91	8.58	16.49	10.21	5.60	15.81
2000–2001	18.93	16.36	35.29	10.42	7.79	18.21	23.19	21.50	44.98
2001–2002	21.12	20.63	41.75	15.32	15.99	31.30	24.91	26.04	51.02
2002–2003	19.00	19.03	38.03	24.85	24.99	49.84	17.16	15.65	32.81
2003–2004	20.78	15.80	36.58	25.04	24.29	49.33	13.07	6.93	20.65
2004–2005	24.04	16.80	40.83	23.20	18.27	41.47	13.34	4.07	17.97
2005–2006	26.69	14.79	41.48	25.08	17.17	42.25	13.68	2.01	16.62
2006–2007	26.30	9.23	35.53	25.06	11.71	36.77	13.17	4.70	17.93
2007–2008	26.29	11.13	37.42	25.22	12.20	37.43	13.84	5.80	19.75
2008–2009	32.84	22.69	55.53	24.62	14.88	39.50	21.60	13.43	35.58
2009–2010	32.59	25.38	57.98	26.89	21.97	48.86	26.71	16.13	43.36
2010–2011	31.13	25.92	56.79	29.80	23.07	52.87	28.82	15.36	44.35
2011–2012	37.91	28.33	66.35	32.12	24.16	56.28	33.35	19.95	53.44
2012–2013	5.38	38.03	43.41	30.71	49.81	80.60			

Source: Reserve Bank of India. Public distribution system—Procurement, off-take and stocks. Sept. 16, 2013.

Note: Total stocks Include coarse cereals. For 2010–2011, procurement data is up to August 18, 2011, off-take for the period up to April 30, 2011 and stocks as on August 1, 2011.

Appendix 12.5

Important Cooperatives Acts

1. Multi-unit Cooperative Societies Act (1942) governs the working of cooperative societies whose objects and area of operation extend to more than one state.
2. National Cooperative Development Corporation (NCDC) Act, (1962) — By repealing earlier Acts, NCDC 1962 Act was enacted to replace the earlier NCD Board.
3. RBI Act 1934—The Agri-Credit Department of the Reserve Bank of India (RBI) was established in April 1935 under the RBI Act, 1934, to maintain expert staff and coordinate operations of the bank in connection with agri-credit and its relation with cooperative banks. We now have the RBI (Amendment) Act, 2006.
4. RRBs—These banks were set up under the Regional Rural Banks Act, 1976.
5. Agricultural Refinance and Development Corporation Act 1963—The Act has been amended in several ways by the Agricultural Refinance Corporation (Amendment) Act, 1975.
6. Central state warehousing corporations—Warehousing corporations were established under the Agricultural Produce (Development and Warehousing) Corporation Act, 1956. They are now regulated under the Warehousing Corporation Act, 1962. Recently, the Warehouse Corporations (Amendment) Act, 2005, has been passed by Parliament.

Source: ICAR. Agricultural legislations. 2010.

Appendix 12.6

Some of the Leading Government Organizations in the Agro-Based Sector (Indicative list)

Numerous organizations have been discussed in the earlier chapters. This Appendix gives readers an idea about a selected few of these to help them understand the approach taken by these organizations in facilitating economic function and growth.

Commission of Agricultural Costs and Prices

The Agricultural Prices Commission was set up in January 1965 to advise the government on price policies for major agricultural commodities. The main objective was to balance price structures from the perspective of the overall needs of the economy, and at the same time, protect the interests of the producers and the consumers. Since March 1985, the Commission has been known as the Commission for Agricultural Costs and Prices.

Minimum support prices (MSPs) for major agricultural products are fixed by the government each year after taking into account the recommendations of the Commission for Agricultural Costs and Prices (CACP). The Commission analyses a wide spectrum of data, covering the costs of cultivation/production, trends and spread of input use, production and productivity of the crop concerned, market prices, both domestic and global inter-crop price parity, the emerging supply-demand situation, procurement and distribution, terms of trade between agriculture and non-agriculture sectors, and so on. Since the price policy involves certain considerations of long-run consequences, the Commission also looks at the yield-raising research being conducted by institutions like ICAR. Data are gathered through field work and expert opinions. The commission supports 25 crops for MSP.

For further information refer to: About CACP. http://cacp.dacnet.nic.in/r2inf/aboutcacp.htm

Indian Council of Agricultural Research (ICAR)

The Indian Council of Agricultural Research (ICAR) is an autonomous organization under the Department of Agricultural Research and Education (DARE), Ministry of Agriculture, GOI. It was established on July 16, 1929 as a registered society under the Societies Registration Act, 1860, in pursuance of the report of the Royal Commission on Agriculture, and was then called Imperial Council of Agricultural Research.

The Council is the apex body for co-ordinating, guiding and managing research and education in agriculture, including horticulture, fisheries and animal sciences, in the entire country. There are 99 ICAR institutes and 53 agricultural universities under its fold.

It has played a pioneering role in ushering in the Green Revolution and in subsequent developments in agriculture in India through its research and technology development. According to sources, it has enabled the country to increase its production of food grains by four times, horticultural crops by six times, fish by nine times (marine by five times and inland by 17 times), milk by six times and eggs by 27 times since 1950–1951. The contribution is quite significant for the national food and nutritional security. It has played a major role in promoting excellence in higher education in agriculture. It is engaged in the cutting edge areas of science and technology development, and its scientists are internationally acknowledged in their fields.

For further information refer to: ICAR. About us. 2010. http://www.icar.org.in/en/aboutus.htm

National Bank for Agriculture and Rural Development

NABARD was set up as an apex development bank with a mandate for facilitating credit flow for promotion and development of agriculture, small-scale industries, cottage and village industries, handicrafts and other rural crafts. It also has the mandate to support all other allied economic activities in rural areas, promote integrated and sustainable rural development and secure prosperity of rural areas. In discharging its role as a facilitator for rural prosperity, NABARD is entrusted with

- providing refinance to lending institutions in rural areas;
- bringing about or promoting institutional development; and
- evaluating, monitoring and inspecting client banks,

Besides this pivotal role, NABARD also

- acts as a coordinator in the operations of rural credit institutions;
- extends assistance to the government, the RBI and other organizations in matters relating to rural development;
- offers training and research facilities for banks, cooperatives and organizations working in the field of rural development;
- helps state governments in reaching their targets of providing assistance to eligible institutions in agriculture and rural development; and
- acts as a regulator for cooperative banks and RRBs.

For further information refer to: NABARD: https://www.nabard.org/english/Home.aspx

Food Corporation of India (FCI)

Food Corporation of India was setup under the Food Corporation Act, 1964, in order to fulfil the following objectives of the Food Policy:

- Effective price support operations for safeguarding the interests of farmers
- Distribution of food grains throughout the country through the public distribution system
- Maintaining satisfactory levels of operational and buffer stocks of food grains to ensure national food security

FCI's objectives are

- to provide farmers with remunerative prices;
- to make food grains available at reasonable prices, particularly to vulnerable sections of the society;
- to maintain buffer stocks as measures of food security; and
- to intervene in the market for price stabilization.

For further information refer to: FCI: http://fciweb.nic.in/

Agriculture and Processed Food Products Export Development Authority (APEDA)

APEDA was established by the GOI under the Agricultural and Processed Food Products Export Development Authority Act passed by Parliament in December 1985. The Authority replaced the Processed Food Export Promotion Council (PFEPC).

The following functions have been assigned to the authority.

- Develop industries relating to the scheduled products for export by way of providing financial assistance or otherwise for undertaking surveys and feasibility studies, participation in enquiry capital through joint ventures and other reliefs and subsidy schemes
- Register persons as exporters of the scheduled products on payment of such fees as may be prescribed
- Fix standards and specifications for the scheduled products for the purpose of exports.
- Carry out inspection of meat and meat products in slaughter houses, processing plants, storage premises, conveyances or other places where such products are kept or handled for the purpose of ensuring the quality of such products.
- Improve packaging of the scheduled products.
- Improve marketing of the scheduled products outside India.
- Promote export-oriented production and development of the scheduled products.
- Collect statistics from the owners of factories or establishments engaged in the production, processing, packaging, marketing or export of the scheduled products or from such other persons as may be prescribed on any matter relating to the scheduled products, and publish the statistics so collected, or any portions thereof, or extracts therefrom.
- Provide training in various aspects of the industries connected with the scheduled products.
- Work on such other matters as may be prescribed.
  For further information refer to: About APEDA. http://www.apeda.gov.in/apedawebsite/about_apeda/About_apeda.htm

Marine Products Exports Development Authority (MPEDA)

MPEDA was set up in 1972 for increasing exports by specifying standards, processing, marketing, extension and training in various aspects of the fisheries sector.

For further information refer to MPEDA: http://www.mpeda.com/inner_home.asp?pg=overview

National Dairy Development Board

NDDB is a premier government body that was responsible for the dairy revolution in India under the leadership of Dr. V. Kurien. The organization created history by spreading the cooperative movement in India in first the diary field and then in edible oilseeds and oil.

For further information refer to NDDB: http://www.nddb.org/

Spices Board of India

The Spices Board (Ministry of Commerce, GOI) is an organization for the development and worldwide promotion of Indian spices. The Board is an international link between Indian exporters and importers abroad. The Board has been spearheading activities for excellence of Indian spices, involving every segment of the industry. The Board has made quality and hygiene the cornerstones of its development and promotional strategies.

The Board's activities include promotion of exports of spices and spice products; maintenance and monitoring of quality of exports; development and implementation of better production methods, through scientific, technological and economic research, and so on.

Spices Board India: http://www.indianspices.com/html/spices_board_intro.htm

Tea Board of India

Tea is one of the industries that by an Act of Parliament come under the control of the union government. The genesis of the Tea Board India dates back to 1903, when the Indian Tea Cess Bill was passed. The bill provided for levying a cess on tea exports, the proceeds of which were to be used for the promotion of Indian tea both within and outside India. The present Tea Board, set up under Section 4 of the Tea Act, 1953, was constituted on April 1, 1954. It has succeeded the Central Tea Board and the Indian Tea Licensing Committee, which functioned, respectively, under the Central Tea Board Act, 1949, and the Indian Tea Control Act, 1938, which were later repealed. The activities of the two previous bodies had been confined largely to regulation of tea cultivation and export of tea, as required by the International Tea Agreement then in force, and to promotion of tea consumption. (Tea Board of India, 2013)

Acknowledgement

The authors would like to place on record the support received from Ms. Rammyaa M, research scholar with the Centre for Logistics and SCM, Loyola Institute of Business Administration, in terms of research support for this chapter.

References

Agricultural and Processed Food Products Export Development Authority (APEDA). About APEDA. http://www.apeda.gov.in/apedawebsite/about_apeda/About_apeda.htm

Commission of Agricultural Costs and Prices. About CACP. http://cacp.dacnet.nic.in/r2inf/aboutcacp.htm

Department of Food & Public Distribution. Website. http://dfpd.nic.in

Food Corporation of India (FCI). Objective. n.d. http://fciweb.nic.in/articles/view/268

Food Corporation of India (FCI). Finance details section: Food subsidy due & released, 2013. http://fciweb.nic.in/finances/view/6

Food Corporation of India (FCI). RTI Manuals Schemes, 2013. http://fciweb.nic.in/articles/view/328

Food Corporation of India (FCI). Stock Position of Foodgrains with Food Corporation of India as on 01.05.2013. 2013. http://fciweb.nic.in/upload/Stock/5.pdf

Government of India. Essential Commodities Act, 1955.

Indian Council of Agricultural Research (ICAR). About us, 2010. http://www.icar.org.in/en/aboutus.htm

Indian Council of Agricultural Research (ICAR). Agricultural legislations, 2010. http://www.icar.org.in/files/Agril-Legislation.pdf

Marine Products Export Development Authority (MPEDA) Overview. n.d. http://www.mpeda.com/inner_home.asp?pg=overview

Ministry of Agriculture. Minimum support prices. Directorate of Economics and Statistics, Aug. 3, 2012. http://eands.dacnet.nic.in/msp/MSPStatement_(2012.03.08)_latest.pdf

Ministry of Consumer Affairs, Food and Public Distribution. Department of Food & Public Distribution website. http://dfpd.nic.in

Ministry of Consumer Affairs, Food and Public Distribution. National Food Security Bill, 2011. Bill No. 132, 2011. http://dfpd.nic.in/fcamin/FSBILL/food-security.pdf

Ministry of Finance, *Economic Survey, 2011–2012.* 2013. http://indiabudget.nic.in/survey.asp

National Agricultural Co-operative Marketing Federation (NAFED). Useful websites, 2013. http://www.nafed-india.com/hindi/useful-websites.asp

National Dairy Development Board. Website. http://www.nddb.org/

Reserve Bank of India. Minimum support price for non-foodgrains according to crop year, September 16, 2013. http://www.rbi.org.in/scripts/PublicationsView.aspx?id=15147

Reserve Bank of India. Public distribution system—Procurement, off-take and stocks, September 16, 2013. http://www.rbi.org.in/scripts/PublicationsView.aspx?id=15148

Tea Board of India. Website, 2013. http://www.teaboard.gov.in

World Trade Organization (WTO). Website 2013. http://www.wto.org

Case 1 Bayer CropScience: Science for a Better Supply Chain

September 14, 2005

Jadcherla, Andhra Pradesh

Suresh Babu, Territory Manager for Jadcherla, was a worried person. Standing in the middle of a cotton field, he could clearly see the telltale signs of an extensive bollworm attack. This was the second year that farmers in Andhra Pradesh had planted BT cotton, supposedly resistant to bollworm attacks. Acreage under BT cotton had increased substantially since last year. Owing to premium pricing, a major section of farmers had opted for BT cotton seeds from illegal third-party sources. The quality of these seeds was always under doubt. Suresh Babu had taken a big risk by giving a sales plan of Spintor that matched last year's sales volume. Spintor was highly effective in controlling the bollworm pest and was a premium brand in the market. The big risk taken seemed to be turning in his favour a few days ago as bollworm attacks were reported to be at the same level as last year. Yet, Suresh could not believe his bad luck. A few feet away, the farmer was filling up the spraying machine with a mix of water and a cola-based soft drink. A rumour had spread through the district that cola-based soft drinks were particularly effective in resisting bollworm attacks. Within a few days, cola-based beverages were selling at a premium in the market. This meant that Spintor sales would not be anywhere near the quantities that Suresh had stocked at dealers and distributors.

Kaithal, Haryana

Sampat Singh, Regional Manager, had just finished calling the Logistics and Distribution Department (LDD) of Bayer CropScience at the Mumbai head-quarters. He felt a growing anxiety as he recalled the day's events. Earlier in the day he had talked to major dealers offering them a 10 per cent discount if they lifted a specified quantity of Sherpa Alpha. Sherpa Alpha was a generic product sold in a highly competitive market. By offering this discount, Singh was sure about capturing at least 25 per cent of the market. He had planned his moves well in advance. The discounts were announced only Wednesday morning and he planned to deliver the quantities by next Monday. This way the competitors would not have enough time to react even if they came to know of the development through the grapevine. As an experienced sales manager, he knew that speed of execution was a prime factor in the success of such moves. It was absolutely essential that the shelf space at dealers be blocked before others could do so. The amount of Sherpa Alpha lying in the godowns was not enough to meet the huge demand. Realizing this, Singh had already been in touch with LDD and arrangements had been made to divert excess stock from nearby regions. Arrangement had also been made with the plant in Chandigarh to produce the excess requirement on a war footing. LDD was reasonably sure that all demand could be met by Monday morning. "All's well that ends well," thought Sampat Singh as he uttered a silent prayer.

Background

Bayer CropScience Limited was a leading firm in the fields of crop protection, non-agricultural pest control, seeds, and plant biotechnology. For the year ended December 2004, its annual turnover of ₹7865 million constituted a major portion of the annual turnover of ₹19816 million of the Bayer group in India. (Table A1.1 provides details about different Bayer group companies in India.) Bayer CropScience had a market share of 20.2 per cent in 2004, down from 22.8 per cent in the previous year. Its main competitors were established players like Syngenta, Dow Chemicals, Du Pont, Monsanto, and Rallis.

The firm converted six zonal sales offices into four zonal profit centres in a recent restructuring exercise. The objective was to inculcate a spirit of entrepreneurship through a decentralized structure, empowerment, and accountability. (The organization structure is shown in Figure A1.1.) The general managers at the zonal level enjoyed substantial decision making powers and had complete responsibility over all zonal functions. At the zonal level, the distribution and finance functions reported to the general manager with dotted relationship reporting to the head office (Figure A1.2).

Table A1.1   Bayer Group of Companies in India

Bayer CropScience Limited, erstwhile Bayer (India) Limited, constituted the core cropscience company and had production facilities at Thane, Himmatnagar, and Ankleshwar.
Bayer Diagnostics India Limited, with headquarters and production facilities in Vadodara, had been active in India since 1976 and had a leading position in the market for medical diagnostic devices.
Bayer Pharmaceuticals Private Limited was established in July 2000 and was a wholly owned pharmaceutical subsidiary of Bayer Industries Limited. The company handled pharmaceutical sales and marketing operations.
Bayer Material Science Private Limited was one of the largest producers of polymers and high-performance plastics. The company had a plant in Cuddalore, near Chennai, which manufactured thermoplastic polyurethanes.
Proagro Seed Company Private Limited, headquartered in Hyderabad, was involved in activities in bioscience to develop hybrid seeds for crops catering to specific farmers' needs across India. The company, incorporated in 1977, had emerged as a national player having the advantage of technology transfer, excellent R&D facilities, and presence in hybrid as well as field crops.
Bilag Industries Private Limited was one of the largest manufacturers of synthetic pyrethroids in the world.
Bayer Polychem (India) Limited, was a 100 per cent subsidiary of Bayer CropScience Limited, and handled non-cropscience businesses which were earlier part of Bayer (India) Limited.

Pesticide Market

The pesticide market was made up of insecticide, herbicide, and fungicide segments. Sales in all these segments were highly seasonal (Figure A1.3) and depended crucially on a good monsoon. A good monsoon resulted in well-nourished crops, which in turn attracted a variety of pests. Different pests could attack the same crop during different stages of a crop life cycle (Table A1.2). Thus different types of pesticides were effective during different stages of crop growth. Certain pesticides were prescribed as prophylactic applications to ensure that the crop was resistant to pest attacks. Curative application of pesticides focused on saving the crop once a pest attack had taken place.

Bayer CropScience sold 53 different pesticides in various pack sizes. Most of these products were established brands having high recall among farmers. (Details about two such products are given in Table A1.3) The technical superiority of its products allowed Bayer CropScience to command a premium in the market.

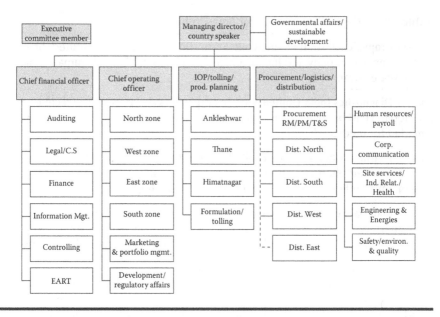

Figure A1.1   Organisational structure of Bayer CropScience.

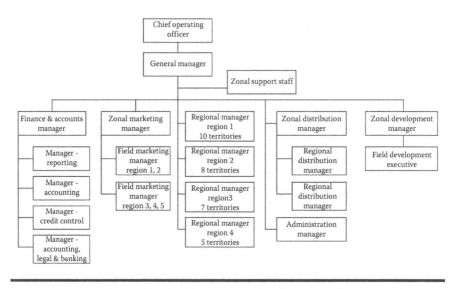

Figure A1.2   Zonal organisational structure of Bayer CropScience.

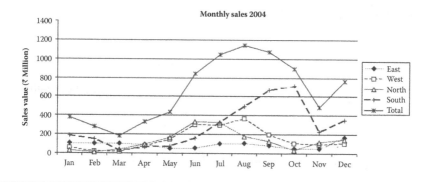

Figure A1.3   Seasonality.

Table A1.2   Pest Attacks on Cotton

Days	Pest	Application	Bayer Product
0–30	Sucking Pest	Prophylactic (50%)	Confidor
30–45	Sucking Pest	Prophylactic (50%)	Confidor-Metasystox
45–60	Bollworm	Prophylactic (20%)	Thiodan-Hostathion-Sherpa alpha-Bilcyp-Decis-Cybil
60–75	Sucking Pest + Bollworm		Thiodan-Hostathion-Spark-Decis-2.8ec-Larvin-Spintor-Decis100-Confidor-Metasystox
75–90	Mixed infestation	Curative (90%)	Thiodan-Hostathion-Spark-Decis-2.8ec-Larvin-Spintor-Decis100-Confidor-Metasystox
90–120	Mixed infestation	Curative (90%)	Thiodan-Hostathion-Spark-Decis-2.8ec-Larvin-Spintor-Decis100-Confidor-Metasystox

Table A1.3   Product Details

Product	Spintor	Sherpa Alpha
Type	Insecticide	Insecticide
Technical	Spinocyte	Alphamethrin (Synthetic pyrethroid)
Crop	Cotton	Cotton
Pests	Bollworm	Initial Stage of Bollworm
Pack Sizes and MRP	75 ml–₹888 250 ml–₹2766	250 ml–₹103 500 ml–₹193 1 ltr–₹377 5 ltr–₹1850
Distributor Margin	25%–30% of MRP	25%–30% of MRP
Bayer Gross Margin	16% of MRP	32% of MRP

Planning Process

The planning cycle in a year was divided into the kharif season (April to November) and the rabi season (December to March). The planning for the kharif season was done in April when sales managers forecasted the demand for a particular product. This was arrived at by first forecasting the acreage under cultivation for different crops. Then the percentage of total acreage to be covered by a particular product was determined. The forecasted sales of the product was arrived at by considering several factors like recommended dosage, number of sprayings, etc. Sales targets could be modified to reflect region-specific targets for market share, stock availability, and strategic considerations. Bayer CropScience divided the 53 products into four segments according to their gross margins (see Table A1.4). The sales plan reflected the increasing focus on A and B categories.

Once finalized, the sales plan was converted to a monthly sales plan using the liquidation trend of sales of the previous year and uploaded in the SAP system. At any point in time, rolling sales forecasts (RFCs) were available in the SAP system for m (current month), $(m + 1)$, $(m + 2)$ and $(m + 3)$ months. RFCs could be modified by generating a mid RFC around the fifteenth of a month. Additional RFCs could be generated according to the need (Table A1.5). The SAP system determined the master production schedule (MPS), which was further fine-tuned following discussions with distribution. Once firmed up, MRP was run to determine the production plan for different production facilities and purchase requirements. (Different monthly planning activities undertaken by the Industrial Operations Planning (IOP) Department are given in Table A1.6.)

Table A1.4 Categorization of Products

Category	Gross Margin	Objective
A	30%–40%	Invest in branding
B	10%–20%	Milking
C	5%–9%	Phase out
D	<5%	Phase out

Production

The production process involved two distinct stages. In the first stage, technical (active ingredients) were produced in batches in large integrated plants at Thane in Maharashtra and Ankleshwar in Gujarat. The Ankleshwar plant was the sole production hub for worldwide sales of several technical. Certain technical were sourced from the associate firm Bilag Industries and some were imported. Technical were stored in bulk in warehouses constructed following strict norms for storage of hazardous materials.

In the second stage, the end products were formulated by dissolving the active ingredients, solvents, and emulsifiers, and packaged in different sizes. Almost 40 per cent of the volume was formulated at Ankleshwar. Another 40 per cent was subcontracted to a firm at Kharar in Chandigarh. The rest was formulated at Himmatnagar in Gujarat and other subcontracting locations, referred to as tolling locations. In January 2004, the state of Jammu and Kashmir announced its industrial policy which offered excise duty waiver (currently 16 per cent) and a tax holiday of 10 years for new industrial units. Bayer CropScience was considering shifting a major portion of its formulation activities from different locations to Jammu to avail the benefits of the tax cut.

There were four production units at Ankleshwar (see Table A1.7). The EC unit contained five vessels for dilution of technicals, one having 5 kilolitre (kl) capacity, three having 6 kl capacity, and one having 12 kl capacity. One vessel having 6 kl capacity was dedicated to herbicide production. Hence, at any point, a maximum of five different products could be formulated. After formulation, the material from any vessel could be directed to one of seven filling stations. The filling stations were dedicated according to pack sizes (see Table A1.8).

Depending on vessel capacity, a minimum batch size of 2.5–5 kl was required for effective stirring. A smaller batch was considered uneconomical as the cost of running the motor used for stirring was invariant of the actual amount inside the vessel. Changeover from one product to another required flushing the entire system with a solvent. The solvent could be reused either in a future flushing or as an ingredient to the formulation process. Strict norms were enforced for checking the quality of flushing. A presence of no more than 800 parts per million of an active

Table A1.5 Month-wise Value of Projected Sales Plan as per First RFC and Variations: 2004 (₹ million)

East	Jan	Feb	Mar	Apr	May	Jun	Jul	Aug	Sep	Oct	Nov	Dec	Total
Plan	111.0	87.8	93.7	53.2	28.2	49.9	82.1	102.3	77.3	47.7	49.1	133.4	915.7
Additional	-3.8	13.2	8.2	42.2	22.3	6.8	21.4	6.0	5.6	3.4	0.8	35.8	162.0
Total	107.2	101.0	101.9	95.4	50.5	56.7	103.5	108.3	82.9	51.0	49.9	169.3	1077.7
% variation	-3.4	15.0	8.8	79.3	79.1	13.5	26.1	5.9	7.3	7.1	1.6	26.9	17.7
West	Jan	Feb	Mar	Apr	May	Jun	Jul	Aug	Sep	Oct	Nov	Dec	Total
Plan	43.5	20.2	12.6	72.3	119.9	275.4	314.8	366.3	129.3	89.5	86.4	74.5	1604.7
Additional	12.8	-0.2	0.7	3.2	23.4	25.3	-20.5	-2.8	69.0	12.0	5.3	28.9	157.1
Total	56.3	20.0	13.2	75.5	143.3	300.7	294.3	363.5	198.3	101.5	91.7	103.3	1761.8
% variation	29.5	-0.9	5.2	4.4	19.5	9.2	-6.5	-0.8	53.4	13.4	6.2	38.7	9.8
North	Jan	Feb	Mar	Apr	May	Jun	Jul	Aug	Sep	Oct	Nov	Dec	Total
Plan	31.4	50.5	44.6	71.0	131.2	328.8	211.7	258.6	58.1	52.0	111.6	123.9	1473.4
Additional	-3.7	-42.5	-5.3	23.9	31.2	4.0	110.7	-82.1	68.6	-21.5	3.3	21.5	108.1
Total	27.7	8.0	39.3	94.9	162.4	332.9	322.5	176.5	126.7	30.5	114.9	145.4	1581.5
% variation	-11.6	-84.2	-12.0	33.6	23.8	1.2	52.3	-31.7	118.1	-41.3	2.9	17.3	7.3

South	Jan	Feb	Mar	Apr	May	Jun	Jul	Aug	Sep	Oct	Nov	Dec	Total
Plan	165.1	157.8	17.5	50.4	68.1	150.1	267.6	458.1	501.6	531.1	227.7	217.4	2812.5
Additional	21.3	-6.5	10.6	15.2	8.7	6.4	57.0	37.2	165.7	183.4	0.8	131.8	631.5
Total	186.4	151.3	28.0	65.6	76.8	156.5	324.6	495.4	667.3	714.5	228.5	349.2	3444.0
% variation	12.9	-4.1	60.5	30.2	12.8	4.3	21.3	8.1	33.0	34.5	0.3	60.6	22.5

All 4 Zones	Jan	Feb	Mar	Apr	May	Jun	Jul	Aug	Sep	Oct	Nov	Dec	Total
Plan	350.9	316.3	168.4	246.9	347.4	804.3	876.2	1185.3	766.3	720.2	474.8	549.3	6806.2
Additional	26.7	-36.0	14.1	84.5	85.6	42.5	168.6	-41.6	308.9	177.3	10.2	218.0	1058.8
Total	377.6	280.3	182.4	331.4	433.0	846.8	1044.8	1143.7	1075.2	897.5	485.0	767.2	7865.0
% variation	7.6	-11.4	8.4	34.2	24.6	5.3	19.2	-3.5	40.3	24.6	2.2	39.7	15.6

Table A1.6   Monthly Planning Cycle

Day of Month	Activity
4	Receipt of RFC
4–6	Generate MPS
6	Meeting between IOP and Distribution
6–7	Release Production Plan
8–9	Release Requisition for Procurement
12	Release Import Plan
16–17	Mid-Month Review
28	Production Plan for next 10 days

Table A1.7   Formulation Units

Unit	Type	Material	Monthly Capacity
WP I	Herbicide	Powder	100 t
WP II	Insecticide	Powder	90 t
Ezeetab	Insecticide	Tablets	1.2 m
EC	Insecticide/Herbicide	Liquid	750 kl

Table A1.8   Filling Stations

Filling Station	Container Type	Pack Size	Capacity per Shift (units)
Bulk	Flexible	20–200 ltr	300
	Flexible	5 ltr	1300
	Tin	5 ltr	1700
Small pack	PET	250/500 ml	9000
	PET	500 ml	8000
	PET	250–1000 ml	9000
	PET/Co-EX	250–1000 ml	9000

ingredient was essential for changing over from one insecticide to another or from an insecticide to a herbicide. For changing from herbicide to insecticide, the norm was zero parts per million. Changeovers took two shifts on average. The formulation units worked on a single shift basis in general and three shift basis during peak season (April to September).

The tolling unit at Kharar was operated by Punjab Pesticides Industrial Cooperative Society Limited (PPIC) and had 14 formulation lines. It supplied formulated products to several manufacturers apart from Bayer CropScience. Several formulation lines were dedicated to Bayer CropScience products during peak season. Working on a single shift basis it could formulate and pack 150–200 kl in a month. The plant could produce a maximum of 750 kl in a month working on a two-shift basis.

Distribution

Bayer CropScience did not have retail presence; instead relied solely on a network of dealers, distributors, and preferred dealers to service the demand. The sales organization was made up of four zones, 25 regions and 169 territories. The South Zone, headquartered in Hyderabad, was by far the biggest (Table A1.9) and looked after the sales in Andhra Pradesh, Karnataka, Tamil Nadu, and Kerala. Andhra Pradesh accounted for almost 25 per cent of national sales. SKUs were transported from any one of the 13 manufacturing/tolling locations to 25 warehouses of which 17 were managed by carrying and forwarding agents (C&FA). (Some of the prominent manufacturing locations and headquarters are shown in Figure A1.4.) South Zone warehouses were located in Hyderabad, Guntur, and Kurnool in Andhra Pradesh, Hubli and Bangalore in Karnataka, Trichy and Coimbatore in Tamil Nadu, and Kottayam in Kerala. Seventy-five per cent of transportation from the manufacturing location to a warehouse used single source-single destination-multi product-full truckloads. The rest involved movement from a single source to multiple destinations on a full truckload basis. From the warehouses, 80 per cent of the material was moved to the distributors in full truckload milk runs. The rest was sent as parcels to a single destination. Distributors managed the movement of

Table A1.9   Zonal Transfers (2004) (₹ Million)

	East	West	North	South
Sales	1077.7	1761.8	1581.5	3444.0
Transfer within zone	46.74	26.35	37.44	37.15
Transfers outside zone	6.68	29.56	87.14	5.78
Branch returns	4.40	0.91	3.75	1.40

Figure A1.4   Prominent locations.

material to dealers using auto-rickshaws and public transport. In Andhra Pradesh, for example, the dealer telephoned the distributor and placed his order. The orders received during the day were dispatched between 9.30 PM and 3.30 AM. A preferred dealer (PD) was appointed in certain satellite markets. A PD was primarily a dealer who was serviced directly from the warehouses. PD could also act as a distributor and sell to other small dealers and big farmers who would in turn supply to co-farmers. In all, Bayer CropScience sold through a network of 2,500 distributors and 35,000 dealers.

Trade Promotions

Individual farmers bought their pesticide requirements from dealers located in nearby markets. Such dealers generally stocked other agricultural inputs like seeds and fertilizers besides pesticides. Trade promotions constituted an important marketing tool in stimulating demand in the market. The promotional schemes were territory-specific and could be categorized into two distinct classes. Liquidation discounts involved SKU-specific discounts offered to liquidate stocks with dealers. Such discounts were floated during the season and were intended to create a "hungama" (excitement) in the market. In contrast, placement discounts were based on the quantity of pesticide stock that a dealer ordered before the start of the season. For example, a dealer may be offered a cash discount of 10 per cent on demand draft and 9.5 per cent on cheques in January for stocks that were likely to be sold during the kharif season. The cash discount would reduce by 1.5 per cent each month. The discount could also be in the form of promotional schemes like gold coins. Such promotional schemes were commonly employed by almost all major players in the pesticide market.

The main objective of trade promotions was to garner shelf space. An effective way to "book" shelf space was to induce the dealer to park a part of his working capital with Bayer CropScience. A dealer could transfer a lump sum amount to Bayer CropScience in January without specifying the SKU-wise requirement. The actual order was then placed at the beginning of the season in April and stocks delivered subsequently. The attractiveness of a promotional scheme depended on whether the dealer paid cash or took material on credit. In 2005, dealers in Punjab and Haryana had surplus cash and could afford to commit funds well in advance. In contrast, Andhra Pradesh was primarily a credit market.

Inventory Management

Availability of the right pesticide in the right market at the right time was key to profitability. The nature and quantum of pest attack was difficult to predict. The extent of a pest attack could be significantly affected by temperature and rainfall. Indian agriculture relied heavily on monsoon rains. Monsoon activity was difficult to predict in aggregate, more so at local level. (The sales forecast accuracy for the Jadcherla territory for 2004 is given in Table A1.10.) The inherent uncertainty meant that stock built up at a particular location in anticipation of a pest attack could suddenly become redundant. Stocks returned from distributors to warehouses were termed sales returns. Sales returns averaged 3.4 per cent in 2004, down from 6–7 per cent in earlier years. Such stocks could then be moved to another location and were classified as intra-zonal or inter-zonal transfers (Table A1.9) depending on the destination. The company did not separately track the intra- and inter- zonal transfers arising from a sales return from those arising out of other reasons. Stocks returned by a zone from its warehouses to manufacturing units constituted branch returns. These stocks were either near expiry or expired stocks. Sales returns exhibited significant variability from one location to another and also from one season to another.

Table A1.10   Sales Forecast Accuracy for Jadcherla Market (2004)

Product	Pack	Forecast (Kg. or L)	Sales (Kg. or L)	Product	Pack	Forecast (Kg. or L)	Sales (Kg. or L)
Antracol	100 g	56	34.8	Metasystox	100 ml	56	12.6
	250 g	77	35.5		250 ml	84	26
	500 g	203	25.0		500 ml	91	38.5
	1 Kg	175	10.0		1 L	42	51
Bayrusil	500 ml	35	98.5	Planofix	100 ml	112	65.4
	1 L	175	178.0		250 ml	294	107.5
	5 L	210	125.0		500 ml	122.5	119
Bilcyp	100 ml	7	28.8		1 L	35	48
	250 ml	60	51.5	Quintal	100 g	39.2	22.2
	500 ml	210	57.0		250 g	39.2	30
	1 L	179	106.0		500 g	21	8.5
Bilzeb	500 g	70	13.8		1 Kg	28	4
	1 Kg	70	18.0	Raft	250 ml	385	250
Confidor	50 ml	49	19.8		500 ml	280	58.5
	100 ml	28	12.6		1 L	70	16

Product	Size			Product	Size		
Decis	100 ml	7	16.2	Sevin	100 g	21	6
	250 ml	28	33.0		500 g	28	31.5
	500 ml	14	49.5	Sherpa Alpha	250 ml	21	8.5
Derosal	100 g	154	34.8		500 ml	70	36.5
	25 Kg	450	225.0		1 L	98	44
Folidol Dust	5 Kg	2100	1955.0	Spark	100 ml	21	29.4
	10 Kg	4200	2280.0		250 ml	56	57.5
	25 Kg	35000	26200.0		500 ml	63	86.5
Hostathion	100 ml	14	72.0		1 L	28	112
	250 ml	56	725.5	Spintor	75 ml	140	7.5
	500 ml	126	1451.5		250 ml	126	8.5
	1 L	8561	5235.0	Thiodan	100 ml	280	45.6
	5 L	14145	4355.0		250 ml	14	90
	20 L	1120	2020.0		500 ml	1050	945
Metacid	100 ml	4	43.2		1 L	6510	6861
	250 ml	14	74.5		5 L	9625	6765
	500 ml	35	118.5	Whip Super	250 ml	350	604.5
	1 L	140	111.0		500 ml	2660	3585

Bayer CropScience tracked the inventory lying in different manufacturing and warehouse locations using the SAP platform on a real-time basis. However, stock level information was available only till the distributor level. This information was used in deciding to call back stocks from one location for redeployment in another location. Redeployment entailed certain risks. In one example, the company decided to redeploy stocks following less than adequate rainfall in a particular territory. Five days after redeploying ₹41 million of stocks, substantial rainfall changed the odds of a pest attack. A decision was taken the day after the rains to rebuild the inventory in that territory. Of the ₹47 million worth of stocks rebuilt, only ₹10 million was the same product as earlier stocked. Redeployment of stocks was possible to the extent that the crop sowing date differed from one territory to another. Redeployments were mostly to nearby territories and within a zone. Interzonal redeployments were rare.

Most pesticides had a shelf life of two years. The company tracked the expiry date of stocks located in different locations and classified stock as "near expiry" if the expiry date was less than three months away. After a season was over, dealers showed an aversion to keeping unsold stocks even if the stocks were not classified as near expiry. Unsold stocks were returned by dealers and could only be repositioned at the beginning of the next season, that too after offering a discount. Old stocks could be reprocessed any time before the expiry date. Reprocessing allowed the shelf life to be extended by two more years and entailed a reprocessing cost amounting to 28 per cent of the original manufacturing cost. The company employed first-in-first-out (FIFO) rule in all inventory related decisions. In a recent initiative, the zones were being asked to return all unsold stocks after a season for reprocessing. All expired stocks had to be incinerated. In 2004, the company had to write off ₹85 million of stocks. A further ₹70 million constituted material loss during reprocessing. It did not track the revenue loss arising from non-availability of stocks.

A further complication involved availability of different pack sizes for a particular product. On average, each of the 53 products had four pack sizes, resulting in more than 200 SKUs. The pesticide price per ml depended on the pack size and could vary considerably (Table A1.3). The IOP Department tried to schedule the production of bigger pack sizes during the season and smaller pack sizes during the off season. This was done since, for a specific production quantity, the filling time for smaller pack sizes were more than that for a bigger pack size.

Inventory planning was based on determination of the stock coverage representing the ratio of stocks to sales. The coverage ratio was tracked on a zonal and national basis and could be tracked on an aggregate basis and also on a specific SKU basis. A target inventory norm of 1.5 months coverage was considered ideal. In practice, the stock coverage varied from 1.5 to 3 months and could be set depending on the product and market conditions.

Science for a Better Supply Chain

September 14, 2005

Mumbai, Maharashtra

Susan D'Costa, planning executive at LDD, was updating a report on the forecast accuracy for presentation to top management. Forecast accuracy had been hotly debated in the headquarters. Earlier in the day, the report seemed to be almost complete when she received data regarding the sudden demand for Sherpa Alpha which was well above the sales forecast. More disturbing was the news about Spintor that came from the Hyderabad office. For a long time, D'Costa had been emphatically stating, "We need a better forecast." She felt time had come to reinforce the need for a better forecast.

As she finished updating the forecast data, D'Costa wondered whether there were alternative forecasting techniques for the pesticide market. Could a better forecast be a competitive tool? Could Bayer CropScience align its supply chain to be responsive to market requirements? Such a move could actually reduce the reliance on a sales forecast. But it may need substantial changes in the manufacturing and logistics processes apart from a change in the organizational mindset. Similarly, promotion schemes needed to be aligned with the supply chain strategy. On her desk was a company brochure prominently displaying the mission statement adopted by Bayer worldwide: "Bayer: Science for a better life." Perhaps the need of the hour in the Indian pesticide market was a science for a better supply chain.

Acknowledgements

Written by Professors Saral Mukherjee and G. Raghuram.

Cases of the Indian Institute of Management, Ahmedabad, are prepared as a basis for class discussion. Cases are not designed to present illustrations of either correct or incorrect handling of administrative problems.

This case has been reproduced with permission from Indian Institute of Management, Ahmedabad. Copyright © 2007.

Case 2 Marico: Disintermediating the Copra Supply Chain

"They have arrived!" announced the eighteen-year old son of Ranganathan as he finished giving directions on the cell phone to the two-member Marico buying team. Even though the visitors were a good two kilometres away, a palpable sense of excitement filled the assembled extended family members. Ranganathan was an affluent coconut farmer-converter in the remote village of Deepalapalli in Tamil Nadu, about 90 km from Kanjikode near the border with Kerala. Apart from the coconuts from his own farm, he procured coconuts from nearby farmers and dried the copra in the cemented conversion yard adjacent to his house. He had been supplying copra to Marico for the last sixteen years but had made his first direct sale only three months ago. Two months ago, in February 2007, he had visited Marico's buying office in Kozhikode and toured the coconut oil factory at Kanjikode, where he had learnt about the testing procedures for copra. During the visit he had met the buying team and invited them to visit his house. After his return, he had opened a savings bank account with ICICI Bank since it offered an Internet banking facility. A brand-new desktop computer adorned the living room. Since Ranganathan was computer illiterate, the task of tracking online account statements and copra transaction related emails using the dial-up connection fell on the eldest son. The son was much more technology savvy than his father and had already become comfortable with the process of submission of bids through SMS.

For the Marico copra buying team consisting of Sasi and Jayanth, the visit to Ranganathan's house was an important milestone. The immediate need for the visit was to understand the special requirements of farmer-converters. For example, at noon, Jayanth had accepted the bid placed by Ranganathan and had specified the acceptance quantity as 7.5 tons while all other successful bidders were allocated

369

in multiples of 10 tons. To a casual observer it might have seemed that Jayanth had discriminated against Ranganathan. In reality, the buying team made it a point to accept all reasonable bids placed by Ranganathan. Every such order was for 7.5 tons as the buying team knew that Ranganathan owned a modified tractor of 7.5 tons capacity. Other traders typically used hired trucks of 10 tons capacity.

More importantly, the visit signalled the direct sourcing of copra from farmer-converters rather than from converter-traders or middlemen. It was a significant chapter in the story of disintermediation which had started sixteen years ago with the shifting of the buying office from Mumbai to Kozhikode. What was not clear was how far the story was from its conclusion.

Company

The company was incorporated in 1988 as Marico Foods Ltd. and subsequently changed its name to Marico Industries Limited in 1989. By 2007, Marico was a leading Indian firm in consumer products and services in the global beauty and wellness space. It marketed well-known brands such as Parachute, Saffola, Sweekar, Hair & Care, Nihar, Shanti, Fiancee, HairCode, Silk-n-Shine, Mediker, Revive, Kaya, Sundari, Manjal, Aromatic, and Camelia. These brands and their extensions occupied leadership positions (mostly first or second) with significant market shares in coconut oil, premium refined edible oil, hair oil, post-wash hair care and conditioner, anti-lice treatment, niche fabric care, and skin care including soap categories (Table A2.1).

Marico's branded products were marketed in Bangladesh and other SAARC countries, the Middle East, and Egypt. The overseas sales franchise of Marico's consumer products (whether as exports from India or as local operations in a foreign country) was one of the largest among Indian companies. According to company estimates posted on the company's website www.maricoindia.com, over 70 million consumer packs reached approximately 130 million consumers in about 23 million households every month through a widespread distribution network of more than 2.5 million outlets in India and overseas. Marico generated a turnover of about ₹15.6 billion (about USD 380 million) during 2006–2007. Profits were over ₹1 billion. The turnover and profits exhibited a compounded annual growth rate of 19 per cent during the period 2002–2003 to 2006–2007.

Coconuts, Copra, and Coconut Oil

India produced about eleven billion coconuts/year in 2005–2006, up from a little over three billion coconuts in 1950–1951. Kerala and Tamil Nadu were the major producers of coconuts, followed by Andhra Pradesh, Karnataka, and a few coastal states (Table A2.2). Coconuts were harvested most of the year, except during

Table A2.1   Categories and Brands

Categories	Brands
Coconut Oil	Parachute, Nihar
Premium Refined Edible Oils promoted as 'Good for Heart' brand	Saffola
Premium Refined Edible Oils	Sweekar
Hair Oils	Hair & Care, Nihar, Shanti (Amla and Badam)
Post Wash Hair Care (Creams, Gels and Creams+Gels) in Egypt	Fiancee, Hair Code
Post-Wash Hair Conditioner	Silk-n-Shine
Anti-lice Treatment	Mediker
Fabric Care	Revive
Skin Care Treatment	Kaya
Ayurvedic Skin Care Products	Sundari
Beauty Soaps in Bangladesh	Aromatic, Camelia
Herbal Bath Soap (presence in Kerala)	Manjal

Source: Company data.

heavy monsoons. The peak harvesting season was March–July. Half of the coconut production was used in its fresh form for coconut water and edible purposes. About 42 per cent was converted into copra and the balance was converted into desiccated powder for use in cooking and flavouring.

Copra is the dried kernel of the coconut and it is mainly used for two purposes: edible item (15 per cent) and milling (85 per cent) into coconut oil. All the varieties of copra, edible or milling, came from the same nut. How the nut shaped up during its growth and the process of conversion to copra determined whether the copra would be for direct edible use or for milling. Edible copra was generally in the form of a cup. It was more compact in shape than milling copra and whiter in colour.

Copra conversion started with the manual dehusking of a coconut, typically done in the growing area after plucking. The dehusked coconut was then transported to yards of converters, who could be large farmers or more often independent players. The coconut was then manually split into two and the outer shell was removed. This had to be done skilfully so that the kernel could be extracted without

Table A2.2   Coconut Production and Prices*

States	Coconut Production (billion units)					
	1950–1951	2001–2002	2002–2003	2003–2004	2004–2005	2005–2006
Kerala	2.03	4.1	4.0	3.7	3.8	3.91
Tamil Nadu	0.46	4.0	3.9	2.9	3.3	3.85
Andhra Pradesh	0.31	0.9	0.9	0.9	1.1	1.22
Karnataka	0.37	1.2	1.0	0.9	0.6	0.63
Others	NA	1.1	1.1	1.0	1.1	1.25
Total	**3.17**	**11.2**	**10.9**	**9.3**	**9.9**	**10.85**

Source: Company Data.

Note: Average weight of a coconut, with husk, immediately after plucking, was 750 grams and after removal of husk was 400 grams. Dehusked coconuts could generally be bought at ₹7500 per ton. The husk was used for extracting fibre (used for weaving products like wall hangings, door mats, etc) and as fuel. It sold at ₹0.15 to 0.40 per kg.

* One acre of land could accommodate about 70 coconut trees and each 'cut' yielded about 12 to 13 coconuts per tree. The minimum spacing between cuts was 40 days. There were about seven cuts in a year resulting in an average yield of 85 to 90 coconuts per tree per year. Coconut production fell in 2003–2005 owing to droughts in Tamil Nadu.

damage. The water in the coconut was just drained. The kernel was then either sun dried or smoke dried for five to six days to produce copra. Copra production was a batch process with batches processed daily. In 2004–2005, Kerala produced the largest quantity of copra (Table A2.3).

Coconut oil yield by crushing copra was about 63 per cent (Table A2.4). Coconut oil was used as edible oil and hair oil, and in the production of soaps, shampoos, paints, biscuits, ice cream, confectioneries, etc. As much as 55 per cent of the oil produced in Kerala was used for edible purposes. In other states, about 15 per cent was used for edible purposes, resulting in an overall edible share of 41 per cent in 2005–2006. In 2006–2007, out of a total edible oil market of 128 lakh tons, coconut oil accounted for 3.2 lakh tons, out of which Marico's share was 48,000 tons (15 per cent by volume and 23 per cent by value). Annual per capita edible oil consumption increased from 9.53 kg in 2002–2003 to 11.48 kg in 2006–2007.

Table A2.3   Copra Production and Price*

Production Share (%)	1991–1992	2004–2005
Kerala	90	79
Tamil Nadu	6	14
Others	4	7
Total	**100**	**100**

Source: Company data.

Note: Demand for edible coconut being relatively more inelastic, copra availability was hit during times of low coconut production. The price of edible copra was generally higher than milling copra. The price of edible copra rose in the festival and marriage season. In times of low demand, edible copra was sold for milling. The main driver for copra prices was the price of coconut oil. The trader in the terminal market obtained the base price of copra (called padthar) as oil price per ton ∗ 60.5 per cent + oil cake price per ton ∗ 35 per cent − ₹1200 for processing charges. The trader then determined the copra price per ton by deducting (i) 2–4 per cent of padthar for low quality copra and moisture content over 6 per cent (ii) ₹400 as vandikkaran's commission, and (iii) a further 1 per cent as his own expense.

* Out of 100 dehusked coconuts of about 40 kg, an average of 12 to 13 kg of copra could be extracted. The balance was the shell (testa) and water, equally distributed in weight. The shell was used as fuel, and to a small extent for handicrafts. It sold at ₹1.8 to 2.0 per kg. During processing for copra, water was just drained and fetched no value. Sun drying was the favoured way for copra production in low rainfall areas. In high rainfall areas like Kerala, copra was produced by smoke drying.

Actors in the Supply Chain

Farmers

The copra conversion chain (Figure A2.1) began with the farmer of nuts. The farmer would either make copra himself or sell the nuts to a copra converter (CC). The incidence of farmers making copra themselves was higher during the summer months when sun drying was possible. The ability of a farmer to convert coconut into copra was limited by his financial strength. Small and marginal farmers did not have the financial capacity to wait for the conversion process to be completed before selling copra.

Because of small farm sizes, per farmer volumes were significantly lower in Kerala than Tamil Nadu, making Kerala relatively more complex in terms of the role of intermediaries and logistics.

Table A2.4   Coconut Oil Production and Use*

Demand Share (%)	2002–2003	2003–2004	2004–2005	2005–2006
Loose Coconut Oil	33	23	22	21
Branded Coconut Oil	30	39	38	36
Value Added Coconut Oil	1	2	2	2
Edible	33	36	37	41
Industrial	3	Insignificant	Insignificant	Insignificant
Total	**100**	**100**	**100**	**100**
Total Demand (lakh tons)	**3.47**	**2.62**	**2.67**	**3.00**

Source: Company data.

* Coconut oil was obtained by crushing copra. Newer millers used expellers while older technologies included rotaries and traditional rotaries called chekku. While expellers gave 63 to 65 per cent oil, rotaries gave from 62 to 63 per cent oil and chekkus gave 58 to 60 per cent oil. Most mills in Kerala traditionally used rotaries. Milling copra yielded about 63 per cent oil, 31 per cent cake, and 6 per cent water. Oil cake sold at approximately 20 to 22 per cent of copra price per ton.

The application-wise demand for coconut oil is provided: Industrial demand had come down significantly, partly because of the drop in production after 2002–2003 and substitution by cheaper imported palm kernel oil.

Copra Converters (CC)

Converters were the intermediaries in the copra supply chain who converted coconuts into copra by either smoke drying or sun drying them. In Kerala, copra conversion was a traditional backyard practice and was prevalent throughout the state. However, organised CCs were typically within 50 km of terminal markets.

CCs in Tamil Nadu could be categorised based on conversion capacity, as small (less than 8000 nuts/day), medium (8000 to 30,000 nuts/day), and large (more than 30,000 nuts/day). There were about 500 small, 400 medium, and 75 large CCs in Tamil Nadu. In the Kangeyam area of Tamil Nadu, a CC would rely on not only his own conversion facility but also on outsourced conversion and purchased

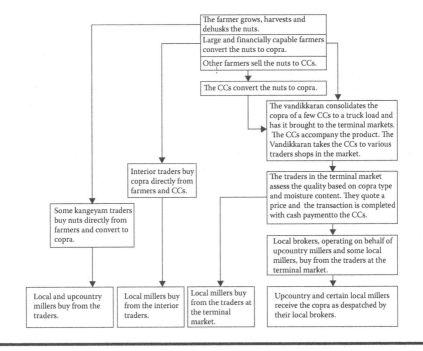

Figure A2.1   Various actors in the copra supply chain. (From company data.)

unsorted copra. Unsorted copra was available at ₹29.5/kg and CCs sold quality copra to Marico at ₹33.5/kg (CIF) and best quality edible copra at ₹40/kg.

The Vandikkaran

A CC did not often possess adequate knowledge about market conditions to negotiate with the trader and sell the produce himself. Additionally, every CC might not be capable of sending a full truck or tempo load to Kozhikode. So CCs contacted a broker called the vandikkaran. The vandikkaran arranged for a truck load of copra from several CCs to reach Kozhikode. Each CC marked his bags with initials for identification. CCs also accompanied the load. In Kozhikode, the vandikkaran took CCs around the market and helped negotiate the best price. The trader gave the vandikkaran a commission on the copra he bought from CCs. This commission was a fixed amount that varied from trader to trader; it averaged ₹400/ton.

Traders in Terminal Markets

The main terminal markets were Kozhikode, Alleppey, and Badagara, all in Kerala. Traders in the terminal markets purchased copra from CCs and sold it to millers.

They also bought copra on a speculative basis if they saw demand rising in the near future. The ability to invest in the trade varied from trader to trader.

Terminal markets witnessed a lot of activity in the morning when the copra arrived. When the vandikkaran or CC approached a trader, he would put the copra in a rope bag for visual inspection followed by a moisture check. He would also ascertain the proportion of edible, milling, and inferior quality copra in the lot. Before supplying to a miller, the trader performed several operations such as fumigation, drying, and sorting copra to meet the miller's requirement. Sorting was done by worker unions and typically cost ₹100/ton for milling copra. The trader could either employ his own employees or use the services of a labour union called kamali to carry out activities such as filling the ready copra into sacks, stitching and marking of sacks, drying, weighing, and loading of copra. Stacking of bags in the truck was performed by another union called Atti. In all, the trader paid ₹500/ton for the different activities including rent for the drying yard. Gunny bags cost a further ₹300/ton.

Local Brokers

Many millers in the upcountry coconut oil markets did not have a buying office in terminal markets. These millers contacted brokers in terminal markets to buy copra for them. Even local millers availed of the services of the brokers. A commission of ₹5/50 kg bag of copra was paid by the seller.

Interior Traders

Interior traders were local traders of copra. They did not operate in the terminal market. They invested money in buying copra from local farmers and CCs and supplied to the local millers. They were not concentrated in one place like traders in a terminal market and hence their volumes tended to be higher.

Converter Traders

A recent trend in Tamil Nadu was direct sourcing of nuts by some of the traders from the Pollachi, Udamalpet, and Karnataka areas. They then converted the nuts to copra and sold it to both local and upcountry millers. This was aided by the low labour cost of conversion. The share of direct sourcing in the Kangeyam area increased from 12 per cent in 2001–2002 to 40 per cent in 2006–2007, with some volatility in the intervening years. In fact, during the 2004–2005 droughts in Tamil Nadu, this share was nonexistent.

Millers

Local millers in Kerala bought copra from local farmers, CCs in the terminal market, and interior traders. A small miller depended on gate buying (buying the copra that came to his gate). A bigger miller would buy from the terminal markets and interior traders apart from some gate buying.

Local millers in Tamil Nadu bought copra from Kangeyam and Pollachi. Both Kangeyam and Pollachi did not have an organised market. The dealings were with traders, either directly or through brokers.

Upcountry millers bought copra from the terminal markets and/or larger traders either directly or through brokers. Marico was one of the few upcountry buyers which had a big presence in the interior markets.

Markets for Coconut Oil

Local demand for coconut oil in Kerala was satisfied by local millers spread all across the state. Demand in Tamil Nadu was primarily met by millers in the Kangeyam area. There existed small local brands in these markets.

Upcountry demand for coconut oil was mainly met by branded oil makers like Marico, Shalimar, Hindustan Unilever, Dabur, etc. In certain pockets such as Akola, Jalgaon, Kolkata, etc., there existed local millers. Oil traders in various parts of the country also met coconut oil demand by purchasing oil from the main oil trading centres in Kerala and Tamil Nadu.

Industrial demand for coconut oil was met by brokers in Cochin, Thrissur, and Kangeyam.

The main market for coconut oil was the oil exchange in Cochin organised by the Cochin Oil Millers Association (COMA). Trade at this exchange took place only through registered brokers. Even though the volume of trade in this exchange was about 5 per cent of the total coconut oil demand of India, prices set in this exchange acted as the benchmark.

Futures Market

Copra and coconut oil prices at the terminal markets exhibited significant volatility (Table A2.5). The futures market provided a mechanism to hedge against this volatility. Futures trading in coconut oil was possible on (i) the First Commodities Exchange of India (FCEI) from October 2001, (ii) the National Multi-Commodity Exchange of India (NMCE) from 2006, and (iii) the Multi Commodity Exchange of India (MCX) from 2006. While copra futures had been traded intermittently on FCEI and NMCE, in recent times none of the exchanges hosted copra futures transactions. Coconut oil was relatively more amenable to futures trade than copra as it was much easier to set standards and enforce quality for coconut oil.

On FCEI, one of the more successful coconut oil futures exchanges, monthly futures for the four subsequent months were open for trading on any given trading day. Trading was organised through an online platform. Each unit of trade was for two tons of coconut oil. At any point in time, the maximum open position for an operator was restricted to 200 units. The volume of trade on FCEI had grown from 2344 tons

Table A2.5   Spot Prices of Copra and Coconut Oil

	Year	Apr	May	June	July	Aug	Sep	Oct	Nov	Dec	Jan	Feb	Mar
	1995–1996	2042	2069	2057	2160	2178	2598	2641	2837	2669	2725	2760	2721
	1996–1997	2594	2481	2577	2558	2797	3171	3465	3802	4226	4298	4472	3745
	1997–1998	3378	3281	3301	3412	3391	3343	3263	3726	3711	3355	3113	2940
	1998–1999	2925	2653	2829	2790	2829	2950	3264	3284	3525	3373	3484	3622
Average monthly copra spot prices at Calicut market (₹ per quintal)	1999–2000	3676	3488	3566	3665	3995	3868	4008	4124	3931	3685	3175	2861
	2000–2001	2703	2599	2360	2170	2295	2169	2139	2133	2185	2115	2270	2172
	2001–2002	1910	1952	2085	2125	2153	2116	2103	2314	2775	2664	2598	2507
	2002–2003	2555	2627	2716	3169	3065	3110	3024	3496	3789	3643	3904	3873
	2003–2004	3702	3601	3434	3610	3838	4121	4531	4755	4485	4419	4236	4416
	2004–2005	4261	4205	4535	4506	4588	4483	4547	4732	4903	4782	4948	4470
	2005–2006	3927	3756	3616	3392	3341	3145	3178	3123	3210	3140	3222	3168
	2006–2007	3138	3130	3070	2957	3223	3289	3575	3577	3649	3586	3506	3247

	Year	Apr	May	June	July	Aug	Sep	Oct	Nov	Dec	Jan	Feb	Mar
Average monthly coconut oil spot prices at Cochin market (₹ per quintal)	1995–1996	3131	3109	3064	3164	3259	3864	4055	4167	3964	4140	4329	4101
	1996–1997	3952	3778	3919	3890	4123	4578	5023	5528	6095	6137	6390	5430
	1997–1998	5032	4845	5010	5017	4955	4956	4743	5308	5260	4809	4330	4099
	1998–1999	4169	3870	4177	4058	4140	4374	4891	4725	5219	4991	5197	5296
	1999–2000	5562	5189	5320	5220	5650	5593	5883	5889	5660	4941	4288	3836
	2000–2001	3665	3360	3255	3013	3164	3009	2859	2970	2984	2826	3099	3011
	2001–2002	2864	2794	3165	3041	3186	3070	2998	3074	3725	3774	3619	3465
	2002–2003	3632	3887	4039	4648	4670	4606	4497	5145	5663	5386	5642	5651
	2003–2004	5381	5309	4951	5463	5730	6163	6564	6940	6756	6644	6069	6195
	2004–2005	6038	6011	6675	6706	6925	6790	6713	6895	7290	7227	7296	6578
	2005–2006	5786	5557	5476	5193	5215	5152	4853	4882	4779	4605	4791	4855
	2006–2007	4692	4849	4913	4563	4729	4940	5448	5481	5488	5323	5197	4813

Source: Company data.

between October 2001 and March 2002 to 123,998 tons between April 2005 and March 2006. However, operators took delivery of only 1236 tons during 2005–2006.

NAFED

A minimum support price (MSP) was fixed for copra and revised annually by the government. The National Agricultural Cooperative Marketing Federation of India (NAFED) was the nodal agency for market intervention. Whenever market prices were below MSP, NAFED bought copra at MSP through its affiliated co-operative societies such as KERAFED in Kerala and TANFED in Tamil Nadu. The copra bought by NAFED was sold back in the market (normally in other states) after a few months. Given the high volume of copra that NAFED bought or liquidated, its actions had a significant impact on copra prices. Speculation about the timings of market interventions led to added uncertainty in the market.

Copra Supply Chain of Marico

Marico was a major player in procuring copra with an estimated 15 per cent share of the milling copra market. The buying activity was critical for Marico not only for cost management but also for quality assurance. Marico had buying teams led by managers who reported to the commercial function of the domestic consumer products SBU (Figure A2.2). Forecasting was based on the inputs of field survey personnel who travelled to sample coconut farms across various locations in south India to assess productivity, production, and overall farm health.

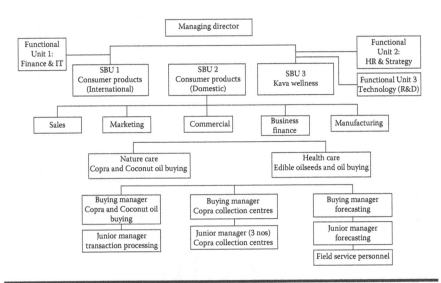

Figure A2.2　Marico organisational structure.

Initially Marico sourced a larger part of copra from Kerala; in recent years the share of Tamil Nadu had become more significant. In 2005–2006, 60 per cent of the copra was sourced from Tamil Nadu (Table A2.6). Kerala had traditional terminal markets and unionised brokers, traders, and converters who exercised their own control on the transactions. As stated by the buying team, "Proximity to factories and not having 'all the eggs in one basket' made Marico continue Kerala as a source." Marico attempted to gain some control in the procurement chain by setting up copra collection centres in Kerala from 2003–2004.

Marico set up a coconut oil factory at Kanjikode in Kerala in May 1993. Before that, Marico sent copra for crushing and processing to a Mumbai factory, which was later shut down. Two factories were operational in Goa and Puducherry. Table A2.7 provides profiles of the three factories while Figure A2.3 provides a map of the factory and buying locations. The average transportation cost of copra to the factories from Kangeyam (the main sourcing point in Tamil Nadu) is given

Table A2.6   Marico's Copra Buying

State-wise Share (%)	2001–2002	2002–2003	2003–2004	2004–2005	2005–2006
Tamil Nadu	53	49	35	52	60
Kerala	38	45	63	46	38
Andhra Pradesh	0	0	0	0	1.2
Karnataka	5	5	1	0	0.1
Others	4	1	1	2	0.7
Total	100	100	100	100	100

Source: Company data.

Table A2.7   Production and Storage Capacities at Factories

	Kanjikode	Goa	Puducherry
Start of Operations	May 1993	September 1997	March 2002
Copra crushing capacity (tons per day)	90	90	120
Copra storage capacity (tons)	300	400	1100
Coconut oil storage capacity (tons)	1250	1000	2000

Source: Company data.

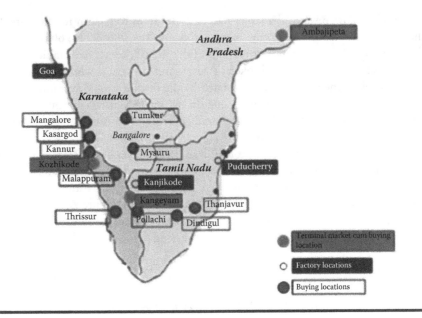

Figure A2.3   Map of Marico operations. (Company data superimposed on map from mapsofindia.com.)

Table A2.8   Average Transportation Cost to the Factories from Kangeyam

Destination	Transport Cost
Kanjikode	₹4300 for 10 tons (200 bags per truck, each of 50 kg copra)
Goa	₹19500 for 10 tons and ₹23000 for 16 tons
Puducherry	₹4500 for 10 tons

Source: Company data.

in Table A2.8. Marico had the option of hiring warehouses for storing copra over and above the copra storage capacity available at the factories. In the recent past, not more than 3000 tons of copra had been stored in hired warehouses.

Quality was a critical issue for Marico. The attempt was to ensure greater discipline at the copra source by having contracts that passed the risk of insufficient quality to the seller. Table A2.9 provides the copra quality specification sheet, along with a list of contractual obligations. Quality was also the consideration why Marico did not source coconut oil as a commodity, while other players did so.

**Table A2.9   Copra Quality Specification Sheet:
"Dala" Superior Quality Full Dry Milling Copra**

Characteristic	Acceptance Limit
Moisture	Max 5%
Inferior	
Rubbery/Immature	Max 5%
Burnt	Max 3%
Green Copra/Mouldy	Max 5%
Insect/Worm Eaten	Not allowed
Smoky	Max 15%
Free Fatty Acid	Max 0.60%
Cut Pieces	Max 2%

Note: 1) Total 'inferior' should not exceed 15 per cent.
2) For delayed deliveries, a debit note would be raised on the party pro rata for the delay period. Delivery beyond 10 days from date of contract would be liable to rejection, except at the option of the buyer.
3) All the contracts were for delivery at factory inclusive of any cess, tax, freight, gunny bags and any other expenses. The goods could be dispatched loose or in gunny bags as per the requirement of the factory and appropriate debits would be made for supplies in loose.
4) Weight and quality observed at factory would be final.
5) Any rejected material would have to be replaced with acceptable quality (as per the specifications above) at party's expense within five days of the rejection date to the factory. Otherwise it would fall under delayed deliveries.
6) The contracted rate was inclusive of good quality gunny bags—clean, dry and free from dust/foreign particles.
7) "Dala" is a category of Copra.

After processing, packaging and distribution, Marico sold coconut oil under the brand name of Parachute. Parachute was the world's largest packaged coconut oil brand. Parachute was of edible grade but customers frequently used it as hair oil. Available in pack sizes from 50 ml to 2 litres, the 200 ml flip-top pack had a maximum retail price of ₹34. The demand for Parachute did not exhibit significant seasonality.

Disintermediation Initiative

Early Initiatives

The disintermediation initiative was a long drawn affair that began in 1991. At that time, the buying office was located at the corporate headquarters in Mumbai. Marico would enter into contracts with Mumbai based brokers who would in turn contact brokers in the terminal markets in Kerala. Copra was sent to Mumbai on FOB basis. There were significant variations in quality between the copra (?) bought in the market and that reaching the factory. It was impossible to raise a debit note for quality or quantity related discrepancies. Prices and payment terms were dictated by brokers and supply disruptions were common.

The first step to solve these issues was the opening of a buying office in Kozhikode in 1991 staffed by a resident manager. Marico could now enter into contracts with traders in the copra market at Kozhikode, bypassing two layers of brokers. The terminal markets had strong labour unions. Not only were the labour charges high, there were separate unions for handling, loading, unloading, drying, stacking etc. In order to reduce dependency on the terminal markets, Marico started contacting individual traders and gradually started sourcing directly from them. The traders were mainly small aggregators who sourced copra from interior villages. To strengthen the vendor development initiative, Marico formed two buying teams in 1994. One team focused on the Kozhikode market while the other focussed on North Kerala, specifically Kasargod and Cannanore areas. As a result of these initiatives, ten vendors were identified in the North Kerala region and were provided dryers to dry copra during the monsoon. At this stage, Marico sourced copra on FOB basis from the terminal markets and on CIF basis through direct supply.

The 90s saw an explosive growth in coconut cultivation in Tamil Nadu. Starting from 1994, Marico focused on developing the vendor base in Tamil Nadu. The market perception was that copra from Tamil Nadu was of poor quality with respect to oil content. However, certain parts of Tamil Nadu were ideally suited for copra conversion. Coconut trees required adequate rainfall for growth but the copra conversion process was aided by dry weather. While Kerala received abundant rainfall, certain neighbouring pockets like Kangeyam in Tamil Nadu received almost no rainfall for eleven months in a year. Marico focussed its vendor development effort in such dry areas in Tamil Nadu. These vendors were provided moisture meters so they could objectively judge the moisture content. They were provided continuous buying support even when their volumes were low. Because of the underdeveloped nature of the market, Marico relied on brokers who represented multiple traders and copra converters.

In 1998, Marico completely discontinued sourcing from the terminal markets, at that time constituting almost 45 per cent of the sourcing volume. Instead, Marico would contract with individual traders in Kerala. These traders supplied full truckload quantities directly to the factories on CIF terms. This eliminated not only the transaction fee at the exchange but also unnecessary loading and unloading at the exchange. Marico continued its reliance on brokers in Tamil Nadu.

The brokers received commission from traders but not from Marico. Given the large copra requirement, the contract prices entered into by Marico were not significantly different from the market-traded prices.

At this stage, the Marico buyers were involved in daily negotiations with a large number of copra traders. Each one would tell long drawn stories of supply unavailability and try to misdirect the buyers regarding the reasonable price of copra. Marico buyers sensed cartelisation among the copra traders. To deal with this phenomenon, efforts towards diversifying the supplier base gathered momentum. During 1999–2003, Marico focussed on developing copra traders and converters in Pollachi and Thanjavur regions of Tamil Nadu and certain pockets in Karnataka and Andhra Pradesh. The vendor base reached 80 by 2003.

In the meantime, significant changes were afoot on the information technology domain. Marico opted for a big bang implementation of an ERP system from SAP in 2001–2002. Subsequently, all Marico copra and coconut oil suppliers were connected to Marico through a web-based portal called Marico Connect. MINET, an online portal, was started in 2002 with the aim of gaining visibility of secondary sales. It was used for communication between the sales personnel and the distributors and was integrated with SAP. The dotcom boom created awareness about the Internet and cyber cafes sprang up in towns and cities. New Age banks such as ICICI Bank started offering Internet banking on a national level. The stage was set for the next level of disintermediation.

Copra Bull Run

The need for disintermediation was felt acutely during the copra bull run of 2003–2005. From a price of ₹18000–20000/ton in 2001–2002, copra prices reached a high of ₹52000/ton in February 2005. This was due to a drastic fall in supply because of severe droughts in Tamil Nadu in 2002–2003 and cyclones in Andhra Pradesh. By this time, Marico had become a dominant player in the branded coconut oil market and volume requirements had shot up considerably. The copra markets were characterised by increased competition from buyers, wide market access for buyers, and high networking among suppliers. Speculative traders entered the market attracted by the high price volatility. Traders would frequently refrain from bringing copra into the market when they expected a price rise. Price rises heightened the expectation of further price rises. The earlier initiatives of developing a large vendor base and direct sourcing helped Marico to a large extent. Still, this episode underlined the need for decreasing the supply disruption risk.

Copra Collection Centres

Copra conversion was an age old practice in Kerala, which was done primarily on a small scale in households. The low volumes meant that there was a need for an aggregator who could then supply truckloads of copra. Marico aimed at dislodging the traders from this role of aggregator through opening up company-owned copra collection centres (Figure A2.4). The first copra collection centre was set up in May 2003 at Perambra in Kerala. However, the initial low volumes meant delays in transportation, resulting in rejections at the factory because of moisture related quality issues. Marico also faced resistance from local traders who initiated a misinformation campaign. Operationally, Marico had to grapple with issues in cash management at the collection centre.

The second collection centre was opened in July 2003 in Makkaraparamba, about 40 km from the first location at Perambra. This area was dominated by millers and Marico faced less resistance from the trade. The converters had been supplying directly to the millers and there was no practice of bringing the copra to the market. This resulted in a significant price difference compared to the prevailing market price. Marico was able to set up successfully a collection centre with mobile collection facility in this locality. An important lesson learnt by Marico was the need for customising the collection centre for local needs and practices. For example, a drying unit was set up in Perambra while mobile collection was employed in Makkaraparamba.

A copra collection centre typically employed a supervisor and a helper. The Marico owned centre was managed by an agent whose commission averaged ₹150/ton. The agent assessed the quality of copra on offer and took the decision to buy or reject.

Figure A2.4   Copra collection centre. (From company photo.)

The copra was then bagged and transported to the assigned factory within 4–5 days. The physical handling of copra was outsourced. Total expenses including salary, rent, electricity, communication, and handling charges amounted to ₹450/ton.

Farmers and CCs selling to the collection centre were small and marginal. They neither possessed sales tax registrations nor had Internet banking facilities. Payments were through cheques and disbursed by the agent. The collection prices were dictated by prevailing local prices, frequently published in local newspapers.

The copra collection centre was a strategically significant move since it not only aligned with the overall disintermediation initiative but also proved to be an assured supply channel. The value of assured supply was appreciated during the copra bull run when the supply from copra collection centres were found to be more stable than other sources. Marico set an ambitious vision in 2003–2004 to increase sourcing from copra collection centres so that half of the requirement of the Kanjikode factory could be met from this source. At that time, the Kanjikode factory processed 9590 tons of copra annually. The gradual increase in sourcing from copra collection centres is shown in Table A2.10. All copra collection centres were in Kerala and more were planned to be set up in Kerala and Tamil Nadu. By 2011, 25 per cent of all copra sourced would be through copra collection centres (Table A2.11).

Table A2.10 Sourcing from Copra Collection Centres

Year	Number of Copra Collection Centres	Quantity (tons)
2003–2004	2	1674
2004–2005	5	4158
2005–2006	9	7873
2006–2007	15	11556

Source: Company data.

Table A2.11 Percentage Share of Sourcing through Copra Collection Centres

	Share of Kerala Buying	Share of Tamil Nadu Buying	Share of Total Buying	Kerala Share of Total Buying	Tamil Nadu Share of Total Buying
Nov 2006	40%	No Sourcing	15%	38%	60%
2011 (Projections)	76%	3%	25%	33%	65%

Source: Company data.

Reverse Auctions

In 2002, Marico decided to discontinue the existing process of daily negotiations with copra traders. Instead, it introduced a reverse auction based procedure. Three one-hour slots were specified every day during which the buying team would accept quotes from copra traders. The communication between traders and the buying team was restricted to the quote: quantity and price data only. At the end of the one-hour period, Marico selected the lowest bidders and informed them about the accepted quantity and the destination factory. The reverse auction based procedure offered Marico several advantages as shown in Table A2.12.

The auction process was not well received by the trading community. It tried to break the auction process by offering to sell copra after auction hours at prices lower than those accepted by Marico. Marico steadfastly refused to renege on any high priced contracts and similarly refused to enter into any contracts after auction hours irrespective of the attractiveness of the price. Unable to break the auction system, Kerala-based traders collectively refused to sell copra to Marico. This blockage lasted from August 10 to September 5, 2002. During this period Marico continued to adhere to the auction system and sourced copra from Tamil Nadu based vendors. Slowly, the trading community started to place bids during auction timings.

Web Based Auctions

Marico next wanted to replace the telephonic auction based system with a web based auction system. A related transformation was the move from a demand draft driven advance payment system to a copra-receipt based e-payment system. These changes were necessitated by the fact that a significant part of the activities of the buying team involved transaction related activities. Vendors would telephone the buying team to know whether their truck had arrived in the factory or whether the payment had been released. Payment to the vendor was in the form of demand drafts issued by the Mumbai office before the receipt of material at the factory. The drafts were couriered to the Kozhikode office, which then couriered them to the vendor on confirmation of material receipt at the factories. A separate debit note had to be generated in case a quantity or quality related discrepancy was reported. Since the vendor came to know of the discrepancy only on receipt of the debit note, further rounds of telephonic calls were the norm.

The biggest hurdle in implementing web-based auctions was the fact that almost all traders were computer illiterate. Marico decided to gradually increase the sophistication of the transaction related procedures between 2003 and 2005. Initially, Marico asked the vendors to send an email containing transaction related details such as purchase order number, dispatch quantity, bill number, truck number, etc., for each transaction. However, the vendors were not prepared to send such emails. Marico facilitated the process by opening a large number of email accounts with *www.rediffmail.com* on behalf of the vendors. It also tied up with

Table A2.12   Comparison of Buying Processes

Dimension	Old Buying Process	Reverse Auction
Time	Buying at the seller's convenience, any time of the day or sometimes at night	Buying only in fixed time slots: 10–11, 12–13, 15–16 hrs
Information	Buying team had to respond immediately, and hence act on old information	Buying team had time to collect information and then respond to offers
No. of offers	No limit on number of offers	Only three offers from a supplier in a trading session
Influence on buyer	Traders usually talked to the buyer before offering and created a context. Thus the seller was able to influence the buyer's mind	Traders were allowed to quote price only. Any market talk happened after trading hours
Price Discovery	No comparison of quotes was possible, hence no mechanism for price discovery	Buying team had the option of collecting and comparing the quotes before responding. Hence, the process facilitated price discovery. This was felt more in a bearish market
Quantity Discovery	No mechanism for capturing total quantity on offer	Discovery of quantity on offer helped in pricing and timing decisions
Differentiation	Big brokers could take larger risks to maximise their benefit	Buying team was able to identify vendors for development and support them with volumes. This also helped restrict the big vendors.

Source: Company data.

local computer training institutes for imparting basic computer literacy and email courses. Standard operating procedures for email checking were prepared in Malayalam and Tamil. Several traders still needed significant hand-holding support from Marico. For example, traders would frequently call the buying office when they forgot the passwords. Marico maintained records of traders' passwords and helped them access email accounts. It also showed the extent to which the trading community trusted Marico. As time progressed, the traders became more comfortable in handling email accounts themselves.

To incentivise the email based system, Marico introduced the FTP (Fast Track Payment) scheme. Under this scheme, a vendor would be paid the day after receipt of copra at the factory. To avail this scheme, selected vendors needed to open an account with certain specified banks. Payment was then made electronically instead of issuing a demand draft. This allowed vendors to rotate their money faster than the earlier case which involved approximately nine days from copra receipt to receipt of demand draft. This benefit was offset by a 0.3 per cent discount on the bill value. Additionally, all FTP vendors had to send an email with dispatch details. FTP vendors were to be given priority service at the factory. By 2006, 95 per cent of vendors had opted for the FTP scheme and Marico stipulated that a vendor needed to send an email containing dispatch details even if the vendor was not on FTP. The 5 per cent vendors who did not opt for FTP included very large players who faced no working capital constraints.

Transformed Supply Chain

By July 2006, Marico had integrated web-based auctions with email based transaction management and e-payment based financial transactions in the portal *www.e-marico.com*. A vendor could now place bids, review his transaction summary, view details of purchase orders, enter shipping details, check the arrival of the truck at the factory, check payment status, bank account statements, and debit or credit notes – all from the comfort of home or a cyber café. The buying team members were provided cell phones that allowed viewing and accepting the bids placed while on the move. Once the buyer accepted a bid on his cell phone, specified the quantity and the destination factory, the information was transferred to the SAP system which automatically generated the purchase orders. The buying team meanwhile could spend their time visiting markets and concentrate more on price discovery than transactional chores.

As Marico kept on transforming the supply chain, newer challenges emerged. Traders would typically congregate at a local cyber café where the cyber café owner would help the traders with the transactions. It was soon apparent that the cyber café owner provided another 'service' of sharing the bidding information of one vendor with another. The response of Marico was to enable placing of bids through SMS. For example, a vendor could send the SMS "MLO 10 3350" to 57575 indicating a bid of 10 tons of copra at ₹3350 per quintal.* The SMS option for placing bids liberated the vendor from being close to a cyber café for placing bids. On a typical day, 11 out of 32 bids placed were through SMS.

The transformed supply chain was a significant departure from the past and stood out against other competitors. The sourcing profile of Dabur (which accounted for 1.3 per cent by volume and 1.6 per cent by value of coconut oil in 2006–2007), a main competitor of Marico, is provided in Box A2.1.

---

* 1 quintal = 100 kg = 0.1 ton.

BOX A2.1   COMPETITOR PROFILE: DABUR INDIA LIMITED

Dabur India Limited was a marketer of value added coconut oil and one of the major organised coconut oil buyers. It bought coconut oil only, not copra. Buying was largely restricted to Kangeyam with average monthly coconut oil purchases of 400 kl. It bought only through brokers and the buying price was benchmarked against prevailing market prices. Delivery was on CIF terms and quality was inspected at the point of receipt. The payment terms specified were 15 days from the date of receipt.

Glossary	
CIF—Cost, Insurance and Freight	SAARC—South Asian Association for Regional Cooperation
ERP—Enterprise Resource Planning	
FTP—Fast Track Payment	SBU—Strategic Business Unit
FOB—Free On Board	SMS—Short Message Service

Source: Company data.

Road Ahead

As dusk settled in the village, Sasi and Jayanth started the journey back to Coimbatore. They had visited several other copra converters and traders that day but none was as special as Ranganathan. He represented the face of the new supplier base for Marico. Yet that future face of the copra supplier was in a state of flux. Ranganathan had mentioned that several smaller converters had approached him for supplying copra. These converters did not want to supply directly to Marico as they did not want to obtain the sales tax registration required for becoming an accredited Marico supplier. Some of them were illiterate and found the Internet based bidding and banking procedures too complicated. Some held an apprehension that a sales tax registration would invite the taxman home. These small converters were more comfortable selling copra to Ranganathan. This practically meant the creation of a new supplier category of a farmer-converter-trader.

Marico was now targeting 25 per cent of its copra sourcing to come from copra collection centres by 2011. To achieve this, new centres needed to be set up in Tamil Nadu, not just in Kerala. However, initial experiences suggested that establishing collection centres in Tamil Nadu would be more challenging given the geographic spread and sales tax related complications. It would also imply a higher level of transportation risk for Marico apart from exposing it to thefts, pilferage, and accidents. More importantly perhaps, the move was likely to see significant resistance from Tamil Nadu based traders.

The policy of cultivating farmer-converters and pure converters by discriminating in their favour while awarding contracts was an ongoing initiative aimed at strengthening the sourcing base. The moot question was whether the farmer-converter represented the final frontier in the disintermediation initiative. Should Marico directly source coconut from farmers and convert to copra either in-house or under subcontract? Should it consider contract farming of coconut? The extreme option was to be vertically integrated from farming of coconut to selling of branded coconut oil. On the other hand, could reintermediation be an option worth considering? For example, would Marico's objective of minimisation of supply disruption be better answered by sourcing coconut oil rather than copra? In the face of an apparent trade-off between cost and risk minimisation, how could Marico reduce supply chain risk at the lowest supply chain cost? Driving through pitch darkness, the one thing clear to Sasi and Jayanth was that the solution had to provide Marico a sustainable competitive edge in future.

Acknowledgements

Written by Professors Saral Mukherjee and G. Raghuram. We thank Mr. Vinod Kamath, Mr. Jayanth K. and Mr. Sasikumar K. of Marico for their support. Research support by Ms. Anjali Dave and Ms. Kruti Mody is acknowledged.

Cases of the Indian Institute of Management, Ahmedabad, are prepared as a basis for class discussion. Cases are not designed to present illustrations of either correct or incorrect handling of administrative problems.

Case 3   Hasmukhbhai K. Nakum: Cold Storage Entrepreneur

It was March 2006. Mr. Hasmukhbhai K. Nakum was thinking about expanding his cold storage business beyond Ahmedabad. Some of the cold storages in Ahmedabad were closing down due to high running costs and poor capacity utilization, but he was contemplating on expansion. He already had four cold storages and each one of them was running profitably. The recent years had been financially good for him, as all his cold storages had come to be known for their professional management. He attributed his success to direct management that enabled cost reduction and focusing on products that provided maximum capacity utilization. With the continuous increase in agricultural production, this business had become an opportunity for Hasmukhbhai.

He wanted to expand beyond Ahmedabad to other cities where the demand was high. He thought he could improve his operating margins and reduce risks by geographical diversification. Based on savings and debt worthiness, he estimated his personal investment to be ₹3.5 crore. His prime concerns were real estate prices and subsidy provided by the Government. He felt that he could compete on his core competency of cost efficiency. He was considering one among the three cities in western India that could be attractive for cold storage business, namely Pune, Mumbai or Surat. He wanted the cold storage to be along highways and/or in proximity to local mandis.

Hasmukhbhai K. Nakum

Hasmukhbhai came from a joint family which was involved in cold storage business since 1978. His uncle managed their family owned cold storage. During his school and college days, he used to observe the functions of the cold storage and learnt the intricacies. He always had the interest of doing something on his own and becoming independent as early as possible. In 1985, their family started two cold storages at Jamkhambhalia and Jamnagar. During this time, Hasmukhbhai completed his intermediate in commerce and was pursuing Bachelors of Commerce. He got motivated by one of his uncles involved in diamond business and learned the basics of business. Then he considered pursuing the family business actively, even by dropping out of college. He calculated the cost and benefit of this decision and decided in favour of business. He dropped out of college and went to Ahmedabad in 1990.

In 1990, he started Hira Cold Storage with his uncle. It started operating with one chamber and an initial capacity of 1500 tons (t). Gradually, they increased its capacity to 8000 t by 2006. Giving into family pressures, he got married in 1995. He had a son in 1998 and a daughter in 2002. In 1999, he started Hapa Cold Storage with his cousin brother. From an initial capacity of 1500 t, it grew to 4500 t during 2002.

In 2003, he decided to go for another cold storage, but independently this time. He purchased the required land. In 2004, the cold storage was started and by 2006, it reached a capacity of 6000 t. This cold storage operated under the name Mother Shree Cold Storage Pvt Ltd (MSCSPL). This unit costed him a total of ₹2.5 crore out which he borrowed ₹1.3 crore from Indian Overseas Bank. The balance sheet for the financial years 2003–2004 and 2004–2005 are given in Table A3.1. The profit and loss account is given in Table A3.2.

In the year 2005, he acquired a cold storage near MSCSPL from Sindhi Bhai and named it Mother Agro Cold Storage. Mr. Sindhi Bhai sold the cold storage since he was not able to professionally manage the cold storage, given its high operating costs. The capacity was 2500 t. The sale price was ₹1.2 crore. For financing both these cold storages, he maintained a 25:75 of own equity and bank debt ratio.

Location Decision for the Cold Storages

When his family was initially operating at Jamnagar, Hasmukhbhai came to know about the demand in Ahmedabad. In the early 1990s, Krishna Cold Storage was the major player and it enjoyed monopoly. It offered eight hours of service. Hasmukhbhai observed that the rent charged was very high and the customers had to pay rent in advance to keep their goods. Hasmukhbhai sensed the opportunity for quality service at Ahmedabad. He also made a rough demand assessment and started searching for a suitable location for his cold storage. The various factors that he considered before zeroing in on Naroda were the following:

■ Proximity to Madhavpura Mandi: Naroda could be easily approached by a subway on Hanuman Road and there were no hassles of traffic signals. Further, one could reach Naroda without entering Ahmedabad city. Thus the goods could reach the store fast and could be a convenience for truck drivers.

Table A3.1   Balance Sheet (in ₹)

Sources of Fund	2004–2005	2003–2004
Issued and Paid Up Capital	28,70,000	22,38,600
(300000 Shares of ₹10 Each)		
Reserves and Surpluses	55,57,500	45,38,800
Loans Funds		
Secured Loans	1,33,75,000	1,04,32,500
Unsecured Loans	32,97,500	24,92,100
Total	**2,51,00,000**	**1,97,02,000**
Application of Funds		
Fixed Assets		
Gross Block	2,33,90,000	1,82,44,200
Less: Depreciation till Date	26,40,000	20,59,200
Net Block	2,07,50,000	1,61,85,000
Investment		
Current Assets, Loans and Advances		
Cash and Bank Balances	1,05,000	81,900
Deposits	5,62,500	4,38,750
Debentures	15,00,000	11,70,000
TDS Receivables	46,500	36,270
Prepaid Insurance	27,500	21,450
Deferred Tax Assets	47,500	37,050
Total	**22,89,000**	**17,85,420**

(Continued)

Table A3.1 (*Continued*) Balance Sheet (in ₹)

Sources of Fund	2004–2005	2003–2004
Less Current Liabilities and Provision		
Expenses Payable	7,500	5,850
Creditors for Capital Expenses	4,00,000	3,12,000
TDS Payable	4,375	3,412
Advances from Customers	1,35,000	1,05,300
Provision for Tax	44,000	34,320
Net Current Assets	**5,90,875**	**4,60,882**
Miscellaneous Expenditure (to the Extent not Written Off)		
Preliminary Expenses	62,500	48,750
Preoperative Expenses		

Source: Mr. Hasmukhbhai K. Nakum.

- Naroda fruit market which was a major customer of cold storage was near this location. A cold store located here would be preferred by traders as it would reduce the transportation cost.
- Kalupur pulses and grocery market was also quite near and could be served well from Naroda. Jaggery was the major commodity. Many of the masala items were also stored.
- Naroda had been riot proof all these years.
- It was close to NH 8. Customers could easily locate and reach the cold store.
- The roads connecting it were broad and well maintained, which would be of real help during the monsoon.

By 2006, there were around 30 cold storages with an average capacity of 2500 t to 3000 t in Ahmedabad. Only ten of these were operating profitably. Of this, four were owned by Hasmukhbhai.

MSCSPL

Selection of Name: Hasmukhbhai said, "I wanted a name which would be remembered easily and have an unaided recall. Mother preserved all nutrients for its offspring." Further, different names were chosen for different cold

Table A3.2   Profit and Loss Account (in ₹)

Particulars	(2004–2005)
Income	
Cold Storage Receipts	2,27,12,832
Other Income	0
Profit on Sale of Shares	3,22,560
Total	**2,30,35,392**
Expenditure	
Manufacturing Expenses (Electricity)	41,28,000
Administrative Expenses	19,15,392
Financial Overheads	45,13,968
Depreciation	99,60,864
Preliminary Expenses Written Off	57,792
Total	**2,05,76,016**
Profit or Loss before Tax	**24,59,376**
Tax and Provision	11,17,898
Deferred Tax Asset	12,01,740
Profit or Loss after Tax	1,39,737
Previous Year's Balance	1,72,156
Current Year's Balance (Transferred to Reserves)	**3,11,893**

Source: Mr. Hasmukhbhai K. Nakum.

storages to avail the government subsidy. Banks would provide 25% back ended subsidy for a single entity, whose capacity was within 5000 t.

Facilities: It had five cold rooms and six floors. The cold rooms comprised small racks. MSCSPL was a multi-commodity storage, primarily including pulses, spices, and fruits. It was a conventional cold storage and used ammonia as the coolant. This technology was economical as compared to Freon based refrigeration. Each cold room could be maintained at a different temperature and stored a particular type of commodity.

Customer Profile: There were five cold rooms in MSCSPL. The items were classified as fruits, pulses and spices. The revenue share was 40%, 40% and 20%, respectively. A few customers accounted for more than 50% revenue in each of the three categories.

Ramdev Spices and Ganesh Traders accounted for nearly 50% of the revenue from spices. BS Traders and Ahmedabad Effel Company accounted for 70% revenue from fruits. These were commission agents and they sold the stored items to wholesalers when the prices were favourable. Murlidhar Traders and Comm Traders accounted for 55% revenue from pulses.

Ramdev Spices operated in Surat, Jamnagar, Pune, Nasik and other locations in the region. All the commission agents involved in fruit trading imported their requirements through Mumbai (JNPT Port) and sold to wholesalers in Ahmedabad, Surat, Pune, Jaipur, Jodhpur and a few other places in the region. Murlidhar Traders had their operations in Ahmedabad, Vadodara and Surat.

Pricing: The rent was charged for a month even if the items were stored for a single day. Commodity-wise rent was fixed based on the three parameters of volume, weight and temperature required.

Box A3.1 provides a visual and calculations for the basis of pricing. The cold rooms comprised smaller compartments called racks, each with a maximum volume of 288 cubic feet and capacity of 7.5 t. There was no minimum requirement for the quantity to be stored. Therefore, even farmers were able to utilize these facilities. However, the major customers were traders and commission agents.

Capacity Utilization: Capacity utilization was the weighted average of the ratio between the actual quantity stored to the maximum that could be stored per unit of time with existing plant and equipment.

There was high seasonality in this business. His average capacity utilization was 80% during 2005. The utilization moved in a cyclic manner with a peak of 100% in summer and low of 50% to 70% towards the end of the rainy season through the beginning of winter.

Figure A3.1 gives the seasonality of revenue for MSCSPL for the calendar year 2005. Table A3.3 gives the seasonality of the main commodities stored by Hasmukhbhai. Figure A3.2 gives the typical harvesting period of spices.

Hasmukhbhai felt that the pricing also had an effect on the capacity utilization. The monthly basis of pricing affected the capacity utilization because even if a customer kept his items for a day, he had to pay for the whole month. A more viable option which Hasmukhbhai was considering was to charge commodities kept for higher duration on the monthly basis, whereas those that were more perishable on

BOX A3.1   BASIS FOR PRICING

Dimensions of a Rack

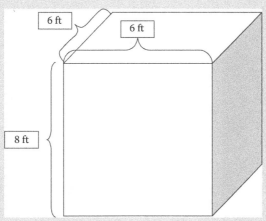

Volume of each rack = 288 ft³
Max weight to be kept at each rack = 7.5 t

Spices—Chilly
- Weight of 1 bag of chilly = 30 kg and the volume of each bag = 8 ft³.
- Number of bags that could be stored in a rack (288/8) = 36 bags = 1.08 t = 288 ft³
- Rent ₹10 per bag per month = ₹360 per rack per month

Pulses—Pigeon Pea
- Weight of 1 bag of pulse = 50 kg and the volume of each bag = 1 ft³.
- Number of bags that could be stored in a rack (7500/50) = 150 bags = 7.50 t = 150 ft³
- Rent ₹3 per bag per month = ₹450 per rack per month

Fruits
- ₹ 20 per carton per month

Source: Mr. Hasmukhbhai K. Nakum.

a fortnightly basis. This could attract such customers who avoid cold storages due to the monthly pricing policy.

Hasmukhbhai felt that capacity utilization can also be improved by optimizing the space and weight by keeping a combination of both heavy and light commodities that required the same temperature (for example chilly and pulses). This required good packing to prevent the exchange of smell and flavour.

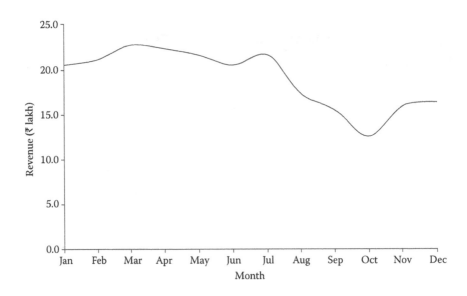

Figure A3.1 Seasonality of revenue, 2005. (From Mr. Hasmukhbhai K. Nakum.)

Table A3.3 Seasonality of Commodities

Commodity	Month of Inflow	Month of Outflow
Jaggery	October to November	March to October
Spices (Tamarind, Coriander, Chilli, Turmeric etc.) – Total 250 Items	December to February	April to November
Pulses and Cereals	March to June	July to December
Fruits		
Pomegranate	September to December	January to June
Apple (Himachal)	September to October	January to June
Kashmir Apple	October to December	January to June
Orange	March to April	January to June
Imported Fruits	January to September	As and When Required

Source: Mr. Hasmukhbhai K. Nakum.

Common Name	Botanical Name	Part Used	Jan	Feb	Mar	Apr	May	Jun	Jul	Aug	Sep	Oct	Nov	Dec
Black Pepper	*Piper nigrum*	Berry	▓	▓	▓									
Cardamom	*Elettaria cardamomum maton*	Capsule									▓	▓	▓	▓
Chillies	*Capsicum annuum*	Fruit	▓	▓	▓									▓
Clove	*Eugenia caryophyllus*	Unopened Flower Bud	▓	▓										▓
Cinnamon	*Cinnamomum zeylanicum*	Bark								▓	▓	▓	▓	
Cassia (Indian)	*Cinnamomum tamala*	Bark	▓	▓							▓	▓		
Ginger	*Zingiber officinale*	Rhizome	▓	▓										
Turmeric	*Curcuma longa*	Rhizome	▓	▓	▓									
Garlic	*Allium sativum*	Bulb			▓	▓								

Figure A3.2   Harvesting period of spices. (From Spices Board India, Spice harvest calendar.)

Common Name	Botanical Name	Part Used	Jan	Feb	Mar	Apr	May	Jun	Jul	Aug	Sep	Oct	Nov	Dec
Mustard	Brassica juncea	Seed		▓	▓	▓								
Kokam	Garcinia indica	Fruit						▓	▓					
Mace	Myristica fragrans	Aril						▓	▓	▓				
Nutmeg	Myristica fragrans	Kernel							▓	▓				
Pomegranate	Punica granatum	Seed	▓	▓	▓									
Tamarind	Tamarindus indica	Mesocarp	▓	▓	▓	▓	▓							
Vanilla	Vanilla planifolia	Pod		▓	▓	▓								
Dill	Anethum graveolens	Fruit & Leaf						▓						
Celery Seed	Apium graveolens	Fruit						▓						
Coriander	Coriandrum sativum	Fruit & Leaf		▓	▓									

Figure A3.2 (*Continued*) Harvesting period of spices. (From Spices Board India, Spice harvest calendar.)

Caraway Seed	*Carum carvi*	Fruit
Cumin Seed	*Cuminum cyminum*	Fruit
Fennel Seed	*Foeniculum vulgare*	Fruit
Fenugreek	*Trigonella foenum graecum*	Seed
Ajowan (Bishop's Weed)	*Trachyspermum ammi*	Fruit
Garcinia	*Garcinia cambogia*	Pericarp
Saffron	*Crocus sativus*	Stigma
Basil	*Ocimum basilicum*	Leaf

Figure A3.2 (Continued)   Harvesting period of spices. (From Spices Board India, Spice harvest calendar.)

Competitive Advantage

■ The service was provided to the customers throughout the day. It was the first such service in Ahmedabad.

■ MSCSPL also maintained and managed a fleet of eight trucks. These vehicles delivered the goods kept by the clients from the cold storage. While returning, they picked up the goods to be preserved from the clients' end. This had two advantages, firstly it created a dependency syndrome on part of the client and secondly the early storage helped the cold store get higher income. The transportation service helped in quicker turnover of the inventory and increased capacity utilization.

■ Rapid expansion of cold storages of MSCSPL helped it enjoy the economy of scale and hence reduced the rent to the lowest levels in the market. Further, it also helped in integrating the functions of all the four cold storages.

■ There were six floors in MSCSPL. Heavier items were stored in the basement and ground floors. It increased the stability and prevented cracks on the cold storage walls. It also helped in easy and quick handling of the items.

■ Unskilled labour was mainly employed. Only the operator was a skilled labour with some years of work experience in an ice factory.

■ No other cold storage was directly managed by its owners since they had several other businesses to manage. For Hasmukhbhai, this was the main business.

■ Hasmukhbhai believed in relationship marketing. He had developed a cordial relationship with all his clients and it helped him win their trust.

■ He often reduced the price for new commodities to experiment their feasibility of preservation which enabled him to gain new markets (for example wheat and more recently, egg). During slack season, he offered 20% to 30% discount to run his cold storage at optimum capacity.

■ Even after two years of operation, there was not even a single complaint on the security of the stored items.

■ He had a long term goal of capturing the market of commodities that were preserved in warehouses. There were high losses in warehouses, whereas quality of the commodities could be better maintained in cold storages with marginal price difference. For example, the cost of preservation of cereals in Central Warehousing Corporation warehouses was ₹2.60 per bag per month whereas it was ₹3.00 per bag per month in the cold storage. With an extra 40 paise per bag, wastage could be prevented.

■ The captured market, capital intensity and the economies of scale were the entry barriers to this business.

Competitor Analysis in Ahmedabad and Gujarat

A perspective on the cold storages in Ahmedabad is given below:

Samarpan Cold Storage (1500 t): It offered multi-commodity storage facility. Low capacity utilization had always been its problem.

Vinodh Cold Storage (3000 t): It was started in 1999 and had 10% of the total market share. Mr. Sivjibhai Patel was the owner of this multi-commodity storage facility.

Shahjanand Cold Storage (1500 t): It dealt mainly with jaggery.

Laxmi Cold Storage (2000 t): It dealt mainly with jaggery.

HC Cold Storage (5000 t): It was started in 1995. It stored fruits only. This cold storage was started by HC Fruits limited by Mr. Hundraj Changomal. He integrated vertically and built this cold storage to reduce cost. It had a market share of 10%. It incurred very high construction cost and ended up spending double the actual capital required for such a project. Basic reason was his inexperience in this field.

Padmavati Cold Storage (2500 t): It was a multi-commodity storage started in 1994 by an NRI, Mr. Nitin Bhawsar.

Alka Cold Storage: It does not exist now. It had a capacity of 1000 t. It was one of the few first cold storages in India. It was a multi-commodity storage.

Cold Storages managed by Hasmukhbhai in Naroda (total market share was 25%): Hira Cold Storage (8000 t), Hapa Cold Storage (4500 t), MSCSPL (6000 t) and Mother Agro Cold Storage (2500 t)

Mr. Hasmukhbhai had the following to say regarding problems faced by cold storages in Ahmedabad and causes of closure of some of them. "More than 45% of the operational cost is towards electricity. The electricity charges in the state are very high. Apart from this, cold storages fall under service industry. So they are levied a duty of 20% over and above the electricity bill. The irony is even industries pay just 10% duty. Cold storages in Gujarat are treated on par with cinema houses. There is an association for cold storages in Ahmedabad and they have taken this issue with many governments but so far there has been no change. The members strongly feel that if cold storages have to pay agricultural electricity tariff, many of the existing cold storages can be turned profitable.

Twenty years back, when the electricity charge was ₹0.20 per unit, the rent was ₹0.50 per kg of potato for a season. Now the electricity charge has gone over ₹5.00 per unit, but the rent remains almost the same at ₹0.60 per kg. It is no wonder that most of the cold storages are not profitable. Also, potato cold storages are operational only during March to September and rest of the period they remain idle. This is another reason for them not being profitable. Other reasons for financial problems and closure are high cost of construction due to inexperience, high cost of manpower and over heads, and competitive pricing."

Regarding MSCSPL's relative better performance, he said, "We dictated the market by decreasing the rent gradually as we grew more and more efficient. At such low rent, it is difficult for many of the cold storages to operate. MSCSPL managed to operate at such a low rent because of its efficient management and economies of scale. Electricity conservation is the key for cold storage management and it can be done in an efficient way only when the owner is directly involved. Manpower management is also an important aspect of cold storage management."

There were 337 cold storages in Gujarat with a capacity of 8.40 lakh t (lt) (Table A3.4). Cold storage capacity was concentrated in Kheda, Anand, Nadiad, Banaskantha, Gandhi Nagar and Ahmedabad districts. The cold storage facilities were mostly for a single commodity like potatoes, oranges, apples, grapes, pomegranates, flowers etc. Junagadh had a large number of small cold storages, primarily for sea food.

Table A3.4   Districtwise Cold Storage: Gujarat (31.12.2003)

District	Number	Capacity (t)
Kheda (including Anand and Nadiad)	73	2,01,286
Banaskantha	40	1,98,207
Gandhi Nagar	25	1,18,178
Ahmedabad	34	67,063
Mehsana	16	52,254
Baroda	20	46,866
Rajkot	15	32,935
Sabarkantha	13	32,236
Junagadh	63	26,057
Sidhpur	5	22,828
Jamnagar	14	22,779
Navsari	1	5,000
Kutch	4	4,376
Surat	4	3,410
Amreli	1	3,000
Surendranagar	3	2,133
Valsad	4	1,374
Bharuch	1	214
Godhra	1	199
Total	**337**	**8,40,395**

Source: Agmarknet. Cold storages in Gujarat. http://agmarknet.nic.in/gjnew.htm

Globally, the most widely stored item in cold storages was carrot, followed by potato. In India, the early cold storages were built, for storing potatoes. Even now, the number of cold storages for potatoes is significantly higher than other commodities.

In 1956, the first cold storage was established in Ahmedabad for storing potato. As per Hasmukhbhai, "During those days, the production was not more than 0.25 lt in Ahmedabad. After the establishment of this cold storage, potato production went up to 10.00 lt in a few years. This increment is solely due to the establishment of the cold storage. The current cold storage capacity in Ahmedabad is about 0.60 lt. Cold storages for potato developed in Kheda district even though there was not much of production. People of Kheda were wealthy and made investments in cold storage projects. Potatoes grown in Deesa taluk of Banaskantha were stored in these storages for a long time. The current cold storage capacity in Kheda (including Anand and Nadiad) is 2.00 lt. The quantity of potato produced is 1.50 lt.

Later, cold storages started developing in Deesa (Banaskantha district) and now there are 40 cold storages with a total capacity of about 2.00 lt. The total potato production is around 6.00 lt. Currently, only the excess quantity goes to the Kheda district. So many cold storages are lying idle in Kheda. Single commodity storage resulted in poor capacity utilization due to their seasonality.

In the conventional cold storages, temperature can be controlled but not the humidity. However, humidity control is very vital in preserving fruits and vegetables effectively. This is possible only in controlled atmosphere (CA) cold storages. There is only one CA storage in Gujarat and it is located in Borsad (Anand district). The location of this storage, being in the interior is inconvenient for the customers. The cost in this type of cold storage is high when compared to conventional storages, making it infeasible for farmers and traders. Only traders who are involved in exports are using its services."

Expansion into Other Cities

Having experienced profits in the cold storage business and considering the demand for cold storages, Hasmukhbhai wanted to expand his business into other cities. He was considering three cities and wanted to choose one among the three. He had no problem with buying an old cold storage and expanding it. He wanted to purchase land with future expansion in view. He wanted to have the land in at least two clusters in the vicinity, since he wanted to avoid inflow/outflow congestion on expansion. The construction would be done phase-wise. A gradual increment in capacity would lead to early utilization of the store and its operation would generate funds to cover the interest part of the loan. He also knew that land value would increase.

For example, the lands for his operations in Ahmedabad were in a cluster, near one another. The land value for MSCSPL was ₹10 l and it had increased during these years. This land was in two clusters. The expansion had been carried out gradually from 1500 t to 6000 t in one and half years.

He felt that the maximum capacity should not exceed 10000 t since it would make customer service difficult. He also wanted the minimum capacity to be 5000 t. Otherwise, there could not be an efficient utilization of labour and this could inflate the operational costs of cold storage. Always one standby machine had to be maintained for the efficient running of any cold storage. Therefore even for 1000 t capacity, two machines and five operators were needed. MSCSPL managed 6000 t with just six people which was by far the most efficient management of people.

Considering his financial status, he was planning a 60:40 debt equity ratio with total investment not exceeding ₹3.5 crore.

Mr. Hasmukhbhai wanted to be in the western region, not only due to managerial convenience, but also because many of his customers may be operating in locations within the region. This region, including Gujarat and Maharashtra, was doing well in agriculture production (Table A3.9 through Table A3.12). He also felt that the major growing metros would offer the best scope. He was considering three possible locations, Surat, Pune and Mumbai for the expansion.

Hasmukhbhai wanted to derive lessons from the overall status of cold storage industry in India, as given in the section, "Status of Cold Storage Industry in India" on page 424. The investment required for construction of a 5000 t capacity cold storage in the three candidate cities, along with Ahmedabad, is given in Table A3.5. His perceptions of the relative advantages and disadvantages of the three cities are given below.

Surat: There were four cold storages catering to the needs of the entire city. The main player was Surat Agro (Table A3.6). There was not much competition from other players. The neighbouring districts of Navsari, Valsad and Bharuch also had six cold storages (Table A3.4). The preservation knowledge among the traders and farmers was primitive. Hasmukhbhai felt that there was a need for three to four cold storages with a capacity of 5000 t each. Even the cost at some places was higher than the suburbs of Mumbai. He felt that managing one more cold store at Surat would not be a problem since the city was not new to him. Since the Agricultural Produce Market Committee (APMC) act was amended in Gujarat, the cold storage owners would be able to procure the goods from farmers directly without paying APMC cess. So, there would also not be any problem of inspection by APMC officials. A major challenge in the initial years could be due to his customer profile, not many of whom operated from Surat.

Pune: This region had high production of onions, grapes and flowers. The consumption demand was growing fast due to migration from Mumbai. Further, because of its proximity to and increasing cold storage rents in Mumbai, many of the traders shifted their base here. The initial investment required to start a cold storage at Pune was less compared to Mumbai. However, being relatively a new place both personally and in terms of customers, operation during the initial years might be difficult.

Table A3.5   Investment Required for a 5000 t Capacity Cold Storage (₹ l)

	Ahmedabad	Surat	Pune	Mumbai
Land (8000 Sq m)	42	46	54	77
Building and Steel	97	106	116	127
Plant and Machinery	28	29	32	35
Insulation	18	19	21	23
Total Investment	**185**	**201**	**223**	**262**

Source: Data provided by Mr. Hasmukhbhai K. Nakum.

The APMC act was not yet amended in Maharashtra. Many traders procured directly from the farmers and stored their goods in warehouses without paying APMC cess. In order to ensure that the cess was not avoided, the APMC officials frequently inspected all the cold storages. This was a common problem to all cold storages. This severely affected customer loyalty.

There were about 30 cold storages in Pune and its vicinity (Table A3.7). Frosty Refrigeration and Sunrise Cold Storage Company were the major players in the Pune market (Table A3.6). They offered cold chain facility and complete logistics solutions with their refrigerated vehicles. Thus Hasmukhbhai would have to invest not only in a cold storage, but in a cold chain.

Mumbai: The demand in Mumbai was always high and right through the year. Lots of imported perishable food items were getting unloaded every day in JNPT. There were many cold storages in the port vicinity catering to the growing demand. In spite of both domestic and import arrivals, as in other places, there was seasonality (Table A3.8). There were more than 70 cold storages in Mumbai and its vicinity (Table A3.7). Some of the small players like California Humifresh India Private Limited, Trans Agrotech Limited and AA Cold Storage and Ice Factory Pvt. Limited provided complete logistics solutions apart from storage service. The initial investment here was high due to the real estate cost. Hasmukhbhai felt that he would be able to charge lesser rent than most of the existing players, giving him competitive advantage. Many of his existing customers had their operations in Mumbai too. This would be an advantage for him in the initial years. Issues related to APMC existed here too.

Table A3.6   Major Competitors

Name	Capacity (t)	Sector	Products
Surat*			
Surat Agro Cold Storage	4500	Private	Multipurpose
Surat Dist Coop Milk Prods Union	1800	Cooperative	Dairy Products
Meghdoot Ice and Cold Storage	771	Private	Multipurpose
Pooja Cold Storage	579	Private	Multipurpose
Amrit Ice Cold Storage	260	Private	Multipurpose
Pune*			
Sunrise Cold Storage	2991	Private	Multipurpose
Mafco Cold Storage	2674	Public	Multipurpose
Frosty Refrigeration	2500	Private	Multipurpose
Maharashtra Krishi Sheet	2220	Cooperative	Multipurpose
Jai Jinendra Cold Storage	2000	Private	Multipurpose
Jayaree Cold Storage	1480	Private	Multipurpose
Temperature Food Ltd.	1353	Private	Multipurpose
Mumbai*			
Shiv Cold Storage	2675	Private	Multipurpose
Byculla Cold Storage	2604	Private	Multipurpose
Greater Bombay Milk Scheme Cold Storage	2252	Public	Dairy Products
Indian Diary Corp.	1283	Cooperative	Dairy Products
Ahmed A Fazalbhoy P Ltd.	1000	Private	Multipurpose

Source: Agmarknet. *Cold storages in Gujarat http://agmarknet.nic.in/gjnew.htm; **Cold storages in Maharashtra. http://agmarknet.nic.in/mhnew.htm

Table A3.7   Districtwise Cold Storage: Maharashtra (31.12.2003)

District	Number	Capacity (t)
Nagpur	12	46077
Thane	33	35194
Sangli	37	31526
Mumbai	71	23051
Dhule	7	22987
Pune	28	21720
Raigad	16	19345
Aurangabad	9	14246
Nasik	63	14079
Ahmednagar	9	13206
Jalgaon	8	8864
Solapur	11	7451
Amravati	3	5789
Satara	4	5081
Kolhapur	8	4834
Jalna	1	2565
Bhandara	3	2335
Ratnagiri	8	1319
Nanded	2	1164
Akola	3	858
Osmanabad	1	305
Latur	1	203
Sindhudurg	1	59
Beed	1	30
Total	**340**	**282288**

Source: Agmarknet. Cold storages in Maharashtra. http://agmarknet.nic.in/mhnew.htm

Table A3.8   Arrival of Fruits and Vegetables in Markets in Mumbai

Name of Commodity	Begin	End	Peak
Banana	Round the Year		
Citrus lime	August	December	September, October, November
Citrus mosambi	September, February	December, April	October, November, March
Citrus Orange	February, September	May, December	March, April, November
Grapes	March	May	March, April
Guava	November	May	December, April
Mango	March	June	April, May
Pomegranate	November, February	January, April	December, March
Ber	October	January	November, December
Apple	August	November	September, October
Papaya	Round the Year		
Sapota	Round the Year		
Brinjal	Round the Year		
Cabbage	December	March	January, February
Cauliflower	Round the Year		
Onion	October, March	December, May	November, April
Potato	February	May	March, April
Tomato	Round the Year		

Source: National Horticulture Board. http://nhb.gov.in/staticfiles/mumbai.htm

Challenges for Expansion

According to Hasmukhbhai, the following were the challenges that he would face:

- Land availability is decreasing in the vicinity of the metros. Land prices are increasing
- Earlier the subsidy was given on the amount invested but now it is given on the volume of storage space created. It is also limited to ₹50 l. (See section on page 436, "Subsidy Schemes for Cold Storages.") The subsidy does not distinguish entrepreneurs with and without experience, across locations and other cost influencing dimensions. Many have entered this business because of the subsidy, creating a mismatch between supply and demand in certain locations
- Power availability is an issue as also its reliability. This has implications on maintaining the products stored. Further, the cost is high
- The seasonality in arrivals makes it difficult to achieve 100% utilization
- There are also variations in the day to day arrivals caused by speculations, black marketing and other malpractices. This is partly due to insufficient data supplied by APMC on expected arrivals

Mr. Hasmukhbhai had to take a decision keeping in mind the relative costs and benefits involved in these three cities. This would further his cause of aggressive growth.

Agribusiness Supply Chain Management

Table A3.9 Production of Pulses, Food Grains, Spices, Fruits, Vegetables and Flowers in Gujarat and Maharashtra

Districtwise Area, Production and Yield of Total Pulses in Gujarat

District	2001–2002			2002–2003			2003–2004			2004–2005		
	Area	Production	Yield	Area	Production	Yield	Area	Production	Yield	Area	Production	Yield
	00 h	00 t	Kg/h	00 h	00 t	Kg/h	00 h	00 t	Kg/h	00 h	00 t	Kg/h
Vadodara	1171	626	535	1184	857	723	1118	780	698	922	828	898
Dahod	–	–	–	–	–	–	718	590	822	757	616	814
Panchmahal	692	366	529	696	437	624	383	419	1094	378	396	1048
Surat	386	307	795	342	227	664	380	400	1053	401	321	800
Bharuch	751	437	582	721	435	602	482	260	539	486	277	570
Banaskantha	551	262	475	605	127	210	533	381	715	596	260	436
Sabarkantha	794	248	312	766	165	216	695	512	737	459	249	542
Kutch	821	248	302	414	50	124	995	529	532	767	241	314
Jamnagar	188	127	675	139	73	519	296	361	1220	213	184	864
Narmada	–	–	–	–	–	–	268	248	925	193	178	922
Ahmedabad	204	84	412	119	70	593	266	132	496	226	132	584
Kheda	174	125	718	238	183	771	165	168	1018	119	129	1084

Surendranagar	97	61	629	185	51	277	210	149	710	218	128	587
Patan	–	–	–	–	–	–	364	219	602	255	120	471
Rajkot	221	161	728	213	40	184	228	217	952	165	107	648
Mehsana	493	274	556	534	169	317	291	152	522	213	106	498
Junagadh	204	133	652	200	109	545	152	126	829	128	94	734
Amreli	100	63	630	75	29	383	128	88	687	137	89	650
Porbandar	–	–	–	–	–	–	159	133	836	69	77	1116
Valsad	219	136	621	187	99	532	113	88	779	98	70	714
Gandhinagar	25	12	480	13	6	415	92	64	696	62	57	919
Navsari	–	–	–	–	–	–	81	58	716	80	52	650
Dang	82	58	707	86	48	561	92	73	793	61	38	623
Anand	–	–	–	–	–	–	56	30	536	43	31	721
Bhavnagar	131	70	534	139	56	415	61	42	689	52	30	577
Total	**7304**	**3798**	**520**	**6977**	**3272**	**469**	**8326**	**6219**	**747**	**7098**	**4810**	**678**

(Continued)

Source: www.indiastat.com

Table A3.9 (Continued) Production of Pulses, Food Grains, Spices, Fruits, Vegetables and Flowers in Gujarat and Maharashtra

Districtwise Area, Production and Yield of Total Food Grains in Gujarat

District	2001–2002			2002–2003			2003–2004			2004–2005		
	Area	Production	Yield	Area	Production	Yield	Area	Production	Yield	Area	Production	Yield
	00 h	00 t	Kg/h	00 h	00 t	Kg/h	00 h	00 t	Kg/h	00 h	00 t	Kg/h
Kheda	3482	6179	1775	3108	4910	1581	2817	5443	1932	2635	5168	1961
Ahmedabad	2185	2972	1360	1949	2060	1059	2532	4307	1701	2420	4282	1769
Junagadh	1138	2929	2574	667	958	1439	1357	4066	2996	1349	3964	2938
Banaskantha	3185	3757	1180	4303	3019	701	3982	6095	1531	3460	3781	1093
Anand	–	–	–	–	–	–	1791	3390	1893	1879	3590	1911
Sabarkantha	2667	3961	1485	2749	2803	1020	2958	1806	1625	2441	3066	1256
Dahod	–	–	–	–	–	1051	3012	4573	1518	3066	2911	949
Vadodara	2601	2528	972	–	–	–	2607	3055	1172	2417	2599	1075
Surat	1839	2728	1483	1538	2465	1603	1783	2910	1632	1684	2524	1499
Panchmahal	3770	4562	1210	4063	5283	1300	2353	2901	1233	2573	2212	860
Rajkot	861	1178	1368	887	712	802	1348	3516	2608	931	1854	1991
Mehsana	2400	3864	1610	2170	2445	1127	1466	2134	1456	1125	1791	1592
Gandhinagar	257	550	2140	175	371	2095	886	1827	2062	717	1570	2190

Surendranagar	1322	1518	1148	1649	2183	1324	755	821	1088	1581	1472	931
Kutch	717	1310	1827	1056	2193	2077	637	659	1035	722	1390	1926
Bhavnagar	1665	1277	767	2005	1817	906	1556	2099	1352	1607	1729	1076
Navsari	2002	1159	579	2224	1372	617	–	–	–	–	–	–
Amreli	1665	1134	681	1520	1157	761	1495	1049	701	1656	1257	759
Patan	989	1107	1119	1133	1743	1539	–	–	–	1786	–	–
Valsad	1530	1088	711	1767	1147	649	516	280	542	1214	2570	1439
Jamnagar	1480	1083	732	2171	2405	1108	905	1364	1504	922	948	781
Bharuch	851	1032	1212	917	1051	1146	675	299	442	990	1420	1540
Dang	1341	566	422	1621	3085	1903	–	–	–	–	413	417
Narmada	1072	533	497	1212	816	673	–	–	–	–	–	–
Porbandar	1915	408	213	1384	353	255	–	–	–	–	–	–
Baroda	–	–	–	–	–	–	–	2694	2564	–	–	–
Bulsar	–	–	–	–	–	–	1482	1919	1294	–	–	–
Total	**1408**	**51527**	**36605**	**1629**	**65707**	**40338**	**1127**	**36210**	**32131**	**1396**	**46407**	**33254**

(Continued)

Source: www.indiastat.com

Table A3.9 (Continued) Production of Pulses, Food Grains, Spices, Fruits, Vegetables and Flowers in Gujarat and Maharashtra

	Area and Production of Spices in Gujarat							
	2001–2002		2002–2003		2003–2004		2004–2005 (P)	
Spice	Area	Production	Area	Production	Area	Production	Area	Production
	h	t	h	t	h	t	h	t
Mustard	247100	292100	160800	172300	160800	172300	291200	404900
Garlic	9300	57600	8240	46390	37680	151370	28520	144360
Cumin	145100	61300	200049	64275	203011	81999	208141	106975
Fennel	29100	30900	17947	25099	36095	46799	34265	46450
Ginger	1050	16280	1150	19410	1910	22940	1830	26920
Chilli	13600	12100	12110	10470	27110	24570	24640	26520
Turmeric	690	12240	840	14690	950	11180	1020	14120
Coriander	5380	3090	2190	1820	6820	5710	9860	10630
Dill Seed	12297	10003	6407	4489	8269	6807	10076	8567
Fenugreek	7208	8294	6258	7494	5684	6695	4857	7014
Ajwan	2717	1429	946	376	1476	556	4628	3744
Total	473542	505336	416937	366813	489805	530926	619037	800200

Source: www.indiastat.com

Districtwise Area, Production and Yield of Total Pulses in Maharashtra

District	2000–2001			2002–2003			2003–2004			2004–2005		
	Area	Production	Yield	Area	Production	Yield	Area	Production	Yield	Area	Production	Yield
	00 h	00 t	Kg/h	00 h	00 t	Kg/h	00 h	00 t	Kg/h	00 h	00 t	Kg/h
Buldhana	2471	782	316	2633	1257	477	2740	1700	620	2552	1085	425
Nanded	1801	761	423	2022	1620	801	2027	1516	748	1727	1076	623
Amravati	2313	1232	533	2163	1444	668	2151	1489	692	173	1073	544
Jalgaon	1560	553	354	1735	1301	750	1837	1213	660	1767	1068	604
Osmanabad	2047	1145	559	2168	984	454	2151	435	202	2123	1052	496
Yavatmal	2218	1843	831	2229	1670	749	2198	1561	710	1892	1022	540
Latur	2230	671	301	2284	1140	499	2221	1133	510	2176	909	418
Washim	1828	911	498	1886	1244	660	1809	1273	704	1605	874	545
Nasik	901	367	407	857	535	624	839	429	511	900	609	677
Parbhani	1765	577	327	1984	1109	559	1991	1010	507	2016	600	298
Ahmednagar	1226	565	461	1190	471	396	831	294	354	1086	573	527
Wardha	729	675	926	823	929	1129	866	946	1092	758	560	739
Jalna	1520	633	416	1456	922	633	1390	691	497	1369	542	396
Beed	1264	479	379	1051	484	461	979	330	337	1006	542	539
Akola	1583	791	500	1512	917	606	1606	1013	631	1547	537	347
Nagpur	887	484	546	916	473	516	1000	611	611	1015	513	506

(Continued)

Table A3.9 (*Continued*) Production of Pulses, Food Grains, Spices, Fruits, Vegetables and Flowers in Gujarat and Maharashtra

Aurangabad	1165	356	306	1100	521	474	1019	450	442	992	498	502
Satara	742	331	446	761	335	440	545	217	398	775	477	615
Sangli	943	498	528	806	223	277	384	139	362	887	436	492
Hingoli	894	351	393	1078	685	636	943	623	661	861	431	501
Pune	1076	531	493	895	408	456	528	251	475	736	375	510
Dhule	570	208	365	597	282	472	739	463	627	691	369	534
Solapur	920	416	452	670	248	370	410	78	190	783	348	445
Nandurbar	919	323	351	835	427	511	798	531	665	777	322	415
Chandrapur	647	252	389	602	214	355	858	391	456	612	196	320
Kolhapur	292	154	527	279	151	541	277	148	534	258	169	656
Thane	192	87	453	172	98	570	191	112	586	192	123	640
Bhandara	234	113	483	234	130	557	314	169	538	201	76	378
Raigad	129	52	403	112	50	446	121	56	463	122	60	489
Gadchiroli	220	101	459	145	67	462	271	112	413	186	51	274
Gondia	147	72	490	141	76	538	191	104	545	132	50	380
Ratnagiri	74	29	392	68	29	426	74	36	486	75	37	487
Sindhudurg	65	26	400	67	27	403	65	29	446	62	29	473
Total	**35572**	**16369**	**460**	**35471**	**20471**	**577**	**34364**	**19553**	**569**	**33854**	**16682**	**493**

Source: www.indiastat.com

Districtwise Area, Production and Yield of Total Food Grains in Maharashtra

District	2000–2001			2002–2003			2003–2004			2004–2005		
	Area	Production	Yield	Area	Production	Yield	Area	Production	Yield	Area	Production	Yield
	00 h	00 t	Kg/h	00 h	00 t	Kg/h	00 h	00 t	Kg/h	00 h	00 t	Kg/h
Ahmednagar	9687	5084	525	9397	4424	471	8364	2169	259	10337	6685	647
Nasik	6124	4702	768	6017	5420	901	5946	4923	828	6239	6076	974
Jalgaon	4327	4697	1086	4675	6687	1430	4844	6397	1321	4679	5911	1263
Pune	8473	5971	705	8322	4517	543	5426	3238	597	7742	5716	738
Satara	4430	4365	985	4711	3940	836	2998	3036	1013	4954	5460	1102
Sangli	4484	3278	731	4321	2196	508	2949	1549	525	5217	5185	994
Aurangabad	5868	4670	796	5498	5488	998	5450	4763	874	5388	4999	928
Solapur	8563	4336	506	8427	3699	439	7134	589	82	8767	4490	512
Beed	6324	3281	519	6493	4702	724	6350	2735	431	6296	4155	660
Kolhapur	1987	4200	2114	1972	3484	1767	1974	3622	1835	2036	4111	2019
Nanded	4452	3891	874	4638	5320	1147	4556	4569	1003	4213	3972	943
Jalna	4917	4033	820	4896	5103	1042	4662	4330	929	4625	3876	838
Thane	1805	3723	2063	1863	1824	979	1893	4301	2272	1888	3701	1960
Osmanabad	6045	3351	554	6286	3700	589	6469	1393	215	6242	3528	565
Raigad	1616	3753	2322	1590	2919	1836	1619	3947	2438	1603	3138	1958
Dhule	2704	2600	962	2790	2176	780	2987	3406	1140	2897	3113	1075
Buldhana	4535	4381	966	4614	4366	946	4657	4725	1015	4426	2947	666
Latur	4890	3527	721	4761	4294	902	4709	3882	824	4501	2924	650
Parbhani	5095	4054	796	5324	4568	858	5406	3941	729	5289	2449	463

(Continued)

Table A3.9 (*Continued*) Production of Pulses, Food Grains, Spices, Fruits, Vegetables and Flowers in Gujarat and Maharashtra

Sindhudurg	898	2050	2283	894	1779	1990	890	2537	2850	887	2258	2546
Yavatmal	3964	3776	953	3895	3566	916	3749	3141	838	3155	2227	706
Amravati	3502	3245	927	3445	3102	900	3403	3162	929	3141	2153	686
Ratnagiri	1109	2412	2175	1148	1794	1563	1132	2767	2444	1121	2150	1918
Akola	2497	2390	957	2476	2727	1101	2544	2775	1091	2427	1673	689
Nagpur	2216	2118	956	2192	1692	772	2336	2167	928	2223	1597	719
Chandrapur	3101	3233	1043	2651	1679	633	3092	2939	950	2567	1569	611
Nandurbar	2165	1754	810	2065	1621	785	2099	2127	1013	2146	1540	718
Hingoli	2476	2400	969	2636	2744	1041	2396	2539	1060	2129	1492	701
Washim	2714	2421	892	2721	2223	817	2559	2206	862	2189	1442	659
Gadchiroli	1771	1900	1073	1702	924	543	1850	2008	1085	1736	1365	787
Bhandara	1846	2324	1259	2070	2187	1056	2153	2989	1388	1921	1362	709
Gondia	2033	2525	1242	1908	1843	966	2002	2809	1403	1799	1073	597
Wardha	1363	1403	1029	1379	1492	1082	1380	1493	1082	1173	971	828
Total	**127981**	**111848**	**874**	**127777**	**108202**	**847**	**119978**	**103174**	**860**	**125953**	**105310**	**836**

Source: www.indiastat.com

Table A3.10   Statewise Production of Fruits in India (000 t)

States/UT	2000–2001	2001–2002	2002–2003	2003–2004	2004–2005
Andhra Pradesh	5003.4	6157.4	7404.8	6871.7	7735.4
Tamil Nadu	6237.7	4342.4	4014.0	3460.1	4467.6
Karnataka	4819.5	4028.9	4008.8	3027.3	4046.9
Gujarat	2268.2	2346.9	2957.5	3586.8	4014.4
Uttar Pradesh	2713.0	2282.8	4313.8	3381.2	2912.8
Bihar	3237.5	2877.0	3038.1	3294.91	2769.5
West Bengal	1656.5	1985.5	1785.6	2111.5	2128.3
Orissa	1284.4	1362.9	1485.5	1352.6	1404.0
Jammu and Kashmir	837.3	1000.9	983.9	1180.5	1217.6
Assam	1293.8	1335.1	1126.5	1181.1	1151.0
Madhya Pradesh	1740.4	1143.8	1112.6	1167.8	1102.6
Kerala	1772.6	1772.6	837.3	1401.8	985.1
Uttaranchal	541.0	376.1	458.1	644.6	788.7
Himachal Pradesh	438.3	263.4	480.4	588.1	720.6
Punjab	479.7	531.7	578.5	628.1	679.5
Tripura	450.8	452.1	459.9	482.1	503.4
Jharkhand	265.1	321.1	321.2	321.2	403.4
Chhattisgarh	154.3	203.1	382.0	401.1	343.2
Manipur	118.7	134.0	137.8	353.3	320.9
Rajasthan	339.3*	200.7	184.8	220.9	238.6
Haryana	232.0	235.2	237.3	257.2	232.2
Meghalaya	186.9	186.9	153.3	199.6	199.6
Arunachal Pradesh	123.1	124.9	82.1	101.3	103.2
Goa	71.5	64.7	72.8	78.7	81.6
Nagaland	290.4	302	65.9	48.8	48.9
Mizoram	66.7	63.4	55.0	42.4	42.5
Andaman and Nicobar Islands*	16.7	16.7	16.7	22.1	23.4
Pondicherry*	26.7	26.7	26.7	19.1	21.1
Sikkim	10.0	10.3	8.1	0.1	12.2
Dadra and Nagar Haveli*	7.1	7.1*	7.1	7.1	7.1

(Continued)

Table A3.10 (*Continued*)   Statewise Production of Fruits in India (000 t)

States/UT	2000–2001	2001–2002	2002–2003	2003–2004	2004–2005
Chandigarh*	1.1	1.1	1.1	1.1	1.1
Lakshadweep*	1.1	1.1	1.1	1.1	1.1
Delhi	1.0	1.0*	1.0	1.0	–
Maharashtra	8680.8	8840.6	–	–	–
Daman and Diu*	3.4	3.4	3.4	–	–
Total	45370.0	43000.9	45203.1	45705.9	49294.8

Source: www.indiastat.com

* Previous year data.

Status of Cold Storage Industry in India

Introduction

India is the largest producer of fruits and second largest producer of vegetables in the world. It was a $70 billion industry in the mid-90s and was expected to double by 2005. The value added food segment was expected to grow at a faster rate, rising from $20 billion to $60 billion during this period. This enormous growth in the food front in India during the last decades of twentieth century, has not only totally eliminated the imports and deficits in terms of food grains, fruits and vegetables, milk and milk products, and fish and fish products but also transformed our Indian agricultural economy from deficit status to that of surplus status in the twenty first century. In spite of that, per capita availability of fruits and vegetables is quite low.

Due to lack of storage facilities, and post-harvest preservation and handling facilities, there is wastage of 25% to 30% of production. Besides this, the quality of a sizable quantity of produce also deteriorates by the time it reaches the consumer. The enormous wastage is mainly because of the perishable nature of the produce which requires a cold chain arrangement to maintain the quality and extend the shelf-life if consumption is not meant immediately after harvest. In the absence of a cold storage and related cold chain facilities, the farmers are forced to sell their produce immediately after harvest which results in glut situations and low price realization. Sometimes farmers do not even get their harvesting and transportation costs. As a result, our production is not getting stabilized and the farmers, after burning fingers in one crop, switch over to another crop in the subsequent year and the vicious cycle continues. Cold storage facilities accessible to them will go a long way in removing the distress sales and ensure better returns.

Table A3.11   Statewise Production of Vegetables in India (000 t)

States/UT	2000–2001	2001–2002	2002–2003	2003–2004	2004–2005
West Bengal	17779.4	18075.3	17376.5	18510.6	18103.2
Uttar Pradesh	13030.4	15044.8	15791.4	14862	15792.8
Bihar	10219.7	8022.9	8288.5	13296.9	13349.1
Orissa	8089.1	7447.4	7126.2	8030.9	8045.6
Tamil Nadu	6011.0	5444.6	4223.3	4672.7	6218.3
Gujarat	3070.8	3278.2	3517.9	3515.2	4867.9
Karnataka	5763.0	4173.2	3707.9	4176.1	4382.9
Maharashtra	5142.0	5128.3	4768.9	4132.1	4044.4
Andhra Pradesh	3147.7	2586.7	2357.9	2882.3	3861.9
Jharkhand	2109.5	1736.3	1300.1	1197.2	3394.9
Haryana	2191.5	2151.9	2051.8	2703.3	2980.4
Punjab	2310.0	2275.6	2319.4	2588.1	2677.4
Madhya Pradesh	3501.9	1817.5	1827.0	2377.0	2659.6
Kerala	2530.9	2541.9	2547.4	2602.9	2490.1
Assam	2693.1	2935.2	2464.4	1958.9	2020.4
Chhattisgarh	1146.3	1355.3	1357.2	1554.1	1266.3
Himachal Pradesh	734.2	639.1	775.7	877.2	1013.5
Uttaranchal	1138.1	737.3	507.5	447.3	951.8
Jammu and Kashmir	757.9	728.9*	332.9	462.9	843.0
Rajasthan	386.4	432.5	358.3	527.6	650.2
Delhi	862.7	747.4*	628.1	626.8	626.8
Tripura	328.1	353.2	360.3	352.2	373.4
Meghalaya	303.6	265.9	338.9	270.5	270.4
Nagaland	253.6	286.0	78.5	88.1	88.1
Manipur	67.4	66.1	71.9	86.0	86.0
Arunachal Pradesh	83.7	83.9	81.5	80.9	78.8
Sikkim	59.7	60.0	59.1	75.0	76.5
Goa	76.0	76.0	68.5	74.7	74.7
Pondicherry	54.2	54.2*	63.7	71.9	74.7
Andaman and Nicobar Islands	15.8*	15.8*	16.3	23.3	30.8
Mizoram	47.3	44.1	31.9	24.0	24.0
Dadra and Nagar Haveli	13.5*	13.5*	13.5	13.5	13.5
Chandigarh	1.7	1.7*	1.7	1.7	1.7
Lakshadweep	0.2	0.2*	0.2	0.2	0.2
Daman and Diu	1.1*	1.1*	0.1	0.1	0.1
Total	**93921.5**	**88622**	**84815.4**	**93165**	**101433.5**

Source: www.indiastat.com

* *Previous year data.*

Table A3.12 Statewise Production of Flowers in India (000 t)

State	2001–2002 Loose	2001–2002 Cut	2002–2003 Loose	2002–2003 Cut	2003–2004 Loose	2003–2004 Cut	2004–2005 Loose	2004–2005 Cut
	t	l	t	l	T	l	T	l
Tamil Nadu	156700		135221	–	16155	–	187342	–
Karnataka	138776	9230	151953	–	143286	5591	145890	4134
Andhra Pradesh	121336	2780	72205	87	49130	44	57875	71
Haryana	17890	1127	32500	1200	58333	461	55583	508
West Bengal	31268	6771	33749	7020	43575	8767	44674	8963
Maharashtra	30376		–	–	48538	–	51705	–
Gujarat			30187	–	30187	–	41811	1969
Delhi	25007		25007	–	25007	–	25007	–
Orissa			–	–	78	11	17252	12
Uttar Pradesh	3400	5350	9753	2650	9753	2650	11905	3527
Punjab	2741		3000	–	3000	–	3075	–
Chattisgarh			60	–	60	–	2829	–
Rajasthan	2434		986	–	2161	–	2604	–
Himachal Pradesh		292	999	283	1504	380	2243	182
Bihar	620	21	1757	11	1757	11	1757	11
Madhya Pradesh	862		–	–	–	–	1097	–
Jammu and Kashmir	61		38	7	207	3	922	110
Manipur	33		30	–	701	–	701	–
Uttaranchal	324		254458	–	545	–	558	–
Daman and Diu			–	–	7	–	7	–
Mizoram			–	–	0	1	0	1
Assam			1000	–	–	–	–	–
Jharkhand			2000	347	–	–	–	–
Pondicherry			–	–	–	–	–	–
Nagaland			–	–	–	–	–	–
Sikkim			–	–	–	–	–	–
Tripura			–	–	–	–	–	–
Others	2748	76	–	9	–	9	–	28
Total	**534576**	**25647**	**754903**	**11614**	**579484**	**17926**	**654837**	**19515**

Source: www.indiastat.com

* Previous year data.

Storage Conditions for Foods

In low temperatures, activities of microorganisms get retarded and hence the shelf life of food items increases. Unlike high temperature, low temperature preservation cannot destroy the spoilage agents, but greatly reduces their activities, providing a practical way of preserving perishable foods in their natural state which otherwise is not possible through heating. There are three groups of products:

1. Foods that are alive at the time of storage, distribution and sale, for example fruits and vegetables
2. Foods that are no longer alive and have been processed in some form, for example meat and fish products and
3. Commodities that benefit from storage at controlled temperature, for example beer, tobacco, khandsari, etc.

Living foods such as fruits and vegetables have some natural protection against the activities of microorganism. The best method of preserving these items is to keep the product alive and at the same time retard the natural enzyme activity which will retard the rate of ripening or maturity.

Preservation of non-living foods is more difficult since they are susceptible to spoilage. The problem is to preserve dead tissues from decay and putrefaction. Long term storage of meat and fish product can only be achieved by freezing and then by storing it at temperature below -15°C. Only certain fruits and vegetables can benefit from freezing. Products such as apples, tomatoes, oranges, etc., cannot be frozen and close control of temperature is necessary for long term storage. Some products can also be benefited by storing under CA and modified atmosphere (MA) conditions.

Dairy products are produced from animal fats and therefore nonliving foodstuffs. They suffer from oxidation and breakdown of their fats, causing rancidity. Packaging to exclude air and hence oxygen can extend storage life of such foodstuffs. The storage requirements of some of the important commodities are given in the following table.

Economic Size of Unit and Land Requirements

Cold storage units can be used to store either a single commodity or multiple commodities. Financial viability of a unit depends upon the intended pattern of use and rental rate prevalent in an area. To set up a 5000 t capacity cold storage unit although one acre of land may be adequate, it is always better to have two acres of land to take care of future expansions and waste management. While selecting the site, care should be taken to select a site at an elevation free from inundation and well connected by road and other communication facilities to both production and consumption centres. The land should be of nonagricultural type and the soil at the site should be firm enough to carry the weight of the building and storage racks.

Technology

A cold storage unit has various elements of technology to maintain the desired room environment for the commodities to be stored. They are described below:

Table A3.13 Desired Storage Environment of Fruits and Vegetables

Commodity	Temperature (F)	Relative Humidity (%)
Fruits		
Apple	30–40	90–95
Apricots	32	90–95
Avocado	40–55	85–90
Asparagus	32–35	95–100
Black berry	31–32	90–95
Cherries	30–31	90–95
Grapes	32	85
Lemons	50–55	85–90
Banana	56–58	90–95
Watermelon	50–60	90
Orange	32–48	85–90
Peach	31–32	90–95
Vegetables		
Cabbage	32	98–100
Carrots	32	98–100
Cauliflower	32	90–98
Cucumber	50–55	95
Brinjal	32	90–95
Lettuce	32	85–90
Potato	50–60	90

Source: National Sustainable Agriculture Information Service, Postharvest handling of fruits and vegetables. Aug. 2000. https://attra.ncat.org/attra-pub/download.php?id=378

Table A3.14 Sectorwise Distribution of Cold Storages in India (31.12.2004)

States/UT	Private Sector		Cooperative Sector		Public Sector		Total	
	No	Capacity (t)	No	Capacity (t)	No	Capacity (t)	No	Capacity (t)
Uttar Pradesh and Uttaranchal	1320	79,33,688	87	2,81,480	3	8,000	1410	82,58,813
West Bengal	317	41,05,177	69	2,97,800	0	0	386	44,02,977
Punjab	364	11,92,593	18	39,092	0	0	382	12,31,685
Bihar	220	8,33,382	18	77,200	0	0	238	9,10,582
Gujarat*	324	8,45,581	19	21,543	8	7,739	351	8,74,863
Madhya Pradesh	143	6,31,430	19	98,848	5	2,434	167	7,32,712
Andhra Pradesh*	213	5,56,147	13	9,270	8	1,190	234	5,66,607
Maharashtra	343	4,20,270	53	19,839	29	7,851	425	4,47,960
Haryana	227	3,65,291	4	3,403	6	11,399	237	3,80,093
Chhattisgarh	66	3,60,974	1	29	1	41	68	3,61,044

(Continued)

Table A3.14 (Continued) Sectorwise Distribution of Cold Storages in India (31.12.2004)

States/UT	Private Sector		Cooperative Sector		Public Sector		Total	
	No	Capacity (t)	No	Capacity (t)	No	Capacity (t)	No	Capacity (t)
Orissa	80	2,23,135	24	51,040	0	0	104	2,74,175
Rajasthan	83	2,68,776	9	3,832	1	14	93	2,72,622
Tamil Nadu*	96	1,49,448	13	7,462	4	4,162	113	1,61,072
Karnataka*	79	1,22,566	18	6,689	17	9,594	114	1,38,849
Delhi	74	1,03,180	2	5,201	16	17,680	92	1,26,061
Jharkhand	17	53,210	8	27,415	0	0	25	80,625
Assam	18	68,796	2	6,000	54	1,120	24	75,916
Jammu and Kashmir	15	40,689	3	2,134	1	46	19	42,869
Kerala*	135	27,450	6	1,080	9	1540	150	30,070

Tripura	2	7,750	1	5,000	5	5,700	8	18,450
Himachal Pradesh	8	11,413	2	767	7	6,195	17	18,375
Chandigarh	5	11,216	1	1,000	0	0	6	12,216
Nagaland	1	5,000	1	1,150	0	0	2	6,150
Goa	24	5,875	0	0	0	0	24	5,875
Arunachal Pradesh	1	5,000	0	0	0	0	1	5,000
Meghalaya	1	1,200	0	0	2	20,00	3	3,200
Andaman and Nicobar Islands	1	170	0	0	1	40	2	210
Pondicherry*	2	115	1	50	2	35	5	200
Lakshadweep*	0	0	0	0	1	36	1	36
Total	4179	18,349,522	392	9,67,324	180	86,816	4701	1,94,39,307

Source: Agmarknet. Sectorwise distribution of cold storage in India. 2005. http://agmarknet.nic.in/sectorcold311204.htm

* As on 31.12.2003.

Table A3.15   Commodity-Wise Distribution of Cold Storages (31.12.2004)

State/UT	Potatoes No	Capacity (t)	Multi Purpose No	Capacity (t)	Fruits and Vegetables No	Capacity (t)	Meat and Fish No	Capacity (t)	Milk and Milk Products No	Capacity (t)	Others No	Capacity (t)	Total No	Capacity (t)
U.P and Uttaranchal	1371	8163232	30	87135	04	3580	02	4027	02	801	01	38	1410	8258813
West Bengal	364	4379347	16	22610	00	00	06	1020	00	00	00	00	386	440297
Punjab	344	1097609	36	123486	00	00	00	00	02	10590	00	00	382	1231685
Bihar	187	6,99,780	51	210802	00	00	00	00	00	00	00	00	238	910582
Gujarat*	164	584848	91	247380	02	1063	61	28442	32	13114	01	16	351	874863
Madhya Pradesh	109	553257	42	173740	04	2577	00	00	11	3125	01	13	167	732712
Andhra Pradesh	00	00	178	583198	11	12913	46	42594	17	14919	05	9243	257	662867
Maharashtra	04	2436	168	357644	93	18094	60	47449	72	19571	28	2766	425	447960
Haryana	172	225991	59	152759	00	00	00	00	06	1343	00	00	237	380093
Chhattisgarh	09	27575	56	333369	00	00	00	00	03	100	00	00	68	361044
Orissa	37	123580	42	146060	00	00	25	4535	00	00	00	00	104	274175
Rajasthan	19	65896	62	204811	00	00	00	00	08	1781	04	104	93	272622
Tamil Nadu*	00	00	42	146095	00	00	51	9965	13	4512	07	500	113	161072
Karnataka	03	00	67	140770	01	22	33	5078	22	3380	00	00	123	149250
Delhi	00	00	51	117276	01	53	04	1239	09	4757	27	2736	92	126061

State													
Jharkhand	06	22500	19	58125	00	00	00	00	00	00	00	00	25	80625
Assam	00	00	24	75916	00	00	00	00	00	00	00	00	24	75916
Jammu and Kashmir	05	11281	12	31473	00	00	00	00	02	115	00	00	19	42869
Kerala*	00	00	03	3200	00	00	140	25780	06	1080	01	10	150	30070
Tripura	00	00	08	18450	00	00	00	00	00	00	00	00	08	18450
Himachal Pradesh	05	9748	04	2354	06	6100	01	78	01	95	00	00	17	18375
Chandigarh	01	1000	04	11131	00	00	01	85	00	00	00	00	06	12216
Nagaland	00	00	02	6150	00	00	00	00	00	00	00	00	02	6150
Goa	00	00	01	3633	00	00	10	1701	02	68	11	473	24	5875
Arunachal Pradesh	00	00	01	5000	00	00	00	00	00	00	00	00	01	5000
Meghalaya	00	00	03	3200	00	00	00	00	00	00	00	00	03	3200
Andaman and Nicobar Islands	00	00	01	170	00	00	01	40	00	00	00	00	02	210
Pondicherry*	00	00	00	00	01	15	03	135	01	50	00	00	05	200
Lakshadweep*	00	00	00	00	00	00	01	36	00	00	00	00	01	36
TOTAL	**2800**	**159968080**	**1073**	**3265937**	**123**	**44417**	**445**	**172204**	**209**	**79401**	**86**	**15899**	**4733**	**19545968**

Source: Agmarknet. Commoditywise distribution of cold storages. 2005. http://agmarknet.nic.in/commcold311204.htm

* As on 31/12/2003.

Refrigeration

A refrigeration system works on two principles:

1. Vapour Absorption System (VAS) and
2. Vapour Compression System (VCS)

VAS, although comparatively costlier, is quite economical in operation and adequately compensates the higher initial investment. Wherever possible such a system should be selected to conserve on energy and operational cost. However, it has its own limitations when temperature requirement is below 10°C and many of the fruits and vegetables except seeds, mango, etc., require lower than 10°C for long storage.

VCS is comparatively cheaper than VAS. There are three types of VCS systems available depending upon the cooling arrangements in the storage rooms i.e. diffuser type, bunker type and fin coil type. Diffuser type is comparatively costlier and is selected only when the storage room heights are low. The operational cost of such units is also higher. Bunker type is the cheapest and is preferred when storage room heights normally exceeds 11.5 m. Its operational cost is also low. Fin coil type, although about 5% costlier than the bunker type, is very energy efficient with low operational cost and higher space availability for storage of produce. Such system is used for units with room heights of 5.4 m onwards. In a refrigeration system, refrigerants are used to pick up heat by evaporation at a lower temperature and pressure from the storage space and give up the heat by condensation at a higher temperature and pressure in a condenser. Freon used to be a common refrigerant. As it causes environmental degradation, its use is going to be banned by the year 2008. Therefore, ammonia is being increasingly used and preferred for horticultural produce cold storage units.

Table A3.16   Comparison of Cooling Units (5000 t Capacity)

Item	Unit	Type of Vapour Compression System		
		Diffuser	Bunker	Fin coil
Installed Electrical Load	hp	180	128	124
Installed Electrical Load	kW	134.28	95.49	92.50
Energy Saving	%	–	29	31
Refrigerant Requirement	kg	1,520	2,200	380
Space Requirement for Cooling System	cu m	452	670	36

Although several types of compressors and condensers are available, medium speed reciprocating compressors and atmospheric condensers are preferred because of the relatively lower cost, energy efficiency and ease in maintenance.

While selecting size of the equipment, care should be taken to assess all loads and proper provision should be made to take care of the peak demand during summer loading and aging of the equipment. Heat load factors normally considered in a cold storage design are:

1. Wall, floor and ceiling heat gains due to conduction
2. Wall and ceiling heat gains from solar radiation
3. Load due to ingression of air by frequent door openings and during fresh air charge.
4. Product load from incoming goods
5. Heat of respiration from stored product
6. Heat from workers working in the room
7. Cooler fan load
8. Light load
9. Aging of equipment
10. Miscellaneous loads, if any

CA Storage

CA storage involves altering and maintaining an atmospheric composition that is different from air composition (about 78% N_2, 21% O_2, and 0.03% CO_2) generally, O_2 below 8% and CO_2 above 1% is used. Atmospheric modification should be considered as a supplement to maintenance of optimum ranges of temperature and relative humidity for each commodity in preserving quality and safety of fresh fruits and vegetables. The benefits of CA storage are:

■ Retardation of senescence (including ripening) and associated biochemical and physiological changes i.e., slowing down rates of respiration, ethylene production, softening, and compositional changes.

■ Reduction of sensitivity to ethylene action at O_2 levels <8% and/or CO_2 levels >1%.

■ Alleviation of certain physiological disorders such as chilling injury and some storage disorders, including scald of apples.

■ CA can have a direct or indirect effect on post-harvest pathogens (bacteria and fungi) and consequently decay incidence and severity. For example, CO_2 at 10% to 15% significantly inhibit development of botrytis rot on strawberries, cherries, and other perishables.

■ Low O_2 (<1%) and/or elevated CO_2 (40% to 60%) can be a useful tool for insect control in some fresh and dried fruits, flowers, and vegetables and dried nuts and grains.

Insulation

All the sides of the cold storage room need to be insulated in order to maintain the required temperature inside. Various types of insulating materials are used for insulation of walls and roof. However, the most commonly used insulation material is thermocol.

Utilities

Availability of dependable power supply at the site needs to be ensured. Since electricity is the major operational cost, running the plant solely on diesel generators may render the unit financially infeasible.

Subsidy Schemes for Cold Storages

A. National Horticulture Board

Capital Investment Subsidy for Construction/Modernization/ Expansion of Cold Storage for Horticulture Produce

Projects up to a capacity of 5000 t with an average cost of ₹2 crore (₹4000 per t) would be promoted for wider dispersal, which includes expansion of existing capacity (including CA/MA stores/Pre-cooling units). In case of modernization and rehabilitation, subsidy at 25% of the capital cost will be determined at ₹1000 per t capacity created.

For other storages, subsidy at 25% of the capital cost is to be determined @ ₹2000 per t capacity created to be 25% promoters contribution 50% term loan by banks at PLR+1% through NABARD. Refinance Banks not availing refinance may also finance such projects with the overall operational guidelines of Government of India 25% back-ended capital investment subsidy by National Horticulture Board (NHB) not exceeding ₹50 l per project.

The subsidy would flow from NHB and operated by NABARD, through commercial/cooperative banks, and by National Co-operative Development Corporation (NCDC) where cooperatives seek loan from NCDC. Wherever term loans are not raised from institutional sources and the promoters fund projects entirely through internal resource generation, NHB would provide subsidy directly.

B. Ministry of Food Processing Industries, Government of India

Scheme for Infrastructure Development - Integrated Cold Chain Facilities

The rate of growth in production of perishable produce is roughly twice as much as general agriculture. Accordingly, to keep pace with the need for storage of perishable produce so that there is no wastage and at the same time raw material supply to food processing industries is not adversely hit in the off season, enhanced cold storage capacity is essential with a view to

- Improve viability of existing cold storages
- Enhance total cold store capacity, both stationary and mobile

Consequent upon NHB's capital subsidy scheme for cold storages, the Ministry's assistance is limited to the following types of cold storages

- Cold storage for non-horticulture produce
- Where the cold storage is an integral part of the processing unit or of the common facilities in food park
- Special type of cold storage with CA or MA facility

C. APEDA

Scheme for Infrastructure Development

Assistance to exporters/producers/growers/cooperative organizations and federations for horticulture and floriculture sector for setting up of specialized storage facilities such as high humidity cold storage, deep freezers, CA or MA storage, etc., 25% of cost subject to ceiling of ₹10 l per beneficiary.

D. Government of Gujarat (Assistance under Agro-Industrial Policy)

Scheme for Agri Infrastructure Development (2005–2006)

The state government will offer the following incentives for projects providing common infrastructure facilities in the value chain of agri-produce from farm to markets as decided by Single Window Clearance (SWC) committee

- Back ended interest subsidy as follows
  - 6% pa back ended interest subsidy for first five years from commencement of operations
  - The aggregate interest subsidy will not exceed ₹400 l
  - The interest subsidy will be available on the funds borrowed from financial institutions for capital investments only. No interest subsidy will be available towards working capital loan or any other loan which is not of the nature of term loan meant for acquiring capital assets
- The state government will assist in preparing pre-feasibility studies through Gujarat Infrastructure Development Board
- The state government intends to provide government land including agriculture farms on long lease basis at reasonable rates

SWC

The state government is committed to provide SWC to ensure that entrepreneurs do not have to visit different government offices to obtain the required for setting up industrial units in the state.

For this purpose the state government has constituted an empowered committee consisting of secretaries in charge of Departments of Finance, Industries, Agriculture and Managing Director, Gujarat Agro Industries Corporation Limited with a view to provide SWC to Agriculture and Food Processing Industry. The committee will also take such measure and decisions as required for proactively promoting the development of Agro Industries in the state. In view of the thrust required to this sector, the committee is chaired by Chief Secretary and also include Industries Commissioner as a member. The committee will take appropriate decisions to affect special assistance from Gujarat Government, if required to attract large scale investments in this sector. The decisions of the committee shall be final and no further approval will be necessary.

E. Government of Maharashtra

The Maharashtra State Agricultural Marketing Board (MSAMB), Pune, has launched 'Cold Storage Subsidy Scheme' from 1st of August 2005. Scheme is applicable for promoters from the Maharashtra state. The scheme details as follows:

Pattern of Assistance

Subsidy at 25% of the total project cost with maximum limit of ₹2.5 l per project. The beneficiary will be eligible for getting subsidy from other financial institutions/agencies. The beneficiary can get subsidy for more than one cold storage.

Features of the Scheme

Subsidy is available for capacity up to 100 t.

- The cost of the project is considered at ₹10000 per t.
- The subsidy rate is considered at ₹2500 per t.
- In principle approval from MSAMB for the project is necessary.
- The detail project report of the cold storage must be of the MSAMB.
- The plan and estimates of the project should be according to the norms finalized by the MSAMB.

Table A3.17 Summary of Schemes

Organization	Subsidy
NHB	Maximum of ₹50 l for 5000 t capacity cold storage per beneficiary
Ministry of Food Processing Industries	Integrated cold chain facilities—25% of the cost of plant and machinery and technical civil works in general areas and 33.33% in difficult areas with a common ceiling of ₹75 l
APEDA	25% of cost subject to ceiling of ₹10 l per beneficiary for setting up of specialized storage facilities such as high humidity cold storage, deep freezers, CA or MA storage etc
Government of Gujarat	6% pa back ended interest subsidy for first five years from commencement of operations. The aggregate interest subsidy will not exceed ₹400 l. The interest subsidy will be available on the funds borrowed from financial institutions for capital investments only. No interest subsidy will be available towards working capital loan or any other loan which is not of the nature of term loan meant for acquiring capital assets
Government of Maharashtra	Maximum of ₹2.5 l for 100 t capacity

Source: MSAMB. Cold storage subsidy scheme. 2013.   http://www.msamb.com/ schemes/coldstorage.htm

Eligible Organizations

The eligible promoters under the scheme shall include agriculture produce market committees, co-operative societies engaged in processing/marketing of fruits, vegetables, and flowers, co-operative sugar factories, agricultural producer's co-operative societies.

Acknowledgements

Written by G. Raghuram, N. Vijaya Baskar and Santosh Kumar Mishra, Indian Institute of Management, Ahmedabad. Research assistance provided by Ramesh Reddy Amereddy, Niti Sirohi and Anju Singla is acknowledged. We acknowledge Mr. Hasmukhbhai for his support.

Cases of the Indian Institute of Management, Ahmedabad, are prepared as a basis for class discussion. Cases are not designed to present illustrations of either correct or incorrect handling of administrative problems.

Case 4　Chilli in Soup (A)

Introduction

It was a damp morning on June 10, 2003. The monsoon had arrived in Kerala a few days ago. Mr. C. J. Jose, a tall well-built man in his fifties, the Chairman of the Spices Board of India (SBI), was in his spacious office deeply engrossed in a file. It was 10 am. His phone started ringing. It was a call from the Secretary, Department of Commerce, Ministry of Commerce and Industry, Government of India. The matter was of serious concern. The Secretary seemed to be worried and conveyed to him the following message. "Yesterday evening, the Ministry received a communication from the Charge d'Affaires, European Union (EU), regarding a notification. It came via the existing Rapid Alert System for Food and Feed (RASFF) of the EU. The notification stated that French scientists have detected traces of Sudan 1, a banned carcinogenic dye in a British food product in the last month. Further investigations revealed that the source of Sudan 1 was red chilli powder exported from India. The British firm had used it as an ingredient. The Ministry was requested to inform the competent authority in India of these findings in order to help avoid such problems in the future. My office will be sending a fax of the notification in an hour." Closing the conversation, he said, "Considering the seriousness of the matter, the Ministry would like to be periodically apprised of the actions SBI would be taking in its capacity as the regulatory body for controlling and certifying the quality of spices exported out of India."

Mr. Jose hung up the phone and brooded for some time over possible actions he could take. He called his Personal Assistant, Mr. Raphael, and asked him to cancel all his meetings and appointments for the day unless something was exigent. He then discussed the matter with Mr. S. Kannan, Director (Marketing), over the phone, and immediately decided to summon a meeting at 11 am, of all the departmental heads and a few other officers whose inputs were desirable. (Table A4.1 gives the functions and organisation structure of SBI.) He was expecting the fax to come before the meeting would start. He did all this within a few minutes of receiving the call.

Table A4.1 Spices Board of India: Functions and Organisation Structure

SBI is governed by a 32 member governing body consisting of representatives of the Lok Sabha, Rajya Sabha, State Governments, certain Central Government Ministries, Planning Commission, plantation labour, spice growers, processors and exporters, and specialist organisations such as Central Food Technology Research Institute (CFTRI) and Indian Institute of Packaging (IIP).

Offices

The Head Office of SBI is located at Cochin. SBI has 22 Regional Offices, 13 Zonal Offices and 35 Field Offices. A central Quality Evaluation Laboratory (QEL) is located at the Head Office. A Biotechnology Lab also functions at the Head Office. Indian Cardamom Research Institute, a research wing of the Spices Board has its main station at Myladumpara (Idukki, Kerala) with Regional Stations located at Thadiyankudisai (Tamil Nadu), Sakleshpur (Karnataka) and Gangtok (Sikkim).

Main Functions

1. Export promotion of all spices through support for:-

    a) Technology upgradation

    b) Quality upgradation

    c) Brand promotion

    d) Research and product development

2. Research, development and regulation of domestic marketing of small and large cardamom

3. Research and production development of vanilla

4. Post-harvest improvement of all spices

5. Promotion of organic production, processing and certification of spices

6. Development of spices in the North East

7. Provision of quality evaluation services

Other Responsibilities Related to Export Promotion of Spices

1. Registration of exporters

2. Quality certification

3. Quality control

4. Collection and documentation of trade information

5. Provision of inputs to the Central Government on policy matters relating to import and export of spices

Regulatory Functions

1. Registration of exporters of spices

2. Licensing of cardamom auctioneers

3. Licensing of cardamom dealers

4. Sampling, testing and analysing of samples of all the spices designated for export out of the country

5. Suspending/revoking the certificates of registration of exporters for violation of the conditions of registration relating to food safety

Source: Spices Board India. Website. 2004.

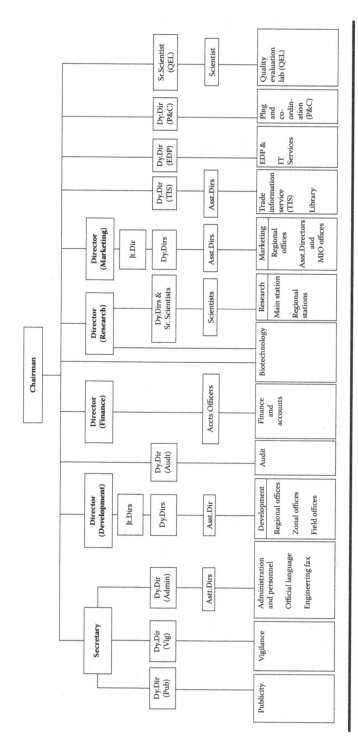

Figure A4.1 Organizational structure. *Abbreviations*: Jt.Dir, Joint Director; Dy.Dir, Deputy Director; Asst.Dir, Assistant Director. (From Spices Board India. Website. 2004.)

Mr. Jose had, in his long administrative service (as an officer of the Indian Administrative Services), faced many a crises in various capacities and always found these an interesting part of life. He was posted as the Chairman of SBI in April 2001. Until the previous day, it had been a relatively easy posting, doing routine regulatory work, besides looking for opportunities to enhance the markets outside India, meeting exporters, attending meetings with similar international bodies regarding the export of Indian spices. But the call from the Ministry in the morning looked like a sign of an impending crisis situation. He started thinking logically about the stakes and the possible fallouts of the incident.

The fax came before the meeting. Three Mumbai based exporters had been identified as the suppliers of the contaminated chilli powder in the RASFF (Box A4.1). During the meeting, Mr. Jose summed up the following issues which had to be addressed by SBI as the licensing authority of exporters of Indian spices:

(i) Assessing the damage done by the event and formulating strategies to control it,

(ii) Defending the brand image of Indian spices in the international market through confidence building exercises, and

(iii) Formulating strategies to avoid future recurrence of such events.

At the end of the meeting, Mr. Jose requested Mr. Kannan to verify the EU claim and collect back-up information in the next few days and report to him. He also requested Mr. K.R.K Menon, Senior Scientist, in charge of the Quality Evaluation Laboratory (QEL), to report on the quality parameters and the tests related to chilli powder that were currently being carried out in the SBI lab.

Mr. Kannan set about his task immediately, keeping in mind the issues above. Examining past correspondence and records, he gathered the following:

1. "Sudan 1 was a red dye that was used for colouring solvents, oils, waxes, petrol, and shoe and floor polishes. Under the colours in Food Regulations 1995 of EU, the red dye was illegal, and was considered to be a genotoxic carcinogen. Its presence, at any level, was not permitted in foodstuffs for any purpose. The ban came following the experimental results on rats, which suggested that the chemical could trigger the formation of malignant tumours. The details about its health risks (Box A4.2) mentioned a significant fact that "the risk of cancer in humans from Sudan 1 has not been proven and any risk from these foods is likely to be very small indeed."

2. As per the RASFF (Box A4.1) dated May 9, 2003, issued by the EU, consignments of chilli powder exported by the following firms to East West

Spices, UK had been detained for the presence of Sudan 1. All were based in Mumbai. The names were:

(i) Gautam Export Corporation
Flat 11-B, 3rd Floor
Koolbreeze CHS Limited, Plot No. K-72
17th Road, Khar (West)
Mumbai-400 052

(ii) Patons Exports Private Limited
10, Koolbreeze CHS Limited,
17th Road, Khar (West)
Mumbai-400 052

East West Spices, UK (the importers) provided the details of how they had traced back the material to the exporters. In order to verify the correctness of

---

BOX A4.1   RAPID ALERT SYSTEM FOR FOOD AND FEED

No 1/22/2003-EP (Agri-V)
Government of India
Ministry of Commerce and Industry
Department of Commerce

Udyog Bhavan, New Delhi
Dated, 10th June, 2003

Chairman
Spices Board
Cochin
Subject:- Rapid Alert Notification

Sir,

1. Kindly find enclosed Rapid Alert Notifications No 2003/111 dated 9th May, 2003 for presence of colour Sudan I in hot red chilli powder exported by Gautam Export Corporation, Mumbai and M/s Patons Exports Private Limited, Mumbai.
Spices Board is requested to take necessary action in the matter.

Yours faithfully,

(KK Singh)
Section Officer

Encl: A/A

(Continued)

BOX A4.1 (*CONTINUED*)   RAPID ALERT SYSTEM FOR FOOD AND FEED

EUROPEAN UNION DELEGATION OF THE EUROPEAN
COMMISSION TO INDIA, BHUTAN, MALDIVES, NEPAL AND SRI LANKA

Minister-Counsellor, Political Affairs and Coordination
Charge d'Affairs a.i.

CK/ck/222
09.06.2003

Mr YYY
Additional Secretary
Department of Commerce
Ministry of Commerce and Industry
Udyog Bhavan
New Delhi-110001

Dear Mr YYY,

According to Regulation (EC) No 178/2002 laying down the general principles and requirements of food law, establishing the European Food Safety Authority and laying down procedures in matters of food safety, I would like to bring to your attention the following notification received via the Rapid Alert System for Food and Feed (RASFF):

Notification:	2003.111 Date: 09/05/2003
Concerning:	hot red chilli powder
Dispatched from:	INDIA
	Via UK
Having been the subject of:	market control
Reason for notification:	colour Sudan 1

Detailed information is attached.
I would be grateful if you could inform the competent authorities in your country of these findings in order to help avoiding such problems in the future. The RASFF contact point (fax number +32.2.222.77.77 or e-mail: sanco-rasff@cec.eu.int) is at your disposal should the authorities of your country have questions about the above-mentioned information.
With kind regards,

Yours Sincerely

Dr ZZZ, Charge d'Affairs a.i.

(*Continued*)

BOX A4.1 (*CONTINUED*) RAPID ALERT SYSTEM FOR FOOD AND FEED

EUROPEAN COMMISSION

HEALTH and CONSUMER PROTECTION
DIRECTORATE-GENERAL

Directorate D – Food Safety: production and distribution chain
D3 – Chemical and Physical risks; surveillance

Brussels, 9 May 2003

RAPID
ALERT
SYSTEM
FOR
FOOD
AND FEED

FOOD

VERY URGENT – TRES URGENT

ALERT NOTIFICATION: 2003/111

ORIGINAL NOTIFICATION

SUBJECT: COLOUR SUDAN 1 IN HOT RED CHILLI POWDER FROM UK

PAGE: Cover Pages (1) + 10

FAX NUMBER: ++32-2-222-77-77

EMAIL: sanco-rasff@cec.eu.int

MARKET CONTROL Product recall/Product withdrawal

Product distributed to France

Manufacturer: East West Spices (UK)

The contact point from UK is kindly requested to provide the possible distribution and details about origin of the incriminated ingredient

---

The contact point from France has communicated to the Commission the following information:

(Continued)

**BOX A4.1 (*CONTINUED*)   RAPID ALERT
SYSTEM FOR FOOD AND FEED**

RAPID ALERT SYSTEM FOR FOOD AND FEED **REGULATION (EC) No: 178/2002 – Art 50/GENERAL INFORMATION:**			
1*	NOTIFICATION TYPE		FOOD
2*	CONTROL TYPE		MARKET
3*	NOTIFYING COUNTRY		FRANCE
4*	DATE OF NOTIFICATION		9.05.2003
HAZARD:			
5*	NATURE OF HAZARD		UNAUTHORISED COLOUR
6*	RESULTS OF THE TESTS		SUDAN I 3.89 G/KG
7*	COUNTER ANALYSIS		
8*	SAMPLING	DATES	30.04.2003
		No OF SAMPLES	TAKING A SAMPLE (2 PACKAGES OF 30 G EACH)
		METHOD	
		PLACE	RETAILER
9*	LABORATORY		
10*	METHODS OF ANALYSIS USED		GC/MS
11*	PERSONS AFFECTED		
12*	TYPE OF THE ILLNESS		
PRODUCT:			
13*	PRODUCT CATEGORY		SPICES
14*	PRODUCT NAME		HOT RED CHILLI POWDER
15*	BRAND/TRADE NAME		
IDENTIFICATION OF THE LOT(S):			
16*	CONSIGNMENT/LOT NUMBER		NOT AVAILABLE
17*	PUBLIC HEALTH CERTIFICATE	NUMBER	
		DATE	
		CVED No	

(*Continued*)

BOX A4.1 (*CONTINUED*)   RAPID ALERT
SYSTEM FOR FOOD AND FEED

18*	DURABILITY DATES	USE-BY DATE	
		BEST BEFORE DATE	
		SELL-BY DATE	
19*	DESCRIPTION OF THE PRODUCT	PRODUCT ASPECT	HOT RED CHILLI POWDER
		No OF UNITS	
		TOTAL NET WEIGHT	
ORIGIN:			
20*	COUNTRY OF ORIGIN		UK
21*	MANUFACTURER	NAME	EAST WEST SPICES, UK
		ADDRESS	
		VET AP-No	
22*	DISPATCHER/ EXPORTER	NAME	
		ADDRESS	
DISTRIBUTION:			
23*	DISTRIBUTED BY	IMPORTER	
		WHOLESALER	
		RETAILER	
24*	DISTRIBUTION TO MEMBER STATES		
	DISTRIBUTION LIST ATTACHED		NO
25*	EXPORTED TO THIRD COUNTRIES		NOT BY THE FIRM
	DISTRIBUTION LIST ATTACHED		NO
IN CASE OF A REJECTION AT THE BORDER:			
26*	POINT OF ENTRY		
27*	TYPE OF CHECK		
28*	COUNTRY OF DISPATCH		

(Continued)

**BOX A4.1 (*CONTINUED*)   RAPID ALERT
SYSTEM FOR FOOD AND FEED**

29*	COUNTRY OF DESTINATION		
30*	CONSIGNEE	NAME	
		ADDRESS	
31*	CONTAINER NUMBER		
32*	MEANS OF TRANSPORT		
MEASURES ADOPTED:			
33*	VOLUNTARY MEASURES		
34*	COMPULSORY MEASURES		MONITORING OF THE DESTRUCTION
35*	JUSTIFICATION		INFRINGEMENT OF § 11-LID 1 OF THE FOOD AND COMMODITY ACT
36*	SCOPE		NATIONAL
37*	DATE OF ENTRY INTO FORCE		
38*	DURATION		
OTHER INFORMATION:			
39*	MINISTRY		
40*	PERSON TO CONTACT		NAME: TEL: FAX: E-MAIL:
41*	OTHER INFORMATION		
42*	CONFIDENTIAL		NO
43*	IF YES, WHICH BOXES (NUMBERS)		
44*	IF YES, REASON		

Source: Spices Board of India.
Attached were consignment notes identifying the following exporters from India:
1. Gautam Export Corporation
2. M/S Patons Exports Private Limited
As sold to East West Spices, UK.

BOX A4.2   HEALTH RISKS OF SUDAN I

Illegal dye: What is the health risk?

There is concern that the dye, Sudan I, has the potential to cause cancer.

However, the Food Standards Agency has said the risk is very low.

Professor Alan Boobis, an expert in toxicology at Imperial College London says there is little reason for the public to be alarmed.

Sudan I was banned from use in food products following experiments on rats, which suggested that the chemical could trigger the formation of malignant tumours.

However, Professor Boobis told the BBC News website that the levels of the chemical fed to the rats bore no relation to the kind of levels that people would be exposed to if they ate contaminated products.

The rats that showed signs of developing tumours were given a daily dose of Sudan I of around 30 milligrams/kg of bodyweight for two years.

Animals given a lower dose – 15 milligrams/kg of bodyweight - showed no signs of cancer-related changes.

The contaminated products would contain Sudan I doses of a much smaller magnitude - micrograms, rather than milligrams.

The animals that did show signs of cancer—known scientifically as tumourigenic changes—started to do so only after many months.

In human terms, this would suggest that if the dye was to have any effect, the symptoms would not start to become apparent for around 20 years.

Genetic damage

The compound triggers cancer growth through what is known as a DNA reactive mechanism.

Within the body it is converted into a form which attacks the DNA of cells, causing damage which can be passed on to the next generation of cells in the affected tissue, ultimately leading to cancer.

However, Professor Boobis said that the risk of eating just one or two contaminated items was trivial.

While not an exact comparison, he likened it to the cancer risk associated with smoking just one cigarette in a lifetime.

Even if a person was to eat a contaminated product every day for several years, the risk of cancer, although higher, was still likely to be very low.

Source: BBC News. Illegal dye: What is the health risk? Feb. 22, 2005.

(Continued)

BOX A4.2 (*CONTINUED*)   HEALTH RISKS OF SUDAN I

"This compound is not a very potent carcinogen in animals," he said. "People should not be unduly concerned about the health effects."

"It is a good idea to remove this substance from the food chain, but this is being done simply as a precaution, not because there is an immediate impact on health."

Federal Institute of Risk Assessment (BfR)

"Dyes Sudan I to IV in food"
BfR Opinion of 19 November 2003

Sudan I was classified as a category 3 carcinogen and as a category 3 mutagen (Annex I of the Directive 67/548/EEC) according to EU criteria. Substances in category 3 give rise to concern because of a possible carcinogenic effect in man but cannot be definitively assessed because of the lack of information.

For Sudan I, the possible intake amounts can be compared in the following way with the dose which led in animal experiments to neoplastic liver nodules.

In the case of an assumed high dye level (3500 mg/kg) and a large consumption amount of chilli powder (up to 500 mg/day), 1750 µg Sudan I can be taken in per day in the worst-case scenario. This corresponds to 29.2 µg/kg bodyweight (at a bodyweight of 60 kg). This amount is below the dose of 30 mg/kg bodyweight by a factor of 1×10^3 at which a statistically significant increase in the incidence of neoplastic liver nodules (NTP, 1982) was observed in animal experiments in rats after chronic administration with Sudan I in feed. In other words, the difference between the amount of Sudan I which can be taken in based on these assumptions in the worst case scenario per day (29.2 µg/kg bodyweight) and the amount at which a statistically significant increase in the incidence of neoplastic liver nodules was observed in animal experiments (30 mg/kg bodyweight) amounts to three orders of magnitude.

Assuming a lower dye level (eg 10 mg/kg) and a large consumption amount of chilli powder (up to 500 mg/day), 5 µg Sudan I can be taken in per day in the worst case scenario. This corresponds to 0.083 µg/kg bodyweight (at a bodyweight of 60 kg). This amount is below the dose of 30 mg/kg bodyweight by a factor of 3.6×10^5 at which a statistically significant increase in the incidence of neoplastic liver nodules (NTP, 1982) was observed in animal experiments in rats after chronic administration of Sudan I in the feed. In other words, the difference between the amount of Sudan I which can be taken in based on these assumptions in the worst case scenario per day (0.083 µg/kg bodyweight) and the amount at which a statistically significant increase in the incidence of neoplastic liver nodules was observed in animal experiments (30 mg/kg bodyweight) amounts to six orders of magnitude.

(Continued)

BOX A4.2 (*CONTINUED*)   HEALTH RISKS OF SUDAN I

Conclusions:

From the above assessments, the conclusion can be drawn that in the case of one-off or occasional consumption of foods which are contaminated with Sudan dyes in concentrations of a few milligrams per kilogram, the risk of cancer is probably very low. This is because the difference between the dye or amine amount, which can be taken in per day in conjunction with a high consumed amount in the worst case scenario and the dose at which carcinogenic effects were observed in animal experiments is roughly five to seven orders of magnitude.

However, this can no longer be assumed for concentrations of several thousand milligrams per kilogram because the difference between the dye or amine amount, which can be taken in per day in conjunction with a high consumed amount in the worst case scenario and the dose at which carcinogenic effects were observed in animal experiments, is now only two to three orders of magnitude.

The estimation that in the case of one-off or occasional consumption of only a few foods contaminated with Sudan dyes, the risk of cancer is probably very low does not mean that there is no risk at all. For carcinogenic substances in category 2, whose action constitutes a clear cancer risk for human beings according to the current state of knowledge, no concentration can be given which could still be considered to be safe (DFG, 2003). Furthermore, the risk of course increases in the case of frequent or ongoing consumption. For precautionary reasons, the intake of substances of this kind should, therefore, be kept as low as possible and any exposure, if at all possible, avoided. This applies even more since intake of carcinogenic amines may also result from other applications like hair dyes (Platzek et al., 1999; SCCNFP, 2002) and the action of several different carcinogenic amines can lead to additive effects.

Furthermore, it should be borne in mind that risks which may be tolerable in the safety at work range must be assessed differently in the food sector if these are avoidable risks as they are in this case.

Source: Federal Institute for Risk Assessment (BfR) Dyes Sudan I to IV in food, Nov. 2003.

the EU claims that the material actually went from the two named exporters (Gautam Export Corporation and Patons Exports Private Limited) to East West Spices, UK, an investigation was conducted by SBI. The investigation showed that the shipping marks and lot numbers of the material sampled corresponded to those used by Volga Spices and Masala Mills Private Limited, at the behest of the named exporters. Volga Spices, also a registered exporter, had processed and supplied the material to Gautam Export Corporation and Patons Exports Private Limited.

Gautam Export Corporation's senior partner Jagdish Advani was a big exporter of spices to Europe. His company used to trade around 5400 tons of spice annually to Europe. Volga Spices was run by the Shaikh family. Imran Shaikh was one of its directors. Patons Exports Private Limited had directors from the Advani family who managed Gautam Export Corporation.

A consignment of 5 tons had been shipped from Gautam Export Corporation to East West Spices (UK), who had in turn exported part of the chilli powder, possibly clubbing with shipments from the other party to an importer in France where the Sudan 1 was detected.

3. Chilli was commercially important for two qualities, its red colour by the pigment Capsanthin and its biting pungency by Capsaicin. The price that chilli powder fetched in the market was determined by its pungency and colour. Sudan 1, a cheap red dye, would brighten up the colour if mixed with crushed chilli (processed from cheap discoloured chilli) adequately during its processing. If Sudan 1 was present in chilli powder it would be the result of a deliberate act of adulteration performed to brighten up its colour, and hence its appearance. Experiments had shown that for good colour enhancement, a couple of kilograms (kgs) of Sudan 1 would be required per ton of chilli powder. This would cost about US$ 5 and could potentially increase per ton chilli price from US$ 600 to US$ 1200.

4. Data related to Indian chilli exports and production (Table A4.2), major importing countries of Indian chilli (Table A4.3), and India's share of chilli imports of some major importing countries (Table A4.4) provide a perspective on the export market scenario of chilli from India."

Mr. Menon, with the help of his subordinates, put together the following information regarding SBI's testing equipment and processes.

1. "The SBI set up the QEL in 1990. The lab was ISO 9002 certified by the British Standards Institution (BSI) in 1994. The QEL adopted the ISO 14001 Environmental Management System for its activities in the lab. The SBI also had two major concepts for quality improvement. The SBI awarded the ISB logo to the consumer packs of spices selectively to exporters who had certified processing and quality control capability and maintained a high level of hygiene and sanitation at all stages. For exports of spices and spice

Table A4.2   Indian Chilli Exports and Production

| Year | Exports | | Production | Annual Average Spot Prices of Indian Chilli in New York | Annual Average Prices of Chilli in India | Export Share of Production |
| | Value | Qty | Qty | | | |
	(₹ m)	('000 t)	('000 t)	(US$/t)	(₹/t)	% (of t)
1998–1999	2,529	68	1,043	1698	4134	6.5
1999–2000	2,547	64	1,056	1587	3300	6.0
2000–2001	2,297	62	1,046	1345	2393	6.0
2001–2002	2,524	70	1,113	1301	2660	6.3
2002–2003	3,151	81	846	800	3358	9.6

Source: Spices Board of India. Indian chilli exports and production, 2003.

Table A4.3   Major Importing Countries of Indian Chilli

Country	1998–1999		1999–2000		2000–2001		2001–2002		2002–2003		Five Year Average Qty
	Value (₹m)	Qty ('000 t)	Value (₹m)	Qty ('000 t)	Value (₹m)	Qty ('000 t)	Value (₹m)	Qty ('000 t)	Value (₹m)	Qty ('000 t)	('000 t)
Sri Lanka	730	20	599	18	618	21	700	25	785	22	21
USA	441	7	649	12	682	13	757	15	868	17	13
Bangladesh	445	11	183	4	33	1	86	3	460	16	7
EU[1]	147	3	205	4	243	5	232	4	247	5	4
Malaysia	61	1	213	5	74	2	107	3	150	4	3
Pakistan	320	15	199	8	87	3	68	3	17	1	6
Total of Above	**2144**	**57**	**2048**	**51**	**1737**	**45**	**1950**	**53**	**2527**	**65**	**54**
Overall Total	**2529**	**68**	**2547**	**64**	**2297**	**62**	**2524**	**70**	**3151**	**81**	**69**
% Share of Above countries	**85**	**84**	**80**	**80**	**76**	**73**	**77**	**76**	**80**	**80**	**79**

[1] The top ten importing countries of EU, which accounted for almost the entire EU imports are elaborated above.

(Continued)

Table A4.3 (Continued) Major Importing Countries of Indian Chilli

Country	1998–1999		1999–2000		2000–2001		2001–2002		2002–2003		Five Year Average Qty
	Value (₹ m)	Qty ('000 t)	Value (₹ m)	Qty ('000 t)	Value (₹ m)	Qty ('000 t)	Value (₹ m)	Qty ('000 t)	Value (₹ m)	Qty ('000 t)	('000 t)
UK	80.0	1.40	87.0	1.42	92.8	1.67	104.7	1.92	98.0	1.80	1.64
Netherlands	16.0	0.30	29.0	0.57	47.1	1.08	33.1	0.75	39.0	0.82	0.70
Italy	22.0	0.39	43.0	0.75	41.3	0.79	36.1	0.62	41.0	0.69	0.65
France	8.5	0.20	14.0	0.30	14.9	0.37	16.0	0.38	13.0	0.29	0.31
Germany	3.5	0.09	11.0	0.34	10.0	0.19	9.5	0.22	11.0	0.24	0.21
Spain	3.5	0.06	4.6	0.07	17.6	0.27	17.5	0.26	20.0	0.34	0.20
Poland	2.8	0.10	2.6	0.11	5.6	0.21	2.4	0.09	4.7	0.16	0.13
Greece	3.7	0.08	12.0	0.24	5.8	0.13	5.0	0.11	5.1	0.10	0.13
Belgium	4.1	0.10	2.4	0.05	3.5	0.06	2.2	0.04	6.9	0.11	0.07
Sweden	3.2	0.02	0.1	0.00	4.2	0.06	5.0	0.05	8.7	0.08	0.04
EU Total	147	3	206	4	243	5	232	4	247	5	4

Source: Spices Board of India. Major importing countries of Indian chilli. 2003.

Table A4.4   India's Share of Chilli Imports of Some Major Importing Countries

Country	1998 Value (US$ m)	1998 Qty ('000t)	1999 Value (US$ m)	1999 Qty ('000 t)	2000 Value (US$ m)	2000 Qty ('000 t)	2001 Value (US$ m)	2001 Qty ('000 t)	2002 Value (US$ m)	2002 Qty ('000 t)
USA (Total)	89	54	90	57	96	59	99	66	125	85
USA (India)	13	9	19	12	18	12	18	14	21	17
India's Share (%)	14.9	15.9	21.0	21.0	18.7	21.1	18.1	20.6	16.4	19.6
EU (Total)			72		75		96			
EU (India)			5.7		5.9		6.1			
India's Share (%)			7.9		7.9		6.4			
Malaysia (Total)	22	23	35	29	32	29	30	33	35	40
Malaysia (India)	0.2	0.4	4.8	5.0	4.5	5.0	2.0	2.7	4.8	6.5
India's Share (%)	0.8	1.5	13.7	17.4	14.3	17.5	6.8	8.3	13.8	16.1
UK (Total)	16	5.3	12	4.7	12	5.1	14	7.5	14	6.2
UK (India)	2.1	1.4	2.5	1.7	2.4	1.7	2.4	1.9	2.6	2.2
India's Share (%)	13.3	25.8	21.3	36.7	20.8	33.8	17.5	25.2	18.9	35
Sri Lanka (Total)			18	20			17		18	25
Sri Lanka (India)			18	20			17		17	24
India's Share (%)			96.3	96.1			99.5		96.2	96.6

Source: United Nations Statistics Division.

products in bulk, it awarded the Spice House Certificates to those processors/ exporters who had a genuine commitment to sustainable quality and export growth and made investments in developing in-house processing facilities, infrastructure and had the necessary competence to ensure consistent quality and reliability. These facilities should cover cleaning, grading, processing, packaging and warehousing.

2. The two important quality characteristics that determined the price of red chilli powder were its colour (measured in ASTA units) and pungency (Capsaicin percentage). Besides that, Aflatoxin (a toxic metabolite produced by certain moulds in food and feed) content was a critical parameter that should not exceed a threshold level (though different for different countries) in the exported chilli powder. Several other parameters could be tested in the SBI lab. (See Table A5.1 for the analytical services offered by SBI including the list of tests, charges and the time taken for getting the report.)

3. The lab had five High Performance Liquid Chromatographs (HPLC) machines for detecting the Aflatoxin content. HPLC was used to find the amount of a chemical compound within a mixture of other chemicals. It could detect up to the level of 5 ppm (5 gm in a ton). An HPLC machine could also be used for detecting Sudan 1. Each machine could analyse 20 samples a day.

4. Usually, the big exporters had their own testing facilities. Most of the requests for testing were received from small exporters who did not have their own facilities. On an average, around 30 samples were analysed each day by HPLC machines for various tests on different spices."

On June 13, 2003, Mr. Jose, Mr. Kannan and Mr. Menon met to discuss the way forward. Mr. Jose asked Mr. Kannan for his views. Mr. Kannan said "I was aware of the unregulated use of banned red dyes to brighten up the color of red chilli powder in the domestic market in order to fetch good prices. But this is the first time it has happened in the export market. Maybe, it has happened in the past too, but was not detected. The quantum of Sudan 1 added by the alleged exporters at 3890 ppm was a deliberate act that needed stern action."

Mr. Kannan further added, "The events after 9/11 had changed the world. Since then the threats of bioterrorism had been lurking in the minds of people in the USA and Europe. Consequently, the food regulatory authorities and the consumers had become more sensitive about any kind of adulteration. Checking had become more frequent and thus increased the chance of detection."

Mr. Jose mentioned that certain sources in Europe had informed him that the EU had been seriously contemplating the imposition of some kind of restriction to ensure the import of Sudan 1-free chilli powder and chilli products to its member countries. It could lead to the requirement of an analytical report from well-established labs certifying the export consignment to be Sudan 1-free. He continued, "If such a situation actually arises,

- How should the SBI act?
- Should the SBI consider mandatory testing of all the export consignments before shipping? In that case, the immediate question would be,
- Is the SBI (and the QEL) equipped to handle the task efficiently and in a reasonably short period of time? or
- Should the SBI license other laboratories which could do the testing and issue the analytical reports? or
- Should the SBI require the exporters to submit analytical reports by equipping their own quality control lab or through other labs? And also,
- How should the SBI monitor the whole process starting from taking the sample to the shipment of the consignment?
- Should they price the test? If so, what should be the price?"

Mr. Kannan informed that during 2002–2003, about 8000 consignments had been shipped out by nearly 150 exporters. While the largest exporter did about 4000 tons, many exporters did even less than 100 tons. Testing of all consignments would increase the work load of the SBI significantly and might have to be done using various other accredited laboratories in the country.

Mr. Kannan also mentioned, "There were also a few private laboratories used by the exporters, but their quality and capacity were not up to the mark. Further, the trade felt that their charges for testing for Sudan 1, at over ₹4000 per sample, are quite high. It would be better if SBI could make the testing mandatory and thus take control of the process. The SBI staff at its offices in the port cities could be given the responsibility of taking samples and sending them to SBI for testing."

Mr. Jose was very concerned about the fact that EU norms would require no trace of Sudan 1, for which testing precision may have to be at the parts per billion (ppb) levels.

Responding to Mr. Jose's concern, Mr. Menon specified that available HPLC machines were not equipped to test for Sudan 1 at the ppb level. A superior machine called LC MS MS would be required for this. It would have a capacity of testing 40 samples per day and would cost around US$ 200,000. The variable cost (material and labour for sample preparation) per test would be about ₹700 excluding the collection charges. Supervisory costs would be in addition and would require one qualified technician per shift.

Mr. Kannan alerted them to the fact that even though 8000 consignments had been shipped over a year, the shipments were not uniform and were subject to high seasonality. Arrivals on a given day could be as high as three times the average.

Mr. Jose emphasised, "In case mandatory testing was to be brought in, it had to be done in a manner that the export time schedules are not affected. Almost all exporters were located at field office locations. Collection, courier of sample and return courier of the analytical report could cost about ₹300 per sample. While mandatory testing by SBI may not require additional human resources at the field offices, it would definitely be required at the Head Office, both for

BOX A4.3   EXCERPTS FROM THE SPICES BOARD ACT, 1986 AND THE SPICES BOARD RULES, 1987

Excerpts from the Act
Chapter II: THE SPICES BOARD

7. **Functions of the board**
   (1) The board may-
      (i)   Develop, promote and regulate export of spices;
      (ii)   Grant certificate for export of spices and register brokers therefor;
      (iii)   Undertake programmes and projects for promotion of export of spices;
      (iv)   Assist and encourage studies and research for improvement of processing, quality, techniques of grading and packaging of spices;
      (v)   Strive towards stabilisation of prices of spices for export;
      (vi)   Evolve suitable quality standards and introduce certification of quality through "Quality Marking" for spices for export;
      (vii)   Control quality of spices for export;
      (viii) Give licences, subject to such terms and conditions as may be prescribed, to the manufacturers of spices for export;
      (ix)   Market any spice, if it considers necessary, in the interest of promotion of export;
      (x)   Provide warehousing facilities abroad for spices;
      (xi)   Collect statistics with regard to spices for compilation and publication;
      (xii)   Import, with the previous approval of the Central Government, any spice for sale; and
      (xiii) Advise the Central Government on matters relating to import and export of spices.

   (2) The board may also-
      (i)   promote co-operative efforts among growers of cardamom;
      (ii)   Ensure remunerative returns to growers of cardamom;
      (iii)   Provide financial or other assistance for improved methods of cultivation and processing of cardamom, for replanting cardamom and for extension of cardamom growing areas;
      (iv)   Regulate the sale of cardamom and stabilisation of prices of cardamom;
      (v)   Provide training in cardamom testing and fixing grade standards of cardamom;
      (vi)   Increase the consumption of cardamom and carry on propaganda for that purpose;

(Continued)

BOX A4.3 (*CONTINUED*)   EXCERPTS FROM THE SPICES BOARD ACT, 1986 AND THE SPICES BOARD RULES, 1987

(vii)   Register and licence brokers (including auctioneers) of cardamom and persons engaged in the business of cardamom;

(viii)   Improve the marketing of cardamom;

(ix)   collect statistics from growers, dealers and such other persons as may be prescribed on any matter relating to the cardamom industry; publish statistics so collected or portions thereof or extracts therefrom;

(x)   Secure better working conditions and the provision and improvement of amenities and incentives for workers; and

(xi)   Undertake, assist or encourage scientific, technological and economic research.

Chapter IV: CERTIFICATE FOR EXPORT OF SPICES

11. No person to export spices without certificate

Save as otherwise provided in this act, no person shall, after the commencement of this act, commence or carry on the business of export of any spice except under and in accordance with a certificate.

Provided that a person carrying on the business of export of spices immediately before the commencement of this act, may continue to do so for a period of three months from such commencement; and if he has made an application for such certificate within the said period of three months till the disposal of such application.

Explanation- The reference in this section to the commencement of this act shall be construed in relation to any spice added to the Schedule by notification under the proviso to clause (n) of section 2 as reference to the date with effect from which such spice is added to the Schedule.

12. Grant of certificate

(1)   An application for grant of certificate shall be made to the board in such form and shall contain such particulars as may be prescribed and shall be accompanied by a receipt evidencing the payment of the prescribed fee.

(2)   On receipt of such application, the board shall,-

(a) if the application is not in the prescribed form or does not contain any of the prescribed particulars, return the application to the applicant; or

(b) if the application is in the prescribed form and contains the prescribed particulars, grant the certificate subject to such terms and conditions as may be determined by regulations.

(Continued)

BOX A4.3 (*CONTINUED*)   EXCERPTS FROM THE SPICES BOARD ACT, 1986 AND THE SPICES BOARD RULES, 1987

13. **Cancellation, suspension, etc, of certificate**
    (1) The board may cancel any certificate on any one or more of the following grounds, namely:-
        (a) that the holder of the certificate has violated any of the terms and conditions of the certificate; and
        (b) that in the opinion of the Central Government it is necessary in the interests of general public to cancel the certificate.
    (2) Where the board, for reasons to be recorded in writing, is satisfied that pending consideration of the question of cancelling the certificate on any grounds mentioned in sub-section (1), it is necessary so to do, the board may, by order in writing, suspend the operation of the certificate for such period not exceeding forty-five days as may be specified in the order and require the holder of the certificate to show cause, within fifteen days from the date of receipt of such order, as to why the suspension of the certificate should not be extended till the determination of the question as to whether the registration should be cancelled.
    (3) No order of cancellation of registration under this section shall be made unless the person concerned has been given a reasonable opportunity of being heard in respect of the grounds for such cancellation.

14. **Appeal**
    (1) Any person aggrieved by an order made under section 13 may prefer an appeal to the Central Government within such period as may be prescribed.
    (2) No appeal shall be admitted if it is preferred after the expiry of the period prescribed therefore:
        Provided that an appeal may be admitted after the expiry of the period prescribed therefore, if the appellant satisfies the Central Government that he had sufficient cause for not preferring the appeal within the prescribed period.
    (3) Every appeal made under this section shall be made in such form and shall be accompanied by a copy of order appealed against and by such fees as may be prescribed.
    (4) The procedure for disposing of an appeal shall be such as may be prescribed:
        Provided that before disposing of an appeal, the appellant shall be given a reasonable opportunity of being heard.
    (5) The Central Government may confirm, modify or reverse the order appealed against.

(*Continued*)

BOX A4.3 (*CONTINUED*)   EXCERPTS FROM THE SPICES BOARD ACT, 1986 AND THE SPICES BOARD RULES, 1987

15. Power to permit export without certificate

The Central Government may, if satisfied that it is necessary or expedient, so to do, in public interest, by notification in the Official Gazette and subject to such conditions, if any, as may be specified therein, permit anybody or other agency to commence or carry on the business of export of spices without a certificate.

Spice House Certificate

The certificate is issued to those processors/exporters who have a genuine commitment to quality, and whose long-term objective is sustained export growth. The Spice House Certificate seeks to identify and recognise processors who have made investments in in-house processing facilities and infrastructure, and have the necessary competence to ensure consistent quality and reliability. These facilities cover all critical areas - cleaning, grading, processing, packaging and warehousing. A fool proof system of quality assurance should be employed at all stages of processing - from raw material selection to final shipping. The processors are also expected to maintain a high degree of sanitation in the plant, while the workers must observe absolute cleanliness and personnel hygiene.

Spice House Certificate holders have quality up gradation as their ultimate objective, but with a basic difference in focus.

The certification programme aims at exporters of spices and spice products in bulk packing. The products covered by the certification programme include whole spices as well as value-added products like spice mixes, ground spices, curry blends, spice oils oleoresins and sterilised, dehydrated, pickled and candied spices.

FORM VII
(See Rule 15A)
SPICE HOUSE CERTIFICATE
SPICES BOARD
(Ministry of Commerce, Government of India, Cochin – 25)

Number.................

M/s...exporter of spices having Exporter Registration No. are hereby granted this Spice House Certificate on the basis of their facilities for cleaning, processing, grading, warehousing, packaging of spices/spice products.

        This Certificate is valid upto the ....day of ....

Place: Cochin   Secretary
Date: (Seal)

(Continued)

BOX A4.3 (*CONTINUED*)   EXCERPTS FROM THE SPICES BOARD ACT, 1986 AND THE SPICES BOARD RULES, 1987

SCHEDULE
FACILITIES TO BE MAINTAINED BY AN EXPORTER OF SPICES FOR GRANT OF A SPICE HOUSE CERTIFICATE

Processing Unit:-

A	Cleaning	Shall have facilities for washing, removal of extraneous matters, stone, dust manual or automatic
B	Processing	Facilities for processing spices, spice mix, spice powders, oils, oleoresins or any other value added item.
	Drying	Shall have facilities for drying spices. The drying yards shall have cemented floors without crevices, provided with skirting all around and nets for preventing entry of birds.
	Details of other processing technology and equipment	A laboratory capable of analysing raw materials and the finished products.
C	Grading	Shall have facilities for grading spices using sieves mechanically or manually operated or for grading the spices by sorting machines or by manual means.
D	Warehousing	Shall have store houses/storage area separate for a raw materials and finished products. Storage premises shall have cemented floors without crevices provided with ceiling and doors to prevent entry of rodents and birds shall be provided with wooden pallets/wire mesh and kept clean to avoid entry of rodents, insects, spiders etc.
E	Packaging	Shall have facilities for packaging spices (manual or automatic)
F	Other Facilities	The unit shall be provided with washing facilities for hands and feet at the entrance of the unit. Toilets facilities separate for ladies and gents, washing facilities with soap. Head gears for workers. The surroundings of the unit shall be maintained free of weeds and dumped waste materials have facilities for disposal of waste material.

(*Continued*)

BOX A4.3 (*CONTINUED*)   EXCERPTS FROM THE SPICES BOARD ACT, 1986 AND THE SPICES BOARD RULES, 1987

Excerpts from the Rules

Chapter IV: CERTIFICATE OF REGISTRATION

15. (1) An application for grant of certificate under section 12 shall be made to the Secretary of the Board in Form I.

   (2) Every application for grant of certificate of registration shall be accompanied by a receipt evidencing the payment of fees of rupees two thousand for a block period of three years ending on the 31st August or part thereof. Renewal fee for such period shall be rupees one thousand.

15. (A) An exporter of spices who has his own or taken on rent or leased premises having facilities for cleaning, grading, processing, warehousing and packing as given in the Schedule annexed to these rules, may apply to the Secretary in Form -IV for grant of a Spice House Certificate. The Secretary, on being satisfied as to the facilities available in the premises owned or taken on rent or lease for a period of not less than three years, shall issue a certificate in Form VII. The certificate shall be valid for the block period of three years for which the Exporter Registration Certificate is valid and it shall be renewed if he maintains such facilities. If on inspection at any time, it is found that the exporter does not have such facilities, the Spice House Certificate issued to him shall be cancelled.

16. (1) Any person aggrieved by an order of the Board made under section 13 may, within sixty days from the date of making of such order, appeal to the Central Government.

   (2) Every appeal made under sub-section (1) shall be made in Form II and shall be accompanied by a copy of order appealed against and a receipt evidencing the payment of rupees twenty five.

   (3) On receipt of appeal under sub-rule (1), the Central Government shall, after giving a reasonable opportunity to the appellant of being heard, pass such order as it may deem fit.

Source: Ministry of Commerce and Industry. The Spices Board Act, 1986; The Spices Board Rules, 1987.

sample preparation and supervision." He was also aware that SBI being a government organisation, recruitment was banned and hence not an option. Recruiting students with the appropriate educational background as "trainees" was a possible way out.

Though he had raised various questions, Mr. Jose wondered whether they were overreacting. Mr. Kannan responded, "It seems that the alleged exporters get their orders through overseas middlemen. It is thus possible that parts of the same shipment could have gone to other processors/customers and the same could be true with other shipments."

Mr. Jose knew that a clear message had to be sent to the industry. He also had the feeling that some of the alleged exporters had political connections. He wanted to make sure that the entire trade did not suffer due to the greed of a few. Further, in terms of regulatory powers, SBI could only suspend (pending investigation) and revoke (after investigation) the exporter's registration for violation of conditions listed in the registration rules. (Box A4.3 gives excerpts from the Spices Board Act, 1986 and the Spices Board Rules, 1987.)

However, violation of food safety norms of the importing country or of the Prevention of Food Adulteration Act was not a listed condition for which registration could be suspended or revoked by the SBI. The regulations also did not empower SBI to draw samples on a compulsory basis from the export consignment. To impose mandatory testing, the regulations would need to be amended.

Acknowledgements

Case prepared by Tathagata Bandyopadhyay (tathagata@iimahd.ernet.in), G. Raghuram (graghu@iimahd.ernet.in), and Neeraj Sisodia (neerajsisodia@yahoo.com), IIM, Ahmedabad.

We acknowledge the research assistance by Vishal Kashyap. We are thankful to Mr. Jose, Chairman and other senior executives of the Spices Board of India for their discussions and data support. Some data have been masked to protect sensitivities.

Teaching material of the Indian Institute of Management, Ahmedabad, is prepared as a basis for class discussion. Cases are not designed to present illustrations of either correct or incorrect handling of administrative problems.

This case has been reproduced with permission from Indian Institute of Management, Ahmedabad. Copyright © 2005.

References

BBC News. Illegal dye: What is the health risk? Feb. 22, 2005. http://news.bbc.co.uk/2/hi/health/4286847.stm

Federal Institute for Risk Assessment (BfR). Dyes Sudan I to IV in food. Nov. 2003. https://www.bfr.bund.de/cm/245/dyes_sudan_I_IV.pdf

Ministry of Commerce and Industry. The Spices Board Act, 1986; The Spices Board Rules, 1987. http://www.indianspices.com/pdf/Act_Rules_English_(SB).pdf

Spices Board of India. Website. 2003. http://www.indianspices.com

United Nations Statistics Division. Accessed on Nov. 19, 2004. http://www.unstats.un.org

Case 5   Chilli in Soup (B)

It was in February 2005, Mr. Jose, Chairman of Spices Board of India (SBI) was reading the depressing news article announcing the largest recall of food products in the UK, due to use of chilli powder imported from India.

His thoughts went back to June 2003, when warnings had come about traces of Sudan I (a banned carcinogenic dye) being found in some consignments of chilli exports from India. The SBI had swung into action immediately suspending for forty five days the certificates of registration of the three firms mentioned in the Rapid Alert System for Food and Feed (RASFF) notifications in Case 4 with effect from June 13 and 20, 2003 (Box A5.1). The suspension had stayed effective, pending the decision regarding the cancellation of their registrations subject to further investigation.

Jose had also called a meeting of the executives of the All India Spices Exporters Forum (AISEF). In the meeting, had voiced his concern about the possible damage this event could trigger, both in terms of short-term and long-term consequences. He had expressed his apprehensions about a possible chain of serious consequences on the exports of Indian spices in terms of brand equity, confidence-building, and new opportunities. He had requested the executives to join hands and help him in controlling the damage to the best possible extent. He had further stated that he was open to any constructive suggestions from their end, and concluded the meeting by saying that the primary and the most difficult task for the Board was to convince the consumers of their trustworthiness.

In the same month, the European Commission (EC) put into effect, on all its member states, emergency measures on the import of hot chilli and hot chilli products. It stipulated that every consignment of hot chilli and hot chilli products should be barred from entering any of its member states unless it was accompanied by an analytical report declaring that it did not contain any trace of Sudan I (Box A5.2).

Jose kept himself abreast of the latest events and was closely following the reports appearing daily in the press. He had also been in touch with European Union (EU)

BOX A5.1   ACTION ON RAPID ALERT NOTIFICATION BY EUROPEAN COMMISSION

As per the Rapid Alert Notification dated May 9, 2003, issued by the EC, consignments of chilli powder exported by the following firms to France have been detained on account of reported presence of a prohibited colourant "Sudan-I".

1. **M/s Gautam Export Corporation**
   Flat 11-B, 3rd Floor
   Koolbreeze CHS Limited, Plot No K-72
   17th Road, Khar (West)
   Mumbai – 400 052

2. **M/s Volga Spices & Masala Mills (Private) Limited**
   10-A, Tata Mill Compound
   Near Elphinston Bridge
   Parel
   Mumbai – 400 012

3. **M/s Patons Exports Pvt. Ltd**
   10, Koolbreeze CHS Limited
   17th Road, Khar (West)
   Mumbai – 400 052

Their certificates of registration as exporters of spices have been suspended for a period of 45 days, effective from June 13 and 20, 2003, pending a decision regarding cancellation of their registration.

SBI has suspended the certificate of registration of the following exporters exporting imported pepper as Malabar Garbled (MG-1) and Tellicherry Garbled Extra Bold (TGEB) for a period of 45 days, effective from June 20, 2003, pending a decision regarding cancellation of their registration.

1. **M/s Nishant Export**
   7/179, D S Road
   Cochin 682 002

2. **M/S Jamnadasmadhvji International Limited**
   Tanna House
   Ground Floor
   11 A, Nathalal Parekh Marg
   Mumbai 400 001

(Continued)

BOX A5.1 (CONTINUED)   ACTION ON RAPID ALERT NOTIFICATION BY EUROPEAN COMMISSION

Press Release

A recent decision of the European Union, published as 2003/460/EC dated June 20, 2003 has put into effect the following emergency measures with respect to hot chilli and hot chilli products:

1. Member States shall prohibit the import of hot chilli and hot chilli products defined in Article 1 unless an analytical report accompanying the consignment demonstrates that the product does not contain Sudan red I (CAS Nr. 842-07-09).
2. The competent authorities in the Member States shall check that each consignment of hot chilli and hot chilli products presented for importation is accompanied by a report as provided for in paragraph 1.
3. In the absence of such an analytical report, the importer established in the Community shall have the product tested to demonstrate that it does not contain Sudan red. Pending availability of the analytical report, the product shall be detained under official supervision.

Hence, all exporters of chilli and chilli products from India should ensure that the analytical report required by the European Commission accompanies consignments of hot chilli and hot chilli products.

Source: Spices Board of India. Action on rapid alert notification by European Union and exporting of imported pepper as Malabar garbled (MG-1). 2003.

officials and kept them informed about the actions SBI had taken in this context. This helped in calming the situation to certain extent.

However, in July 2003, it erupted again. The EC notified Food Standards Agency (FSA), UK, (Box A5.3) about the food products contaminated with Sudan I found in France with an origin in the UK. The contamination was traced to an Essex based firm, East Anglia Food Ingredients. Subsequent investigations by FSA revealed that 25 food products sold in the UK were found to contain chilli powder imported from India and contaminated with Sudan I. FSA issued its first food hazard warnings (later called food alerts) (Box A5.4), asking the local authorities to ensure that the contaminated products were removed from the sales (Box A5.5). In the meantime, the UK implemented emergency measures of EU.

As a follow up to the food alerts in September 2003, the FSA issued Food Guidance Notes (Box A5.6) on February 18, 2004 to companies in the food industry asking them to withdraw and recall contaminated products as soon as they were identified. The FSA also asked all manufacturers who used chilli powder or chilli products (if it was imported from India and supplied to them prior to

**BOX A5.2   EMERGENCY MEASURES BY
THE EUROPEAN COMMISSION**

Official Journal of the European Union   21.6.2003

COMMISSION DECISION
of 20 June, 2003
on emergency measures regarding hot chilli and hot chilli products
(Notified under document number C (2003) 1970)
(Text with EEA relevance)
(2003/460/EC)

THE COMMISSION OF THE EUROPEAN COMMUNITIES,

Having regard to the Treaty establishing the European Community,

Having regard to Regulation (EC) No 178/2002 of the European Parliament and of the council of 28 January 2002 laying down the general principles and requirements of food law, establishing the European Food Safety Authority and laying down procedures in matters of food safety (1), and in particular Article 54 thereof,

Whereas:

(1)   Under Regulation (EC) No 178/2002 the Commission is to suspend the placing on the market or use of a food or feed that is likely to constitute a serious risk to human health, and take any other appropriate interim measure when such risk cannot be contained satisfactorily by means of measures taken by the Member States concerned.

(2)   On 9 May 2003, France sent information through the rapid alert system for food and feed relating to discovery of the dye Sudan red 1 in hot chilli products originating from India. There is no evidence that products of Community origin are concerned by such findings.

(3)   Available experimental data indicate that Sudan red 1 may be a genotoxic carcinogen. It is, therefore, not possible to establish a tolerable daily intake. Sudan red 1 may also exert sensitizing effects by dermal route or inhalation. It has also been classified as a category 3 carcinogen by the International Agency for Research on Cancer (IARC).

(4)   Therefore the findings reported by France point to an adulteration constituting a serious health risk.

(5)   On 5 June 2003, in the light of the possible extent of the problem, France adopted interim protective measures and informed the Commission thereof.

(*Continued*)

BOX A5.2 (*CONTINUED*)   EMERGENCY MEASURES BY THE EUROPEAN COMMISSION

(6)   Accordingly, the Commission must put the matter before the Standing Committee on the Food Chain and Animal Health within 10 working days of the measures having been adopted by France, with a view to the extension, amendment or abrogation of the national interim protective measures.

(7)   Given the seriousness of the health threat, it is necessary to extend the measures taken by France to the whole Community. Moreover, account should be taken of potential triangular trade, especially for products for which there is no official certification of origin. In order to protect public health, it is appropriate to require that consignments of hot chilli and hot chilli products imported into the Community in whatever form, intended for human consumption, should be accompanied by an analytical report provided by the importer or food business operator concerned, demonstrating that the consignment does not contain Sudan red 1. For the same reason, Member States shall carry out random sampling and analysis of hot chilli and hot chilli products at import or already on the market.

(8)   It is appropriate to order the destruction of adulterated hot chilli and hot chilli products to avoid their introduction into the food chain.

(9)   Since the measures provided for in this Decision have an impact on the control resources of the Member States, the results of these measures should be evaluated at the latest after 12 months in order to assess whether they are still necessary for the protection of public health.

(10)   This evaluation should take account of the results of all analyses carried out by the competent authorities.

(11)   The measures provided for in this Decision are in accordance with the opinion of the Standing Committee on the Food Chain and Animal Health.

Article 1
Scope

This Decision applies to the following hot chilli and hot chilli products, in whatever form, intended for human consumption:

- Fruits of the genus Capsicum, dried and crushed or ground within CN code 0904 20 90.

(*Continued*)

BOX A5.2 (*CONTINUED*) EMERGENCY MEASURES BY THE EUROPEAN COMMISSION

Article 2
Conditions for import of hot chilli and hot chilli products

1. Member States shall prohibit the import of hot chilli and hot chilli products defined in Article 1 unless an analytical report accompanying the consignment demonstrates that the product does not contain Sudan red 1 (CAS Nr 842-07-09).
2. The competent authorities in the Member States shall check that each consignment of hot chilli and hot chilli products presented for importation is accompanied by a report as provided for in paragraph 1.
3. In the absence of such an analytical report, the importer established in the Community shall have the product tested to demonstrate that it does not contain Sudan red 1. Pending availability of the analytical report, the product shall be detained under official supervision.

Article 3
Sampling and analysis

1. Member States shall take appropriate measures, including random sampling and analysis of hot chilli and hot chilli products presented for importation or already on the market in order to verify the absence of Sudan red 1. They shall inform the Commission of positive (unfavourable) results through the rapid alert system for food and feed. Negative (favourable) results shall be reported to the Commission on a three-monthly basis. This report shall be submitted during the month following each quarter.[1]
2. Any consignment subjected to official sampling and analysis may be detained before release into the market for a maximum period of 15 working days.

Article 4
Splitting of a consignment

If a consignment is split, a certified copy of the analytical report provided for in Article 2(1) shall accompany each part of the split consignment.

Article 5
Adulterated consignments

Products referred to in Article 1 that are found to contain Sudan red 1 shall be destroyed.

(Continued)

BOX A5.2 (*CONTINUED*)   EMERGENCY MEASURES
BY THE EUROPEAN COMMISSION

Article 6
Recovery of costs
In relation to Article 2(1), (3) and Article 5, costs resulting from analysis, storage and eventual destruction shall be borne by the importers or food business operators concerned.

Article 7
Review of the measures
This Decision shall be reviewed by 20 June 2004 at the latest.

Article 8
Addressees
This Decision is addressed to the Member States.

Done at Brussels, 20 June 2003
For the Commission
David Byrne
Member of the Commission

Source: European Commission. Commission decision of 20 June 2003 on emergency measures regarding hot chilli and hot chilli products, *Official Journal of the European Union*. Accessed Nov. 12, 2005, from http://eurlex.europa.eu/LexUriServ/LexUriServ.do?uri=OJ:L:2003:1 54:0114:0115:EN:PDF
[1] April, July, October, January.

July 30, 2003) to ensure that it had not been contaminated with Sudan I. The guidance added: "These companies should consider sampling relevant batches in order to ensure that their products are not contaminated. The FSA would like to be notified of all results, whether they are positive or negative, in order to obtain as much information as possible about the distribution of the contaminated chilli products."

Since May 2003, the SBI had been receiving a number of RASFF notifications relating to the presence of Sudan I in consignments of chilli powder exported from India. Most of the RASFF notifications reported involvement of the two firms viz, Gautam Export Corporation and Volga Spices & Masala Mills Private Limited. This sudden turn of events in July 2003, worried Jose. The FSA had been regularly issuing food alerts as soon as new contaminated products were identified. The number was increasing every month. In view of the new developments, Jose was seriously contemplating the introduction of a number of regulatory mechanisms, to ensure that the quality of Indian chilli exports satisfied the food safety requirements of the importing countries. For this, he had been in constant touch with

BOX A5.3 FOOD STANDARDS AGENCY

The Food Standards Agency (FSA) is an independent UK Government department set up by an Act of Parliament in 2000 to protect the public's health and consumer interests in relation to food. Although the FSA is a Government agency, it works at 'arm's length' from government because it does not report to a specific minister and is free to publish any advice it issues.

FSA is led by a board appointed to act in public interest and not to represent particular sectors. Board members have a wide range of relevant skills and experience. FSA is headquartered in London and has national offices in Scotland, Wales and Northern Ireland. It is accountable to Parliament through Health Ministers, and to the devolved administrations in Scotland, Wales and Northern Ireland for its activities within their areas.

How FSA works

- It provides advice and information to the public and the Government on food safety from farm to fork, nutrition and diet. It also protects consumers through effective food enforcement and monitoring.
- It supports consumer choice by promoting accurate and meaningful labeling.
- It works with many different people and organizations, including consumers, consumer organizations, manufacturers, retailers, farmers, local authority enforcement teams and the catering industry. It also works with other Government departments and agencies.

Key aims of FSA's strategic plan during 2001–2006

- Reduce food borne illness by 20% by improving food safety through the food chain
- Help people to eat more healthily
- Promote honest and informative labeling to help consumers
- Promote best practices within the food industry
- Improve the enforcement of food law
- Earn people's trust by what it does and how it does it

Key aims of FSA's strategic plan during 2005–2010

- Continue to reduce food borne illness
- Reduce further the risks to consumers from chemical contamination including radiological contamination of food
- Make it easier for all consumers to choose a healthy diet, and thereby improve quality of life by reducing diet-related disease
- Enable consumers to make informed choices

Source: Food Standards Agency. About us. 2005. Accessed Nov. 12, 2005 from http://www.food.gov.uk/aboutus.

BOX A5.4   FOOD ALERTS

Food Alerts are the FSA's way of letting local authorities and consumers know about the problems associated with food and, in some cases, provide details of specific action to be taken.

These are often issued in conjunction with a product withdrawal or recall by a manufacturer, retailer or distributor.

Food Alerts were previously known as Food Hazard Warnings and were issued under four categories (A, B, C, and D), but the new system took effect from October 25, 2004.

Now they are issued under two categories:

1. Food Alerts for Action (replacing A, B and C)
2. Food Alerts for Information (replacing Category D)

A Food Alert issued by the FSA

Chilli powder imported from India containing Sudan I dye used in Rajah Brand seasoning mixes

(July 31, 2003)

Category C: Action as deemed necessary

Heads of Environmental Health Services and Directors of Trading Standards will wish to be aware that contaminated chilli powder containing an illegal dye Sudan I, has been imported into the UK from India and used in products produced for BE International Foods, a division of HP Foods Limited.

Sudan I is not a permitted colour under the Colours in Food regulations 1995. It is considered to be a genotoxic carcinogen and its presence, at any level, is not permitted in foodstuffs for any purpose.

The Rajah brand products listed below have been identified as containing Sudan I. The company has undertaken a full trade withdrawal of the products with any Best before End dates until July 18, 2005, from all their customers.

Any product with a Best Before End date of July 18, 2005, or after this date, is confirmed to be made with chilli powder that is clear of Sudan I.

HP Foods Limited has set up a helpline to address customer enquiries. The Rajah Consumer Services team can be contacted on 0800 0721 122 and the company will arrange collection and reimbursement of the consumer's costs.

Only the products mentioned below are affected; no other Rajah products are affected.

(Continued)

BOX A5.4 (*CONTINUED*)   FOOD ALERTS

The details of the products are as follows:

- Rajah Tandoori Masala
- Rajah Natural Tandoori Masala
- Rajah Jerk seasoning
- Rajah Chilli & Lemon seasoning
- Rajah Oriental Noodle seasoning
- Rajah Chicken seasoning

These products are retailed in 100 g and 400 g packets. These sizes and larger sizes (1 kg and 8 kg) are also sold through wholesalers.

Action to be taken by local authorities were deemed necessary

The products detailed above present a health risk if consumed in sufficient quantity, and do not comply with the food safety requirements specified in the Food Safety Act 1990 due to contamination with Sudan I. In addition, Sudan I is not a permitted colour under the Colours in Food Regulations 1995.

Local authorities should take action if these products are found within their area. Enforcement officers should ensure that they are withdrawn from sale and destroyed, if necessary, using powers under the Food Safety Act 1990.

It is essential that all Local Authorities liaise at county level to ensure that enforcement authorities agree who will take the lead in dealing with this incident.

Given the nature of the problem, and that this type of product is likely to remain for long periods in consumers' homes between purchase and consumption, further local publicity may assist in preventing consumers eating any affected product they may have.

Local authorities are, therefore, asked to consider providing local publicity at the next available opportunity to this issue.

For that purpose, local authorities may wish to use the Food Standards Agency's press release as a guide as to what to say in any local news release.

Source: National Archives (UK). Chilli powder imported from India containing Sudan I dye used in chutney and relish produced by Shaw's (Huddersfield) Ltd. July 31, 2003. http://tna.europarchive.org/20110116113217/http://www.food.gov.uk/enforcement/alerts/2003/jul/sudanichi

BOX A5.5   LIST OF PRODUCTS NAMED IN FOOD ALERTS

Products containing chilli imported from India, mentioned in subsequent Food Alerts are:

- Products produced by Shaws (Huddersfield) Ltd, Huddersfield, UK (August 20, 2003).
- Favourite brand products, packed by Martin Foods, Belfast, Ireland (September 30, 2003).
- Pasand Green Chilli Pickle, India (January 24, 2005).
- Nice N Spice Chilli Powder, imported by Nice N Spice Foods Ltd, UK (January 09, 2004).
- Dragon Brand and O'Kane Food Service Curry Sauce, packed by Dragon Brand Foods, UK (September 30, 2003).
- Deggi Mirch Chilli Powder imported from India by Mark One Ind Spices Ltd, UK, manufactured by Mahashian Di Hatti Ltd, India (April 16, 2004).
- Natco Brand Madras Curry Powder and Curry Powder Mild, distributed by Natco Foods Ltd, UK (April 02, 2004).
- Ritu's Garlic Curry Sauce, manufactured by Cerbos Impex Pvt Ltd, New Delhi, and distributed by Mosscope Ltd, London (December 23, 2004).

Source: National Archives (UK). Chilli powder imported from India containingSudanIdyeusedinchutneyandrelishproducedbyShaw's(Huddersfield) Ltd. July 31, 2003. http://tna.europarchive.org/20110116113217/http://www.food.gov.uk/enforcement/alerts/2003/jul/sudanichi

the representatives of the AISEF. He was not so concerned about the big exporters. Such exporters themselves were conscious about their brand equity in the overseas market and had their own sampling and testing facilities. They also had their consignments tested by renowned laboratories in India and abroad. He was worried more about the smaller exporters who were into trading and had no control on the quality of the supplied materials.

One possibility, he thought, was to get SBI directly involved in the process of sampling and testing of every consignment of chilli powder/chilli products that was to be shipped out of India. This would give SBI complete control in terms of monitoring the quality of chilli products that were shipped out. He envisaged a dire need to study the scenario to bring forth practical suggestions. He entrusted this task to Dr. Thomas, Director–Research and Mr. Kannan, Director–Marketing to

BOX A5.6 FOOD STANDARDS AGENCY GUIDANCE

The FSA regularly issues guidance to food industry representatives and other stakeholders on a range of topics, often as a result of new regulations coming into force.

Guidance Notes on Sudan dyes in chilli imported from India

Wednesday, 18 February 2004

The Food Standards Agency investigations have identified that certain food products sold in the UK have been found to contain chilli powder, chilli products or curry powder imported from India illegally contaminated with the Sudan dyes, which are chemicals that could cause cancer.

Introduction

The Agency was notified in July 2003 by the European Commission (EC) that products contaminated with Sudan I found in France had been produced in the UK and, more recently, of products produced in Italy and imported into the UK.

Sudan dyes are not permitted colours under the Colours in Food Regulations 1995. These are red dyes that are used for colouring solvents, oils, waxes, petrol, and shoe and floor polishes. They are considered to be a genotoxic carcinogen and their presence, at any level in food, is not permitted in food for any purpose.

SBI subsequently suspended the certificates of registration as exporters of spices of the Indian companies listed in Annex 1 from 13 and 20 June 2003, and action is being taken to cancel their certificate of registration, if so required.

Legislation

An EC Decision, which was made on 21 January 2004, requires that cargoes of dried and crushed or ground chilli and curry powder coming into any EU Member State must now be accompanied by a certificate showing that they have been tested and found to be free of Sudan I, Sudan II, Sudan III and Sudan IV.

The Food (Hot Chilli and Hot Chilli Products) (Emergency Control) (England) (Amendment) Regulations 2004 implementing the EC Decision came into force on 27 January 2004.

Any consignment that does not have a certificate will be detained for sampling and analysis.

(Continued)

BOX A5.6 (*CONTINUED*)   FOOD STANDARDS AGENCY GUIDANCE

Action requested by FSA

The Agency has asked companies to withdraw and recall contaminated products as soon as they are identified. The Agency and local authorities are working together to ensure that any products which have been identified as containing contaminated chilli powder, are removed from sale and destroyed. Both retail and catering products have been identified as being contaminated.

In addition, the FSA is asking all manufacturers who use chilli powder or chilli products as part of the ingredients in their products to ensure that it has not been contaminated with Sudan I–IV as specified in the Food (Hot Chilli and Hot Chilli Products) (Emergency Control) (Amendment) (England) Regulations 2004, if it was imported from India and supplied to them prior to January 27, 2004.

These companies should consider sampling relevant batches in order to ensure that their products are not contaminated.

The FSA would like to be notified of all results whether they are positive or negative, in order to obtain as much information as possible about the distribution of contaminated chilli products.

Analysis of chilli powder or products containing chilli powder

Whilst the Commission Decision does not specify the method of analysis for Sudan I–IV, the use of validated methods by official food control laboratories is an important requirement of the EU Additional Measures Directive 93/99/EEC for food. This Directive also requires that official food control laboratories should be accredited to the ISO/IEC Guide 25 (now the ISO/IEC 17025 Standard). In the UK, all chemical analysis food control laboratories are required to have generic accreditation for analytical procedures, including HPLC, as well as for specific methods of analysis.

The FSA is currently carrying out collaborative trials on two methods of analysis, which have been developed to determine the presence of Sudan I and details of these two methods, that have been shown to work, are attached below. These methods are also applicable for testing Sudan II and further work is underway to detect Sudan III and IV.

(Continued)

**BOX A5.6 (*CONTINUED*) FOOD
STANDARDS AGENCY GUIDANCE**

Laboratories carrying out the analysis should comply with these Accreditation Standards and have demonstrated their ability to carry out these particular analyses successfully. In particular, they should be able to demonstrate that they can carry out the separation of Sudan I from potentially interfering substances, most notably capsanthin, a carotenoid found naturally in chillies, and which has an absorption maximum very similar to that given for Sudan I (483 nm for capsanthin and 478 nm for Sudan I).

Disposal of Contaminated Product

Foodstuffs containing Sudan I–IV would not be classified as hazardous waste. On this basis, the waste may go to any permitted facility that is authorized to accept non-hazardous waste of this type, including a transfer station, landfill and incinerator.

Source: Food Standards Agency. Sudan dyes in chilli imported from India: Guidance notes. Mar. 2004. Accessed Nov. 15, 2005 from http://www.food.gov.uk/foodindustry/guidancenotes/foodguid/sudanguidance.

---

study the feasibility of his proposal and submit a report within two weeks. Their task was as follows:

1. To find if the available resources in terms of equipment, human resources, and logistics were adequate to handle the foreseeable demand.
2. To shorten the time taken in the whole process, from receiving the request for sampling and testing from the exporter to sending the analytical report to the exporter.
3. To evaluate the cost for the tests and suggest ways to minimize the same, so that the exporters would find it acceptable.

After analysing the whole situation, Dr. Thomas and Mr. Kannan stated in the report that the plan could be made feasible by extending working hours of the lab by recruiting part-time lab technicians for testing and employing outside agencies for sampling. The part-time lab technicians could be employed from the pool of SBI-trained post-graduate students from the nearby universities. However, the report pointed out that the exporters might have to wait for the report, sometimes, as long as ten days depending upon the workload and availability of human resources. In addition, they suggested that testing charges be fixed at a level slightly above the break-even.

Subsequent to these developments, Jose met the representatives of AISEF several times during the next month and convinced them of the urgency of measures of this kind. Hence, some of the suggestions were considered in finalizing the details of the circular as well as the modalities of the whole operation by SBI.

On October 9, 2003, SBI issued a circular (Box A5.7) making it mandatory for every exporter to get the consignments of chilli powder or any spice product containing chilli, sampled and tested for Sudan I in the SBI Laboratory (Box A5.8) before shipment, with effect from October 23, 2003. The exporters were asked to notify SBI's nearest office (Table A4.1 in Case 4 about SBI) and the head office in the prescribed format (Box A5.7) at least 48 hours before the shipment. The cost of sampling and testing was to be borne by the exporter with the assurance that the present arrangement would be reviewed after three months. It also stated that the exporter could ship the consignment before the analytical reports were obtained, but in case the report showed presence of Sudan I, the shipments would have to be called back at the risk and cost of the exporter.

On November 12, 2003, an addendum to the previous circular (October 9, 2003) was issued. In order to ensure that the chillies exported did not contain Aflatoxin (a toxic metabolite produced by certain moulds in food and feed) above the tolerance limit (Table A5.2), the test for Aflatoxin of whole chillies was made mandatory (Box A5.9).

On January 21, 2004, the EC, amending the earlier decision of June 6, 2003, made it mandatory that import of chilli and chilli products into EU be accompanied by an analytical report showing that the product does not contain Sudan II, III, and IV, in addition to Sudan I.

During February 16 to February 27, 2004, officials of the Food and Veterinary Office (FVO), EC, visited the SBI head office in Kochi to assess the control system in place. They reported, "The laboratory empowered for official analysis does not generate results for Sudan I examination with complete reliability, since its limit for detection of Sudan I is three parts per million (ppm), which is not sufficiently low and furthermore, as Sudan II, III, and IV are not in the analytical scope of the SBI lab." Therefore, from March 19, 2004, the SBI lab started testing for Sudan I and Sudan II, III, and IV.

Before the visit of the FVO, Jose had already started the process of procuring two High Profile Liquid Chromatography (HPLC) MS/MS machines (Box A5.10) with investment of ₹16 million in order to upgrade the testing facilities. These equipment had a detection limit of five parts per billion (ppb) or less. The HPLC MS/MS machines were installed and test procedures were validated in May 2004 and the new machines became operational in June 2004.

In the meantime, the SBI received around 38 RASFF notifications till May 2004. In all, 17 rapid alerts reported Gautam Exports Corporation and Volga Spices & Masala Mills Private Limited as the sources of contaminated chilli powder. Out of the remaining 21, two did not name any Indian source, nine did not quantify the presence of the prohibited substance alleged to be present, and eight

BOX A5.7 SPICES BOARD OF INDIA CIRCULAR: MANDATORY TESTING FOR SUDAN I

To,

All leading exporters of chilli/chilli products
Copy to: The Regional Offices of the Board
No.MD/CHI/01/03/ October 9, 2003

CIRCULAR

As you are aware, we have received a number of Rapid Alert System for Food and Feed (RASFF) notifications since May 2003 relating to the presence of the prohibited carcinogenic dye 'Sudan-1' in some consignments of chilli powder exported from India. We have also received reports about product recalls in a number of countries on the ground that the manufacturers of those products used contaminated chilli powder imported from India. The European Union has since mandated that each consignment of hot chilli and chilli products imported into EU must accompany an analytical report demonstrating that the chilli products do not contain 'Sudan-I'. The European Spice Trade Association has subsequently asked the Spices Board to indicate to them as to what steps are being taken to ensure a higher degree of reliability and safety to spice products exported from India.

In the light of the above, after due consultations, it has been decided that with effect from 23rd October, 2003, all exporters intending to export chilli powder or any other spice product containing chilli, excluding chilli oil & oleoresin, to any destination, should notify Spices Board's nearest office and the head office in the prescribed format, at least 48 hours before shipment, and assist Spices Board or any agency designated by it, to draw samples from the export consignment. The containers will have to be sealed in the presence of concerned official/agency soon after samples are drawn, to ensure that the samples represent material intended to be/actually exported.

Samples will be analyzed in the Board's laboratory or any other designated laboratory for the presence of 'Sudan Red-1'.

If the exporter is confident that the consignment is free of adulterants, shipment may be effected. At his option, the exporter can wait for the receipt of analytical report from the Spices Board prior to effecting shipment. In case the consignments have left and the analysis indicates the presence of contaminants, the buyer will be notified and the shipment will have to be called back at the risk and cost of the exporter.

The cost of the above procedure will be debited to the exporter.

This arrangement will be reviewed after 3 months.

(S. Kannan)
Director (Marketing)
(Continued)

BOX A5.7 (*CONTINUED*)   SPICES BOARD OF INDIA
CIRCULAR: MANDATORY TESTING FOR SUDAN I

Prescribed Format for Details of Proposed Export of Chilli and Products Containing Chilli
(Circular No. MD/CHI/01/03 Dated 9.10.2003)

Notification No. _____   Date: _____
(Serial wise)

1	Name & Address of the Exporter	Tel No Fax No
2	Spices Board Registration No	
3	Name & Address of the processor, if different from the Exporter	Tel No Fax No
4	Details of product(s) to be exported	

Product(s)	Quantity (ton)	Value (FOB) ₹/Kg

5	Address of the place where container is to be stuffed	
6	Lot No & Shipping Marks	
7	Date and Time of Stuffing	
8	Port of Shipment	
9	Port & Country of Discharge	
10	Name of the Vessel	
11	Container No	
12	Invoice No & Date	

Place: Signature:
Date: Name & Designation
                           (with seal)

Source: Kannan, S. Circular MD/CHI/01/03: Mandatory testing for Sudan I. Spices Board of India, Oct. 9, 2003. Accessed Nov. 15, 2005 from http://www.indianspices.com

BOX A5.8   QUALITY EVALUATION LABORATORY OF THE SPICES BOARD OF INDIA

The Quality Evaluation Laboratory of the SBI was established in 1989. It provides analytical services to the Indian spice industry and monitors the quality of spices produced and processed for export in the country. It has facilities to analyse various physical, chemical and microbial parameters including pesticide residues, aflatoxin, and heavy metals, in spices and spice products and also adulterants like Sudan-I-IV in chilli products. The laboratory follows internationally accepted procedures for the various analyses.

The laboratory is certified by British Standards Institution, UK, for the ISO 9002 Quality Management System in 1997 and for the ISO 14001 Environmental Management System in 1999. The laboratory has submitted its application to the Department of Science & Technology for accreditation under NABL. All the activities under the ISO 9002 system established in the laboratory are fully computerized.

The laboratory is equipped with state of the art technology to undertake the analysis as per the requirements of importing countries.

The educational qualification of technical staff working in the quality evaluation & upgradation laboratory of SBI:

1. Senior Scientist: MSc AIC
2. Scientist (QC): MSc (Applied Microbiology)
3. Junior Scientist: MSc (Ag)
4. Junior Scientist: MSc (Applied chemistry)
5. Lab Technician: MSc (MLT)
6. Lab Technician: MSc Chemistry
7. Junior Lab Technician: BSc Chemistry
8. Junior Lab Technician

Major equipment purchased since 1991:

1. UV-VIS Spectrophotometers
2. Atomic Absorption Spectrophotometer
3. Gas Chromatographs
4. High Performance Liquid Chromatographs with various detectors
5. Zoom Stereomicroscopes
6. High Performance Thin Layer Chromatograph
7. Chilled Circulating System
8. Gas Chromatograph - Mass Spectrometer (GC-MS)

(Continued)

> ## BOX A5.8 (*CONTINUED*)   QUALITY EVALUATION
> ## LABORATORY OF THE SPICES BOARD OF INDIA
>
> 9. GPC (Gel Permeation Chromatograph)
> 10. High Performance Liquid Chromatographs with MS/MS Facility
> 11. Chilled Water Circulating System
> 12. Stomacher
> 13. Water Activity Meter
>
> **Validation/check sample programme**
>
> To validate the analytical methods adopted, the laboratory participates in
> Check Sample/Validation programmes organized by international/national
> agencies regularly viz, American Spice Trade Association (ASTA) and
> International Pepper Community (IPC), Jakarta. In addition to this, it con-
> ducts check sample programmes for major parameters (like Aflatoxin, Sudan
> I–IV and Pesticide residues) with various laboratories spread over the major
> importing countries and laboratories attached to spice export units in India.
>
> **Analytical services offered by the quality and evaluation laboratory of
> the SBI**
>
> The laboratory offers analytical service to the exporters, traders, farm-
> ers and research organizations in spices and spice products. During the
> period 2003–2004, 12,147 samples were analysed for various parameters,
> including pesticide residues, Aflatoxin and Sudan I–IV in chilli and chilli
> products.
>
> Table A5.1 gives the details of the analysis on various parameters, showing
> the analytical charges, minimum quantity and number of days required for
> completion of the analysis.

alleged the presence of Sudan I at varying levels, but less than two ppm. In this
context, it needs to be mentioned that SBI found, by experimentation, that even
2,000 ppm* of Sudan per kg would not be sufficient to convert a given variety of
Indian chilli powder to a high colour value category.

During July 31, 2003 to January 31, 2005, the FSA issued 56 food hazard
warnings/alerts on Sudan dyes 1, II, III and IV involving more than 200 products. In the meantime, AISEF had been demanding the withdrawal of the circular
issued on October 9, 2003 by SBI. The big exporters stated that most of their

---

* 1 ppm means 1 g of Sudan I mixed with 1 ton of chilli powder, 1 ppb means 0.01 g of Sudan I
  per ton of chilli powder.

Table A5.1 SBI Quality and Evaluation Laboratory Analytical Services

Sr. No.	Name of Analysis	Minimum Quantity Required for Analysis in Gram	Charges ₹	No. of Days Required[f]
1	Aerobic spore count	100	150	3
2	Agmark specifications	500	[a]	>
3	ASTA specifications	500/250 × 10 Nos[b]	[a]	1
4	Acid insoluble ash	250	250	6
5	Aflatoxin(B1, B2, G1, G2) [HPLC method]	250[e]	1500	6
6	Alcohol soluble extract	250	120	4
7	*Bacillus cereus*	100	350	6
8	Bulk density/Litre weight of spices	1000	100	1
9	Calcium as CaO	250	120	5
10	Capsaicin - HPLC method (%/SHU)	100[e]	350	4
11	Capsaicin - UV difference method (%/SHU)	100[e]	250	4
12	Chromate test (Qualitative)	100	100	2
13	*Clostridium perfringens*	100	500	6
14	Cold water soluble extract	100	120	4
15	Conforms	100	220	5
16	Colour value (ASTA METHOD)	100[e]	180	4
17	Common salt	100	100	5
18	Crocin	10	180	3
19	Crude fibre	100	300	6
20	Curcumin (ASTA method)	100[e]	250	4
21	Enterobacteriacea	100	180	5

(Continued)

Table A5.1 (*Continued*)   SBI Quality and Evaluation Laboratory Analytical Services

Sr. No.	Name of Analysis	Minimum Quantity Required for Analysis in Gram	Charges ₹	No. of Days Required[f]
22	Excreta, mamalian	500/250	100	1
23	Excreta, others	500/250	100	1
24	Extraneous/foreign matter	500/250	100	1
25	E. coli	100	180	4
26	Filth, heavy	100	400	4
27	Filth, light	100[c]	400	5
28	Gingerols and shogaols	100[e]	750	5
29	Heavy metals (Cd, Cr, Cu, Fe, Mg, Mo, Pb and Zn)	100[e]	120[d]	5
30	Heavy metal: Arsenic	100	300	5
31	Heavy metal: Mercury	100	300	5
32	Insect defiled/infested	500/250	100	1
33	Light berries	250	100	1
34	Moisture (ASTA method)	250	100	2
35	Mould (microbiological)	100	120	5
36	Mould (physical)	250	100	1
37	Non-volatile ether extract	100	150	6
38	Oleoresin (EDC extractables)	100	150	5
39	Organochlorine Pesticides Residues (Isomers of BHC, endosulfan and DDT; heptachlor, aldrin, dieldrin, endrin and endrin aldehyde)	200[e]	2000	6
40	Organochlorine pesticide residue—Dicofol	200	800	6

(Continued)

Table A5.1 (*Continued*)   SBI Quality and Evaluation Laboratory Analytical Services

Sr. No.	Name of Analysis	Minimum Quantity Required for Analysis in Gram	Charges ₹	No. of Days Required[f]
41	Organophosphorous pesticides residues - (I) (chlorpyrifos, dimethoate, disulfoton ethion, methyl parathion, phorate, parathion and quinalphos)	200[e]	2000	6
42	Organophosphorus pesticides residues (II) (triazophos, methyl pirimiphos, phosalone, monocrotophos)	200	2000	6
43	Pyrethroid pesticides residues (cypermethrin and fenvalerate)	200[e]	1000	6
44	Pesticide residues, single compound (any of the above)	200[e]	800	6
45	Picrocrocine	10	200	3
46	Pipeline (ASTA method)	100[e]	250	4
47	Residual solvent	200	500	3
48	Safranal	10	180	3
49	Salmonella	100	420	6
50	Shigella	100	350	6
51	Starch	100	300	6
52	*Staphylococcus aureus*	100	420	6
53	Sudan-1 Dye (HPLC method)	100	750	3
54	Sulphite reducing Clostridia	100	600	5
55	Sulphur dioxide	200	300	4

(Continued)

Table A5.1 (*Continued*)   SBI Quality and Evaluation Laboratory Analytical Services

Sr. No.	Name of Analysis	Minimum Quantity Required for Analysis in Gram	Charges ₹	No. of Days Required[f]
56	Thermostable bacteria	100	175	3
57	Total ash	100	150	4
58	Total plate count	100	150	3
59	USFDA specifications	500/250 × 10 nos[b]	[a]	6
60	Volatile oil	200[e]	100	3
61	Vanillin	30	250	3
62	Water soluble ash	100	200	6
63	Whole insects, dead (by count)	250	100	1
64	Yeast (microbiological)	100	120	5
65	Yeast and mould (microbiological)	100	220	5

Source: Spice Board of India. Quality Evaluation Laboratory. Issue No. 9, Oct 21, 2003.

Note: Depends on the number of parameters for individual spices and number of samples.

[a] Charges based on the parameter specified for each spice.
[b] In the case of analysis for ASTA/USFDA parameters, sample size should be a minimum of 500 g for heavier items like pepper, ginger, tumeric etc. and 250 g sample for low density items like chilli and seed spices.
[c] Quantity for light filth analysis for Tamarind concentrate by AOAC method should be a minimum of 600g.
[d] Analytical charges ₹120/- per element.
[e] In the case of analysis of oleoresin samples of spices for the above parameters a minimum of 50 g. sample size is required and in case of coriander and herbs a sample size of 300 g is required.
[f] No. of working days required for analysis.

The samples may be sent to the Senior Scientist (QC), Quality Evaluation Laboratory, Spices Board, Sugandha Bhavan, Palarivattom, Cochin-682 025, along with analytical charges in the form of cash/D.D. drawn in favour of secretary, Spices Board.

Table A5.2   Tolerance Limits for Aflatoxin for Foods and Spices

Sr. No.	Country	Commodity	Aflatoxin B1 (µg/kg)	Total Aflatoxin (µg/kg)
1	EU	Food	2	4
		Spices	5	10
2	Australia	Peanuts, Tree nuts	–	15
3	Bangladesh	In preparation a limit of	100 µg/kg	Used in Practice
4	Canada	Nuts and Nut products	–	15
5	China	Peanuts and Maize	20	–
6	India	All Food products	–	30
7	Indonesia	Spices, Peanuts	–	20
8	Japan	All Foods	10	–
9	Malaysia	All Foods	–	35
10	Pakistan	No Regulations	–	–
11	Sri Lanka	All Foods	–	30
12	UK	Food	2	4
		Spices	5	10
13	USA	All Foods	–	20

Source: Spices Board of India.

BOX A5.9   SBI CIRCULAR: MANDATORY TESTING FOR AFLATOXIN

Attention Chilli/Chilli Product Exporters

No. MD/CHI/01/03/   November 12, 2003

CIRCULAR

Samples drawn from export consignments of chilli powder and other products containing chillies, in terms of our Circular No. MD/CHI/01/03 dated 9th October 2003, would be analyzed for presence of Aflatoxin also.

In view of the need for ensuring that chillies exported out of the country do not contain Aflatoxin above tolerance limit, the scope of the Circular No. MD/CHI/01/03 is extended to all verities of whole chillies to be exported with effect from 17.11.2003.

(S. Kannan)
Director (Marketing)

Source: Kannan, S. Circular MD/CHI/01/03: Mandatory testing for aflatoxin. Spices Board of India, Nov. 12, 2003. Accessed Nov. 15, 2005 from http://www.indianspices.com

BOX A5.10   HIGH PROFILE LIQUID CHROMATOGRAPHY

What is HPLC?

High Performance Liquid Chromatography (HPLC) is a chemistry based tool for quantifying and analysing mixtures of chemical compounds.

With today's advanced instrumentation and chemistry supplies, capabilities have increased – earning the High Performance name.

HPLC is an analytical technique for the separation and determination of organic and inorganic solutes in any samples especially biological, pharmaceutical, food, environmental, industrial, etc. In a liquid chromatographic process, a liquid permeates through a porous solid stationary phase and elutes the solutes into a flow-through detector. The stationary phase is usually in the form of small-diameter (5–10 mm) uniform particles, packed into a cylindrical column. The typical column is constructed from a rigid material (such as stainless steel or plastic) and is generally 5–30 cm long and the internal diameter is in the range of 1–9 mm.

What is HPLC used for?

It's used to find the amount of a chemical compound within a mixture of other chemicals. An example would be to find out how much caffeine there is in the cup of coffee (or tea, or cola) I have to get me moving in the morning.

How is the sample analysed?

Dissolve the sample in a solvent (like water or alcohol), thus the term LIQUID chromatography.

How is the sample amount measured?

A detector measures response changes between the solvent itself, and the solvent & sample when passing through it. The electrical response is digitized and sent to a data system. The graphic below shows basic system components:

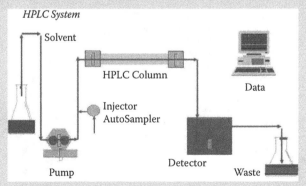

HPLC System

(Continued)

BOX A5.10 (*CONTINUED*) HIGH PROFILE
LIQUID CHROMATOGRAPHY

- **Solvent Delivery System** - pushes the solvent stream through the instrument at constant flow rate
- **Auto sampler** - introduces the sample into the liquid stream of the instrument
- **Column** - a stainless steel tube packed with silicon beads that separates what I'm looking for (the caffeine) from other compounds (like sugar)
- **Detector** - An optical sensor (usually) that detects changes in the characteristics of the solvent stream
- **Data System** - A means of controlling the system components and storing, processing and displaying data

A high pressure pump is required to force the mobile phase through the column at typical flow rates of 0.1–2 ml/min. The sample to be separated is introduced into the mobile phase by injection device, manual or automatic, prior to the column.

Mass Spectrometry (MS)

Principle of Mass Spectrometry

The mass spectrometer is an instrument designed to separate gas phase ions according to their m/z (mass to charge ratio) value.

The "heart" of the mass spectrometer is the analyser. This element separates the gas phase ions. The analyser uses electrical or magnetic fields, or combination of both, to move the ions from the region where they are produced, to a detector, where they produce a signal which is amplified. Since the motion and separation of ions is based on electrical or magnetic fields, it is the mass to charge ratio, and not only the mass, which is of importance. The analyser is operated under high vacuum, so that the ions can travel to the detector with a sufficient yield.

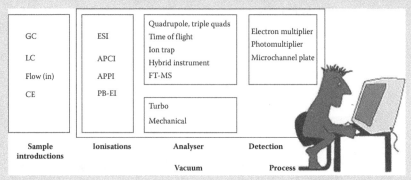

Sample introductions	Ionisations	Analyser	Detection
GC	ESI	Quadrupole, triple quads	Electron multiplier
LC	APCI	Time of flight	Photomultiplier
Flow (in)	APPI	Ion trap	Microchannel plate
CE	PB-EI	Hybrid instrument	
		FT-MS	
		Turbo	
		Mechanical	
		Vacuum	Process

(Continued)

BOX A5.10 (*CONTINUED*)   HIGH PROFILE
LIQUID CHROMATOGRAPHY

In addition to the analyser, the mass spectrometer also includes:

■ A vacuum system
■ Tools to introduce the sample (LC, GC)
■ Tools to produce the gas phase ions from the sample molecules
■ Tools to fragment the ions, in order to obtain structural information, or to get more selective detection
■ A detection system
■ Software and computing.

MS/MS

It is the combination of two or more MS experiments. The aim is either to get structure information by fragmenting the ions isolated during the first experiment, or to achieve better selectivity and sensitivity for quantitative analysis.

MS/MS is done

■ either by coupling multiple analysers (of the same or different kind) or,
■ with an ion trap, by doing the various experiments within the trap.

Principles of LC/MS

LC/MS is a hyphenated technique, combining the separation power of HPLC, with the detection power of mass spectrometry. Even with a very sophisticated MS instrument, HPLC is still useful to remove the interferences from the sample that would impact the ionization.

Closely related to LC/MS are some other techniques, like flow injection/MS, CE or CEC/MS, capillary LC or nano LC/MS.

In all cases, there is a need for an interface that will eliminate the solvent and generate gas phase ions, and then transferred to the optics of the mass spectrometer.

Most instruments now use atmospheric pressure ionization (API) technique where solvent elimination and ionization steps are combined in the source and take place at atmospheric pressure.

When electron impact ionization (EI) is the choice, the solvent elimination and ionization steps are separate.

(Continued)

BOX A5.10 (*CONTINUED*)   HIGH PROFILE LIQUID CHROMATOGRAPHY

The interface is a particle beam type, which separates the sample from the solvent, and allows the introduction of the sample in the form of dry particles into the high vacuum region.

Electron impact is of interest for molecules which do not ionize with API technique, or when an electron impact spectrum is necessary, since it provides spectral information independent of the sample introduction technique (GC or LC, or direct introduction) and instrument supplier.

Source: Waters Corporation. Website. 2005. Accessed Nov. 15, 2005 from http://www.waters.com

customers would accept analytical reports issued only by some designated overseas labs. To them, the analytical reports issued by SBI were useless. In addition, the mandatory testing had been costing them millions of rupees annually. Apart from this, Jose had been under pressure from some influential quarters to revoke the suspension orders placed on the alleged firms. However, he chose not to wilt. However, at the same time, he was not sure how long he would be able to withstand the pressure. Then occurred an event which had the biggest impact on this whole episode.

On February 7, 2005, Premier Foods, UK, informed the FSA that a customer sample of Worcester sauce (Crosse & Blackwell) in Italy had tested positive for Sudan I. On February 10, 2005, it confirmed that Crosse & Blackwell Worcester Sauce and a 2002 batch of chilli powder tested positive for Sudan I. Worcester Sauce was used as a flavouring agent in the manufacture of over 450 supermarket and top brand products. On February 24, 2005, the FSA published a list of 474 food products with public warning not to consume them. This triggered the biggest British food product recall off the shelves of supermarkets and departmental stores (Box A5.11).

The maze of middle-men, legal contracts, and export houses that work the world's commodity markets, made it almost impossible to trace with certainty the source of contaminated products. However, *The Guardian*, a leading newspaper of the UK, could establish the tortuous supply chain. Gautam Exports Corporation supplied a five ton consignment of chilli powder adulterated with Sudan I to EW Spices, an importer of the UK. EW Spices supplied it to East Anglia Food Ingredients. The chilli was then sold to another company called Unbar Rothon, which in turn supplied it to Premier Foods. Premier Foods added the contaminated chilli to Worcester sauce. The contaminated batch sold to Premier Foods had slipped through the lot.

BOX A5.11   NEWS ARTICLES REGARDING
THE PRODUCT RECALL

PRESS TRUST OF INDIA
Saturday, February 19, 2005

Cancer-causing dye found in Indian chilli powder

A total of 359 food products have been taken off shop shelves in Britain after they were found contaminated with an illegal food dye and three companies in India dealing in chilli powder have been identified as the source of the contamination, the Food Standards Agency said.

The Sudan I dye, linked to an increased risk of cancer, was in the chilli powder used by premier foods to make a Worcester sauce and a range of popular pasta sauces, used in other products, the agency said. According to BBC, three companies in India have had their license to trade suspended by the Indian spice board pending further investigation. Sudan I, a red dye used for colouring solvents, oils and waxes, is banned for use in foodstuffs in the UK and across the EU. Since July 2003, all chilli powder imported into the UK has to be certified free of Sudan I. The brands affected include Bertolli as well as own-label sauces made for supermarket chains Sainsbury's and Safeway. Sudan I is normally used for industrial purposes in solvents, petrol and shoe polish and is banned in food. The Food Standards Agency (FSA) has issued a warning advising people not to eat the products but said there was "no need to panic" because of the "very low risk".

Premier foods, in a statement said "the product withdrawal is a purely precautionary measure and the levels at which Sudan I occurs in the products concerned presents no immediate risk to health." Dr. Julie Sharp, of cancer research UK, said the people who had already eaten foods that had been contaminated had no reason to panic. "The risk of cancer in humans from Sudan I has not been proven and any risk from these foods is likely to be very small indeed," she said. The FSA is working with the industry and local authorities to ensure that any remaining affected foods are removed from the sale. It was first discovered in a consignment of the Worcester sauce exported to Italy. But authorities in France first raised the alarm in May when Sudan I was discovered in UK-made chutneys and relishes. As a result, the EU ordered member states to examine imports of ground chilli in addition to random tests on products. A spokesman for Unilever Best foods, which makes Bertolli, said: "The dye was identified by our supplier in Italy during quality assured checks and they notified us."

"The brand is certainly well known but the one we have named is the only product with chilli powder in it."

(Continued)

BOX A5.11 (*CONTINUED*) NEWS ARTICLES REGARDING THE PRODUCT RECALL

THE TIMES OF INDIA
19 Feb 2005 1641 hrs IST

Indian red chilli has UK sweating

LONDON: A carcinogenic spice trail that began in several chilli grinding, packaging and exporting companies in India nearly three years ago has ended in a 15-million-pound food recall across Britain, the largest in its history, and caused panic to spread across at least 15 countries that imported foods contaminated with the banned, cancer-causing chemical colourant SUDAN-1.

The recall of hundreds of food products, ranging from British staples such as shepherd's pie, pasta bake, chicken wings, pizza and pasta sauces, was ordered by the UK's independent watchdog, the Food Standards Authority (FSA) late on Friday. The FSA ordered the destruction of all suspect foodstuffs manufactured over the past 16 months using a five-tonne consignment of red chilli powder from India.

Sources said the powder was thought to have been supplied by at least two western Indian companies. The powder arrived in the warehouses of Unbar Rothon in the English county of Essex and subsequently found its way into a well-known British brand of Worcester sauce manufactured by another company, Premier Foods.

On Saturday, Unbar Rothon admitted to STOI that it was "investigating the matter." But a close-lipped company source insisted, "We have nothing to tell you yet because there is nothing to tell."

The chilli powder, suspiciously red from the addition of SUDAN-1, was added to the Worcester sauce, which was used to flavour popular British and European ready meals, long-life foods, frozen suppers and condiments.

The sauce, with its SUDAN-1 flavouring, also found its way to the US and Canada and several European countries, including Italy.

With panic-stricken Western food standards authorities under enormous public pressure to trace the contaminated food, identify the spice chain, punish and possibly prosecute, the presence of SUDAN-1 in Indian chilli exports is thought to be deeply damaging.

In trenchant criticism and posing a huge question for the future of Indian chilli exports to the West, British food campaigner Joanna Blythman said there was "no excuse" for a substance like Sudan 1 finding its way into food.

(Continued)

BOX A5.11 (*CONTINUED*)   NEWS ARTICLES REGARDING THE PRODUCT RECALL

The FSA told this paper that the dye is normally used in shoe polish, colouring solvents, oils, waxes and petrol. It has been shown to cause cancer in mice and in tests on human liver cells.

On Saturday, food industry experts said the adverse publicity for Indian chilli powder could knock off-balance the recent proud boast that chilli exports are a sunshine part of the spice trade. In November, the Spice Board of India made the self-congratulatory revelation that chilli exports had increased from 38,000 tonnes in April-October 2003 to 54,000 tonnes over the same period last year.

India exports spices to 120 countries. But trade experts have long pointed out the competitive challenges India faces from other spice-exporters countries, including Vietnam, Indonesia and Guatemala, unless it speedily enforces a nation-wide ban on the manufacture and sale of unregistered pesticides and the spurious use of synthetic dyes to bulk up spice quantities.

The Board's data showed that total spice export clearances were exceedingly strong with shipments last September nearing the 30,000 metric tonnes-per-month barrier, a whopping increase on September 2003's total of just 18,000 metric tonnes.

On Saturday, industry pundits were asking how long it could last without active Indian spice controls and image-building.

The British, European and North American SUDAN-1 food alert comes just 18 months after the European Union (EU) enforced stringent emergency measures to regulate the import of chilli powder from India.

The damning Brussels directive said that SUDAN red I "may be a genotoxic carcinogen (and) may also exert sensitizing effects by dermal route or inhalation".

Just days after the EU's Indian chilli red alert, leading British companies using Indian spices in food manufacture or supplying them to other vendors, including Shaws and BE Foods International, admitted they might be forced to seek "alternative sources" for chillies and other spices.

Rashmee Z Ahmed, TNN

(Continued)

BOX A5.11 (*CONTINUED*)   NEWS ARTICLES REGARDING THE PRODUCT RECALL

NDTV PROFIT.COM
Sunday, February 20, 2005 (London):

Indian chilli pulled off supermarket shelves in UK

Red chilli powder manufactured by three Indian companies was pulled off shelves across UK after the search for carcinogenic elements in spices unearthed alarming results.

A total of 359 food products, including fast food items such as pastas and pizzas, have been taken off from supermarket shelves across London as they are found to carry the Sudan I dye.

This dye, which leads to a high cancer risk, was found in commonly used foods as an ingredient in sauces. Sudan I is also said to be used in shoe polish, oils and waxes.

"The risks associated with these foods are extremely small, but it's clearly unacceptable for an illegal food dye to be in any food," said Dr. Jon Bell, Chief Executive, Food Standards Agency.

"If consumers have got any of these products in their cupboards in their houses, then they should throw them away or return them to where they bought them from," he warned.

The trade licenses of the companies supplying this ingredient have been temporarily suspended by the Indian Spice Board.

Looks like the Indian chilli, famed for its superior quality, needs to prove just how hot it is across the European Union.

NDTV Correspondent

THE TIMES OF INDIA
21 Feb 2005, 2323 hrs IST,

UK mulls action against suppliers

LONDON: Seventy-two hours after the biggest food scare in British history, the message from the Food and Drink Federation (FDF), which represents the UK's largest manufacturing sector: long-term "trusted" suppliers of the contaminated chilli powder are to blame because tests are not routinely conducted for the banned dye SUDAN-1.

(Continued)

**BOX A5.11 (*CONTINUED*)   NEWS ARTICLES
REGARDING THE PRODUCT RECALL**

Even as a flurry of stringent legal claims were mooted in a bid to reclaim the millions of pounds being spent on recalling and destroying expensive food, the FDF's comments appeared to mark a new mood to punish, prosecute and penalize long-term spice suppliers to the UK, notably Indian companies.

Experts said on Monday that many established spice exporters to the UK might face the chop unless they demonstrably cleaned up their act. The FDF's comments are considered significant because it represents key British players of an industry that exports nearly £10 billion of food and drink worldwide.

Carcinogenic SUDAN-1's presence in a five-tonne consignment of chilli powder imported from India in 2002, sparked a several-million-pound food recall across the UK and triggered panic across the US, Canada, New Zealand, Italy, France and other European Union countries. SUDAN-1 has been illegal in Britain for a decade but the contaminated, three-year-old Indian import is thought to be the first conclusive proof that Indian spice suppliers have continued to break the rules.

London lawyers said on Monday that they expected a blizzard of product liability claims "up the supply chain", possibly leading all the way back to India. But there was no indication how and if legal claims against Indian companies could be pursued, in practice. The cost of removing an estimated 360 food items from British supermarket shelves, including popular pies, pasta sauces, pizzas, casseroles, sausages and sauces, is estimated to be at least £15 million.

In a knee-jerk fear at the prospect of footing part of the bill, at least some of the food companies that bought the contaminated Indian chilli powder, have begun to distance themselves from all responsibility.

In a game of pass-the-parcel, several British spice importers refused to admit to the initial import of chilli powder from India. One company, East Anglian Foods Ingredients, from Essex, near London, admitted to first handling the chilli powder but claimed it bought it from another source.

The chilli powder was also handled by spice and herb specialists Unbar Rothon, which insisted the "powder came straight in and straight out and we did not process it".

The chillies were then sold to Premier Foods, which fatefully used it to make the Worcestershire sauce that flavoured whole fridge-full of chilled meals, heat-and-eat suppers and snacks. On Monday, Premier Foods claimed it was "misled" by Unbar Rothon and is seeking legal advice.

(*Continued*)

BOX A5.11 (*CONTINUED*)   NEWS ARTICLES REGARDING THE PRODUCT RECALL

But because both Unbar Rothon and East Anglia Food Ingredients deny preliminary responsibility, legal experts said the blame could ultimately be laid to rest on the shoulders of Indian suppliers.

They pointed out that the contaminated chilli powder arrived in Britain despite the Indian Spice Board's acceptance of European Spice Association requirements, which specify that imports be "free from adulteration".

Late on Monday, a spokesperson for the British Retail Consortium said the chilli scare was affecting small convenience stores and large supermarket chains alike. Even fast-food chain McDonalds is affected because a low-fat Caesar salad dressing is seasoned with products flavoured with that initial, five-tonne Indian consignment.

Meanwhile, the FDF's deputy director-general Martin Paterson also controversially appeared to offer a loophole to suppliers of suspect products with the admission that checks for the banned dye SUDAN-1 were not routine on foodstuffs containing imported chilli powder.

SUDAN-1 is normally used as a colouring in solvents, oils, waxes, petrol, and shoe and floor polish.

Rashmee Z Ahmed, TNN

The Guardian
Tuesday February 22, 2005

Watchdog under fire as cancer-dye foods top 400

More than 400 well-known processed foods are to be removed from sale because they are contaminated with an illegal red dye, which can cause cancer, it emerged yesterday.

The Food Standards Agency added 59 more products through the day to its list of 359 adulterated foods first published last Friday.

The agency also warned that the crisis, already the largest food alert since BSE, was likely to get worse, as it came under attack for failing to prevent the lapse in food safety and for taking too long to make the information public.

The FSA's chief executive, Dr. Jon Bell, admitted that the carcinogenic dye, Sudan 1, had been found in Crosse and Blackwell Worcester sauce only by chance.

(*Continued*)

BOX A5.11 (*CONTINUED*)   NEWS ARTICLES REGARDING THE PRODUCT RECALL

It had been used to colour a batch of chilli powder used as an ingredient in the sauce. The sauce in turn was sold on to hundreds of food companies for manufacture into famous brands of food and supermarket ready meals.

Dr. Bell also acknowledged that the adulteration of chilli powder may have been going on undetected for years. The French authorities first raised the alarm about imports of chilli with Sudan 1 in 2003.

The contamination of the Crosse and Blackwell Worcester sauce made by Premier Foods was not picked up by supermarkets, food manufacturers or enforcement authorities in the UK, but by a laboratory in Italy on January 28.

Premier Foods told the FSA of its concerns on February 7 but it was not until February 18 that an initial list of products affected was posted on the FSA website. Supermarkets meanwhile are understood to have received information a few days before the public.

The conservative health spokesman, Chris Grayling, accused the FSA of taking its "eye off the ball".

"I think it is all the more regrettable that consumers were left for a period of up to 10 days still buying these products, unaware that there was an issue," he said.

The full scale of the contamination is still being uncovered. The FSA said that over 300 food companies were involved in the effort to trace how far the Worcester sauce had spread.

Those identified so far include leading supermarkets Tesco, Asda, Sainsbury's, Morrisons, Waitrose, M&S, and top brand owners such as crisp makers Walkers, Unilever, Heinz, McDonald's and Schweppes. The FSA said it could not guarantee that there was not more adulterated chilli in circulation.

Premier Foods declined to comment yesterday. It said on Friday that it had certificates from its suppliers that guaranteed that the chilli it used was free of Sudan 1. The FSA could not say who had issued these certificates or whether they had come from a properly accredited laboratory.

An FSA spokesman said Premier Foods obtained the chilli powder from Essex-based spice and herb specialist Unbar Rothon which in turn received it from East Anglian Food Ingredients, also in Essex. Both these companies are brokers who may not have handled the chilli directly themselves. Unbar Rothon declined to comment last week but EAFI said: "Where contamination could be established we recalled stock."

(Continued)

BOX A5.11 (*CONTINUED*)   NEWS ARTICLES REGARDING THE PRODUCT RECALL

Sudan 1 is an azo dye, known to be carcinogenic in animal tests. It is not permitted for use in food in the EU. It has also been found as an adulterant in samples of palm oil imported into the UK.

The FSA said Sudan 1 presented no risk of immediate illness but could increase the risk of cancer, although the risk "generally is likely to be very small". The French food authority which uncovered the Sudan 1 in chilli powder from the UK said it "cannot exclude the possibility of a risk to human health even at low doses... all measures should be taken to ensure that the consumer is not exposed to this substance".

FSA
Thursday 24 February 2005

Sudan I timeline

Sudan I was first discovered in adulterated chilli products in May 2003. Since then the Food Standards Agency has been working to ensure that it is kept out of the food chain. Read our timeline to find out what happened between May 2003 and the major recall of products in February 2005.

Friday 9 May 2003
France informs the European Commission that testing has detected the illegal dye Sudan I in hot chilli products originating in India.

Wednesday 30 July 2003
UK implementation of European Commission emergency measures comes into force covering the import of hot chilli and hot chilli products from India. All consignments must be accompanied by a certificate showing that they have been tested and are free from Sudan I. If there is no certificate, they are to be tested at the port of entry to the EU. Member states are asked to carry out random sampling and analysis of consignments at import and products already on the market.

Thursday 31 July 2003
The Food Standards Agency issues its first food hazard warnings (now called food alerts) on Sudan I found in products containing hot chilli. Local authorities are asked to ensure that 25 products are removed from sale.

(*Continued*)

**BOX A5.11 (*CONTINUED*)   NEWS ARTICLES
REGARDING THE PRODUCT RECALL**

Monday 15 September 2003
The Agency issues guidance to industry asking companies to withdraw and recall contaminated products as soon as they are identified. The FSA also asks all manufacturers who use chilli powder or chilli products to ensure that it has not been contaminated with Sudan dyes if it was imported from India and supplied to them prior to 30 July 2003.

The guidance adds: 'These companies should consider sampling relevant batches in order to ensure that their products are not contaminated. The FSA would like to be notified of all results whether they are positive or negative, in order to obtain as much information as possible about the distribution of contaminated chilli products.'

Wednesday 18 February 2004
The Agency issues updated guidance to industry, asking companies to withdraw and recall contaminated products as soon as they are identified. The FSA also asks all manufacturers who use chilli powder or chilli products to ensure that it has not been contaminated with Sudan dyes if it was imported from India and supplied to them prior to 27 January 2004. The guidance repeats the Agency's request for companies to sample relevant batches in order to ensure that their products are not contaminated and to notify the Agency of all results.

31 July 2003–31 January 2005
In the 18 months since it first became aware of the adulteration of chilli products with Sudan dyes, the Agency issues 56 food hazard warnings/food alerts on Sudan dyes, involving more than 200 products.

Monday 7 February 2005
Premier Foods informs the FSA that a customer sample of Worcester Sauce in Italy had tested positive for Sudan I (Crosse & Blackwell Worcester Sauce). Premier Foods advises the Agency that it will carry out further tests to check the results.

Wednesday 9 February 2005
Premier Foods informs the FSA that five products could be affected.

(Continued)

BOX A5.11 (*CONTINUED*) NEWS ARTICLES REGARDING THE PRODUCT RECALL

Thursday 10 February 2005
Test results are received by the FSA and it is informed that Crosse and Blackwell Worcester Sauce and a 2002 batch of chilli powder have tested positive for Sudan I.

Friday 11 February 2005
Premier Foods informs the FSA that it is compiling a list of customers for the five products.

Monday 14 February 2005
Premier Foods provides the FSA with a list of more than 160 customers in the UK. Hundreds of products are potentially affected.

Tuesday 15 February 2005
The FSA meets with representatives of the food industry and requires full disclosure by companies of their affected products, their removal from sale, and appropriate publicity to inform consumers.

Wednesday 16 February 2005
The FSA meets again with food industry representatives and repeats its demand for detailed product information.

Thursday 17 February 2005
The bulk of the information on the affected products is received in the evening via the British Retail Consortium.

Friday 18 February 2005
Further product information continues to arrive. A press release is issued by the FSA and a list of some 360 products placed on the Agency website at 1.30 pm.

Monday 21 February 2005
Sixty products are added to the list, a second press release is issued, a further meeting with retailers is held, and a deadline of Thursday 24 February is set for the removal from sale of all contaminated products.

Tuesday 22 February 2005
Nine additional products are added to the list of food products contaminated with the illegal dye Sudan I.

(Continued)

BOX A5.11 (*CONTINUED*)   NEWS ARTICLES REGARDING THE PRODUCT RECALL

Thursday 24 February 2005
The Agency announces that over the past week it has coordinated with local authorities in further inspections of companies and factories allegedly involved in the supply of foods contaminated with the illegal dye Sudan I. The investigation is the largest of its kind ever undertaken in the UK. The information will form part of the Agency's wider investigation to establish how this failure in the industry's legal duty to provide safe and fit food happened. The Agency's advice to local authorities is that they will need to consider enforcement action on the basis of the evidence they collect.

Thursday 24 February 2005
The Agency adds 146 more products to the list of food products contaminated with the illegal dye Sudan I, taking the total to 474. The Agency says that the vast majority of products contaminated with Sudan I have been removed from sale.

Tuesday 8 March 2005
The Agency is advised of a further 43 products that companies have withdrawn from the food chain as a result of Sudan I contamination, bringing the total to 580.

The Times
February 26, 2005

Premier Foods faces £100 m bill for Sudan 1

PREMIER FOODS, the company at the centre of the chilli powder contamination scandal, faces a potential £100 million bill after experts gave warning that insurers were unlikely to cover the cost of Britain's biggest product recall, *The Times* has learnt.

Carcinogenic contaminants such as Sudan 1, the colourant used in the chilli powder, are excluded from standard product recall insurance cover, the world's largest insurance brokers said yesterday.

David Palmer, director of Aon Crisis Management, the broker, said: "This particular incident isn't a covered incident, as far as I'm aware. Policies cover contaminants that manifest themselves within 120 days, not years later."

(Continued)

**BOX A5.11 (*CONTINUED*) NEWS ARTICLES
REGARDING THE PRODUCT RECALL**

While another broker said that it was possible to buy specialized policies from offshore insurers that cover a company against being forced to recall a product at the behest of a regulator, he added that "this isn't the type of policy the average small-to-medium food manufacturer would have".

AIG, the American insurer, is believed to have underwritten Premier's product-recall policy. The US group declined to comment yesterday.

The Food Standards Agency, the watchdog, has ordered the recall of more than 470 products that contain Worcester sauce supplied by Crosse & Blackwell. Premier Foods used Sudan 1 contaminated chilli powder in the manufacture of the sauce, which it supplied to a number of supermarkets and food producers, which in turn used it as an ingredient in their own products.

Brokers said that a large-scale product recall could cost up to £50 million. However, because of the unprecedented scale of the Sudan 1 contamination, the total bill could come to as much as £100 million, they added.

Premier's market capitalization is about £686 million.

A number of supermarkets, including Tesco, Somerfield, Kwik Save and the Co-op, said yesterday that they would recoup their recall costs from their supplier, which was Premier.

Premier is expected to bear the brunt of the recall bill but has said that it will attempt to pass on the cost to its small, privately owned suppliers.

"We believe the responsibility for any financial costs associated with the recall will rest with our suppliers and their insurers," the company said in a statement on February 18.

It later added: "In any event, the company is insured against such eventualities."

However, when asked yesterday whether its own product recall cover insured the group against carcinogenic contaminants, Premier declined to comment.

Christine Seib

BOX A5.12   GROUP TESTING

Suppose 1,000 people are subjected to a blood test for testing the presence of syphilitic antigen. This can be administered in two ways. (i) Each person can be tested separately. In this case, 1000 tests are required. (ii) The blood samples of 10 (any suitable number) people can be pooled and analysed together. If the test is negative, this one test suffices for the 10 people. If the test is positive, each of the 10 persons must be tested separately, and in all, 11 tests are required for 10 people. Assume that the syphilis prevalence rate is 0.001 and the chance that the test would be positive is same for all people. Further, the chance of a person being positive in no way influences the others' being detected positive. Also assume that the cost of testing a single sample is ₹1,000. Then on the average, the total cost of testing 10 people in (ii) above can be computed as follows.

$$\text{Total Cost (on the average, in ₹)} = 1{,}000 * (0.999)^{10} + 11{,}000 * (1-(0.999)^{10})$$
$$= 990 + 11{,}000 * (1-0.99)$$
$$= 990 + 110$$
$$= 1{,}100$$

So on the average the cost per person would be ₹110

Suppose instead of 10, if 20 blood samples are pooled, then the average per person cost would be calculated as above.

$$\text{Total Cost (on the average, in ₹)} = 1{,}000 * (0.999)^{20} + 21{,}000 * (1-(0.999)^{20})$$
$$= 980.20 + 11{,}000 * (1-0.9802)$$
$$= 1{,}396$$

Per person average cost (in ₹) = 1,396/20 = 69.8

If we pool 100 blood samples the same calculation shows

Per person cost (in ₹) = 104.26.

Thus it shows that, for a given prevalence rate if group sizes vary (number of samples to be pooled) the cost per sample varies. In fact, the 'per person' average cost if plotted against group size, gives a curve that looks like the following. Thus, a group size exists for which the cost attains minimum value. In this case it is found to be 32.

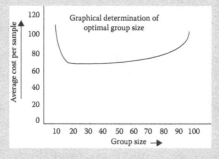

Source: Dorfman, R. The detection of defective members of large populations, *Annals of Mathematical Statistics* 14, 436–440, 1943; Feller, W. *An Introduction to Probability Theory & Its Applications*, New York: Wiley, 1950, p. 189.

BOX A5.13   RANDOM SAMPLING VS. SYSTEMATIC SAMPLING

Samples are to be taken to determine the Aflatoxin level in a lot of, say, 25 tonnes of chilli powder. Suppose the chilli powder is packed into two thousand five hundred 10 kg packs. The total weight of the aggregate sample should be 10 kg and the number of incremental samples should be 100. Suppose a further sample is collected in one of the following ways:

1. 100 packs are selected at random without replacement and from each pack, 100 grams of chilli powder are taken out from a few randomly selected locations.
2. First, select a pack at random from the first 25 packs and then every 25th pack is selected. For example, if the first pack selected is 10th, then the next 99 packs will be 35th, 60th,...., 2485th.

(We assume here throughout that the number of incremental samples is less than the number of packs which is the case in almost all situations in practice.) In sampling literature, the first scheme is known as random sampling and the second scheme as systematic sampling.

Note, the estimate of average Aflatoxin content in the lot per 100 g = (sum of Aflatoxin contents in the incremental samples)/100

It can be shown:

(i) In case of random sampling, standard error of the estimate

$$= \text{lot SD}/\sqrt{100} = \text{lot SD}/10$$

(Lot SD = Standard deviation of Aflatoxin content in all 25,000 possible units of 100 g chilli powder each)

(ii) In case of systematic sampling, standard error of the estimate

$$= (\text{lot SD} / \sqrt{100} * \sqrt{(1 + (100 - 1) * R}$$

where R represents a measure of heterogeneity which is precisely the correlation coefficient between all pairs of elements in a systematic sample. R lies between $-1/(100-1)$ to 1. R is negative in case the elements within a systematic sample tend to be extremely different. In this case, systematic sampling will naturally be better than random sampling. On the other hand, if R is positive then the elements within a sample are fairly homogeneous and hence systematic sampling yields a higher standard error than random sampling.

Source: Scheaffer, R.L., Mendenhall, W., and Ott, R.L. *Elementary Survey Sampling*, Boston: Duxbury Press, 1995, p. 256.

BOX A5.14   SAMPLING PROCEDURES CONFORMING TO EU NORMS

In super-cession of previous directions issued by the SBI, the following sampling procedure would apply to drawing of samples under the mandatory pre-shipment inspection put in place by SBI under the SBI (Registration of Exporters) Amendment Regulations 2004:

Separate set of samples are to be drawn from each lot (of 25 tons) of the commodity offered for sampling provided that if the lots offered for sampling at any given time are different in terms of origin, variety, type of packing, packer, consignor or markings, they have to be considered as separate lots. For example,

(a) If the commodity offered for sampling distinguished by origin, variety, type of packing, packer, consignor or markings exceeds a maximum weight of 25 tons, each 25th ton should be considered as a separate lot.

(b) If the commodity offered for sampling consists of different varieties of products, each variety should be considered to be a separate lot if the respective weights are 25 tons or less.

(c) If the total weight of any given variety for example, is 35 tons, the first 25 tons will form the first lot and the remaining 10 tons will form the second lot.

Sampling Frequency

From a given lot, the quantity of material to be drawn for constituting the aggregate sample will be determined on the basis of sampling frequency determined by the following formulae:

Sampling Frequency (SF)

$$= \frac{\text{Weight of a lot (in kg)} * 0.1}{\substack{\text{Required weight of the aggregate sample} * \\ \text{Weight of the unit of packing}}}$$

The weight of the material to be drawn from each unit (incremental sample) (carton, bag, retail pack or any other packing unit) will be 100 g.

Thus if the offered and identified lot has a weight of 18 tons and if the material is packed into 50 kg packs, the sampling frequency would be

$$= \frac{\text{Weight of the lot (in kg)} (18000 * 0.1)}{\substack{\text{Required weight of the aggregate sample (in kg)} (10) * \\ \text{Weight of the unit of packing in kg} (50)}} = 3.6$$

(Continued)

BOX A5.14 (*CONTINUED*) SAMPLING PROCEDURES CONFORMING TO EU NORMS

In this case, incremental sample of 100 grams has to be drawn from every third or fourth bag and number of bags to be opened for making the aggregate sample would be 100. If on the contrary, the total weight of the lot is 18 tons and the unit weight is 25 kg (material packed in 25 kg bags), the sampling frequency would be

$$\frac{18000 * 0.1}{10 * 25} = 7.2$$

Every seventh or eighth bag should be opened. From each bag, 100 grams should be collected and the total weight of the aggregate samples would be 10 kg. If the total weight of a differentiated lot is 10 tons consisting of 800 cartons of 12.5 kg each and each carton contains 120 packs of 100 g each (500 grams being the allowance given for the weight of the empty carton), the sampling frequency will be 10000 * 0.1/8 * 12.5 = 10.

In this case, every tenth carton should be opened and from each carton, two packs of 50 g each shall constitute the incremental sample.

How to determine the required number of incremental samples and required weight of aggregate samples?

The number of incremental sample to be taken is determined by the total weight of the differential lot/sub-lot and is to be calculated as per the following table:

Lot Weight	No of Incremental Samples	Total Weight of the Aggregate Sample
100 kg or less	10	1
More than 100 kg up to 200 kg	15	1.5
More than 200 kg up to 500 kg	20	2
More than 500 kg up to 1 ton	30	3
More than 1 M.T up to 2 tons	40	4
More than 2 tons up to 5 tons	60	6
More than 5 tons up to 10 tons	80	8
More than 10 tons up to 25 tons	100	10

(Continued)

BOX A5.14 (*CONTINUED*)   SAMPLING
PROCEDURES CONFORMING TO EU NORMS

On the contrary, if the individual packing is more than 100 g, instead of picking out the full pack, 100 g each may be drawn from every 8th carton. However, in such cases some discretion can be exercised provided the minimum requirements are met.

Thus depending upon the weight of the lot/sub-lot being sampled, the aggregate weight of the samples drawn would vary between 1 kg and 10 kg.

Preparation of laboratory samples

From such a homogenized aggregate sample, three laboratory samples should be prepared, each consisting of 200 g. One sample of 200 g should be given to the exporter and two samples are to be sent to SBI.

Lots of less than 100 kg

In case the lot weight is less than 100 kg, incremental samples should be drawn from 10% of the bags/cartons/packing units.

Separate set of samples for each separate lot of 25 tons

As stated earlier, in case the total weight of the identified lot is more than 25 tons, a separate set of samples should be drawn from every additional 25 tons or less.

Features, which characterize separate lots

Irrespective of weight, material which can be distinguished on the basis of origin, variety, type of packing, packer, consignor or markings should be treated as separate lots for sampling.

The above-prescribed sampling procedure has come into force for all consignments.

Source: Spices Board of India. *Spices Market Weekly,* Aug. 21, 2005. Ministry of Commerce and Industry, Govt. of India.

The sequence of events brought several issues to the forefront.

(i) What actions should Jose consider in response to this biggest recall of food products?

(ii) The event clearly makes it a point that traceability of food chains should be given utmost importance in the near future. What would be the strategy of SBI to cope up with this added dimension of the event?

(iii) Is it possible to use group testing (Box A5.12) for Sudan dyes to minimize the cost?

(iv) As Aflatoxins are found in a highly heterogeneous manner in a lot, sampling the lots for measuring Aflatoxin should be done carefully. Such being the case, what is more appropriate: random sampling or systematic sampling of the sacks from the lot (Box A5.13 and Box A5.14)?

(v) During the shipment of the chilli powder/product, due to faulty packaging, the presence of moisture in the container would aid the growth of moulds (if present in the product) and thus, enhance the probability of rejection once it reaches the shore. What would be the possible remedies to this problem?

Acknowledgements

Prepared by Professor Tathagata Bandyopadhyay and Professor G. Raghuram, Indian Institute of Management, Ahmedabad.

Assistance and data support provided by Mr. Jose, Chairman, and other senior executives of the Spices Board of India are gratefully acknowledged. Some data has been masked to protect sensitivities.

Case 6   Adani Wilmar Limited

It was August 2000. Sitting in his cabin, Mr. Pakarashi, the Logistics Manager of Adani Wilmar Limited (AWL), was working on the distribution network for the new brand of edible oil to be launched by the company.

Edible oil was still an important commodity in India. Since there was little difference in the raw material and processing costs for different companies, one of the major areas where one could get a competitive advantage was in managing the supply chain. Transportation accounted for approximately 70 per cent of the logistics cost. This meant setting up an optimal distribution network that focused on transportation costs.

AWL was setting up a refinery of 600 tons per day (tpd) capacity at Mundra, a port on the Gulf of Kutch, Gujarat. This was the largest refinery in the country, and was expected to be fully operational by September 2000. (The National Dairy Development Board (NDDB), a cooperative, currently owned the biggest edible oil refinery with a capacity of 250 tons per day.) Refined stocks would have to be distributed to dealers in the markets selected initially in western and northern India. Issues under consideration were the location of warehouses, allocation of dealers to warehouses, and transportation mode choice from the refinery to the warehouse.

Background

The Adani group started as a trading company, mainly focusing on export of commodities. After a fast growth over 10 years, it had a turnover (value of goods traded) of ₹35 billion in 1999–2000, with net profit of ₹1.2 billion. It had recently entered into the infrastructure sector, with the building of Mundra port through a 50–50 joint venture with the government of Gujarat (represented by the Gujarat Port Infrastructure Development Company Limited, promoted by the Gujarat

515

Maritime Board and Gujarat Industrial Investment Corporation Limited). The port was built with an investment of ₹3.9 billion and was designed for a through-put of 1.7 million tons per annum (mtpa). The port started operations in October 1998 and handled a cargo of 293,000 tons in 1999–2000 and 122,000 tons during April to June, 2000. A further investment of ₹4 billion was being made to increase the port capacity, along with a 57 km railway siding for evacuation.

The group had formed a 50–50 joint venture with Wilmar Trading Private Limited (WTPL) of Singapore in June 1999 to enter into the edible oil business. WTPL had a turnover of US$ 2.1 billion in 1999. It was the world's second largest player in edible oil trade, having its own plantations in Malaysia and Indonesia. It also owned a fleet of vessels to transport the products to different countries. Under this joint venture, crude edible oil was to be sourced and imported from Indonesia and Malaysia, refined in Mundra, and marketed in India for domestic consumption. The aim of the joint venture was to exploit the synergies offered by port ownership (Adani Group) and understanding of the international edible oil market (WTPL). In fact, WTPL had approached Adani for the joint venture, as the latter had suitable infrastructure in the form of a private captive port in Mundra.

Market and Competition

AWL was planning to market the refined edible oil in western and northern India, since this would give the company an advantage in logistics because of the ease of servicing these regions from the Mundra port. The company would find it difficult to service the eastern India markets, since higher road freight (in comparison to ocean freight) would not make it competitive, especially if a competitor sought to access this market through an east coast port. For example, road freight from Mundra to Kolkata was around ₹3000/ton, while ocean freight was approximately ₹500/ton from Singapore to west coast India. (Ocean freight to the east coast was only marginally less.) For the same reason, reaching the western and northern India markets through an east coast port (even though the edible oil was being sourced from countries located to the east of India) would not be competitive.

Consumption patterns of edible oils in India differed from one region to another depending on taste and preferences of the people. For example, in the north region, a large variety of oils such as soya, mustard, rapeseed, and sunflower were used, while in the west, groundnut, cottonseed, and soya oils were consumed. In the south region, coconut oil and to a lesser extent sesame oil was the preferred medium. Mustard oil was the primary medium in the east. Edible oil consumption was price elastic. (Geographical demand patterns for various edible oils are shown in Box A6.1.)

The company was planning to have various edible oils in its portfolio. In the short term, it was looking to focus on following distribution: 50 per cent soya, 20 per cent cottonseed, and 30 per cent others (palmolein, rapeseed, etc.) because the demand for these oils was mainly in northern and western India.

BOX A6.1   EDIBLE OIL CONSUMPTION PATTERNS AND GROWTH

Per capita consumption of oil in India has been increasing over the years from 6.8 kg/year in 1991 to about 10 kg/year in 1999–2000. Though this growth was high in recent years, the average was lower than that in most other countries of the world. Annual per capita consumption was 24.4 kg in USA and 29.4 kg in Greece in 1997, while the world average was estimated to be more than 12.0 kg/year.[a]

Edible oil consumption is price elastic, tending to increase with a decrease in price.[b] This indicates that the consumption of oil is set to increase with increase in income, though its share in total consumption would decrease, since its elasticity is only 0.90. In the base scenario of per capita income growing by 4 per cent annually, an average Indian's yearly edible oil requirement is slated to rise from 9.81 kg in 1999–2000 to 16 kg by 2015.[c] If the per capita income growth is going to be 6 per cent, edible oil consumption would go up to 20.60 kg, i.e. twice the current level over the next 15 years.

In recent years, the Indian edible oil market has seen a lot of changes, especially in integration with the world markets. A study showed that the correlation coefficient between Indian and world edible oil prices was as low as 0.65 till 1995, when edible oil was put on OGL. After 1995, the Indian prices generally moved in tandem with the world prices, the differences being mainly because of varying import duties.

India being a vast country, there are specific regional preferences, depending largely upon the oils available in the region. For example, people in the West and South prefer groundnut oil, while those in the East and North use mustard seed or rapeseed oil. Likewise several pockets in the South have a preference for coconut or sesame oil.

Preferences have also changed over time. Groundnut accounted for about three-fourths of consumption in the early 60s followed by rapeseed and mustard. Demand increased with increase in population and income. Initially, increased demand was met through domestic production. However with emphasis on food grain self-sufficiency, area under oilseed cultivation shifted to wheat and rice. India started importing edible oils. With increasing prices and decreasing supplies of these oils, the consumer was exposed to other oils like soya bean and sunflower. Groundnut oil now accounts for only about one-third of total oil consumption in India. Soya bean oil and sunflower together account for one–fourth of oil consumption.

(Continued)

BOX A6.1 (*CONTINUED*) EDIBLE OIL CONSUMPTION PATTERNS AND GROWTH

Edible oils are consumed in three forms: non-refined (or filtered) oils, vanaspati (partially hydrogenated edible oil mixture), and refined oils. Non-refined oils are mainly made from groundnut and mustard seeds. Traditional customers prefer the strong and distinctive taste of these oils. Inhabitants of the northern plains have a preference for hard fat and use vanaspati for special items like sweets. Its production is about 1.3 mtpa for the oil year (November 1998 to October 1999). Vanaspati has the ability to absorb a heterogeneous variety of oils which do not generally find direct marketing opportunities because of consumers' preference for traditional oils such as groundnut oil, mustard oil, sesame oil, etc.

Newer oils like Soya bean, sunflower, rice bran, cottonseed, and oils from tree and forest sources have found their way to the edible pool largely through the vanaspati route. As a result of price, availability, and even health consciousness, refined edible oils have gained increasing acceptance. Through technological means such as refining, bleaching, and de-odourization, all oils could be made practically colourless, odourless, and taste free. These oils could be made from a variety of sources like cottonseed, sunflower, palm oil (or its liquid fraction palmolein), Soya bean, rice bran, etc. whose availability is in plenty. The share of raw oil, vanaspati, and refined oil in the total edible oil market is, respectively 42.0 per cent, 13.4 per cent, and 42.7 per cent (November 1997 to October 1998).[d]

[a] http://www.mpopc.org.my/newsapril2000.htm
[b] Rakesh Jain, "Background Paper on Edible Oil Industry in India," www.madhyabharat.com/sopa/theme
[c] http://www.indiancommodities.com/shb.htm
[d] http://fcamin.nic.in/sugar_edbl.htm#Consumption Pattern

The market for edible oils was estimated to be 13 mtpa[1] in 2000. In the organized sector the main players were ITC, NDDB, Ruchi Soya, and Marico. In recent times, the market share of national brands like Godrej Sunflower, Sweekar, Saffola, etc., had decreased and regional brands had gained from them, thanks to their lower price. (Sales of various players in the edible oil market are shown in Table A6.1.) However the Edible Oils Packaging (Regulations) Order, 1998 made it obligatory for edible oil companies to sell only packed oil. Given that packing machines would require significant investment, players in the unorganized sector would in future be at a disadvantage compared to the established brands of organized sector.

AWL chose to focus on importing crude edible oil because of differential duty favourable to crude oil over refined oil. The idea behind this duty structure

Table A6.1   Sales of Leading Edible Oil Companies (₹ crore)

Industries	1995–1996	1996–1997	1997–1998	1998–1999
Ruchi Soya Industries	347.03	499.54	703.5	1210.21
ITC Agro-Tech	364.36	552.36	562.94	851.78
Marico Industries	309.37	393.12	468.15	521.69
Godrej Foods	140.88	217.25	233.56	399.96
Madhya Pradesh Glychem Inds.	116.99	129.41	119.40	354.08
K.S. Oils	140.20	90.30	141.08	246.50
Vippy Industries	96.12	124.05	157.76	211.81
Navcom Industries	187.5	167.15	186.70	186.7
N.K. Industries	110.79	189.95	182.99	182.99
Jayant Vegoils and Chemicals	–	–	151.24	181.05
Chawla Brothers	–	–	90.21	132.42
Vimal Oils and Foods	81.45	110.41	111.70	111.70
Sarvottam Industries	29.84	65.04	109.42	109.42
Rishi Oil and Fats	67.75	82.25	105.00	105.00
S.M. Dyechem	64.53	37.39	44.57	99.37
Surya Agroils	96.53	72.89	91.54	91.54
Gujarat Ambuja Exports	13.40	22.05	76.76	91.41
Amrit Enterprises	21.44	21.44	40.16	89.37
Amrit Banaspati	54.61	64.35	68.30	86.37
Prestige Foods	91.78	57.31	89.70	79.92

Source: CMIE Database, August 2000.

(Table A6.2) was to encourage import of crude oil in order to boost the capacity utilization of domestic refineries. On the other hand, the duty structure did not favour further backward integration to include crushing by importing oilseeds, though seed crushing units were running at 30 per cent capacity utilization. The high duties on oilseeds were the result of lobbying by the powerful domestic farmer lobby, which felt that lower duties would affect oilseed prices.[2] Given the transient nature of import duties over the past five years, AWL was open to either importing refined oil or oilseeds if it made economic sense. The cost of a refinery and a seed crushing plant was estimated at ₹1.5 billion.[3]

Table A6.2   Import Duty Structure

Type of Oil	1994–1995	1995–1996	1996– December 1999	December 1999–June 2000	After June 2000
Refined edible oils other than coconut, RBD palm, RBD palm kernel	65%	30%	20%	15%	35%
Other refined edible oils	65%	30%	20%	15%	45%
Crude palm oil for manufacture of vanaspati	65%	30%	20%	15%	15%
Other crude oils	65%	30%	20%	15%	25%

Source: Company data.

Operations

A state-of-the-art 600 tpd refinery was being set up in Mundra with technical know-how from the Wilmar group. AWL intended to run the refinery at full capacity. After refining, almost half of the oil would be sold as bulk. In that case, the buyer would take care of the logistics. The other half would be sold through a distribution network to be set up by the company in the regions mentioned earlier.

Before setting up the refinery, AWL was importing and trading in both crude and refined edible oil. Crude edible oil was sold to refiners in Gujarat and Rajasthan. Refined edible oil was sold to oil companies in bulk. Most of the oil was handled at the Mundra port. A large part of the imported oil was sold on high seas. Even after setting up the refinery, trading in crude and refined edible oil imports was expected to continue. (Total imports were expected to be about 0.5 mtpa, while refining would be around 0.2 mtpa.)

Distribution Network

For packed stocks, the company was thinking of setting up a network, which would consist of three parallel channels dealing in consumer and bulk packs. The consumer packs were 1 litre poly pouch, 1 litre bottle, 5 litre bottle, 5 litre jerry can, and 15 litre jerry can. The bulk packs were 15 litre tin and 15 kg tin.

The three parallel channels were:

Distributors: They would deal in all consumer packs up to 15 litre jerry cans. They would supply to retailers.

Institutional buyers: They would be customers with large demand, such as canteens, restaurants, messes, etc. They would be serviced directly from company depots, bypassing the distributors and retailers.

Super-stockists: They would be traders dealing in 15 litre and 15 kg tins. The responsibility of the company would end once the stock reached the super-stockists. Since these stocks would be traded, the price would be determined daily, based on demand and supply. The super-stockists would sell to smaller institutional buyers, typically not serviced directly by the company.

The distribution network would be serviced by warehouses. About 300 tons of edible oils per day would have to be distributed in a timely and cost-effective manner. Considering this, the company had begun setting up warehouses in Ahmedabad, Rajkot, Udaipur, Jaipur, Delhi, Karnal, Ludhiana, Ghaziabad, Kanpur, and Varanasi. In other places like Nashik, Jalgaon, Nagpur, Indore, Gurgaon, Solan, and Chandigarh where the off-take was not expected to be high, it planned to outsource warehouse management to carrying and forwarding agents (C&FAs). (Table A6.3 and Table A6.4 give product-wise expected demand through these warehouses for consumer and bulk packs.) The company did not see any significant cost savings in outsourcing, as long as there was a reasonable throughput. On the other hand, it preferred managing the warehouses because of the control it could exercise. These locations were being considered based on geographic intuition of the market spread and reasonable transport availability.

The mode of transport and routing had to be decided upon. Various modes of transport that could be considered were:

■ Road (Standard trucks)
■ Railways (Containers, i.e. multimodal transport)

In the case of road transport, freight rates were such that direct movement from Mundra to a warehouse location might not always be the best. For example, the freight rate from Mundra to Kanpur was more than the sum of the freight rates from Mundra to Indore and from Indore to Kanpur. The rates were influenced by several factors including return load availability. This would be important in deciding on location of warehouses and the routing of loads from Mundra to different locations. One of the possible architectures suggested by Mr. Pakarashi was the "hub and spoke".

Rail transport would be by containers, handled by Concor, the container transport company of the Indian Railways. The rates provided by Concor were from the refinery to warehouse locations that could be serviced by it. Concor would

Table A6.3 Forecast of Sales through Own Warehouses (tons/month)

Branch	Sunflower		Soya Bean		Cottonseed		Rapeseed		Vanaspati		Total	
	Cons	Bulk	Cons	Bulk	Cons	Bulk	Cons	Bulk	Cons	Bulk	Cons	Bulk
Ahmedabad	80	150	30	800	350	1500	0	0	50	300	510	2750
Rajkot	20	50	20	200	150	500	0	0	25	125	215	875
Udaipur	25	25	50	400	0	0	20	40	25	100	120	565
Jaipur	35	50	100	800	0	0	30	60	50	300	215	1210
Kanpur	30	40	35	350	0	0	30	30	50	200	145	620
Varanasi	15	20	30	300	0	0	15	50	50	200	110	570
Ghaziabad	30	40	35	350	0	0	30	50	50	200	145	640
Delhi	250	200	50	500	0	0	50	50	50	250	400	1000
Karnal	20	20	20	175	0	0	20	20	50	150	110	365
Ludhiana	30	30	30	125	0	0	30	50	50	250	140	455
Total	**535**	**625**	**400**	**4000**	**500**	**2000**	**225**	**350**	**450**	**2075**	**2110**	**9050**

Source: Company data.

Table A6.4   Forecast of Sales through C&FAs (tons/month)

Branch	Sunflower		Soya Bean		Rapeseed		Total	
	Cons	Bulk	Cons	Bulk	Cons	Bulk	Cons	Bulk
Nashik	30	40	0	50	0	0	30	90
Jalgaon	0	30	0	50	0	0	0	80
Nagpur	30	40	30	100	0	0	60	140
Indore	30	40	0	100	0	0	30	140
Gurgaon	20	40	20	100	0	50	40	190
Solan	10	20	30	150	0	50	40	220
Chandigarh	10	40	20	50	0	0	30	90
Total	**130**	**250**	**150**	**1350**	**0**	**300**	**280**	**1900**

Source: Company data.

Note: Bulk refers to sale through super-stockists.

probably move the containers by road from Mundra to the inland container depots in Kandla or Ahmedabad and then dispatch them as part of a trainload to the warehouses. However, there was some uncertainty as to from when Concor would be able to offer its service.

Apart from timeliness and cost, transit losses would influence mode choice and route. Transit losses could occur wherever there was direct handling of the packs. The company had experienced losses of about 0.1 per cent in handling, especially if the work was not properly supervised. The average selling price would be about ₹30/litre.

To understand the trade-offs and implications of some of the choices in the distribution network, Pakarashi decided to do an analysis of the UP market.

Distribution in UP

The UP market was the second largest in size for AWL, after Gujarat. The total monthly offtake was expected to be about 2200 tons. The district-wise demand forecast along with dealer locations is given in Table A6.5. To service these dealers, seven potential warehouse locations were considered. Keeping in view service time considerations for the secondary movement, dealers to be serviced by a warehouse were restricted to a distance of 500 km. Table A6.6 gives dealer locations, which could be serviced by a warehouse, along with distances. The secondary transportation cost along with other elements of the logistics cost is given in Table A6.7. Figure A6.1 gives a map of UP showing the potential warehouse locations and districts. The transportation cost from Mundra to various warehouse locations (including those in UP), from other warehouse locations to those in UP, and by container movement through Concor are given in Table A6.8. Given the remoteness of Mundra and the nature of truck transport markets, direct road transport from Mundra to warehouses was not always the cheapest. Ahmedabad and Indore were more active markets. If the transportation would be through these two cities, Mr. Pakarashi anticipated some additional costs in handling and coordination. Figure A6.2 gives a map of India showing Mundra and warehouse locations of interest.

Mr. Pakrashi needed to select warehouse locations out of the seven proposed, and assign dealers to warehouses. It would also be important to ensure that there were no imbalances in the workload of the warehouses, if they were to be managed by the company. Smaller volume warehouses could be candidates for outsourcing to C&FAs. He also needed to decide on mode choice and routing from Mundra.

AWL was open to postponing the packing to locations closer to demand points. While this would enable better servicing of demand, the refined oil would need to be transported in tankers, which had a higher freight rate. Currently, the packing facility was to be located in the refinery itself.

Table A6.5   Dealer and Districtwise Demand Forecast

Dealer Location	Districts Served	Demand (tons/month)	Dealer Location	Districts Served	Demand (tons/month)
Agra	Agra	38	Gorakhpur	Gorakhpur	45
	Mathura	21		Deoria	65
	Total	59		Basti	40
Aligarh	Aligarh	45		Total	150
	Etah	29	Haldwani	Nainital	34
	Total	74		Chamoli	10
Allahabad	Allahabad	72		Almora	18
	Total	72		Pithoragarh	12
Azamgarh	Azamgarh	46		Total	74
	Mau	37	Jaunpur	Jaunpur	47
	Total	83		Pratapgarh	32
Badaun	Badaun	54		Total	79
	Total	54	Jhansi	Jhansi	20
Bareilly	Bareilly	62		Lalitpur	10
	Pilibhit	28		Hamirpur	20
	Rampur	33		Jalaun	17
	Total	123		Total	67
Bijnor	Bijnor	34	Kanpur	Kanpur	63
	Total	34		Unnao	30
Dehradun	Dehradun	22		Total	93
	Uttarkashi	5	Kheri	Kheri	35
	Tehri Garhwal	13		Shahjahanpur	24
	Garhwal	15		Total	59
	Total	55	Lucknow	Lucknow	38

(Continued)

Table A6.5 (*Continued*)   Dealer and Districtwise Demand Forecast

Dealer Location	Districts Served	Demand (tons/ month)	Dealer Location	Districts Served	Demand (tons/ month)
Farrukhabad	Farrukhabad	34		Barabanki	37
	Etawah	24		Rae Bareli	32
	Total	58		Total	107
Fatehpur	Fatehpur	26	Maharajganj	Maharajganj	23
	Banda	26		Siddharthanagar	24
	Total	52		Total	47
Firozabad	Firozabad	21	Mirzapur	Mirzapur	24
	Mainpuri	18		Sonbhadra	16
	Total	39		Total	40
Ghaziabad	Ghaziabad	59	Moradabad	Moradabad	90
	Muzaffarnagar	62		Total	90
	Meerut	76	Saharanpur	Saharanpur	51
	Bulandshahr	62		Haridwar	25
	Total	259		Total	76
Ghazipur	Ghazipur	35	Sitapur	Sitapur	38
	Ballia	33		Hardoi	38
	Total	68		Total	76
Gonda	Gonda	49	Sultanpur	Sultanpur	37
	Bahraich	38		Faizabad	44
	Total	87		Total	81
			Varanasi	Varanasi	70
				Total	70

Source: Company data.

Table A6.6   Distance of Dealer Locations from Potential Warehouse Locations (km)

Dealer	Bareilly	Ghaziabad	Gorakhpur	Jhansi	Kanpur	Lucknow	Varanasi
Agra	210	220	–	220	285	365	–
Aligarh	165	115	–	305	275	370	–
Allahabad	480	–	300	400	195	240	125
Azamgarh	–	–	120	–	370	290	100
Badaun	50	110	–	–	250	220	–
Bareilly	0	235	–	435	325	245	–
Bijnor	100	150	–	–	–	–	–
Dehradun	340	215	–	–	–	–	–
Farrukhabad	140	240	–	240	150	230	–
Fatehpur	345	470	–	270	80	100	245
Firozabad	180	140	–	255	225	305	–
Ghaziabad	235	0	–	445	390	480	–
Ghazipur	–	–	140	–	390	350	75
Gonda	365	–	140	420	200	120	280
Gorakhpur	–	–	0	–	340	265	210
Haldwani	105	255	–	–	430	350	–
Jaunpur	–	–	160	450	265	260	60
Jhansi	435	445	–	0	220	300	–
Kanpur	325	390	340	220	0	80	320
Kheri	130	370	265	320	200	120	420
Lucknow	245	480	265	300	80	0	300
Maharajganj	–	–	50	–	380	300	260
Mirzapur	–	–	290	480	275	320	80
Moradabad	95	140	–	490	420	340	–
Saharanpur	330	170	–	–	–	–	–
Sitapur	160	395	265	385	155	85	385
Sultanpur	410	–	175	380	220	140	160
Varanasi	–	–	210	–	320	300	0

Source: Company data.

Table A6.7   Logistics Costs (₹/month)

	<200 tons/ month	*200–400 tons/ month*	*400–600 tons/month*	*>600 tons/ month*
Rent	10,000	20,000	30,000	35,000
Manpower	15,000	20,000	25,000	27,000
Electricity	2,000	3,000	4,000	5,000
Documentation	2,000	3,000	4,000	4,500
Equipment	2,000	3,000	4,000	4,500
Miscellaneous	4,000	6,000	8,000	9,000
	35,000	**55,000**	**75,000**	**85,000**

Source: Company data.

Notes: Primary transportation cost: as in Table A6.8
Secondary transportation cost: ₹1.50/ton/km
Warehousing cost
Inventory holding cost was ₹0.23 per kg (assuming 15 days inventory, and 18 per cent annual cost, and ₹30 per kg).
C&FA charges were ₹0.15 per kg

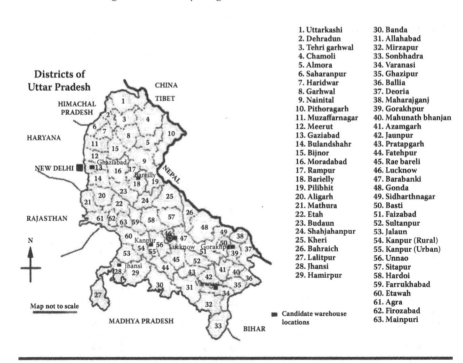

1. Uttarkashi
2. Dehradun
3. Tehri garhwal
4. Chamoli
5. Almora
6. Saharanpur
7. Haridwar
8. Garhwal
9. Nainital
10. Pithoragarh
11. Muzaffarnagar
12. Meerut
13. Gaziabad
14. Bulandshahr
15. Bijnor
16. Moradabad
17. Rampur
18. Barielly
19. Pilibhit
20. Aligarh
21. Mathura
22. Etah
23. Budaun
24. Shahjahanpur
25. Kheri
26. Bahraich
27. Lalitpur
28. Jhansi
29. Hamirpur
30. Banda
31. Allahabad
32. Mirzapur
33. Sonbhadra
34. Varanasi
35. Ghazipur
36. Ballia
37. Deoria
38. Maharajganj
39. Gorakhpur
40. Mahunath bhanjan
41. Azamgarh
42. Jaunpur
43. Pratapgarh
44. Fatehpur
45. Rae bareli
46. Lucknow
47. Barabanki
48. Gonda
49. Sidharthnagar
50. Basti
51. Faizabad
52. Sultanpur
53. Jalaun
54. Kanpur (Rural)
55. Kanpur (Urban)
56. Unnao
57. Sitapur
58. Hardoi
59. Farrukhabad
60. Etawah
61. Agra
62. Firozabad
63. Mainpuri

Figure A6.1   District map of Uttar Pradesh showing warehouse locations.

Table A6.8 Relevant Primary Transportation Cost (₹/ton)

	Ahmedabad	Udaipur	Jaipur	Indore	Jabalpur	Raipur	Bareilly	Ghaziabad	Gorakhpur	Jhansi	Kanpur	Lucknow	Varanasi	Concor (ex Mundra)
Mundra	400	750	975	750	1450	1200	1550	1300	1900	1500	1600	1700	1800	
Ahmedabad		180	800	400		1000	1200	900	1600	850	1100	1200	1500	
Udaipur			550				900	700						
Jaipur								500			700			
Indore					500	700	850	650	1100	600	750	800	1000	
Jabalpur						300			650		500		500	
Raipur									950				850	
Bareilly									500				550	
Ghaziabad							300		800		450		700	950
Gorakhpur													250	
Jhansi									550		250	300	500	
Kanpur									400			125	350	1135
Lucknow									300	300	150		300	
Varanasi									250					

Source: Company data.

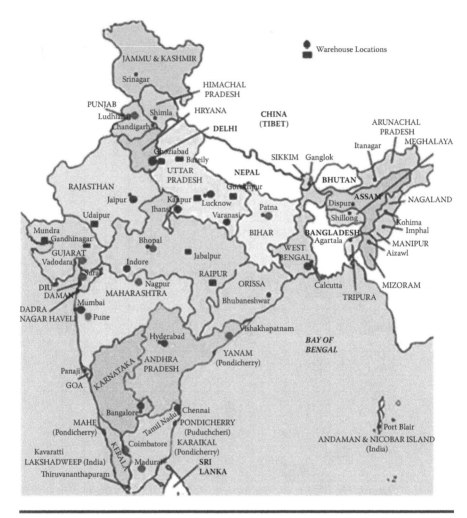

Figure A6.2   Location of AWL facilities.

Acknowledgements

Written by Sarang Deo, Sanjay Kumar Singh, Professor G. Raghuram, and Sanjay Choudhari. Sarang and Sanjay were PGP II students at Indian Institute of Management, Ahmedabad during 1999–2001. We acknowledge the inputs provided by Mr. Pakarashi, Logistics Manager of AWL.

Cases of the Indian Institute of Management, Ahmedabad, are prepared as a basis for class discussion. Cases are not designed to present illustrations of either correct or incorrect handling of administrative problems.

This case has been reproduced with permission from Indian Institute of Management, Ahmedabad. Copyright © 2001.

Notes

1. *Source:* www.indiancommodities.com
2. Rakesh Jain, "Background Paper on the Edible Oil Industry in India," www. madhyabharat.com/sopa/theme.
3. www.expressindia.com, accessed on January 25, 2000.

Case 7 Woolworths Limited, Australia

It was December 2001. The Supply Chain Executive Team (SCET) of Woolworths Limited was called for an important meeting. This meeting was headed by the General Manager, Supply Chain, Michael Luscombe. The main focus was on Project Refresh, which had commenced in August 1999 and was being driven by this team. This project was a plan to 'renew and reinvigorate' the company, leveraging initiatives in the supply chain. Since the commencement of Project Refresh, sales had improved by $2 billion and stripped $0.5 billion from Woolworth's costs. Luscombe and his team members were pleased with these numbers. Most of the benefits were expected to come from the Supermarkets Division, which accounted for about 85% of the revenues.

The SCET reported to the Supermarkets Executive Team and also the Corporate Support Group. It also had to coordinate with the Supermarket Buying and Marketing Team. Table A7.1 gives the composition of these groups and teams. It also provides an insight into the different roles held by the Senior Management Team (related to supermarkets) over the recent five years.

A brief description of Project Refresh is given in Box A7.1, with excerpts from the company's annual reports of 1999, 2000 and 2001. The project was envisaged as a three level initiative, with increasing sophistication, to yield an expected savings of over $9 billion over 9 years, starting 1999. The initial success of Project Refresh made Luscombe and his team optimistic about advancing to level 2 initiatives. They also decided to rechristen this initiative as Project Mercury, with an increased focus on the end-to-end supply chain and information technology (IT). The team was keen on identifying business process outsourcing (BPO) partners for this initiative. Appendix 7.1 give an overview and significant details of the project, as would be relevant for a BPO partner to work with the company.

One of the most important decisions being considered by SCET was the restructuring of the Distribution Centre (DC) network. Woolworths was operating 31 DCs

Table A7.1a   Management

	I. Corporate Support Group (2001)	
	Name	*Designation*
1	Bradley, Steve	General Manager, Corporate IT
2	Brookes, Bernie	Chief General Manager, Supermarket Buying and Marketing
3	**Corbett, Roger**	**Group Managing Director/Chief Executive Officer**
4	Howard, Judy	General Manager, Human Resources
5	Jeff, Rohan	General Manager, Corporate Services
6	**Luscombe, Michael**	**General Manager, Supply Chain**
7	Mcmorron, Dick	Chief General Manager, General Merchandize
8	Onikul, Naum	Chief General Manager, Supermarkets Operations
9	Reid, Gary	General Manager, Business Development
10	Wavish, Bill	Finance Director
	II. Supermarkets Executive Team (2001)	
	Name	*Designation*
1	Flood, Tom	General Manager, Supermarkets Operations
2	Foran, Greg	General Manager, Merchandizing Logistics, General Merchandize and Private Label
3	McFadzean, Tony	General Manager, Liquor
4	Pokorny, Peter	General Manager, Fresh Foods
5	Sidler, Hans	General Manager, Petrol
6	Winn, Penny	General Manager, Supermarkets Retail Support

(Continued)

Table A7.1a (*Continued*)   Management

III. Supply Chain Executive Team (2001)		
	Name	*Designation*
1	Hill, Paul	National Supply Chain Performance Manager
2	Hope-Johnstone, Craig	National Supply Chain Operations Manager
3	Kochanowicz, Daniel	National Supply Chain Strategy Manager
4	**Luscombe, Michael**	**General Manager, Supply Chain**
5	McLaughlin, Charles	National Transport Manager
6	Ramsay, Peter	Divisional Program Office Manager
IV. Supermarket Buying and Marketing Team (2000)		
	Name	*Designation*
1	Aylen James	Senior Business Manager, Grocery 2
2	Brookes, Bernie	Chief General Manager, Supermarket Buying and Marketing
3	Custance, Gavin	Senior Business Manager, Inventory
4	Dhnaram, Greg	Senior Business Manager, State Liaison
5	Dunn, Ian	Senior Business Manager, Trade Development and Relations
6	Hillen, Bevan	Senior Business Manager, Deli/Bakery
7	Hunt, David	Senior Business Manager, Merchandizing Support
8	Johnston, Murray	Senior Business Manager, Grocery
9	MacDonald, Ian	Senior Business Manager, Perishables
10	McAtamney, Jon	Senior Business Manager, Produce
11	McEntee, Pat	Senior Business Manager, Meat
12	Mintzis, Liz	Senior Business Manager, General Merchandize and Cleansing
13	Nahmani, Avner	Senior Business Manager, Liquor and Tobacco
14	O'Brien, Grant	Senior Business Manager, Marketing
15	Pokorny, Peter	General Manager, Fresh Foods

Source: Woolworths Ltd. Financial report to shareholders. 2000 and 2001.

(Continued)

Table A7.1b Career Progression of Woolworths' Senior Management Team (Related to Supermarkets)

Name	1997	1998	1999	2000	2001
Bradley, Steve			Corporate Manager Information Technology	Corporate Manager Information Technology	Corporate Manager Information Technology
Brookes, Bernie	General Manager QLD Supermarkets	General Manager QLD Supermarkets	Chief General Manager Supermarkets Buying and Marketing	Chief General Manager Supermarkets Buying and Marketing	Chief General Manager Supermarkets Buying and Marketing
Clark, Grant					Supermarkets Manager Region 6 (NSW, ACT)
Corbett, Roger	Managing Director Retail	Chief Operating Officer	Group Managing Director Chief Executive Officer	Group Managing Director Chief Executive Officer	Group Managing Director Chief Executive Officer

Cornell, Ian	Chief General Manager Supermarkets	Chief General Manager Supermarkets			
Flood, Tom	General Manager WA Supermarkets	General Manager WA Supermarkets	General Manager VIC Supermarkets	General Manager VIC Supermarkets	General Manager Supermarkets Operation
Luscombe, Michael			General Manager Supply Chain	General Manager Supply Chain	General Manager Supply Chain
Onikul, Naum	General Manager VIC Supermarkets	General Manager VIC Supermarkets	General Manager Supermarkets Operation	Chief General Manager Supermarkets Operation	Chief General Manager Supermarkets Operation
Roberts, Trevor	General Manager Distribution	General Manager Distribution	General Manager Distribution		
Wavish, Bill			Chief Financial Officer	Finance Director	Finance Director
Winn, Penny		Store Operations Manager Big W	National Manager Ezy Banking	National Manager Ezy Banking	General Manager Supermarkets Retail Support

Source: Woolworths Ltd. Annual reports, 1997–2001.

BOX A7.1 PROJECT REFRESH

In the 1999 Annual Report (published in October 1999), Chairman, John Dahlsen wrote:

In August 1999, we announced Project Refresh, an exciting and important program which identified the developing trends in the retail environment and the need for changes designed to deliver increased customer focus and shareholder wealth.

Project Refresh will give management the opportunity to build on the organization's market position, by a rigorous focus on customer needs, a more responsive management team, better supplier relationships and best practices in cost control, cost measurement and tracking.

Critically, Project Refresh will see the divestment of non-core assets and the focusing of attention and funds on the core businesses. It will result in important structural changes and the need to continue the process of review and re-invigoration of the organizations' core value proposition—a Customer Centric Business.

The 2000 Annual Report stated the following:

In August last year, we announced the implementation of Project Refresh. It is a significant and far reaching initiative which covers all aspects of our business as we re-examine everything we do.

Whilst the supermarkets' central Shared Service Buying function is the most recognizable change for our suppliers, Project Refresh also covers supply chain, IT, human resources, organizational redesign and cost of doing business programmes.

The initial benefits will be extended to all areas of our businesses and will become part of our operational and corporate culture. Every effort is being made to achieve our gains without redundancies, by retraining and redeployment. The members of the teams working on Project Refresh are drawn from every state in Australia with around 30 people working on the various initiatives at any one time.

Supermarkets Shared Services

The change to central shared service buying and marketing was the result of a restructure of the business from a state based buying and marketing structure to a single functionally based business. Some of the main building blocks for the new structure—which we have called 'Shared Services'—were developed by examining best practices in supermarkets operations from around Australia and the world.

(Continued)

BOX A7.1 (*CONTINUED*) PROJECT REFRESH

The new structure replaces a system of six state buying and marketing operations and enables us to buy most products centrally, while still maintaining state buying in some fresh foods categories. It has the capacity to considerably increase efficiency, eliminate duplication, improve supplier relations and drive supply chain economies.

The new shared services function allows us to buy more effectively, to offer more consistent quality fresh food and increase value through better buying and reduced pricing. Our decisions will also be better informed because of the establishment of cross-functional teams, made up of people with experience in buying, marketing, stock control and store management. The new structure will further encourage a strong service culture, enabling business managers to provide quicker service and better support to store managers and their teams.

Supply Chain

As the year progressed, the considerable strategic advantages of our major new warehousing and distribution facilities became more apparent. We saw improved in-stock positions, particularly at the Christmas and Easter peak trading periods. Following a comprehensive review of our supply chain management, we have formed a single supply chain structure which applies both technology and logistics as one function for the total company. It is expected that this will bring significant advantages to the company, in stock flow, stock turns and in-stocks. Much more remains to be done in this area and the financial rewards we will generate from a total supply chain management strategy are considerable.

The 2001 Annual Report stated the following:

This far reaching initiative covers all aspects of our business as we examine everything we do with a view to saving costs, increasing sales and improving effectiveness. In particular, the areas of cost savings cover:

- Examination of all line item expenses
- Improvements to the total supply chain and its over-arching IT
- Functional reorganization to a national or shared service basis whilst increasing regional and in-store empowerment.

(Continued)

BOX A7.1 (*CONTINUED*)   PROJECT REFRESH

Earlier this year we announced that we expect annual savings as a result of this initiative to reach $185 million, of which $50 million had been realized. In the global environment which influences all aspects of our business and against which we continuously benchmark ourselves, change and cost saving are perpetual. Initially, as a result of Project Refresh and increasingly as a consequence of continuous change, we now expect annual savings over the next five years to increase to approximately 1% of sales. That is further savings exceeding $200 million per annum.

Supermarkets Buying and Marketing

The Project Refresh changes began in August 1999 with a review of how we conduct our supermarkets buying and operations. This resulted in a decision to change our organizational structure from a divisionally based group of businesses in each state to a single, functionally based business. The first concrete realization of this structural change was effected in June 2000, when supermarkets moved to a national shared service environment for buying and marketing.

In line with our objectives of improving the ranging, layout, competitive pricing and promotional offer to our customers, this was the first major approach to a centralized management environment in supermarkets. It laid the foundation for a streamlined management structure in finance, human resources, operations and supply chain.

In its first full year of operations, supermarkets buying and marketing put in place new measures to improve communication throughout stores, including the in-house television programme WOW TV, and Merchant of the Year competitions to encourage competition between store teams to build the best store displays and gain the best incremental sales.

Supplier Relations

The quality of our relationship with our supplier partners is an important ingredient in providing a continually improving offer to our customers.

Our supplier focus groups that have been run each quarter demonstrate an improving supplier relationship and a genuine focus on sales with our suppliers. Suppliers have been pleased with the consistency and uniformity of our interaction. Feedback shows improving supplier relationships with a genuine mutual focus on increasing sales.

(Continued)

BOX A7.1 *(CONTINUED)*   PROJECT REFRESH

Later in the year, we will be hosting a dinner to thank our suppliers and present the Suppliers of the Year Awards. These awards provide us with a great opportunity to recognize the outstanding teamwork between Woolworths and its many suppliers.

Supply Chain

Our supply chain is defined as the 'end to end' linkage of the activities of buying and marketing and store operations. In simple terms, the supply chain strategy focus is on being store friendly. This will ensure that the appropriate products get on the shelf faster and cheaper whilst using less inventory.

A high level strategy for the Woolworths supply chain has been based upon Woolworths' strong existing supply chain network, and on world's best practice, developed during study tours of major retailers in Europe and the USA. Implementation plans for the strategy are being developed, and will be phased in over the next five years.

A significant amount of supply chain work in the financial year revolved around the development of IT systems to support inventory management in the Supermarkets Division. Four major projects underway are IT systems to support:

- Ordering or merchandize into our DCs
- Ordering of merchandize into our stores
- Better visibility of stock holding across the company
- Better management of stock delivered directly to our stores from suppliers

Other projects undertaken in the year under review include:

- Development, and implementation of supply chain integration plans for the converted former Franklins stores
- Rationalization of the Victorian distribution facilities, including closure of one site and a major upgrade to the Mulgrave Distribution Centre
- Centralization of the Dick Smith Electronics (DSE) distribution facilities from a multi-site to a one-site operation, based at Chullora in Sydney
- Development of integration plans for the Tandy Electronics supply chain into the DSE business
- Extension of the BIG W Warwick, Queensland distribution facility to handle increased volumes as this business expands
- Brismeat operation – design, tender and commencement of works to refurbish and extend the Ipswich Meat Facility.

(Continued)

BOX A7.1 (*CONTINUED*)   PROJECT REFRESH

Merchandize Logistics

During the financial year, the Woolworths supply chain merchandize logistics department was formed, with the aim of working with our suppliers to improve the flow of their merchandize through the supply chain. This is an area with potential to remove substantial costs, for both suppliers and for Woolworths.

Woolworths Transport

The Woolworths National Transport Department was formed as part of the new supply chain structure. This group has identified a number of strategic initiatives that will be positive for the Woolworths group and work has commenced on implementation of these initiatives. The National Transport Department has also provided a national approach to transport contract management, with performance based service level agreements, which are providing immediate benefits to the company.

IT

Supply chain system improvement was the main focus of IT. We have a major development programme underway over the next three years, that enables and supports our overall supply chain strategy. The main achievements in the last year were:

- The implementation of a new warehouse replenishment system (StockSMART) to better control the inventory levels of our everyday stock
- The pilot of a promotional warehouse replenishment system
- The pilot of an AutoStockR
- The development of more sophisticated systems to control the ordering and delivery of DSD

We are confident that in the next year, as we roll out the pilot systems, we will gain significant improvements in our in-stock positions, as well as a reduction in our overall stock levels.

We have rolled out new stores back-office infrastructure to over 600 supermarkets. This provides a standard, Microsoft Windows Platform, in every supermarket which will be the basis for many new in-store IT applications in the coming years.

Source: Woolworths Ltd. Financial reports to shareholders for years 1999–2001.

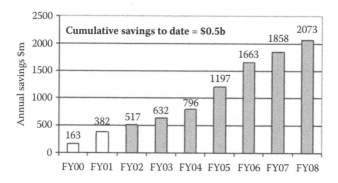

Figure A7.1 Cumulative savings. (From Woolworths Ltd. Company results. Aug 26, 2006.)

to cater to the supply needs of its Supermarkets Division. The retail outlets were receiving stock from multiple DCs, and did not leverage synergy in supply. The replenishment at the DC level was based on manual processes. Luscombe and his team had to take several decisions. First, whether to consolidate the 31 DCs to a smaller number, by having multiple product categories (ambient and chilled) supplied from the same DC. If this happened, they would then need to identify the number of DCs and their locations. This would also imply that third party managed DCs would now be managed by the company. Initiatives to improve DC operations, including IT, would become critical.

Company Overview

Woolworths was one of the largest retail chains in Australia, retailing food, groceries, liquor, petrol, and general merchandize including consumer electronics. It had also entered into the wholesale business. By the end of 2001, it was operating 1359 stores in Australia and 33 in New Zealand. It was headquartered near Sydney in Bella Vista, New South Wales, Australia. Table A7.2 gives some major events in the growth of Woolworths since the opening of the first store in 1924 in Sydney.

Table A7.2 Major Events

Year	Events	Location
1924 (December 5)	First store opened: Woolworths Stupendous Bargain Basement	Old Imperial Arcade, Pitt Street, Sydney
1933	Expanded to 23 stores	
1936 (April 1)	Bought eight stores from Edments Ltd	
1955	200th store was opened First supermarket was opened	Civic Center, Canberra Beverly Hills, Sydney
1960	Acquired Rockmans chain of women's clothing stores Acquired Flemings supermarket	Sydney
1970	First discount department store, Big W was opened	
1981	Acquired Dick Smith Electronics consumer electronics chain	
1985	Acquired the Australian stores of the American company, Safeway Inc (126 stores)	Eastern Australia
1987	Woolworths and Safeway supermarket launched "The Fresh Food People" campaign	
1989	Acquired by Industrial Equity Limited (IEL)	
1993	Woolworths Limited floated on the Australian Stock Exchange at $2.45 a share	
1996	First Plus Petrol outlet was opened	Dubbo, NSW
1997	First convenience store, Metro was opened	Sydney
1998	Internet HomeShop service began	Sydney

(*Continued*)

Table A7.2 (*Continued*)   Major Events

Year	Events	Location
1999 (August 11)	First Ezy Banking service to customers was launched	Queensland
1999 (August 31)	Launched Project Refresh to reorganize the business	
2000	Sold Rockmans chain of women's clothing stores	
2000 (December 11)	Acquired Internet grocery retailer GreenGrocer.com.au	
2001 (April 10)	Acquired 224 Tandy Electronics stores from Canada based InterTan Inc	
2001 (June 4)	Purchased 67 of Franklins' 282 supermarket stores from the Hong Kong-based Dairy Farm Group	

Source: Woolworths Ltd. Media releases.

The first store was called Woolworths Stupendous Bargain Basement. Nominal capital was 25,000 pounds and attracted just 34 shareholders. The original concept was to establish a store that sold everyday needs of general merchandize at low prices. Woolworths successfully focused on its low price approach to general merchandize right through until the late 1950s when it diversified into food retailing through the supermarkets format. By 1960, Woolworths was the first Australian retailer to operate in all Australian states and territories. The 1960s and 1970s saw further diversification with the acquisition of the Rockmans women's clothing chain and the establishment of the BIG W discount department store chain. In the early 1980s, Woolworths acquired the Dick Smith Electronics chain. Then in 1985, Woolworths bought the Australian Safeway store group from Safeway of the US to become the largest food retailer in Australia.

At the end of the 1980s, Woolworths became the subject of a takeover bid from Industrial Equity Limited (IEL) that was successfully completed in 1989. Woolworths remained a wholly-owned subsidiary of IEL until 1993 when it was re-floated on the Australian stock exchange at $2.45 a share. Woolworths diversified into petrol retailing in 1996, giving discounts to attract customers. The Internet HomeShop service was introduced in 1998 and Ezy Banking followed in 1999. The same year, Woolworths launched Project Refresh with the task of reorganizing many parts of the business.

Chisholm Manufacturing, a processed foods and small goods manufacturer and Rockmans, the women's clothing chain, were disposed of in 2000, as Woolworths sought to focus on its core businesses of supermarkets and general merchandize. This process continued in 2001 with the sale of Crazy Prices and the acquisition of Tandy Electronics, and 67 Franklins supermarkets.

Woolworths Retail Activities

Woolworths Limited was made up of a number of businesses, all providing the customers with quality, value and everyday low prices. The company was operating in Australia through several retail banners. Table A7.3 lists the various retail brands under the Woolworths umbrella. The brands were operated under three divisions: Supermarkets, General Merchandize and Wholesale. The numbers of brandwise outlets are given in Table A7.4. Not all the outlets carried liquor or had a petrol pump.

Supermarkets

- Woolworths: The company's premier supermarkets chain operated in every Australian state and territory except Victoria
- Safeway: In Victoria, Woolworths was named Safeway
- Food For Less: Woolworths also operated a number of smaller supermarkets under this brand in some areas
- Flemings: Group of four supermarkets located in Sydney (the remnants of a chain purchased in the 1960s)

At the end of the financial year 2001, Woolworths had 604 supermarkets distributed right around Australia. The distribution of supermarkets was closely aligned with that of the population.

Chisholm Manufacturing was another wholly owned business of Woolworths in Supermarkets Division. All the bulk ham, bacon, sausages and small goods sold in all the supermarkets were packed and supplied through Chisholm Manufacturing.

Plus Petrol

In 1996, Woolworths entered the petrol market, with wholly owned Plus Petrol outlets in Dubbo, NSW. With canopies adjacent to Woolworths supermarkets, it allowed Woolworths to meet more of its customer needs in one location and also helped to increase store sales. The petrol business returned a profit of $4.6 million in fiscal 2001. At the end of fiscal 2001, Woolworths had 166 petrol outlets.

Liquor

The total liquor business was organized under four brands:

- Woolworths Liquor: Liquor stores, either attached or located within Woolworths supermarkets (known as Safeway Liquor in Victoria)
- BWS: Liquor stores located away from the company's supermarkets
- Dan Murphy's: Known as 'liquor supermarkets' and one of the company's best growth performers
- First Estate: Fine wine stores

The total business generated sales of more than $1 billion per annum. Woolworths' share of the national liquor market was around 10% compared with Coles Myer's estimated market share of 13%.

Ezy Banking

Ezy Banking, Woolworths' banking joint venture with the Commonwealth Bank in 1999, was viewed as a separate business from the supermarkets group but was implemented to drive traffic through the stores and thus generated incremental sales. It was another example of meeting customers' everyday needs in one location and it also provided Woolworths with an opportunity to learn more about its customers. In just two years, in 2001, 610,000 accounts were opened exceeding the five year target of 500,000 accounts.

General Merchandize

The General Merchandize division was much smaller than the Supermarkets Division, accounting for only 13% of sales. The division comprised:

BIG W

This discount department store chain sold a wide range of general merchandize. Its direct competitors were Target and Kmart, both members of the Coles Myer Group. Through the implementation of Every Day Low Prices (EDLP), improved supply chain management, and other benefits coming out of Project Refresh, Woolworths was able to drive impressive performance improvements from BIG W.

Consumer Electronics

- Dick Smith Electronics: Sold hobby electronic products as well as computer products
- Dick Smith Powerhouse: Innovative and interactive store which sold consumer entertainment products
- Tandy Electronics: Sold computers, communications and electronic goods

Table A7.3   Retail Brand

	Supermarkets		General Merchandize				
	Food and Groceries	*Liquor*	*Petrol*	*Discount Stores*	*Convenience Stores*	*Consumer Electronics*	*Wholesale*
	Woolworths	***Woolworths Liquor***	Plus Petrol (in Woolworths, Safeway and Big W)	Big W	Metro	Dick Smith Electronics	Australian Independent Wholesaler (AIW)
	Safeway	***Safeway Liquor***		*Rockmans		Dick Smith Powerhouse	
	Food For Less	Dan Murphy's		**Crazy Prices		Tandy Electronics	
	Flemings	First Estate		**Woolworths Variety			
	Franklins	Mac's Liquor					
		BWS					

Note: ***Bold italicized brands*** are served by the DCs under examination.

* Sold in 2000.
** Sold in 2001.

Table A7.4   Brandwise Outlets

Retail Brand	1996 June 23	1997 June 29	1998 June 28	1999 June 27	2000 June 25	2001 June 24
Supermarkets (Food and Groceries)	505	518	542	559	585	604
Woolworths Liquor	–	38	38	42	41	130
Safeway Liquor*						
Dan Murphy's*						
First Estate*						
Mac's Liquor						
BWS*						
Plus Petrol	–	12	49	98	137	166
Big W	71	78	82	85	87	90
Rockmans	246	252	257	258	–	–
Crazy Prices	74	85	100	116	134	–
Woolworths Variety	1	1	1	1	1	–
Metro*						
Dick Smith Electronics	107	113	115	119	123	138
Dick Smith Powerhouse	–	1	2	4	6	9
Tandy Electronics	–	–	–	–	–	222
Australian Independent Wholesaler*						

Source: Woolworths Ltd. Financial reports to shareholders for years 1999–2001.

* Data not available.

Table A7.5   Retail Industry Turnover in Australia

Year	Turnover ($b)	Growth (%)
1998	181.0	N/A
1999	185.1	2.3
2000	192.9	4.2
2001	196.6	2.0
Forecasts		
2002	202.5	3.0
2003	207.6	2.5
2004	211.9	2.0
2005	217.3	2.5
2006	223.4	2.8
2007	229.8	2.9
2008	235.8	2.6

Source: IBIS Industry Division Report. http://invest. vic.gov.au/

Woolworth's Metro

These were inner-urban convenience stores located in Sydney and Brisbane, selling a range of pre-prepared meals for the 'time poor' customer.

Wholesale

The Wholesale Division covered Australian Independent Wholesalers (AIW), who traded in Victoria, New South Wales and Queensland, and Statewide Independent Wholesalers (SIW), which traded in Tasmania. The latter was only 60% owned by Woolworths. This division, which was the smallest amongst other divisions, contributed only 3% sales of Woolworths.

Australian Retail Industry

The Australian retail environment was dynamic and constantly evolving. There were 70,000 employing retail businesses in Australia. In 2000-01 they transacted $197 billion of business, growing at 2% to 3% per annum (Table A7.5). The industry was the largest employer in Australia, employing 920,000 people

Table A7.6   Statewise Population and Retail Income Distribution

States	Capital	Area (sq km)	Population ('000)	Population (%)	Retail Income (%)
New South Wales (NSW)	Sydney	801,352	6371.7	33.6	34.2
Victoria (VIC)	Melbourne	227,590	4645.0	24.5	23.8
Queensland (QLD)	Brisbane	1,734,190	3655.1	19.3	19.2
Western Australia (WA)	Perth	2,532,422	1851.3	9.8	9.8
South Australia (SA)	Adelaide	985,324	1467.3	7.7	7.6
Tasmania (TAS)	Hobart	67,914	456.7	2.4	2.2
Australian Capital Territory (ACT)	Canberra	2,349	311.9	1.6	2.2
Northern Territory (NT)	Darwin	1,352,212	210.7	1.1	1.0
Total		7,703,353	18969.6	100	100

Source: City Population. Australia: Urban centers. http://www.citypopulation.de/Australia-UC.html

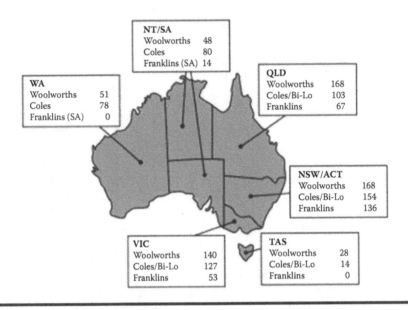

Figure A7.2 Major retailers' supermarkets in Australia, 1998. (From Parliament of Australia, Chap. 2 of Fair Market or Market Failure. 2001.)

(about 12% of the workforce). Over the period 1993–2000, employment grew by over 20%. Australia's biggest retailers in terms of sales, market share and impact on the retail industry were Coles Myer, Woolworths, Foodland, Harvey Norman, Bunning, and David Jones. Appendix 7.2 provides an insight into the Supermarket Division's competitive environment.

The state-wise population and retail income distribution is given in Table A7.6. The major supermarket retailers' outlet across the state, as of 1998, is given in Figure A7.2.

According to Australian Food Statistics 2001 published by the Department of Agriculture, Fisheries and Forestry, supermarkets and grocery stores accounted for around 66% of total food sales (excluding liquor) of almost $64 billion, in 1999–2000. This result showed a 2.6% increase in sales over the year 1998.

[http://www.myfuture.edu.au/services/default.asp?FunctionID=5104&Industr yGroupID=240]

Supermarkets Division

Australia, as a country, was divided into eight administrative states and territories. Figure A7.3 gives a map of Australia, along with its important cities. The state-wise distribution of the 604 Woolworths supermarkets is given in Table A7.7. The supermarkets were organized into ten regions, as seen in Figure A7.4. Large areas of Australia still did not have an outlet due to a very low density of population.

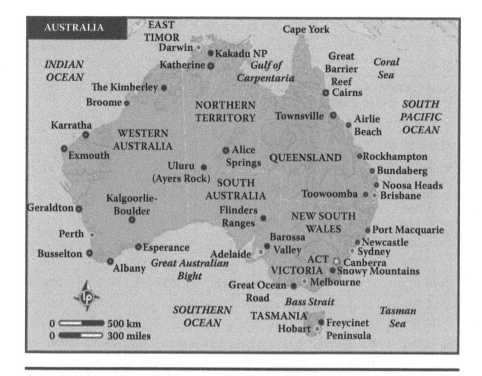

Figure A7.3   Map of Australia. (From Lonely Planet. www.lonelyplanet.com)

The main product categories of the supermarkets are given in Table A7.8. The supermarkets business was viewed as consisting of food, groceries, liquor and petrol. In terms of supplies, the DCs handled chilled and ambient separately. Produce and liquor, to a significant extent, was handled independently through dedicated DCs. For the financial year ended 24 June 2001, the Supermarkets Division had revenue of $17.5 billion, as seen in Table A7.9. The earnings before interest and tax (EBIT) were $600 million. Table A7.10 provides a five year analysis of the financial results of Woolworths.

Woolworths' sales turnover totalled almost $21 billion and EBIT came in at just over $700 million. The company recorded a net profit of $428 million. The market capitalization was recorded at $11.6 billion. The market share was in the range of 28% to 40%. Woolworths was the second largest employer in Australia with a staff of over 140,000 people.

Project Refresh

In 1999, Woolworths achieved an operating profit of $312 million, at 4% over the year 1998, and also a 9.6% growth in sales. Though this performance was very good by Australian standards, and placed the company in the top half of

Table A7.7   Statewise Distribution of Supermarkets

States	1997 June 29	1998 June 28	1999 June 27	2000 June 25	2001 June 24
New South Wales (NSW) and Australian Capital Territory (ACT)	162	174	178	192	199
Queensland (QLD)	106	111	111	112	115
Victoria (VIC)	130	133	145	149	151
South Australia (SA) and New Territories (NT)	43	45	45	51	53
Western Australia (WA)	48	50	52	52	57
Tasmania (TAS)	29	29	28	29	29
Total	**518**	**542**	**559**	**585**	**604**

Source: Woolworths Ltd. Financial report to shareholders. 2001.

Australian listed companies, there was a need to grow shareholder value faster. The management team was determined to have a more dynamic and outward-looking approach, and so, in August 1999, launched 'Project Refresh'. This was an exciting and important programme which identified the developing trends in the retail environment and the need for changes designed to deliver increased customer focus and shareholder wealth.

The major objectives of Project Refresh were:

■ Focus on customer needs
■ Cost control, cost measurement and tracking
■ Create greater shareholder value through strategic capital management
■ Create a new management structure that would allow decisions to be made faster and by the most appropriate person, whether at store, regional or national level and would also reduce the costs of unnecessary duplication
■ Focus on supply chain, inventory management, buying and marketing, IT, human resources, organizational redesign and cost of doing business
■ Provide better supplier relationships
■ Refocus on core businesses of supermarkets and general merchandize retailing
■ Seek buyers for Rockmans and Chisholm Manufacturing businesses

In short, Project Refresh initiative covered all aspects of business with a view to saving costs, increasing sales and improving effectiveness.

Project Refresh, since its inception in 1999, concentrated on a number of initiatives, including a significant business restructuring programme as well as numerous

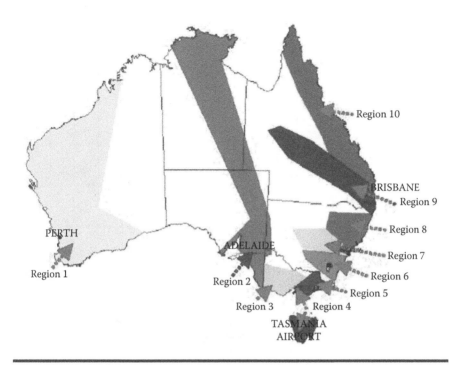

Figure A7.4   Regional distribution of supermarkets. Region 1: WA. Region 2: SA, VIC, NSW, NT, WA. Region 3: VIC. Region 4: VIC, TAS. Region 5: VIC, NSW. Region 6: NSW, ACT. Regions 7–8: NSW. Region 9: QLD, NSW. Region 10: QLD. (From Woolworths Ltd.)

cost reduction programmes. These initiatives were collectively known as Refresh Level I initiatives.

While the benefits of Refresh Level I continued to flow, the company shifted its major focus to Refresh Level II. This was called Project Mercury, an integrated, end-to-end supply chain improvement programme spanning over five years in both logistics and enabling technology. (Woolworths Ltd. Financial reports to shareholders for years 1999–2001.)

Project Mercury

This project, started in October 2001, was considered the most important initiative driving change in Woolworths history, and $1 billion would be spent on the upgrade of supply chain and associated IT. This supply chain redesign programme was initiated with a business objective of improving the process of delivering products to the customers. In achieving this, the programme aimed to generate a competitive advantage for its supermarkets. It would impact logistics, buying and marketing, supermarket operations, and IT functions.

Table A7.8 Supermarkets Product Categories

Chilled/Frozen		Ambient						
Meat	Perishable (Other than meat)	Deli/Bakery	Produce (F&V)	Food			Liquor	Petrol
				Other Food	Non Food			

	Purchase within Australia		Share of Australian Produce
	$ m	%	%
F & V	1280	97.5	25.0
Meat	1140	100.0	14.3
Beef	480		
Lamb	200		
Pork	95		
Chicken	185		
Small goods	177		
Dairy & Eggs	627	94.0	9.2

Source: Woolworths Ltd.

Table A7.9   Financial Summary of Supermarkets ($m)

		1997 June 29 (53 weeks)	1998 June 28 (52 weeks)	1999 June 27 (52 weeks)	2000 June 25 (52 weeks)	2001 June 24 (52 weeks)
Sales ($m)	Food, Liquor and Groceries	12,583.8	13,374.5	14,247.0	15,251.3	16.772.3
	Petrol	–	–	316.4	472.5	747.1
	Total Supermarkets	12,583.8	13,374.5	14,563.4	15,723.8	17,519.4
Cost of doing business (%)		–	–	21.8	21.9	21.4
EBIT to sales (%)		3.1	3.3	3.0	3.3	3.5
EBIT ($m)	Food, Liquor and Groceries	397.6	453.2	451.5	534.0	614.0
	Petrol	–	(3.9)	(2.9)	(1.0)	4.6
	Total Supermarkets	397.6	449.3	448.6	533.0	618.0

Source: Woolworths Ltd. Financial report to shareholders. 2001.

Table A7.10 Five Year Analysis

	2001 52 weeks	2000 52 weeks	1999 52 weeks	1998 52 weeks	1997 53 weeks
Profit and loss					
Sales[1] ($m)					
Food, liquor and groceries	16,772.3	15,251.3	14,247.0	13,374.5	12,583.8
Petrol	747.1	472.5	316.4	–	–
Total supermarkets	17,519.4	15,723.8	14,563.4	13,374.5	12,583.8
BIG W	2,069.8	1,913.9	1,788.0	1,644.5	1,516.8
Consumer electronics	418.0	338.2	298.6	241.9	223.1
Wholesale	697.8	675.3	520.7	388.8	151.5
Total trading operations	**20,705.0**	**18,651.2**	**17,170.7**	**15,649.7**	**14,475.2**
Discontinued operations[2]	210.1	337.6	356.6	351.4	324.5
Total group	**20,915.1**	**18,988.8**	**17,527.3**	**16,001.1**	**14,799.7**
EBIT[3] ($m)					
Food, liquor and groceries	614.0	534.0	451.5	453.2	397.6
Petrol	4.6	(1.0)	(2.9)	(3.9)	–
Total supermarkets[4]	618.6	533.0	448.6	449.3	397.6

BIG W	83.4	74.3	62.2	56.2	48.0
Consumer electronics	30.8	26.0	20.8	15.7	14.6
General merchandise	114.2	100.3	83.0	71.9	62.6
Wholesale	5.0	2.9	(2.1)	(9.7)	(4.2)
Total trading operations	**737.8**	**636.2**	**529.5**	**511.5**	**456.0**
Net property income	33.1	24.8	33.0	22.3	15.1
Head office overheads	(59.0)	(50.2)	(40.5)	(33.9)	(45.1)
Total unallocated[5]	(25.9)	(25.4)	(7.5)	(11.6)	(30.0)
Continuing operations	711.9	610.8	522.0	499.9	426.0
Discontinued operations	(5.3)	10.8	17.4	16.3	23.0
Total group	**706.6**	**621.6**	**539.4**	**516.2**	**449.0**
EBIT to sales (%)					
Supermarkets	3.53	3.39	3.08	3.36	3.16
General merchandise	4.59	4.45	3.98	3.81	3.60
Wholesale	0.72	0.43	(0.40)	(2.49)	(2.77)
Total	3.38	3.27	3.08	3.23	3.03

(Continued)

Table A7.10 (Continued)   Five Year Analysis

Profit and loss ($m)	2001 52 weeks	2000 52 weeks	1999 52 weeks	1998 52 weeks	1997 53 weeks
Profit and loss ($m)					
Sales	20,915.1	18,988.8	17,527.3	16,001.1	14,799.8
Cost of goods sold	15,561.0	13,983.4	12,790.3	11,710.4	10,856.4
Grass profit	5,354.1	5,005.4	4,737.0	4,290.7	3,943.4
Grass profit margin	25.60%	26.36%	27.03%	26.82%	26.64%
Branch and administration expenses	(3,737.7)	(3,548.3)	(3,400.0)	(3,072.4)	(2,866.4)
(excluding rent, depreciation and amortisation)	17.87%	18.69%	19.40%	19.20%	19.37%
EBITDAR	1,616.4	1,457.1	1,337.0	1,218.3	1,076.9
EBITDAR margin	7.73%	7.67%	7.63%	7.61%	7.28%
Rent	(600.0)	(546.7)	(527.7)	(479.4)	(442.1)
EBITDA	1,016.4	910.4	809.3	738.9	634.8
Depreciation	(300.7)	(282.8)	(265.0)	(219.9)	(184.1)
Amortisation of goodwill	(9.1)	(6.0)	(4.9)	(2.8)	(1.7)
EBIT	706.6	621.6	539.4	516.2	449.0
Interest	(13.1)	(27.8)	(45.5)	(42.8)	(41.2)
WINS distribution	(47.7)	(26.1)	–	–	–
Net profit before tax	645.8	567.7	493.9	473.4	407.8
Taxation	(217.4)	(203.6)	(181.3)	(172.7)	(149.5)
Normal net profit after tax	428.4	364.1	312.6	300.7	258.3

Adjustment for change in company tax rate	–	(8.4)	–	–	–
Abnormal items after tax	–	(60.1)	(55.3)	(21.1)	–
Outside equity interests	(0.4)	(0.1)	(0.3)	(0.2)	(0.3)
Operating net profit attributable to the members of Woolworths Limited after WINS	428.0	295.5	257.0	279.4	258.0
Balance sheet ($m)					
Funds employed					
Inventory	1,731.8	1,648.3	1,652.6	1,562.4	1,488.3
Accounts payable	(1,666.4)	(1,571.8)	(1,281.1)	(1,202.7)	(1,101.1)
Net investment in inventory	65.4	76.5	371.5	359.7	387.2
Other assets	320.0	443.5	424.7	342.6	301.6
Other creditors	(855.5)	(798.8)	(653.1)	(536.8)	(485.2)
Fixed assets	2,587.7	2,194.1	2,216.3	1,890.2	1,589.3
Total funds employed[6]	2,117.6	1,915.3	2,359.4	2,055.7	1,792.9
Net tax balances	(49.0)	(64.4)	(28.3)	(52.6)	(85.0)
Provision for dividend	(155.4)	(137.8)	(115.2)	(102.6)	(101.1)
Net assets employed	1,913.2	1,713.1	2,215.9	1,900.5	1,606.8
Net debt[7]	(387.6)	(82.2)	(731.3)	(527.9)	(381.1)
Total equity	**1,525.6**	**1,630.9**	**1,484.6**	**1,372.6**	**1,225.7**
Woolworths income notes	(583.0)	(583.0)	–	–	–
Outside equity interest	(3.7)	(3.3)	(3.2)	(2.9)	(2.5)
Shareholders funds	**938.9**	**1,044.6**	**1,481.4**	**1,369.7**	**1,223.2**

(Continued)

Table A7.10 (Continued) Five Year Analysis

	2001 52 weeks	2000 52 weeks	1999 52 weeks	1998 52 weeks	1997 53 weeks
Cash flow ($m)					
EBITDA	1,016.4	910.4	809.3	738.9	634.8
Movement in net investment in inventory	34.6	276.1	4.6	28.3	52.5
Other operating cash flows	8.7	56.7	(11.5)	(22.3)	59.8
Net interest paid	(25.0)	(33.6)	(53.7)	(42.8)	(41.1)
Tax paid	(225.7)	(142.7)	(174.7)	(193.1)	(46.2)
Operating cash flow	809.0	1,066.9	574.0	509.0	659.8
Gross capital expenditure	(537.4)	(420.8)	(764.9)	(685.0)	(699.4)
Proceeds on disposal	173.1	111.0	145.7	157.7	129.2
Other investing cash flows	(185.0)	16.8	(32.0)	3.2	(19.6)
Free cash flow	259.7	773.9	(77.2)	(15.1)	70.0
Movement in gross debt	211.5	(519.8)	290.9	221.0	53.2
Woolworths income notes	–	583.0	–	–	–
WINS distribution	(47.7)	(24.7)	–	–	–
Dividends paid	(212.1)	(173.5)	(154.1)	(153.3)	(96.1)
Share buybacks	(349.4)	(548.4)	–	–	–
New shares issued	44.0	26.9	22.7	23.3	12.3
Net cash flow	(94.0)	117.4	82.3	75.9	39.4
Shareholder value					

ROFE⁸ (Pre-tax return on funds employed)					
Before abnormals	35.04	29.08	24.43	26.83	26.33
After abnormals	35.04	24.69	20.52	25.11	26.33
Du Pont analysis (abnormals excluded)					
EBIT to sales	3.38	3.27	3.08	3.23	3.03
Debt burden⁹	91.40	91.33	91.56	91.71	90.83
Tax burden¹⁰	66.34	64.14	63.29	63.52	63.34
Asset turn¹¹	4.23	3.99	3.99	4.18	4.44
Financial leverage¹²	4.99	3.77	3.08	2.94	2.82
Return on investment¹³	43.19	28.92	21.88	23.15	21.88
Earnings per share					
Ordinary share price closing	10.85	6.18	5.0	5.28	4.31
Market capitalization ($m)	11,235.2	6,550.8	5,764.2	6,019.2	4,842.5
Weighted average shares on issue	1,065.8	1,125.0	1,146.2	1,132.4	1,109.4
Normal basic EPS	40.16	32.36	27.25	26.54	23.26
Total basic EPS¹⁴	40.16	26.27	22.42	24.67	23.26
EPS pre goodwill amortisation	41.01	32.89	27.67	26.78	23.41
Interim dividend	12.0	10.0	8.0	8.0	8.0
Final dividend	15.0	13.0	10.0	9.0	8.0
Total dividend	27.0	23.0	18.0	17.0	16.0
Payout ratio (before abnormals) %	66.37	66.88	66.28	64.34	69.53
Payout ratio (after abnormals) %	66.37	82.40	80.63	69.26	69.53
Price/earnings ratio (times)	27.0	23.5	22.3	21.4	18.5
Price/cash flow ratio (times)	14.28	6.50	10.00	11.70	73.00

(Continued)

Table A7.10 (Continued) Five Year Analysis

	2001 52 weeks	2000 52 weeks	1999 52 weeks	1998 52 weeks	1997 53 weeks
Growth rates (% increase)					
Sales	10.14	8.34	9.54	8.12	11.34
Sales per equivalent week	10.14	8.34	9.54	10.20	9.23
Same store sales	6.31	4.74	3.99	4.28	4.39
Sales per square meter	6.22	4.39	1.81	3.61	1.58
EBITDA	11.63	12.49	9.53	16.40	15.38
EBIT	14.04	15.24	4.49	14.97	14.60
NPBT	13.76	14.94	4.33	16.09	13.09
NPAT	44.84	14.98	(8.02)	8.29	10.45
Normal EPS	24.12	18.75	2.68	14.10	6.60
Financial strength					
Interest cover ratio	11.62	11.53	11.85	12.06	10.90
Fixed charges cover	2.40	2.40	2.30	2.30	2.19
Sales to inventory[15]	12.38	11.51	10.90	10.49	10.34
Gross capital expenditure to EBITDA (%)	52.88	46.22	94.52	92.71	110.17
Operating cash flow per share	0.76	0.95	0.50	0.45	0.59
Gearing[16] (%)	20.26	4.80	33.01	27.80	23.72
Current assets to current liabilities (%)	80.71	90.37	109.85	110.81	106.43

	71.90	64.88	67.86	65.20	64.53
Total liabilities to net tangible assets[17]	71.90	64.88	67.86	65.20	64.53
Productivity					
Stores (number)					
Supermarkets					
NSW and ACT	199	192	178	174	162
Queensland	115	112	111	111	106
Victoria	151	149	145	133	130
South Australia and Northern Territory	53	51	45	45	43
Western Australia	57	52	52	50	48
Tasmania	29	29	28	29	29
Total supermarkets	604	585	559	542	518
Freestanding liquor	130	41	42	38	38
Plus Petrol	166	137	98	49	12
General merchandise					
BIG W	90	87	85	82	78
Dick Smith Electronics	138	123	119	115	113
Powerhouse	9	6	4	2	1
Tandy	222	–	–	–	–
Crazy Prices	–	135	117	101	86
Rockmans	–	–	258	257	252
Total	**1,359**	**1,114**	**1,232**	**1,186**	**1,098**

(Continued)

Table A7.10 (Continued)   Five Year Analysis

	June 2000	Opened/acquired	Closed/sold	June 2001
Stores (movement)				
Supermarkets				
New South Wales	192	10	3	199
Queensland	112	3	–	115
Victoria	149	5	3	151
South Australia and Northern Territory	51	3	1	53
Western Australia	52	5	–	57
Tasmania	29	1	1	29
Total Supermarkets movements	585	27	8	604
Freestanding Liquor	41	90	1	130
Plus Petrol	137	29	–	166
General merchandise				

	2001 52 weeks	2000 52 weeks	1999 52 weeks	1998 52 weeks	1997 53 weeks
Big W		87	3	–	90
Crazy Prices/Variety		135	9	144	0
Dick Smith Electronics		123	15	–	138
Dick Smith Power House		6	3	–	9
Tandy		–	223	1	222
Total store movements		**1,114**	**399**	**154**	**1,359**
Area (sqm)					
Supermarkets	1,317,840	1,254,744	1,206,202	1,149,431	1,105,518
General merchandise	602,718	614,515	619,333	589,029	552,645
Total	1,920,558	1,869,259	1,825,535	1,738,460	1,658,163

(Continued)

Table A7.10 (*Continued*)   Five Year Analysis

	June 2000	Opened/ acquired	Closed/sold	June 2001
Sales per square meter				
Supermarkets (excluding petrol)	12,727.1	11,811.5	11,635.8	11,382.7
General merchandise	4,127.6	3,369.1	3,202.6	3,148.3
Total	10,028.4	8,947.3	8,778.4	8,638.3

Source: Woolworths Ltd. Financial report to shareholders. 2001.

Notes to statistics

[1] Sales for prior periods have been restated to exclude WST.

[2] Discontinued operations include Chisholm Manufacturing and Crazy Prices sold in 2001 and Rockman sold in 2000.

[3] EBIT for the periods 1998 to 2000 are as previously reported Ie excluding individually significant non-recurring items (previously described as abnormal items).

[4] Supermarket EBIT for prior periods has been restated to reflect IT costs previously reported as unallocated.

[5] Unallocated expense represents corporate costs relating to the Woolworths group as a whole, and profits derived by the group's corporate property division including the disposal of development properties. These amounts are not identifiable against any particular operating segment and accordingly they remain unallocated, as required by Accounting Standard MSB 1005.

6 Funds Employed is net assets excluding net tax balances, provision for dividends and net debt.

7 Net debt is gross debt less cash on hand, cash at bank and cash on short term deposit.

8 Return on Funds Employed (ROFE) is EBIT as a percentage of average funds employed for the year.

9 Debt burden is net operating profit before income tax expressed as a percentage of EBIT before abnormal items.

10 Tax burden is normal profit after income tax expressed as a percentage of normal profit before income tax.

11 Asset turn is Total Sales divided by average Total Assets for the year.

12 Financial leverage is average Total Assets divided by average Shareholders Funds for the year.

13 Return on investment is net profit after income tax, divided by average Shareholders Funds for the year.

14 Total basic earnings per share is net profit after income tax attributable to Members of the Company after WINS distribution, divided into the weighted average number of ordinary shares on issue during the year. The weighted average number of ordinary shares on issue has been calculated in accordance with Accounting Standard AASB 1027. Fully diluted EPS is not significantly different from basic EPS.

15 Sales to inventory is total sales for the period divided by average inventory.

16 Gearing is net repayable debt divided by net repayable debt plus total equity.

17 Total Liabilities excludes deferred income tax liability and provision for dividend and includes outside equity interests.

Supply chain strategy was developed after evaluating systems and logistics features of leading global retailers following which the company determined an appropriate and optimum solution for Woolworths. This solution would address the following key design considerations:

- Common integrated systems required to support supply chain operations
- Store supply chain costs (from the supermarkets' back dock to the shelf)
- DC location and numbers
- DC function (cross-docking and flow-through)
- Composite supply chain (integrating cold and ambient)
- Transport management (primary and secondary freight)
- Process improvement across the network
- Improve buying and supply chain systems to help make cost savings in these areas
- Make distribution process easier and much efficient for the stores

The Mercury Programme would be the umbrella term for all projects across the business that impacted the way in which products were supplied from vendors to customers. It would be the consolidation of many projects already underway within supply chain, buying & marketing, and supermarket operations, with the support of IT and human resources.

One of the important decisions was about consolidating the 31 DCs into a more responsive and cost efficient supply system. Figure A7.5 gives the current DC network. Table A7.12 gives the specific locations of the DCs and product types being supplied. The proposed DC network and locations are given in Figure A7.6. This network would consist of nine regional DCs (RDCs) and two national DCs (NDCs). Figure A7.7 gives the before and after structure of flows between DCs and outlets. The rationale for this proposal is given in Appendix 7.1.

The NDCs would distribute to over 700 stores nationally. The most significant change and challenge for NDC was to convert its operations from servicing just over 238 NSW stores to servicing nine RDCs and over 700 stores nationally.

For example, the Sydney NDC was one of the two NDCs in Australia. The other was Mulgrave Grocery in Melbourne. Sydney NDC would range ambient products such as slow moving grocery and general merchandize lines. It would service Sydney and Melbourne Metro stores directly from the NDC and service all other stores via their respective RDCs.

Delivery frequencies would change from daily store deliveries to once or twice a week. For example, Yennora DC would move 700,000 cartons in a seven day operation. Sydney NDC would move 1.3 million cartons, previously a six day operation.

The RDCs, such as Sydney and Wyong RDCs, would service supermarkets with fast moving lines. They would be a composite facility, providing fast moving grocery, general merchandize, frozen, locally manufactured and all chilled and fresh products to stores. Two-third of the goods supplied would be from Sydney or Melbourne NDCs. Reducing the 31 DCs into 11 multi-temperature sites would reduce DC operating costs by $14 m in FY08.

Figure A7.5   DC network map showing number of DCs per zone. (From company data.)

Table A7.11   DC Network, Location and Type

Location	Zone	Type			Total
		Ambient	Produce	Chilled/Frozen	
Perth(WA)	Central West	3	1	1	5
Adelaide(SA)	Central West	1	1	2	4
Melbourne(VIC)	South	3	1	1	5
Devonport(TAS)	South	2	1	1	4
Sydney(NSW)	East	5	2	1	8
Brisbane(QLD)	North	3	(1)*	1	5
Townsville(QLD)	North		1		1
Total		17	7 (8)	7	31 (32)

Source: Company data.

* Under consideration.

Table A7.12 DC Locations

DC	State	Name	Location	Type
1899	NSW	Moorebank DC	Moorebank	Produce
1904	NSW	Helles Avenue DC	Moorebank	Ambient
1944	NSW	Yennora	Yennora	Ambient
1979	NSW	Sydney RDC– Ambient	Minchinbury	Ambient
1905	NSW	Hume AIW–Liquor	Alexandria	Ambient
1911	NSW	Aiw Warwick Farm	Warwick Farm	Ambient
1947	NSW	Sydney RDC–Fresh	Minchinbury	Produce
1910	NSW	Versacold Arndell Prk Sfd DC	Arndell Park	3rd Party
2899	QLD	Everton Park DC	Everton Park	Ambient
2920	QLD	Acacia Ridge DC	Acacia Ridge	Ambient
2908	QLD	Beenleigh Road DC	Acacia Ridge	Ambient
2953	QLD	Versacold Satellite Warehouse	Murarrie	3rd Party
2919	QLD	Townsville Produce DC	Bohle Townsville	Produce
3902	VIC	Hume DC	Broadmeadows	Ambient
3919	VIC	Mulgrave Produce DC	Mulgrave	Produce
3989	VIC	Clayton DC	Clayton	Ambient
3911	VIC	Melbourne NDC	Noble Park	Ambient
3908	VIC	Versacold Laverton DC	Laverton North	3rd Party
5903	SA	Gepps Cross Produce DC	Epps Cross	Produce
5910	SA	Adelaide RDC–Ambient	Gepps Cross	Ambient
5911	SA	Adelaide RDC–Liquor	Gepps Cross	Ambient
5918	SA	Croydon DC	Ridleyton	Chilled
4899	WA	Miles Road DC	Kewdale	Ambient
4901	WA	Miles Road DC	Kewdale	Ambient

(Continued)

Table A7.12 (*Continued*)   DC Locations

DC	State	Name	Location	Type
4905	WA	Perth Produce DC	Perth Airport	Produce
4916	WA	Bunbury DC	Boyanup Road	Ambient
4903	WA	Versacold Spearwood RDC–Freezer	Spearwood	3rd Party
7180	TAS	Derwent Park DC	Derwent Park	Chilled
7380	TAS	Breadalbane DC–SIW	Breadalbane	Ambient
7385	TAS	Prospect DC–SIW	Prospect	Ambient
7191	TAS	Devonport Produce DC	East Devonport	Produce

Source: Company data.

Note: All 3rd Party were for chilled.

Supply Chain

Project Mercury's mission was to implement the world's best practices in supply chain management to provide a better shopping experience for the customers each and every time. The supply chain strategy would not be just about DCs and transport, but would involve every part of the supermarket's business. This would identify different areas for improvement such as:

- The 'flow through' of products from suppliers to stores with less handling and lower cost along the way (Figure A7.8)
- Product movement through the supply chain 'ready to fill' directly onto store shelves
- The right amount of stock arrivals at the store when the stores need it
- Inventory management in the supply chain and forecasting stock requirements at DCs and stores
- The efficiency and effectiveness of the transport network (Figures A7.9 and A7.10)

The SCET, in partnership with the buying and marketing team, were working with the suppliers to achieve these improvements.

Systems to Support Supply Chain Strategy

The SCET identified possible benefits of implementing IT systems throughout the supply chain. Hence the team was in a process of considering inclusion of following systems to support supply chain strategy as a part of Mercury Project.

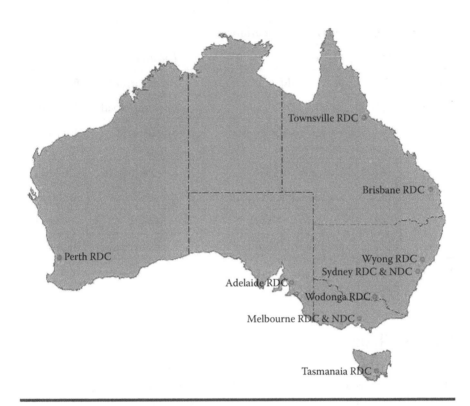

Figure A7.6 Map of proposed DC network. (From company data.)

Table A7.13 Location of Proposed DCs

States	Cities	NDC/RDC
Queensland	Brisbane Townsville	RDC RDC
New South Wales	Sydney Wyong	NDC/RDC RDC
Victoria	Melbourne Wodonga	NDC/RDC RDC
Tasmania	Devonport	RDC
South Australia	Adelaide	RDC
Western Australia	Perth	RDC

Source: Company data.

Abbreviations: NDC, National Distribution Centre; RDC, Regional Distribution Centre.

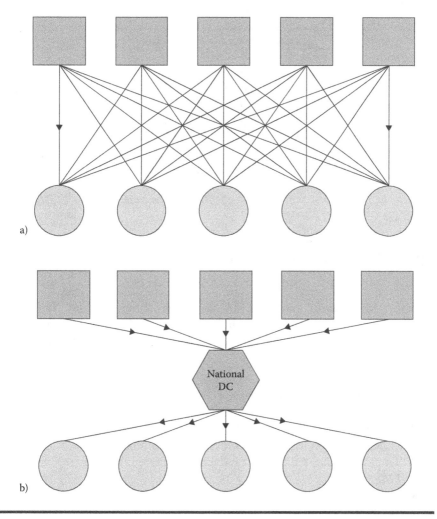

Figure A7.7   Structure of flows between DCs and outlets: a) before consolidation; b) after consolidation. (From company data.)

IT

IT systems would play a major part in implementing supply chain strategy. The aim was to ensure the maximum service level to the customer with minimum inventory levels and operating costs. All components of the supply chain (including suppliers) would be linked electronically with much greater coordination between their operations. Appendix 7.1 provides a description of the IT systems.

IT would deliver maximum benefit to the business from the use of technology for the lowest cost. The systems associated with IT were Warehouse Management System (WMS),

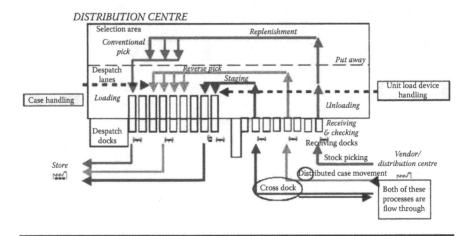

Figure A7.8   Flow through. (From company data.)

Transport Management System (TMS) and Replenishment System. These were already in use for different operations in DCs and stores. A revamped form of these systems would be executed as a part of Project Mercury. Broadly, following programmes would be under IT systems:

- Automatic store replenishment with AutoStockR
- DC replenishment with StockSMART
- Computer-based ordering and receiving for direct store deliveries (DSD)
- Tracking stores stock on hand with Perpetual Inventory
- Better managing the flow of products into and out of DCs with enhanced TMS
- Operations of DCs through WMS

In short, the following programmes in revised form would be executed as part of the Mercury Project.

- WMS
- TMS
- Replenishment System

WMS

WMS, a database-driven computer programme, would be an integral part of Project Mercury which controls the movement of stock, people, forklifts, etc. within the DCs.

Supply Chain Transport
Making it easier to receive deliveries and clear the backdocks

Michael Luscombe
General Manager Supply Chain

We've come a long way – a delivery being unloaded in 1956.

In the Supply Chain Division we manage the transport of stock across the supply chain, from vendors to distribution centres and to stores.

Our transport operations are managed in several different ways. For example in Victoria and Hobart, we own and manage our own transport operations. In other areas, we have commercial arrangements in place with external transport companies to deliver products from our distribution centres to stores.

It all amounts to a massive transport operation. In fact, the combined size of the fleet we use to service our 700 supermarkets is one of the largest in Australia. The size of the transport task provides us with the opportunity to improve efficiencies in many areas.

We are constantly working towards improving our transport operations to ensure that we provide the best possible service to our customers, the stores. One of the ways we are doing this is by taking on more responsibility for the management of our transport. By taking on this responsibility we are able to coordinate how our transport operations are run and therefore provide better service to stores.

Here are some of the things we are doing to improve our transport service.

Improving service to stores
In the future, transport to metropolitan stores will be quite different to what it is today.

At present, supermarkets typically receive large loads of product from our distribution centres for each department. For example, on any given day, metropolitan supermarkets may receive deliveries from up to six trucks a day. On average, one of these trucks will carry a load of fresh produce, one will carry chilled and frozen items, three will carry loads of dry groceries and one will carry a load of general merchandise.

As many of you well know, disseminating stock into the store is not always easy because sometimes there just isn't enough room for it on the shop floor which means the stock needs to be stored on top of the capping area, in the backdock or in the coolroom. Delivering large quantities of one type of product adds to this problem.

In the future we will move away from delivering large loads of similar product to delivering the various products to the stores as and when they need it. We will do this by using trucks that can carry composite loads. A composite load is a mixture of product categories on the one truck. Smaller quantities of products from different categories will come off the truck on Roll Cages and then be put straight out on display, ready for purchase by the customer. This means there will be less stock in the backdock.

To enable composite loads, future trucks will have trailers which can transport different products on the same load. Trailers will have separate areas for produce, chilled and frozen items and dry grocery and general merchandise. We will be trialing composite loads soon in South Australia.

We are focussed on assisting stores in keeping back rooms as clear as possible through the timely removal of returnable items like Roll Cages, pallets and bins. The returnable items are loaded onto empty trucks leaving the store. This not only helps reduce the clutter in back rooms but is a much better way of utilising our trucks.

We are also working hard towards improving on time delivery to stores through improved communication procedures between distribution centres, transport operators and stores.

Focusing on safe operations
We only do business with reputable companies who value safety as highly as we do and who have similar safety standards. The transport companies we do business with must have,

- drivers who are skilled and take part in regular training
- high quality equipment that is well maintained
- responsible driving programs including fatigue management programs
- environmental policies and standards, such as vehicle disc brakes to cut down noise levels and low emissions from trucks with modern refrigeration units.

WOOLWORTHS NEWS

Figure A7.9   Supply chain transport. (From Woolworths Ltd.)

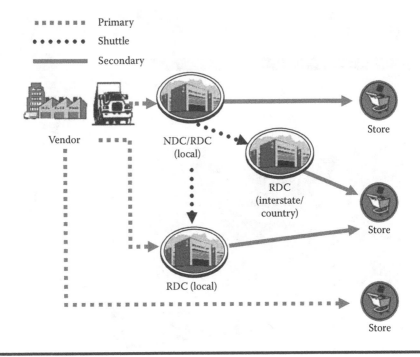

Figure A7.10   Transport network. (From company data.)

This was also responsible for running the DCs operating processes and information flows such as receiving stock from vendors, picking store orders and dispatching these orders to stores. WMS would provide significant benefits to DCs in terms of speeding up picking times and pick accuracy.

A typical warehouse of Woolworths supermarkets would receive products such as groceries, perishables/frozen food, bakery, deli, fresh produce, meat and liquor with more than 100,000 lines. The number of cartons received per year would be 680 million and the number of pallets received per year would be 8.5 million.

Woolworths supermarkets were facing the following operational issues which led to implementation of WMS under Project Mercury:

Receiving

- Full screen receiving with invoices from the drivers
- No handheld receiving at the docks
- No advanced shipment notice receiving
- No Express Receiving-Lower Productivity

Picking

- Picking with labels
- No Hands free picking
- Less accurate and productive especially Non CLS picking

Replenishment of Pick Slots

- Triggering of replenishments not based on real-time balance on hand

Loading

- Manual paper based loading
- Consolidation without using the system
- No visibility of loading status

Miscellaneous

- Insufficient pick slots for the products
- Inefficient management of despatch lanes
- No updating of put-away priorities
- No optimization of secondary freight
- No concept of umbrella warehousing
- No proper management of warehouse mark-outs
- Produce warehousing managed by different system
- Limited cross-docking and flow-through—mostly manual
- Paper based picking for interstate stores and warehouse transfers
- No capability of rejecting stock before put-away was performed
- No advance shipment notice from vendors—no EDI documents to and from the warehouses

This system would be one of the components of end-to-end supply chain strategy which supported RDCs and NDCs. The key considerations for redesigning WMS would be safety, accuracy and cost. The new features of WMS would be:

- All product categories in composite multi-temperature DC
- Voice-enabled processes (picking, forklift drivers)
- Cross-dock processes (NDC and vendor)
- Dynamic pick locations (multiple slots, rotating slots)
- RF loading across multiple temperature zones
- Random weight processes

WMS strategy would also be set up for pilot site selection and rollout. Following would be the business benefits of implementing WMS:

■ A single integrated WMS would be implemented to support the future network of multi-temperature distribution centres (RDCs)
■ QA management would allow quality assurance of pallets received at the DC and rejection of low quality produce products
■ Products would be maintained with quality suffix and inter item transfers would be performed
■ Increased picking accuracy
■ Increased labour productivity
■ Reduced cost of printing and distributing picking documents
■ Reduced cost of re-keying order amendments, picking confirmations
■ Hands free and eyes free—makes picking easier (helpful for items like frozen foods which were handled through gloves)
■ Real time stock updates
■ Would enable business to inquire and maintain the cross dock information by transport order, thus enhancing the visibility of the incoming cross dock Unit Load Devices (ULDs) from NDC and the cross dock ULDs that had arrived in RDC but were not yet loaded, thus giving the RDC a better planning capability
■ Reduced manual intervention required to manage cross dock ULDs in the warehouse
■ Automation of pallet handling (put away, replenishment, empty pallet handling) in ambient component of the facility
■ Would satisfy a 2011 design year OM (Other than Meat) throughput of approximately 225 million cartons per year across all categories

DCs would also consider 'voice picking–hands and eyes free' technology that eliminates need for paper during the order picking process. It would be the best method for improving picking in DCs. It was expected that accuracy of picking would increase substantially. This would be a step towards store friendly principles.

Company Data

TMS

Primary Transport

Primary transport means movement of product from vendors:

■ to the NDC/RDC network
■ into an interstate DC for full truck load (FTL) deliveries from vendors
■ DSD

Shuttle Transport

This was a movement of product between DCs (local, intrastate and interstate DCs).

Secondary Transport

This was a movement of product from the RDCs or NDCs to the stores and the return of ULDs, i.e. roll cages, dollies, crates, etc.

TMS, a web-based software would help manage primary freight (movement of goods from vendors to DCs). It would keep track of the shipments in the transportation process. This would enable the operators to plan accurately and direct the movement of trucks to get the stock arriving in the DCs at the right time for the lowest cost. This system was used by Walmart, Tesco and other large retailers overseas.

TMS would be introduced in Project Mercury's supply chain strategy after identifying the following benefits:

■ Improved delivery to DCs
■ Utilization of Woolworths' volumes to consolidate shipments
■ Greater control over flow-through and cross-docking
■ Improved visibility of product movement in the supply chain
■ Lower inventory

The SCET would manage the transport of stock across the supply chain, from vendors to DCs and to stores. The transport operations, serving 700 supermarkets would be one of the largest in Australia.

CAPs, a Vehicle Scheduling and Routing System would help plan the transport to stores in case of access restrictions to stores due to curfews or physical restrictions. The system would map out the best possible routes and sequences for trucks to take.

Replenishment System

Various replenishment programmes would be initiated as part of the Mercury initiative:

■ Automation of the existing replenishment system
■ Fresh Food, Produce and DSD integration into automated replenishment system
■ Integration of DC forecasting process and store replenishment process

Following would be the major replenishment systems considered for implementation:

■ StockSMART–Improved DC level forecasting including promotions
■ AutoStockR–Store forecasting and replenishment
■ WOWLink–Improved information to vendors
■ DSD Conversion Project
■ Self-Ready Trays
■ Roll Cages

A major task would be to build the link between store and DC replenishment to support an integrated solution.

Following would be the future considerations of replenishment system:

- Store level promotional forecasting/replenishment
- Mix of aggregated store forecasts (fast movers) and DC-level forecasts (slow movers) for DC replenishment
- Introduction of system forecasting and replenishment into fresh food categories
- Closer relationships with key vendors to support flow-through

StockSMART (DC Forecast Based Replenishment)

StockSMART would be an automated forecasting and replenishment system for stock in DCs. It would be implemented in all DCs for grocery and general merchandize items. Service levels would be expected to improve by over 2% while 3 days' worth of inventory would be removed from the DCs.

AutoStockR (Store Forecast Based Replenishment)

AutostockR or Automated Stock Replenishment would be a computer assisted ordering system that would order stocks for departments in supermarkets, including liquor. This system would be a key foundation of the Mercury Project. AutostockR would provide a forecast for each item in every supermarket across the country. This forecast would enable supply chain to gain greater vision of store requirements, improving DC service levels and the way in which stock would deliver to supermarkets. AutostockR would be running in over 100 supermarkets. The potential benefits of this system would be improved productivity, reduction in store out of stocks, unnecessary inventory in stores, and double handling in stores.

WOW Link

WOW Link would be an electronic resource, via internet to monitor Woolworths information from across the buying and supply chain areas. This would provide Woolworths and its trading partners access to:

- Forecast Replenishment plans
- DCs inventories
- Details of vendor's purchase orders
- Upcoming promotions
- Service level information
- Allocation details and Woolworths' product records
- A vendor's logistics performance (via scorecards and alert messages)

Developed in collaboration with a pilot team of 8 trading partners, WOW Link would be a three phased system. The initial phase would provide a basic one way flow of information between Woolworths and its trading partners. The second phase would enable some interaction between the two in terms of exchanging a range of information. The third phase would provide full interaction and collaboration between Woolworths and its trading partners.

DSD Conversion Project

Woolworths' 15% of volume was supplied direct to stores from the vendors. Another valuable component of the Mercury Programme would be the DSD Conversion Project. This project was aimed at converting 50% of DSD vendors over to warehouse supply through Woolworths' DCs. Benefits would be experienced not only for Woolworths but also for the vendors. There would be an increase in delivery frequency to the stores, the removal of congestion at stores' back docks and reduction in the number of invoices to be processed.

Shelf Ready Trays

Shelf Ready Trays were packed trays wrapped with film or a cardboard sleeve that could directly be fitted on the shelf for display. These would be identified as efficient shelf replenishment methods which would improve productivity as well as occupational health and safety at store level and reduce the end-to-end supply chain costs.

Roll Cages

Roll Cages were ULDs for transporting products from the DCs to the stores, rather than using a pallet. Roll Cages reduced the number of times stock was touched at the store and thus reduced the handling costs. Roll Cages would be introduced in Woolworths DCs and stores in Perth and WA. The SCET would proceed with the next stage of the Roll Cage implementation to other parts of the country.

The rationalization of DCs, combined with new cross-dock and flow-through processes, supported by the new WMS, would utilize the site advantages and further reduce both cost and stock levels, and the cost of transport from DCs to stores. Reducing the volume of DSD and introducing electronic store delivery would reduce costs by utilizing DC's infrastructures as well as eliminating administration costs. For stores, the introduction of phased replenishment store restocking capabilities, along with store ready ULDs (eg. shelf ready trays and roll cages) would reduce overall costs. (myfuture: http://www.myfuture.edu.au/)

Looking Forward

Michael Luscombe was looking for a BPO service provider for his company. He expected that the BPO partner should not only see from the IT perspective but should also have a strategic insight for analyzing the financials of the company. The BPO service provider should be able to visualize the implications of restructuring the distribution network along with implementation of various IT systems. Most important was the position of the company in the competitive world of other retail companies in terms of service and cost.

Appendix 7.1

Mercury Programme

The Mercury Programme would be considered as a business transformation for Woolworths supermarkets.

Following were some of the objectives of Mercury Programme:

- Changes to business to support the Supply Chain Development Programme (SCDP) are wider than just "Supply Chain" Operations
- Significant changes required across supply chain, buying & marketing, IT, stores, vendors
- Co-ordinate cross-divisional activities and make it a total business strategy
- Currently stores receive stock from multiple DCs
- A Regional Distribution Centre consolidates these sites into a single point of supply for a group of stores
- These new Distribution Centres process all merchandize categories, requiring multiple temperature environments
- $14M result from fractionalizing DC labour cost
- Together with composite transport, these facilities provide improved service to stores by enabling multiple deliveries for all product categories
- 100% control of DC network
- Transport benefits result from reduced number of distribution nodes

As a part of Mercury programme WMS, TMS and Replenishment systems would be implemented. The respective objectives were:

WMS Project Objectives

Develop WMS Business Requirements required to support NDCs and RDCs of the future - for all categories.

- New Processes: Cross Dock, Carton Flow Through, Voice Picking, optimized system directed work sequencing, etc.
- User requirements as defined by current WMS business users
- Requirements to support existing processes (interface and integration)

Develop WMS strategy for existing DCs

- What should be done with existing DCs from a WMS perspective
- Pilot Site selection

TMS Project Objectives

- Execution: Support the management and execution of all freight within the organization
- Visibility: Provide visibility of freight movement to buyers, vendors, carriers, DC's, stores and transport operations
- Optimization: Optimize freight movement for both inbound (from vendors) and outbound (to stores)
- Automation: Remove much of the administrative overhead currently associated with managing transport
- Reporting and Monitoring Decision Support: Provide comprehensive reporting and monitoring capability for tracking operational performance and cost.
- Make decisions regarding product flows to support the new NDC/RDC flow through network

Replenishment Project Objectives

Key business requirements to be fulfilled by the future replenishment capability:

- Support store service level requirements while reducing store inventory and out-of-stocks
- Support delivery flexibility and one touch initiatives to minimize store labour
- Standardize processes/systems for store and DC replenishment across all categories
- Warehouse demand forecasting based on aggregated store demand
- Responsive to merchandizing strategies such as promotions, events, allocations, seasonality
- Key input for transport planning
- Support of flow through (Cross Dock/Reverse Pick) processes
- Provide tool to centrally manage service levels and presentation level policies
- Support alternate source of supply to provide disaster contingency
- Providing medium and long range forecasts to vendors
- Provide visibility and measurement of replenishment activities

Network Model Project

Leading retailer of Australia executed network optimization project and came to a conclusion on the number of DCs to be operated.

Project Brief:

- Determine the location and number of RDC and NDC facilities.
- Capital costing of proposed DCs infrastructure.
- Determine the optimal range configuration and capacity in each facility.
- Determine the annual throughput levels guidelines on range in facilities (taking into account flow through/XD).
- Be available for iterative use in the event that input data becomes more accurate or identified constraint conditions deem it appropriate.
- The model is required to provide multi period outputs, to be utilized as an ongoing tool as required.

The Network Modelling Project delivered:
The expected number, location and size of the RDCs and NDCs:

- The estimated cost of the network
- Insights into 2008 DC network and flows
- A network data set
- Issues for other project teams
- Next steps for network modelling process

Key Assumptions:

- Five product categories: Ambient, Chilled, Frozen, Meat and fish, Produce
- Management of peaks and range changeovers incorporated into network flows and DC design projects
- All produce, chilled, and meat and fish ranged in RDCs
- All national frozen lines ranged in NDCs
- All local SKUs (including frozen) ranged in RDCs
- Each market area serviced by only one RDC
- RDCs are allocated the fastest XX Ambient SKUs
- Some of which will be ranged and some of which will be flow-through

Analysis to date:

- 500, 2000, 3500 and 5000 fastest ambient SKUs allocated to RDCs
- RDCs at Sydney, Melbourne, Adelaide, Perth, Brisbane, Devonport, Townsville, "Yass", "Newcastle", "Tamworth", "Seymour", "Albury"
- NDCs at Sydney, Melbourne, Brisbane, "Albury", "Tamworth"
- Other sites (Darwin, Overseas, Perth NDC) were considered, but low volumes did not warrant a full analysis

Note: Town name only to be taken as approximate location

- Costs included: DC operating (picking, flow-through, cross-dock), transport (primary, secondary), capital costs
- Costs excluded (for now): Inventory, capital costs of XD, capital cost of FT automation,
- DC fixed costs

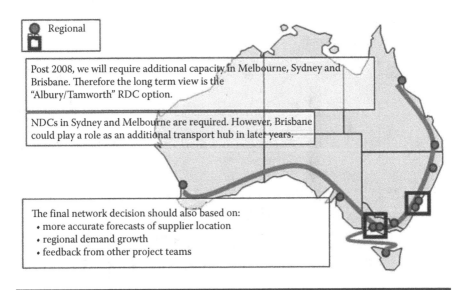

Figure A7.11   Suggested optimal solution, post-2008.

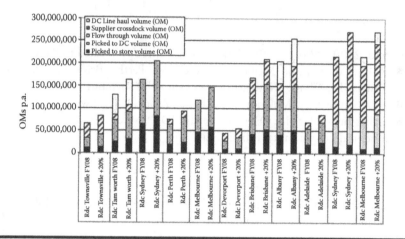

Figure A7.12 This model of 2008 and later projections offers flexibility to cater to at least 20% more volume than FY08.

Factors Involved in Developing an Optimal Network

■ Where are the suppliers?
■ Where is the demand?

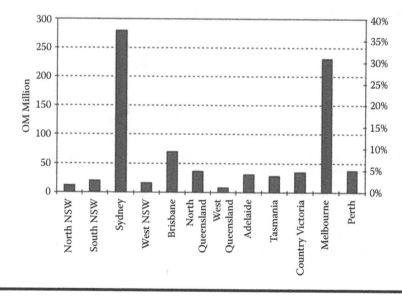

Figure A7.13 Two-thirds of all goods are supplied/made out of Sydney or Melbourne. (Based on vendor survey of top 50% of each category.)

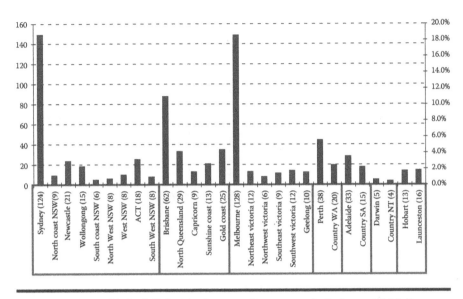

Figure A7.14   A third of all demand comes from stores in Sydney and Melbourne.

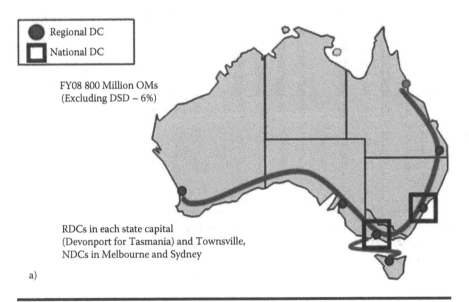

Figure A7.15   Initial base case network. a) Without Townsville operating costs are $40 M higher.

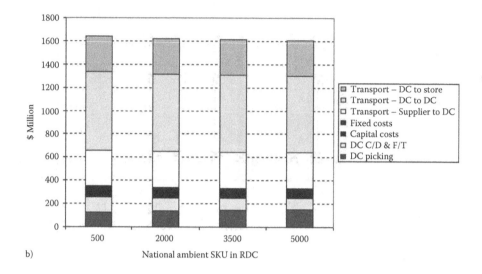

b)

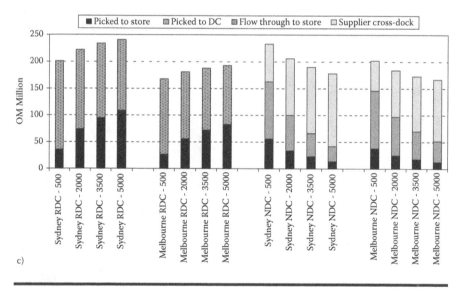

c)

Figure A7.15 (*Continued*) Initial base case network. b) and c) Total costs do not vary greatly as RDC size varies. We are viewing the entire supply chain from factory to store.

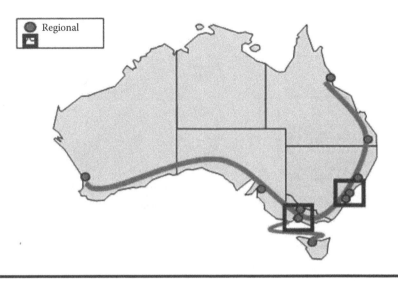

Figure A7.16   Initial optimal network, based on current data and costs and current assumptions.

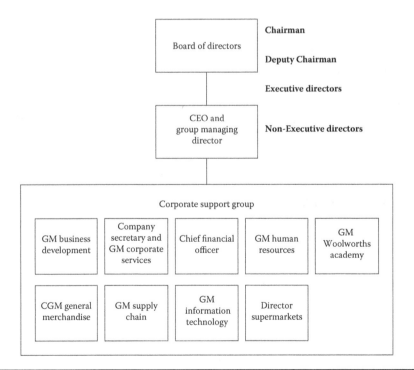

Figure A7.17   Organizational structure. CEO = Chief Executive Officer; GM = General Manager; CGM = Chief General Manager.

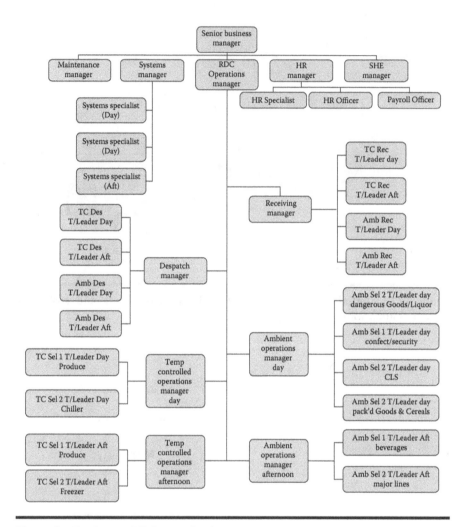

Figure A7.18 Larger RDC organisation structure.

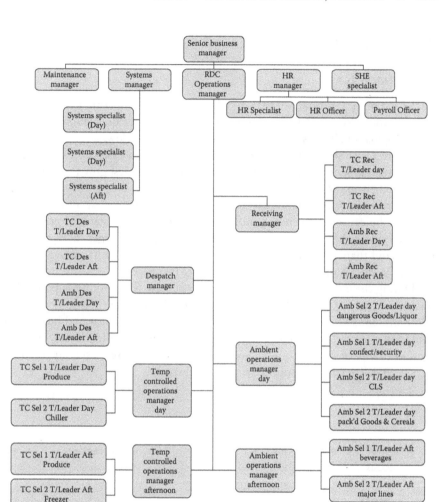

Figure A7.19   Smaller RDC organisation structure. (From company data.)

Appendix 7.2

Supermarkets

At the end of the last financial year, Woolworths had 604 supermarkets distributed right around Australia. The distribution of supermarkets is closely aligned with that of the population. This is clearly shown in Figure A7.20 below.

Figure A7.21 shows the growth in store numbers, state by state, over the last five years. New South Wales and Victoria have been the fastest growing states while store numbers in the other states have been relatively static.

Surprisingly, Queensland has not been one of the fastest growing states as might be expected given the shift in population to that state. However, this may be explained by Woolworths', until recently, having a larger market share in Queensland than any other state apart from Tasmania, which has seen almost no increase in store numbers over the period. Refer Figure A7.22.

During the last fiscal year an agreement to acquire 67 Franklins supermarkets was entered into[2]. The supermarkets, including refurbishment costs, were to cost $360 million, add an expected $1.5 billion to sales in a full year and be earnings neutral in the current fiscal year. These figures will be a little higher now that the number of supermarkets acquired has increased to 71. All, except six, supermarkets have been delivered and re-opened under Woolworths.

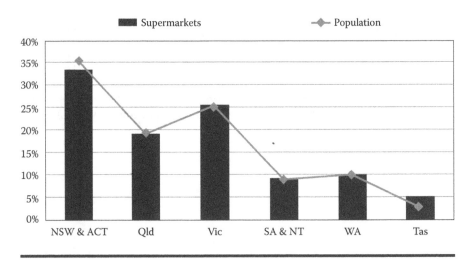

Figure A7.20   Supermarket and population distribution.

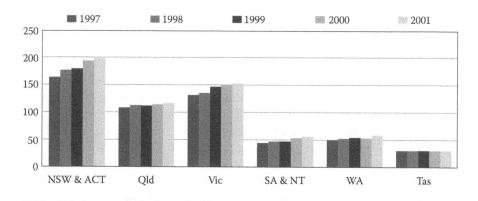

Figure A7.21 Growth in store numbers by state.

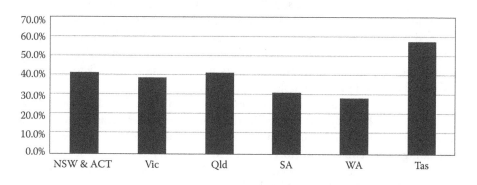

Figure A7.22 Market share by state. (From AC Nielson ScanTrack. Total defined grocery, Oct. 10, 2001.)

Woolworths also operates a number of smaller supermarkets under the Food for Less banner in New South Wales and Queensland. Some of the smaller Franklins supermarkets are now Food for Less stores. These smaller stores offer mostly non-perishable grocery items and usually include frozen food, dairy and delicatessen areas. The stores average 1000 square metres in size and are generally located in secondary or suburban strip shopping centres.

Growth in the retail division will be driven through increasing store numbers and bolt-on acquisitions for both supermarkets and liquor. Under a five year plan supermarket numbers will increase by 15–25 per annum. Some of these will be green field developments while others will come from acquisition of independent operators.

Retail Strategy and Supply Chain Management

Woolworths' retail strategy can be defined as every day, low price (EDLP). The EDLP strategy is sometimes referred to as the Wal-Mart approach. Wal-Mart is the world's largest and most successful supermarket retailer. EDLP means seeking the best prices from suppliers and maintaining minimum profit margins on every product sold—constantly. It is not about 25% off store wide sales, or rotating specials on selected items, and then maintaining higher prices the rest of the time. With EDLP, Woolworths only has specials when the supplier provides the cost reduction.

A critical aspect of supply chain management is relationships with suppliers. Under the old state based buying function, relationships with suppliers were fragmented and varied. The move to a single, national buying function provided an opportunity for all supplier contracts to be re-negotiated, delivering cost savings to Woolworths and its suppliers and stronger relationships. Where Woolworths did not always rate highly with its suppliers previously, it is now rated No.1.

While buying and marketing have moved to a national shared services platform, distribution, by necessity, is a zone based function. Distribution centres are maintained in each zone with transportation from distribution centres to stores being undertaken by transport contractors in each state except Victoria, where Woolworths maintains its own fleet—a legacy from Safeway.

Two high velocity distribution centres are operated in NSW and Victoria, these being the Minchinbury and Hume Distribution centres. These centres receive and distribute to Woolworths supermarkets the 4,500 fastest moving grocery items, excluding fresh, chilled and frozen foods. Minchinbury and Hume service their own states and some of the needs of Tasmania. This amounts to 180 stores for Hume and more than 200 for Minchinbury.

The distribution centres are hi-tech, fully automated, dedicated facilities where the primary objective is to move product through the centre as quickly as possible, thus minimising the investment in inventory. Bulk buys of goods do not go into storage but will be held on the distribution floor as 'cross stock'. In other words, the goods are moved from the arrivals bays to the departure bays within two to three days or preferably less.

Other goods move into a high rise warehouse, serviced by robotic cranes for storage for periods of three to six weeks, before being moved to the stock picking area prior to distribution to the supermarkets. Cross stock goods remain on pallets while stored goods are removed from pallets in the stock picking area for distribution in smaller quantities.

At the present time it is necessary to maintain sizeable goods inventories in distribution centres to avoid stock-out problems in stores. The existing inventory management systems employ computerisation and bar code and scanning technology but the ability for stores to re-order from check-out scanners and suppliers to deliver, just in time, is some way off. This ability should be established within three to four years and lead to significant annual supply chain savings.

Lower volume product is processed through separate distribution centres that are less dependent on technology and more labour intensive. These centres handle 9,000 slower moving items including some general merchandise. There are separate distributions facilities again for fruit and vegetable produce and frozen and chilled foods are handled by P&O Cold Stores.

There is also a national distribution centre at Moorebank in NSW that handles 4,000 lines of general merchandise goods and imported goods for the supermarkets that are not particularly time sensitive.

Plus Petrol

The first Plus Petrol site was opened in Dubbo in 1996. The business has steadily expanded to the point where it generated 4% of the Supermarket division's sales but less than 1% of EBIT, last year. Woolworths views the petrol business as an adjunct to its retail offering, creating incremental store sales. Petrol retailing fits with Woolworths' core competency being a high volume, low margin business and, with canopies located adjacent to supermarkets, it allows Woolworths to meet more of its customers' needs in the one location.

The petrol business returned a profit of $4.6 million in fiscal 2001, before allowing for the cost of the discount given to Woolworths shoppers and the value of incremental sales. While no figures are provided, Woolworths says incremental sales exceed the cost of the discount. An independent analysis undertaken by Jebb, Holland, Dimasi showed that there is a significant incremental sales benefit for stores with adjacent canopies.

Moreover, more volume from existing and new outlets will improve the economics of the petrol business. With currently 212 canopies, Woolworths says it is close to achieving breakeven on a standalone basis.

The number of Plus Petrol outlets has rapidly increased as Figure A7.23 shows. At the end of fiscal 2001, Woolworths had 166 petrol outlets and expects to have 250 by year end. Some of the growth will come from 69 ex-Liberty Oil outlets to be added to the chain under a lease arrangement. Woolworths has already taken delivery of 35 of the outlets.

Liquor

Woolworths stand-alone liquor business has been much slower to develop. Until fiscal 2001, stand-alone stores numbers had been constant at around 40 but 90 stores were added last year to bring the total to 130 by year end. The growth came mostly from a buying spree that saw Woolworths acquire the 13 store Booze Bros chain in South Australia, the 11 store Toohey Bros chain in NSW and the 45 store Liberty Liquor chain was acquired during the year at a cost of $72 million.

The total liquor business has been reorganised under four brands: Dan Murphy, destination outlets; Woolworths Liquor, attached to supermarkets; BWS,

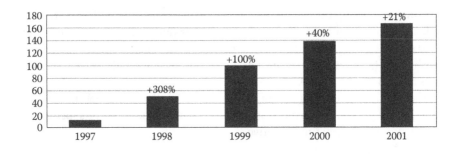

Figure A7.23   Plus Petrol outlets.

neighbourhood stores; and First Estate, fine wine stores. The total business is understood to generate sales of more than $1 billion per annum and this may increase to as much as $1.5 billion this year as more stores are added.

One constraint to growing this business is the availability of liquor licences. In Victoria, Woolworths exceeds the cap on the number of licences that can be held. The cap is set at 8% of the licences issued but the legislation is due to be reviewed in 2003.

Woolworths share of the national liquor market is understood to be around 10% compared with Coles Myer's estimated market share of 13%.

Ezy Banking

Ezy Banking, Woolworths banking joint venture with the Commonwealth Bank, is viewed as a separate business from the Supermarket group but was implemented to drive traffic through the stores and thus generate incremental sales. It is another example of meeting customers' every day needs in one location and it also provides Woolworths with an opportunity to learn more about its customers.

Woolworths has found that Ezy Bank customers have changed their shopping habits from, perhaps coming to the store once a week, to visiting several times a week. Customers spend less at each visit but over a week are spending more than they did previously. Providing banking services also helps Woolworths learn more about its customers habits and needs. This information will become critical in the battle to win and maintain market share in the future. Refer section 4.5.

Woolworths runs Ezy Banking under an alliance with the Commonwealth Bank to produce a co-branded product for its customers. Woolworths has no capital risk associated with Ezy Banking. In this respect the model that has been pursued is different from that chosen by other retailers to provide banking services. For example, Tesco in the United Kingdom, is the bank, it holds a banking licence.

At the moment Ezy Banking product range extends to a savings and transaction account, a credit card, and rewards scheme. The rewards are vouchers redeemable in the supermarkets and provided by suppliers. The product range will be extended and may include insurance and home loans, over time.

In the meantime, in just two years 610,000 accounts have been opened exceeding the five year target of 500,000 accounts.

e-Commerce

HomeShop is Woolworths internet shopping service. During fiscal 2001 Woolworths acquired a 38% interest in Greengocer.com to expand its internet offering and recently moved to 100% ownership.

HomeShop utilises a store based model, in other words, customer orders are supplied and delivered from their local store. Greengrocer.com utilises a central picking model, operating from purpose built facilities. Owning both operations provides Woolworths with back office synergies and the opportunity to experiment with different operating models.

The capital invested to date is not significant.

Financial Performance

The recent financial performance of the whole Supermarket division is presented in Table A7.14 below. Note EBIT margin is fine as would be expected for this division but return on funds employed is substantial and exceeds that of the other divisions.

The supermarket division enjoyed strong revenue growth over the last year, including the contribution from Plus Petrol (without this, revenue growth would have been only 10%). Sales growth was facilitated by an improved product offering and reduced costs, which also enabled Woolworths to increase market share of ABS measured Food Liquor and Grocery to 25.6% from 24.6%. Project Refresh allowed costs to be reduced by 0.52%, of which 0.38% was passed on to customers and the balance to shareholders.

Supermarkets in all states performed strongly with South Australia benefiting from the acquisition of the Booze Bros liquor chain ($12 million). Western Australia from the acquisition of Advantage Supermarkets ($21 million) and Tasmania from the realigning of Roelf Vos and Purity branding with Woolworths.

A five year summary of sales growth for the division (excluding petrol) is presented in Figure A7.24 along with the key drivers; sales per square metre and store numbers. Note growth in sales per square metre has been steadily increasing since 1999. This is the result of improving productivity, and in fiscal 2001, the demise of Franklins and vastly increased liquor store numbers, with higher sales per square metre value. These factors all combined to take the growth rate to almost 5% last year.

Table A7.14   Recent Financial Performance–Supermarket Division

$ millions	Full Year 27/06/1999	Interim 9/1/2000	Full Year 25/06/2000*†	Interim 7/1/2001	Full Year 24/06/2001
Sales	15,399	8,432	15,724	9,322	17,519
Increase			2.1%	10.6%	11.4%
Contribution on consolidation	84.8%	80.2%	82.8%	80.9%	83.8%
EBIT	515	284	533	317	619
Margin	3.3%	3.4%	3.4%	3.4%	3.5%
Increase			3.6%	11.5%	16.1%
Contribution on consolidation	95.4%	79.2%	85.7%	79.2%	87.5%
Funds Employed	1443.3	1322.1	1295.5	1080.2	1312.6
ROFE	35.7%	42.9%	41.1%	58.6%	47.1%
Increase			-10.2%	-18.3%	1.3%
Contribution on consolidation	61.2%	52.1%	67.6%	54.4%	62.0%

* Adjusted for WST.
† Continuing operations.

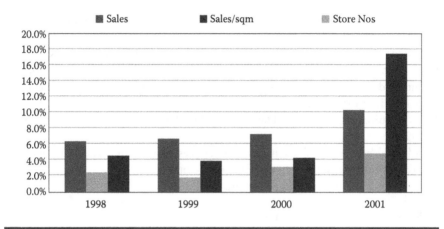

Figure A7.24   Divisional growth summary. (From company data.)

Major Participants and Market Shares

The industry is dominated by a small number of firms. The two major participants are Woolworths Limited and Coles Myer Limited, then some way behind comes Metcash Trading Limited and Foodland Associated Limited, which are wholesalers and distributors to whom independents are aligned under banner groups, and then other independents. The Franklins supermarket group, owned by Hong Kong based Dairy Farm International Holdings Limited, had a similar national presence to Metcash Trading until it exited the industry late in the first half of this year. Its parent Dairy Farm determined that return on investment was insufficient to justify a presence in this industry and its stores were sold-off, mainly to Woolworths (71), Coles Myer (20) Foodland (36) and more than 100 to independent grocers aligned with Metcash's IGA distribution group.

Summary details on the other participants are as follows:

■ Coles Myer Limited

The Coles and Bi-Lo supermarket chains are subsidiaries of Coles Myer Limited. Coles Myer is Australia's largest private employer with 162,000 employees. Coles Myer has 608 Coles and Bi-Lo supermarkets The Coles supermarkets are positioned at the 'full service' end of the market and compete directly with Woolworths and Safeway supermarkets, while Bi-Lo is positioned at the 'no-frills' end.

The supermarket division of Coles Myer also includes liquor retailing through Liquor and fast food retailing through its Red Rooster stores. Coles Myer is also Australia's largest general merchandiser through its Myer Grace Bros, Kmart and Target stores.

Coles Myer has a market capitalisation of approximately $9.6 billion.

Metcash Trading Limited

Metcash is around 75% owned by South African company, Metro Cash and Carry Limited. Metcash is the successor to the former Davids Limited, and has maintained its position as Australia's largest grocery wholesaler and distributor operating in all states and territories except Western Australia. Metro Cash and Carry acquired its interest in Davids in 1998 following the company's floatation in 1994.

Metcash's IGA distribution business services more than 1,100 independently owned IGA aligned supermarkets. Employees total 5,220.

Market capitalisation is approximately $880 million.

- Foodland Associated Limited
  - Foodland is the major distributor and sole independent wholesaler in Western Australia. Apart from the provision of wholesale grocery products, Foodland provides financial incentives and marketing support to FAL franchise and other independent supermarkets. Foodland is the fourth largest retailer listed on the Australian Stock Exchange. With a market capitalisation of approximately $1.2 billion and employs 15,000 people.
- Independents
  - Independent retailers form the remainder of the sector. The National Association of Retail Grocers of Australia (NARGA) is the main representative body for this group and its figures indicate that there are more than 3,000 independent supermarkets and grocers around Australia.

The independents vary in size from corner stores to full size supermarkets and many operate within the banner groups aligned with the wholesalers, Metcash and Foodland. The number of non-aligned independents is not accurately known but is thought to be around 50% of the total NARGA membership.

The vertically integrated structure of the supermarket operations of Woolworths and Coles Myer provides a number of advantages in purchasing, warehousing and distribution, and pricing over the independent operators. Centralised buying allows opportunistic purchases of larger volumes than normal before known price increases or as part of joint promotional exercises or pricing campaigns with the producer. This also allows greater scale to co-ordinate promotional activities and exploit generic advertising advantages.

Centrally co-ordinated store orders allow larger warehouse pick-up runs enabling efficiencies in distribution to be exploited. Average store orders for independents are much smaller and as a general rule smaller stores are more labour intensive and costly to run. Economies of scale allow larger stores to utilise labour, energy and floor space more efficiently.

Finally, Woolworths and Coles Myer supermarkets often have the privilege of being anchor tenants in major shopping centres which means they pay substantially less rent per square metre than other tenants and often enjoy more flexible terms and conditions under their lease agreement with the landlord. In addition, prime positions for extended hours trading and high volume customer flows are secured.

With intense competitive pressures such as these, rationalisation of the independents will continue. Only larger independents with the capital to refurbish stores and to incorporate the new technologies used by their major competitors are likely to survive. However, unlike metropolitan areas, less competition in rural and regional Australia is likely to protect many smaller independents.

Market share for each of the participants can be measured in different ways. Chart 4.4 presents market share as measured by the AC Nielson survey, on a moving annual turnover basis to 28 October 2001. The market share figures are sourced from AC Nielson's ScanTrack service, which uses scanned supermarket sales for Woolworths, Coles Myer and Franklins and warehouse withdrawal data for Metcash, Foodland and other independents.

The product range covered is broad including packaged groceries, dairy and frozen food but excludes fresh meat and vegetables, delicatessen and bakery goods, milk and potato chips. The survey also excludes convenience stores, route trade and direct store delivery, about one-third of all grocery categories according to Woolworths.

Peer Comparisons

A comparison of Woolworths domestic and international peers, based on Standard & Poor's statistics, is provided in Table A7.15 below.

Acknowledgement

Prepared by Professor G. Raghuram, Indian Institute of Management, Ahmedabad, and G. Kuberkar.

All data, except those mentioned as 'company data', has been sourced from the Internet. Support rendered by TCS is gratefully acknowledged.

Cases of the Indian Institute of Management, Ahmedabad, are prepared as a basis for class discussion. Cases are not designed to present illustrations of either correct or incorrect handling of administrative problems.

Table A7.15 Peer Comparison–Average of Last Three Fiscal Years

Criteria	Woolworths (AUDm)	Coles Myer (AUDm)	Kroger Co (USDm)	Safeway Inc (USDm)	Sainsbury Plc (GBRm)	Carrefour SA (EURm)	Ahold Koninklijke NV (EURm)
Credit Rating	A-/Stable	A- BBB-BBB/Stabl	A/Negative	A+/Stabl	BBB/Stabl	WatchNeg/Positive/- e/A-2/A-1 A-1	e/A-2/A-2* -
EBIT/Int	2.3	2.0	2.5	4.0	4.8	6.0	2.5
Return on Capital	15.8%	12.1%	23.7%	21.6%	12.2%	14.7%	10.7%
Op Income/ Sales	6.5%	7.5%	7.2%	9.4%	8.5%	6.5%	8.2%
Free Op Cashflow/ Debt	2.1%	1.1%	1.6%	5.4%	10.3%	(54.7%)	2.4%
Total Debt/ Capital	74.8%	69.6%	103.2%	67.4%	31.5%	51.9%	56.0%
Sales	18442.3	22397.9	34858.7	25014.8	15734.7	48053.0	37505.0
Equity	1301.7	2716.9	728.9	3178.7	4843.7	7029.3	7907.7
Total Assets	4534.5	7670.9	13956.7	12883.0	11389.0	31,721.7	26,293.7

Source: Standard & Poor's CreditStats: Retail Companies and supermarkets, November 2001. http://www.ybmarkets.co.uk/

* Since downgraded to BBB +/Stable/A-2.

References

City Population. Australia: Urban centers. http://www.citypopulation.de/Australia-UC.html

IBIS Industry Division Report. http://invest.vic.gov.au/

Parliament of Australia, Joint Committee, The retailing sector, Chapter 2 of *Fair Market or Market Failure*. 2001. http://www.aph.gov.au/Parliamentary_Business/Committees/Joint/Former_Committees/retail/report/~/media/wopapub/senate/committee/retail_ctte/report/c02_pdf.ashx

Woolworths Ltd. Financial reports to shareholders for years 1999–2001. http://www.woolworthslimited.com.au/

Woolworths Ltd. Media releases. http://www.woolworthslimited.com.au/

Case 8   Akshaya Patra, Gandhinagar

Akshaya Patra's nutritious meal gives a lot of pleasure to small children. And it provides a foundation for successful life.

—Narendra Modi, Chief Minister of Gujarat

It was September 18, 2010. Jaganmohan Krishna Dasa, President of The Akshaya Patra Foundation (TAPF), Gandhinagar, was reviewing their activities and performance in the school mid-day meals programme. They were being asked to increase the number of beneficiaries and move into other domains.

The review was also to help him conceptualise the terms and conditions for serving food to the patients in the Civil Hospitals of Ahmedabad and Gandhinagar. This was in response to the offer made by the Government of Gujarat (GoG) to TAPF on August 15, 2010. It was a bid to serve healthy and hygienic food to the patients of the Civil Hospitals. A milestone achieved rarely by any other Non-Governmental Organisation (NGO) encompassing mid-day meals program, TAPF was awarded the offer merely based on the success rate it had acquired by serving nutritious and delicious food to nearly 2.30 lakh children in Gujarat. The Gandhinagar kitchen worth ₹210 lakhs had the capacity to feed around 1,50,000 children. A similar kitchen in Vadodara which was opened a couple of months earlier had the capacity to feed around 80,000 children.

At 7:00 am, the Gandhinagar kitchen was alive with workers with cap-covered heads and gloved hands brewing rice and curry in the behemoth cauldrons. Some of them moved the cooked rice for packing into containers; some others stirred the curry, while rest of them managed filling the containers with the mechanized meal in accordance with the delivery requirement. Specially insulated vehicles were parked just outside near a platform waiting to be loaded, where the containers were placed route-wise for delivery.

Back in one of the schools, the children waited eagerly for the blue van to arrive. As soon as the food arrived, they scampered to get their plates and formed a queue to wash them. All of them gathered to sit cross-legged in a shed with their plates in front of them, waiting to be served their share. For some of them, it was the only wholesome meal they could savour in a day. They continuously kept chanting *"Hare Krishna Hare Ram"* (a mantra preached by International Society for Krishna Consciousness (ISKCON), the parent body of TAPF) as an act of thanking the deity before eating. United as they seemed, each of them touched the food only after the last child was served.

Mid-Day Meal Scheme In Gujarat

Mid-Day Meal (MDM) scheme introduced in 53 *talukas* of Gujarat in 1962, covered the age group of 6–11 years. In 1984, the GoG extended the programme to the entire state, covering all the students in that age group. The scheme was discontinued during 1990 and 1991, and replaced by the Food for Education Programme. This programme provided 10 kilograms (kg) of food grains free-of-cost every month to children with 70 per cent attendance in school. The MDM scheme, however, was reintroduced from January 15, 1992.

In 1995, the Government of India (GoI) launched the National Programme - Nutritional Support to Primary Education (NP-NSPE) scheme where all the children of the primary schools (run by GoI/local bodies) from Standard (Std) I to V were given 100 grams and Std VI to VII were given 150 grams of free food grains (wheat/rice) on the basis of enrolment. Gujarat was one of the pioneer states to introduce the concept of providing hot cooked mid-day meals to all children from Std I to VII of primary schools run by State Government and Municipal Corporations (Table A8.1 and Figure A8.1 present coverage comparison and administrative structure of MDM in Gujarat). The total budget for 2009–2010 was ₹461.84 crores (GoI–₹301.84 crores and GoG–₹160.00 crores).

The GoG entrusted the work of running the MDM scheme in Ahmedabad Municipal Corporation and Gandhinagar *taluka* to TAPF in June 2006. The Vadodara Municipal Corporation and Kalol and Mansa *talukas* of Gandhinagar district were assigned to TAPF in October 2009.

Akshaya Patra

An initiative of ISKCON, the program commenced in the year 2000. It served nutritious food to 1,500 children in five schools in Bangalore. The original idea was to extend the concept of *prasadam*, usually food items given to devotees, from just in their temple to a more meaningful and needy segment like poor school children. By August 20, 2010, they were feeding about 12,28,580 children in around 8,261 government schools in India across nine states.

Table A8.1   Coverage Comparison of MDM in Gujarat

Particulars	2006–2007	2007–2008	2008–2009	2009–2010
No of primary schools covered	31,682	32,577	32,453	32,762
No of MDM centres	29,991	30,731	30,999	32,275
No of children	38,26,586	39,50,173	40,01,502	39,30,051
Std I–V	31,08,015	30,94,651	31,09,903	30,29,254
Std VI–VII	8,08,571	8,55,522	8,91,599	9,00,797
No of days of meal provision	207	214	208	147
No of employees	86,421	86,643	87,485	88,442

Source: Commissionerate of Schools & MDM, Gujarat State, Education Department, Government of Gujarat, 2010.

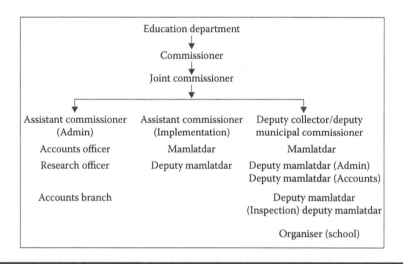

Figure A8.1   Administrative structure of MDM in Gujarat. (From Commissionerate of Schools & MDM, Gujarat State, Education Department, Government of Gujarat, 2010.)

On June 14, 2007, TAPF got underway to serve mid-day meals to 65,000 underprivileged students in Gujarat covering government schools of Ahmedabad and Gandhinagar. Initiated as a temporary kitchen, it soon developed into a fully-mechanized facility. By June 2010, there were two TAPF kitchens in the state, both centralized, one in Gandhinagar and another in Vadodara. TAPF had enrolled more than 1,30,000 school children in 529 schools of Ahmedabad/Gandhinagar and 80,000 children in 286 schools of Vadodara for the mid-day meal. In three years, they had experienced a jump of over 320 per cent serving close to 2,11,000 children in the state (Table A8.2 outlines the reach of TAPF in Gujarat). The number of beneficiaries in the Ahmedabad/Gandhinagar schools usually varied from 95,000 to 96,000, implying a 74 per cent utilization of the MDM scheme in the schools served by TAPF.

The two service areas of Gandhinagar and Vadodara were looked after by a unit called TAPF, Gandhinagar. It was also was responsible for a rural MDM at Nathdwara, Rajasthan. TAPF, Gandhinagar reported to the Head Office (HO) in TAPF, Bangalore (Figure A8.2 illustrates the organisational structure of the TAPF in 2010). TAPF, Gandhinagar had a total of 260 workers (Refer to Table A8.3 for the staff details).

Supply Chain Activities: From Grains to Meal

Procurement, Sourcing, and Storage

TAPF received rice from the Gujarat State Civil Supplies Corporation (GSCSC) which in turn received it from Food Corporation of India. Fortified *atta* was

Table A8.2   TAPF's Reach in Gujarat State

District	Area	Schools	Children	Routes	Vehicles
Gandhinagar	Kalol *taluka*	128	33,370	7	7
	Mansa *taluka*	116	23,938	7	7
	Daskroi *taluka*	33	9,011	2	2
	Gandhinagar city and rural	186	48,374	12	12
	Ahmedabad city schools	66	16,359	5	5
	Total	**529**	**131,052**	**33**	**33**
Vadodara	Vadodara city schools	100	40,000	8	8
	Vadodara *taluka*	186	40,000	10	10
	Total	**286**	**80,000**	**18**	**18**
	Total	**815**	**211,052**	**51**	**51**

Source: TAPF, August 2010 and Google maps accessed on October 04, 2010.

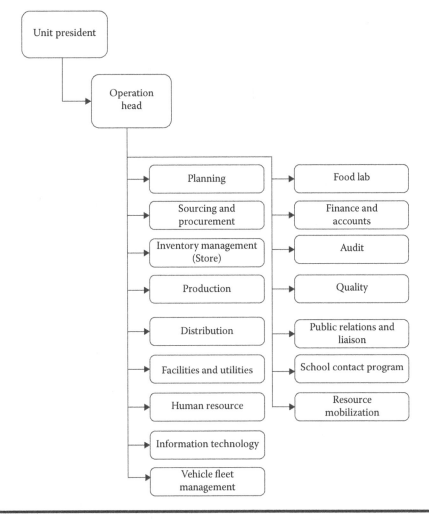

Figure A8.2   Organisational structure of TAPF in Gandhinagar. (From TAPF, September 2010.)

procured from the local mills in Ahmedabad which were outsourced by GSCSC. The GoG made provisions in advance for the next month for required food grains like rice to TAPF. Grain and *atta* requirements were calculated on the basis of actual attendance of each month for the following month. While GSCSC was expected to deliver to the MDM centres, this did not always happen as per required schedule. TAPF felt more comfortable getting the supplies directly from the GSCSC, for which charges were reimbursable. As of 2010, the charges were ₹75 per quintal in Gujarat. The GoG would provide funds the following month after the receipt of the details for the current month from TAPF. The flow of funds

Table A8.3 Staff Details

Sr. No.	Title	Departments											Grand Total
		Production	Quality	Maintenance	Materials	Store	Transportation	House Keeping	Security	Finance	HR	Marketing	
A	**Management Staff**												
1	Head/Senior Manager												0
2	Manager	1									1		2
3	Deputy Manager									1			1
4	Assistant Manager												0
5	Senior Executive				1		1					1	3
6	Officer/Executive												0
7	Junior Executive		1							2		2	5
8	Coordinator/Supervisor	3		1	2	2		1					9

#	Category											Total
9	Executive Assistant/Office Boy	1	1		3						2	7
B	**Work Men Category**											
1	Drivers						28					28
2	Cook	5										5
3	Assistance Cook											0
4	Distribution (Route Boys)						42					42
5	Vessel cleaner	34										34
6	Vegetable Cutting Staff	10										10
7	Electrician/Technician									6		6
8	Helpers/Casual Labour	73				8	11		8	6		108
	Total	127	1	7	3	8	81	12	8	6	2	260

Source: TAPF, September 2010.

seemed to be a concern as bills were pending for the last three years which totalled to approximately ₹15 lakhs.

Vegetables, pulses, oil, and spices were purchased locally as per requirement from Kalupur and Madhupura *mandis* in Ahmedabad at subsidized rates. Prices negotiated with the vendors were 10–12 per cent lower than the market price. For instance, the market price of *moong mogar* was ₹70 per kg but vendor's supply price inclusive of delivery was ₹64 per kg. In June 2010, turmeric, chilli, coriander, and cumin seeds, etc., were centrally procured from different markets in India to have savings in purchases. *Mutter dal* was procured from the retail chain of Reliance for its superior quality (Reliance imported *Mutter dal* from Australia). Among the various raw materials, *moong dal*, oil, turmeric, cumin seeds, and tomatoes were the high-value items.

On account of the inflation in 2009, central procurement in bulk was done by the TAPF HO to hedge against price increases of items like potato, *mutter dal*, chilli, turmeric, and coriander. These items were kept in cold storages and released to different TAPF units. However, their forecasting seemed to be fallacious as the prices came down drastically in 2010. Therefore, local purchasing was preferred to centralized procurement as the rates of these items varied across the country. Further, the storage facility in the Gandhinagar kitchen had space constraints.

Inventory at the storehouse mainly comprised food grains, provisions, and vitamin supplements. Inventories received as government grants were determined at the lower of the market price or regulated price for accounting purposes and a purchase requisition was raised by the store department when the stocks reached their reorder level (refer to Table A8.4 for the minimum stock, maximum stock, reorder level, and shelf life of some of the items). In case the lead time of the items exceeded the normal delivery time and the stocks hit their minimum level, emergency purchases were made to continue operations smoothly. Accordingly, the purchase department collected samples and quotations from different suppliers. (Table A8.5 provides a detailed grocery rate comparison at Gandhinagar). The samples were checked by the quality control executives and the supervisor-in-charge of operations. The criteria for the selection of vendors were governed by market rate, offer rate, quality of samples based on the raw material specification, and the past performance of the supplier's commitment for delivery on a given date, time, requirement, and quality as per the sample.

There was a reasonable level of automation in the preparation of meals. The necessary equipment and spare parts were procured from different vendors as per specific requirements (Table A8.6 shows the spare parts purchased, their purpose, and the frequency of procurement). To maintain hygiene at the storehouse, the Pest Control of India was contracted to carry out fly management operations every day, rodent check three times in a month, and fumigation once a year.

Production

TAPF developed the menu keeping in mind the local tastes and preferences of the children. The menu was primarily designed to fulfil the norms of providing a

Table A8.4   Stock Norms

Doc. No.:	ST
Issue Status:	01
Revision Status:	00
Revision Date:	
Page:	1 of 1

The Akshaya Patra Foundation, Gandhinagar

Sr. No.	Material Name	Classification	Minimum Stock Level			Shelf Life (Months)
			Min Stock (Kg)	Max Stock (Kg)	Reorder Level (Kg)	
1	Moong Dal	Dal	1,000	5,000	1,500	6
2	Moong Mogar	Dal	1,000	5,000	1,500	6
3	Tuver Dal	Dal	1,000	5,000	1,500	6
4	Peas	Pulses	300	2,000	500	6
5	Kabuli Channa	Pulses	400	1,000	500	4
6	Desi Channa	Pulses	400	1,000	500	4
7	Kapasiya Oil	Oil	500	6,500	2,000	18
8	Ajawain	Spices & Condiments	5	60	20	12
9	Cloves	Spices & Condiments	4	25	8	24
10	Dhaniya Powder	Spices & Condiments	30	300	60	4

(Continued)

Table A8.4 (Continued) Stock Norms

						Doc. No.:	ST
The Akshaya Patra Foundation, Gandhinagar						Issue Status:	01
						Revision Status:	00
						Revision Date:	
						Page:	1 of 1
			Minimum Stock Level				
Sr. No.	Material Name	Classification		Min Stock (Kg)	Max Stock (Kg)	Reorder Level (Kg)	Shelf Life (Months)
11	Garam Massala	Spices & Condiments		15	100	25	4
12	Haldi Powder	Spices & Condiments		25	300	65	4
13	Hingh	Spices & Condiments		4	60	10	6
14	Kali Mirch	Spices & Condiments		4	50	10	24
15	Rai	Spices & Condiments		20	200	50	24

16	Red Chilly Powder	Spices & Condiments	30	300	60	12
17	Red Chilly (Sabut)	Spices & Condiments	3	50	10	12
18	Salt	Spices & Condiments	250	1,500	400	12
19	Tej Patta	Spices & Condiments	2	50	10	6
20	Tamarind	Spices & Condiments	25	100	40	4
21	Jaggery (Gud)	Spices & Condiments	100	1,000	300	3
22	Jeera	Spices & Condiments	45	300	60	20
23	Sugar	Spices & Condiments	100	1,000	200	6

Source: TAPF, September 2010.

Table A8.5 Grocery Rate Comparison for Gandhinagar

Material	Quantity	Suppliers							Order Finalize		
		REQUIREMENT FOR GANDHINAGAR (JULY & AUGUST)									
		RAJ TRADING AHMEDABAD	MAHESHWARI AHMEDABAD	PRAMUKH GANDHINAGAR	ADC ST AHMEDABAD	RAFIK KIRANA GANDHINAGAR	MAJITHAI BARODA	AVANI AHMEDABAD	ORDER FINALIZE	INVOICE NAME	RATE
MUTER DAL									MUTER DAL		
9000 KG 28/7/10	18000 KG	X	22.5	22.31	21.91	22.51	X	X	9000 KG 28/7/10 — 18000 KG	PRAMUKH-TRADERS & SUPPLIERS	22.31
9000 KG 02/08/10			22	22					9000 KG 02/08/10		
SALT 22/07/10	2000 KG	X	X	X	X	X	X	X	SALT 22/07/10 — 2000 KG	SHREE HARIHAR	1.9
CHANA DAL 27/07/10	5000 KG	X	30	29.71	29.8	29.91	X	X	CHANA DAL 27/07/10 — 5000 KG	MAHESHWARI TRADERS	29.8
JAGGERY 26/07/10	800 KG	35	X	35	22.5	35.51	X	X	JAGGERY 26/07/10 — 800 KG	PRAMUKH-TRADERS & SUPPLIERS	35
SUGAR 22/07/10	300 KG	29	X	29.21	29.25	29	X	X	SUGAR 22/07/10 — 300 KG	RAFIK KIRANA	29
DELIVERY ON 02/08/10									DELIVERY ON 02/08/10		
MOONG CHHILKA DAL	4000 KG	X	67.5	68.21 / 64	64.71 / 68.55	68.5	X	X	MOONG CHHILKA DAL — 4000 KG	PRAMUKH-TRADERS & SUPPLIERS	64
DESHI CHANA	1000 KG	X	27	25.91	25.8	26	X	X	DESHI CHANA — 1000 KG	PRAMUKH-TRADERS & SUPPLIERS	26.4
PALM OIL	12000 KG	721	X	725	X	728	X	X	PALM OIL — 12000 KG	RAJ TRADING	718
(15KG NEW TIN)	800 TIN	(718 FINAL)	X						(15KG NEW TIN) — 800 TIN		

400+400 – 02/08 & 12/08

Item	Quantity							Supplier	Price
AJWANI	200 KG	188	X	185	185	190	X	PRAMUKH-TRADERS & SUPPLIERS	185
CLOVES	25 KG	315	X	347	350	350	X	RAJ TRADING	315
DHANIYA POWDER	300 KG	60	X	67	68	68	62	MAJITHIA MASALA	62
HALDI POWDER	600 KG	195	X	219	206	220	214	MAJITHIA MASALA	214
GARAM MASALA	100 KG	110	X	130	125	131	85	THE AHMEDABAD COOPERATIVE DEPARTMENT STORE (ACDS)	125
HING	200 KG	120	X	121	122	129	110	AVANI ENTERPRISE	110

DELIVERED ON 05/08/10

Item	Quantity							Supplier	Price
KALI MIRCH	30 KG	245	X	297	290	300	X	RAJ TRADING	245
RAI	200 KG	32	X	33.25	34	33.5	X	ACDS	34
RED CHILLY POWDER	600 KG	X	X	73	74	74	78	MAJITHAI MASALA	78
RED CHILLY	50 KG	60	X	65	65	65	X	RAJ TRADING	60
TAMARIND	500 KG	X	X	49	56	50	X	PRAMUKH-TRADERS & SUPPLIERS	49

Source: TAPF, August 2010.

wholesome and nutritious meal as specified by the GoI under the MDM scheme. This implied a minimum of 450 calories and 12 grams of protein consisting of *dal* with vegetables and rice, that is a minimum of two items on every working day to students studying in standard I–VII. The calorific value and protein content were analysed by third parties who prepared reports every month on raw materials and cooked food (Refer to Tables A8.7 and A8.8 for the Gandhinagar menu and the calorific and protein value of the cooked meals).

Table A8.6　Spare Parts Order Frequency

Spare Parts	Purpose	Frequency
General Spares		
Heat Proof Grease	Greasing *Roti* Machine Bearings	Monthly
Ball Valve (1.5 Inch)	Plumbing	Monthly
Screws (1.5 Inch)	General	Monthly
Bearing (Skf-1209)	*Roti* Machine	Quarter Yearly
Bearing (Zkl-1209)	*Roti* Machine	Quarter Yearly
Pedestal Bearing (Skf P-210)	*Roti* Machine (Small)	Quarter Yearly
LPG Regulator	Gas Line-*Roti* Machine	Bi Monthly
Burner Assembly	*Roti* Machine	Bi Monthly
Door Closer	General	Quarter Yearly
SS Valve (1 Inch)	Plumbing	Monthly
Gi Valve (1 Inch)	Plumbing	Monthly
Gas Pipe (2 m)	Gas Line - *Roti* Machine	Monthly
Hose Pipe (2 m) Big	Steam Line - Cauldrons	Half Yearly
Hose Pipe (2 m) Small	Steam Line - Cauldrons	Half Yearly
Penetrating Lubricant	*Roti* Machine	Monthly
Gear Oil (140 Ep)	*Roti* Machine Gear Wheels	Weekly
Chain Lock Half Pin	*Roti* Machine Chain	Weekly

(Continued)

Table A8.6 (*Continued*)   Spare Parts Order Frequency

Spare Parts	Purpose	Frequency
General Spares		
Chain Lock (Gear Box)	*Roti* Machine Chain	Weekly
Chain Lock (*Tava*)	*Roti* Machine Chain	Monthly
Copper Gas Pipe	Gas Line - Roti Machine	Monthly
Tava (*Roti* Machine)	*Roti* Machine	Monthly
Oil Pump Cuppler Rubber	*Roti* Machine Oil Pump	Monthly
Trolley Wheel	Trolleys	Monthly
Electrical Spares		
CFL Bulb (11W)	General	Quarter Yearly
Bulb (100W)	General	Quarter Yearly
Switch Socket (6a)	General	Monthly
3 Pin Plug	General	Monthly
Tube Light Starter	General	Monthly
1-Way Switch	General	Monthly
4 Core Flexible Wire	General	Half Yearly
MCB Switch (32 Amp)	Electric Panel	Quarter Yearly

Source: TAPF, September 2010.

As required by GoI, the Memorandum of Understanding (MoU) specified the average daily composition of a meal, with the aggregate being 180 grams (refer to Table A8.9 for the average daily composition). This average was monitored by the government and had to be achieved over the meals served in a month. However, there were three reasons because of which the average may not be achieved: (i) the choice of items in the menu constituting a wholesome meal may not have the ingredients in the same proportion as stated by the GoI, (ii) the number and the specific weekdays of holidays in a month due to which the aggregate and the proportion of the ingredients would vary, and (iii) the actual consumption habits of the beneficiaries differed and most of them did not consume the specified aggregate.

Also, it was difficult to arrive at the composition of items like *Khichdi* since it comprised of both rice and *dal*. The *Sabji* provided in the menu differed based on

Table A8.7 Daily Menu

As Suggested by GoG		As suggested by the government in the MoU, the required daily allowance was 450 calories and 12 grams of protein.
Day	*Menu*	
Mon	Lapsi, Sukhadi	
Tue	Khichdi, Sabji	
Wed	Dal Dhokli	
Thu	Plain Rice, Dal	
Fri	Muthiya, Handvo	
Sat	Vegetable Pulav	

Note: As followed by TAPF, Gandhinagar.

Table A8.8 Calorific and Protein Value

Day	*Menu*	*Energy (Kcal/100g)*	*Protein (Grams/100g)*
Monday	Mutter Dal, Thepla, Mutter Pulav	153+194+378 = 725	7+5+9 = 21
Tuesday	Plain Rice, Roti, Sabji	156+371+159 = 686	3+9+2 = 14
Wednesday	Jeera Rice, Dal Dhokli, Sukhadi*	143+86 = 229	5+4 = 9
Thursday	Plain Rice, Alu Chana Dal, Thepla	378+156+156 = 690	9+3+8 = 20
Friday	Khichdi, Roti, Mixed Sabji	371+82+159 = 612	9+6+2 = 17
Saturday	Vegetable Pulav, Mutter Dal, Sukhadi*	135+153 = 288	5+7 = 12

Source: TAPF, August 2010.

* Calorific and protein value for Sukhadi were not provided.

Table A8.9   Average Daily Composition (Specified by GoI)

Component	Amount (grams)
Wheat/Rice	100
Pulses	20
Vegetables and condiments	50
Edible oil	10
Total	**180**

Source: TAPF, August 2010.

seasonality and availability but the amount in grams complied with the standards specified in the MoU.

TAPF did an initial survey on the schools to be covered by the programme to arrive at an approximate number of beneficiaries. Corresponding to this, the GoG also provided an estimate of the beneficiaries. A list of working days and holidays was made, while deciding on the daily menu and recipe. The number of working days varied with each year and was usually fixed between 200 and 220 days.

On a daily basis, the drivers and helpers got the food requirement figures from the school authorities and entered it into the food list for the next day's production (refer to Table A8.10 for the food list). In cases of high variation of demand during admissions, festivals or post vacation period, the supervisor-in-charge personally visited the school and inquired about the reasons that caused the change from the school authorities and informed the production department.

A production planning report, derived from the food list was prepared to determine the start time for the production, time required for cooking, and the quantities needed for production such as the number of cauldrons, etc. The material requisition was subsequently made based on the number of *dal* and rice cauldrons to be utilized, the number of *rotis* (or *theplas*) to be prepared and the quantities per batch which was submitted to the stores for the issue of raw materials. On any given day, depending on the menu, the Gandhinagar kitchen prepared any three of approximately 2.4 lakh *rotis*, three tonnes of *dal*, eight tonnes of *sabji*, and five tonnes of rice.

Table A8.10 Food List for Thursday September 9, 2010, as obtained on September 8, 2010

The Akshaya Patra Foundation, Gandhinagar

FOOD LIST & BATCH NUMBER DETAILS

Ahmedabad City

			Doc. No.:	DIST-09
			Rev. No.:	00
			Rev. Date:	

Route No. 1 Timing: 8.30 (A.M.) VDNFJFN; L8L
Driver Name: Alpesh Jani/Helper's Name: Bala Bharwad (99747 84860)
Supervisor Name: Paresh Solanki Mo.: 9274324951

(A'bad City) Morning — Thursday

9-9-2010

Sr. No.	School Name	Avg.	Thepla	Big	Medium	Batch No.	Rice Big	Medium	Small	Batch No.	Dal Big	Medium	Small	Batch No.	Vessels
1	Laxminagar	185	200	0	1		0	0	1		0	0	1		3
2	Sabarmati (Hindi - 1)	150	400	0	1		0	1	0		0	1	0		3
3	Sabarmati - 5	200	400	0	1		0	1	0		0	1	0		3
4	Sabarmati (Gujarati - 7)	250	400	0	1		0	1	0		0	1	0		3
5	Sabarmati (Gujarati - 3)	195	400	0	1		0	1	0		0	1	0		3
6	Sabarmati (Hindi - 3)	192	400	0	1		0	1	1		0	1	1		5
7	Dharmnagar Vidhyalay	84	400	0	1		0	1	0		0	1	0		3
8	Keshavnagar (Gujarati - 1)	250	400	0	1		1	0	0		1	0	0		3
9	Mahatma Gandhi (Gujarati - 1)	400	800	1	0		0	2	1		0	2	1		7
10	Vadaj (Hindi - 1)	100	400	0	1		0	1	0		0	1	0		3
11	Vadaj - 1	350	400	0	1		0	1	0		0	1	0		3
12	Vadaj (Gujarati - 3)	320	400	0	1		0	1	0		0	1	0		3

Sr. No.	School Name	Avg.	Thepla	Thepla Big	Thepla Medium	Rice Big	Rice Medium	Rice Small	Rice Batch No.	Dal Big	Dal Medium	Dal Small	Dal Batch No.	Vessels
13	Nava Vadaj (Gujarati - 1)	330	400	0	1	1	0	1		1	0	1		5
14	Nava Vadaj (Gujarati - 3)	280	500	0	1	1	0	1		1	0	1		5
15	Nava Vadaj (Hindi - 1)	280	1200	1	0	2	0	0		2	0	0		5
16	Nava Vadaj (Gujarati - 3)	100	400	0	1	1	0	0		1	0	0		3
	Total:	3666	7500	2	14	6	11	5	0	6	11	5	0	60

9-9-2010

Route No. 2 ~8 GVAZ ov Z Timing:- 8.30 (A.M.) VDNFJFN; L8L
Driver Name: Alpesh Jani/Helper's Name: Bala Bharwad (99747 84860)
Supervisor Name: Paresh Solanki Mo.: 9274324951

(A'bad City) Noon (2:00 P.M.) — Thursday

Sr. No.	School Name	Avg.	Thepla	Thepla Big	Thepla Medium	Rice Big	Rice Medium	Rice Small	Rice Batch No.	Dal Big	Dal Medium	Dal Small	Dal Batch No.	Vessels
1	Gota Housing - 1	150	250	0	1	0	1	0		0	1	0		3
2	Thaltej - 2	450	500	0	1	1	0	0		1	0	0		3
3	Naranpura (Hindi - 3)	550	1600	2	0	3	0	0		3	0	0		8
4	Naranpura Marathi - 1	75	200	0	1	0	1	0		0	1	0		3
5	Narayanpura - 1	200	400	0	1	1	0	0		1	0	0		3
6	Naranpura (Hindi - 1)	375	700	1	0	1	0	0		1	0	0		3
7	Elis (Gujarati - 7)	300	600	1	0	0	1	0		0	1	0		3
8	Elis Bridge - 8	500	500	0	1	1	0	0		1	0	0		3
9	Elis Bridge - 18	260	400	0	1	0	1	0		0	1	0		3
10	Vasana (Gujarati - 3)	300	600	1	0	0	0	1		0	0	1		3

(Continued)

Table A8.10 (Continued) Food List for Thursday September 9, 2010, as obtained on September 8, 2010

						Doc. No.:	DIST-09
						Rev. No.:	00
						Rev. Date:	

The Akshaya Patra Foundation, Gandhinagar

FOOD LIST & BATCH NUMBER DETAILS

Ahmedabad City

Route No. 2 ~8 GVAZ ov Z Timing:- 8.30 (A.M.) VDNFJFN; L8L
Driver Name: Alpesh Jani/Helper's Name: Bala Bharwad (99747 84860)
Supervisor Name: Paresh Solanki Mo.: 9274324951

9-9-2010

Sr. No.	School Name	Avg.	Thepla	Thepla Big	Thepla Medium	Batch No.	Rice Big	Rice Medium	Rice Small	Batch No.	(A'bad City) Noon (2:00 P.M.) Dal Big	Dal Medium	Dal Small	Batch No.	Thursday Vessels
11	Vasana - 6	250	400	0	1		0	1	0		0	1	0		3
12	Vasana (Gujarati - 5)	375	400	0	1		0	0	1		0	0	1		3
13	Labour Child	35	100	0	1		0	0	1		0	0	1		3
14	Vasana (Gujarati - 1)	250	300	0	1		0	0	1		0	0	1		3
15	Vasana - 2	250	350	0	1		0	1	0		0	1	0		3
16	Elis (Gujarati - 17)	200	400	0	1		0	0	2		0	0	2		5
17	Vasana - 4	300	500	0	1		0	1	0		0	1	0		3
	Total:	4820	8200	5	13		7	7	6		7	7	6		58

Source: TAPF, September 2010.

Pre-processing and Cooking

The kitchen was set in motion around 9:30 pm with the precooking activities such as washing and cutting of vegetables. The masala-making process started at 4:30 am followed by cooking at 5:00 am. The total production and cook-to-consume times varied between 6 and 7 hours and 5 and 6 hours, respectively.

Being a centralized kitchen, mechanization facilitated reduced human handling, thereby ensuring quality. Grains, pulses, and spices were washed, sorted, and stored. Vegetables were washed, sorted, peeled, and cut using specifically designed vegetable-cutting machines and then stored. Cutting was done in different patterns either by machine or manually. It was preferred to cut the vegetables into small shapes so that the children could consume them easily. Condiments were sorted, cleaned, ground, mixed, and stored. Oil was stored in cans and used either directly for certain items or poured into tanks for usage on tap or as a spray.

The rice-cleaning machine had a capacity of five tons per hour. It was divided into three sections. The first section separated metal adulterants such as nails, iron filings, etc. The second section filtered out dust, moth, chaff, husk, and broken rice grains while the third section removed stones from the rice.

Steam was the medium of cooking. It was generated using boilers and then passed into the cauldrons for cooking. There were separate cauldrons for cooking rice or rice-based items and for *dal* or *dal*-based items. Rice cauldrons had the capacity to cook 100 kg of rice in 20 minutes for 1,000 children. Each *dal* cauldron had the capacity to prepare 1,200 litres of *dal* in under two hours for 6,000 children. The Gandhinagar kitchen had deployed eight rice cauldrons and five *dal* cauldrons for total production.

One of the innovations of the TAPF kitchens was the *roti*-making machine which could yield up to 40,000 *rotis* per hour. Water was added to flour in vessels to make the dough mechanically. The dough was fed manually to the machine whose first step was flattening the dough through rollers. The flattened dough was cut into round shapes using the rotary cutting die. The *rotis* were further moved into a burner for cooking them on both the sides. The *rotis* were then sprayed with oil through a mesh, to keep them soft and tasty.

It was observed that, in general, children did not consume more than two *rotis* (some did not even consume more than one *roti*). This resulted in insufficient wheat intake. The two *rotis* which totalled 40 grams did not meet the norm for minimum wheat consumption set at 50 grams per day. To help the situation, a new item *Sukhadi*, a popular sweet made from fortified *atta,* was introduced in the menu. TAPF designed and procured a special machine to prepare it. The approximate cost of the machine was ₹1.6 lakhs and it produced *Sukhadi* at the rate of 150 kg per hour. (Figures A8.3 through A8.10 display the equipment used in the kitchen for cooking.)

Figure A8.3 *Sukhadi* machine in Akshaya Patra kitchen.

Figure A8.4 Cauldrons in Akshaya Patra kitchen.

Figure A8.5 *Roti*-making machine in Akshaya Patra kitchen.

Figure A8.6 *Roti*-cutting machine in Akshaya Patra kitchen.

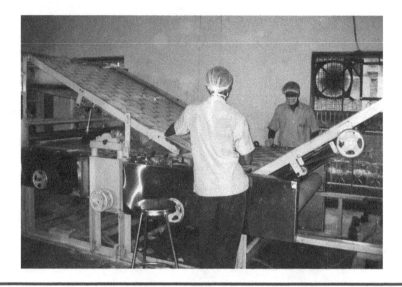

Figure A8.7 *Roti*-making machine in Akshaya Patra kitchen.

Figure A8.8 De-stoning machine in Akshaya Patra kitchen.

Figure A8.9   Boiler.

Figure A8.10   Oil spraying on *rotis*.

Packing

The prepared food was packed into stainless steel containers with tight-fitting lids that prevented pilferage, leakages, and retained heat. The expenditure on containers was borne by TAPF, while other utensils and plates in schools were supplied by the GoG. By 2010, TAPF had spent approximately ₹50 lakhs on containers. The containers were in three different sizes—small, medium, and big, with the capacity to accommodate food (other than *rotis*) for about 50, 100, and 200 beneficiaries, respectively. The requirement for each school was determined in units of containers based on the average beneficiaries, but moderated based on the actual requirements for the next day. However, even this led to supply mismatches since per capita consumption of each school varied, affected by factors such as heavy rains, festivals, local events, etc. In general the supply was more than less.

For packing *rotis*, medium and big containers with the capacity of about 600 and 1,200 *rotis* respectively were generally used. This was because of insufficient number of small containers, whose usage was preferred for *dal, Sabji* or rice. After the *rotis* were produced, they were packed school-wise (and thus not necessarily to its full capacity), using a weighing machine. Since each *roti* weighed 25 grams, the medium and big containers would weigh about 15 kg and 30 kg, respectively.

Each container was labelled after packing. The label comprised the name of the school and route number, and in the case of *rotis*, the equivalent number packed. The packed containers were then placed on a platform route-wise next to the vehicle allotted for that route. When all the containers for a route were packed, a food-packing report was prepared, giving the number of containers and the packing time for each route. (Refer to Table A8.11 for the food-packing report. This report does not have the *Sukhadi* details, an item being introduced recently.)

Distribution

Each vehicle was assigned a driver, helper, and a supervisor. The driver was responsible for ensuring that the vehicle was loaded properly and kept a copy of the food list and the packing report. The containers were loaded into the vehicles in the 'Last In First Out' (LIFO) order, meaning containers for the last school were placed inside and those for the first school were placed near the door. After unloading, the driver/helper verified the required quantity needed for the next day from the school authorities. The supervisor had the responsibility of ensuring that the food reached every school safely as per the requirement and would often make surprise visits to schools.

The vehicles dropped the containers in the schools (about 14 to 18 schools per vehicle) and on their return trip, picked up the empty containers. Since schools had different timings, they were categorized into early morning and late morning deliveries. The vehicles departed at 7 am for early morning delivery around

Table A8.11   Food Packing Report of September 8, 2010

			The Akshaya Patra Foundation, Gandhinagar					Doc. No.:	PDN 03
								Issue status: 01	
								Revision status: 00	
								Date	8.09.10
								Page	1
			FOOD PACKING REPORT						
			RICE/SPECIAL RICE/KHICHDI						
Sr. No.	BOILER No.	ROUTE No.	BIG	MEDIUM	SMALL	TOTAL VESSELS PER ROUTE	PACKING TIME	REMARKS	
1	1,2	1	6	11	5	22	05:20		
2	2,3	2	7	8	4	19	05:25		
3	4	7	6	6	2	14	05:30		
4	5,6	8	4	11	7	22	05:35		
5	6	9	5	7	6	18	05:40		
6	7,6	23	12	5	7	24	05:45		
7	8,9	24	10	8	3	21	05:50		
8	9,1	26	6	10	2	18	05:55		

(Continued)

Table A8.11 (Continued) Food Packing Report of September 8, 2010

							Doc. No.:	PDN 03
		The Akshaya Patra Foundation, Gandhinagar					Issue status: 01	
							Revision status: 00	
							Date	8.09.10
		FOOD PACKING REPORT					Page	1
			RICE/SPECIAL RICE/KHICHDI					
Sr. No.	*BOILER No.*	*ROUTE No.*	*BIG*	*MEDIUM*	*SMALL*	*TOTAL VESSELS PER ROUTE*	*PACKING TIME*	*REMARKS*
9	11	29	6	5	3	14	06:00	
10	11,12	31	3	8	1	12	06:05	
11	12	32	7	2	3	12	06:10	
12	13	14	5	8	2	15	06:15	
13	14	15	6	5	3	14	06:20	
14	14,15	20	3	9	1	13	06:25	

15	15,16	21	6	13	0	19	06:30	
16	16,17	25	11	7	1	19	06:35	
17	18	30	5	8	2	15	06:40	
18	19	5	3	8	4	15	06:45	
19	19,2	10	10	9	0	19	06:50	
20	21	11	5	9	4	18	06:55	
21	22	12	5	9	3	17	07:00	
22	23	16	7	6	1	14	07:05	
23	23,24	17	4	4	3	11	07:10	
24	24	22	5	8	3	16	07:20	
25	25,26	33	11	9	1	21	07:25	

(Continued)

Table A8.11 (Continued) Food Packing Report of September 8, 2010

The Akshaya Patra Foundation, Gandhinagar

Doc. No.:	PDN 03
Issue status: 01	
Revision status: 00	
Date	8.09.10
Page	1

FOOD PACKING REPORT

RICE/SPECIAL RICE/KHICHDI

Sr. No.	BOILER No.	ROUTE No.	BIG	MEDIUM	SMALL	TOTAL VESSELS PER ROUTE	PACKING TIME	REMARKS
26	26,27	3	6	10	0	16	07:35	
27	27,28	4	6	8	2	16	07:40	
28	28	18	5	6	4	15	07:50	
29	29	19	5	3	2	10	07:55	
30	29,3	28	7	6	4	17	08:05	
31	31,32	6	8	14	3	25	08:10	
32	32,33	13	3	9	1	13	08:20	
33	34,35	27	8	4	5	17	08:30	

Subji/Dal/Dal Dhoklay

Sr. No.	BOILER No.	ROUTE No.	BIG	MEDIUM	SMALL	TOTAL VESSELS PER ROUTE	PACKING TIME	REMARKS
1	1	1	6	11	5	22	06:05	
2	1	2	7	7	5	19	06:10	
3	1;2	7	6	6	2	14	06:15	
4	2	8	4	10	8	22	06:20	
5	2	9	5	7	6	18	06:25	
6	2,3	23	12	5	7	24	06:30	
7	3	24	10	8	3	21	06:35	
8	3	26	5	10	3	18	06:40	
9	3	29	6	4	4	14	06:45	
10	4	31	3	8	1	12	06:50	
11	4	32	7	2	2	11	06:55	
12	4	14	5	8	2	15	07:00	
13	4	15	6	5	3	14	07:05	
14	5	20	3	9	1	13	07:10	

(Continued)

Table A8.11 (Continued) Food Packing Report of September 8, 2010

Subji/Dal/Dal Dhoklay

Sr. No.	BOILER No.	ROUTE No.	BIG	MEDIUM	SMALL	TOTAL VESSELS PER ROUTE	PACKING TIME	REMARKS
15	5	21	6	13	0	19	07:15	
16	5,6	25	11	7	1	19	07:20	
17	6	30	5	7	3	15	07:25	
18	6	5	3	8	4	15	07:30	
19	6	10	10	9	0	19	07:35	
20	7	11	5	8	5	18	07:40	
21	7	12	5	9	4	18	07:45	
22	7	16	7	5	2	14	07:50	
23	7	17	4	4	3	11	07:55	

24	7,8	22	5	7	4	16	08:00	
25	8	33	11	9	1	21	08:05	
26	8	3	6	10	0	16	08:10	
27	8,9	4	6	8	2	16	08:15	
28	9	18	5	6	4	15	08:20	
29	9	19	5	3	2	10	08:25	
30	9	28	7	6	4	17	08:30	
31	10	6	8	12	4	24	08:35	
32	10	13	3	9	1	13	08:40	
33	10	27	8	4	5	17	08:45	

Source: TAPF, September 2010.

8:30 am for consumption at 9:30 am. Late morning departures started at 10 am for consumption at 12:30 pm.

Schools were assigned to a specific route keeping in mind that the food delivered to the last school should retain the temperature of at least 60°C. TAPF also tried to ensure that the delivery to the last school did not exceed the cook-to-consume time of six hours. Within these norms, an attempt was made to cover the maximum number of schools on a particular route.

TAPF started with 19 routes in 2007 and extended to 33 routes by 2010, to cater to the increase in the number of schools and beneficiaries. Out of the 33 routes, three routes made early morning deliveries while the rest made late morning deliveries.

In case of unforeseen events like traffic diversion or road blocks, the driver followed an alternate route to cover the schools. If emergencies such as vehicle break down occurred, then alternate vehicles were arranged to ensure uninterrupted distribution. To provide for such contingencies, the departure time built in at least 45 minutes slack.

Vehicles

TAPF owned a fleet of specially designed vehicles procured from Mahindra and Mahindra for maintaining the temperature of the food while carrying it across the required distances. Some vehicles were bought by TAPF while several firms had donated vehicles to TAPF for this cause (Figure A8.11 and Table A8.12) However, there was still a shortage of vehicles and funds for purchase. Hence, more were hired on yearly contracts. TAPF deployed a total of 33 vehicles (21 owned and 12 rented), one for each route. The hired vehicles were deployed on the routes that visited the rural areas. Most of the vehicles used in Gandhinagar in 2010 were Maxx Pickup, costing approximately ₹8.6 lakhs.

The owned vehicles were custom-designed. They had the capacity to accommodate 72 containers (24 big and 48 medium). They had an inbuilt rack system for storing the containers which prevented spillage even on bumpy roads. The vehicles had thermocol insulation which preserved the temperature of the food for up to five hours. The vehicles had a door in the rear which was closed during transit to prevent contamination.

On the other hand, the hired vehicles, being for a temporary period, did not invest in the above additional features. The absence of rack system resulted in lesser capacity of containers (32 big and 23 medium) and the possibility of spillage. Lack of insulation and an open rear made the food susceptible to external temperature and contamination.

TAPF used four categories of vehicles across the country, based on their capacity. (Refer to Table A8.13 and Table A8.14 for the capacity of each vehicle category and the cost structure of the hired vehicles across various locations.)

Figure A8.11   Vehicle types: a) owned, b) hired.

Table A8.12   Vehicle Types and Donors

Vehicle Donated	Company Name	Vehicle Donated	Company Name
Maxx Pickup	Gujarat Industries and Power Co Ltd	Maxx Pickup	GMDC
Maxx Pickup	GSPC	Maxx Pickup	Kalpataru Power Transmission Ltd
Tata Ace	Arvind Mills	Maxx Pickup	State Bank Of India
Maxx Pickup	Udgam School	Maxx Pickup	Gokulesh Petroleum
Maxx Pickup	ONGC	Maxx Pickup	Navneet Publication India Ltd
Maxx Pickup	Gujarat Alkalies And Chemicals Ltd.	Maxx Pickup	Cadila Pharmaceuticals Ltd
Maxx Pickup	GAIL India Ltd	Maxx Pickup	Wagh Bakari Group of Companies Ltd

Source: TAPF, September 2010.

Table A8.13   Vehicle Capacity

Vehicle	Actual Capacity (Containers)	Capacity Utilized (Containers)	Number of Beneficiaries	Total Beneficiary Capacity
Eicher	180	91	4,116	8,147
Tata Ace	45	50	2,005	1,804
Owned Maxx Pickup	72	61	2,754	3,250
Hired Maxx Pickup	55	51	2,579	2,780

Source: TAPF, September 2010.

Table A8.14   Vehicle Information Cost Structure of Hired Vehicles across Locations

Location	Cost of Hired Vehicles
Vrindavan	₹7 per km
Mangalore	₹1,050 per day (₹17.50 per km for TATA 407)
Jaipur	₹420 per day for Maxx Pickup
Hyderabad	₹700 per day + Diesel for TATA Ace (₹11.20 per km)
Puri	₹300 per day + Diesel for TATA Ace
Guwahati	₹800 per day + Diesel for Maxx Pickup
Bangalore	₹1,000 per day including Diesel
Hubli	₹16,000 per month (₹615 per day + Diesel) for TATA 407
Bhilai	₹500 per day + Diesel
Ahmedabad	₹455 per day + Diesel (For Maxx Pickup)
Baroda	₹650 per day + Diesel
Bellary	All TAPF Owned Vehicles

Source: TAPF, September 2010.

Quality Measures

TAPF implemented a set of stringent regulations for controlling the standard of food. These dealt with production, storage, handling, and distribution, ensuring that the food was safe to consume and free of any kind of biological toxins.

Mechanization of the facility had helped them in achieving the highest standards of hygiene by minimizing human handling of food. The parameters defined by them for food safety and hygiene were:

Good Hygiene Practices

- Personal Hygiene: Safety gloves, shoe covers, foot wears, aprons, hand sanitizers, and proper hand washing
- Equipment Hygiene: Washing of equipment with soap solution, cauldron sterilization, and distribution containers sterilization

Good Manufacturing Practices

- Adhering to the Prerequisite Program for food safety and following the production process stated in the Standard Operating Process. (The program and process had been evolved and codified by TAPF to be followed by all kitchens).
- Cooking as per standards of ISO 22000:2005, and monitoring the critical control points through temperature control.

Good Storage Practices

- Maintaining good storage conditions for the raw material: Cool, dry, and clean
- Following the rule of First Expiry First Out (FEFO)
- Pest infestation control

Good Distribution Practices

- Proper packing
- Tight-fitting lids
- Hygienic containers

In terms of procurement, they insisted that the raw materials be packed properly in clean gunny bags or any other food grade/safe plastic materials and not be in contact with any bags of cement, chemicals, soap powders or non-vegetarian items etc. They also ensured that raw materials were delivered in neat and clean, closed or covered vehicles. The acceptance and rejection criteria were based on the quality and safety of the raw material. (Refer to Table A8.15 for the quality checks imposed on some of the raw materials.) The rejected goods were sent back to the supplier, in case quality checks turned out to be negative.

Table A8.15 Raw Material Specifications

Turmeric powder
1 The powder should have characteristic odor and flavor and should be free from mustiness or other foreign odor
2 It should be free from mold, living and dead insects, insect's fragmentation, rodent contamination
3 The powder should be free from any added coloring matter including led chromate and morphologically extraneous matter including foreign starch
4 Moisture not more than 10.0 percent by weight
5 Salmonella Absent in 25 grams
6 Total ash on dry basis not more than 9.0% by weight
7 Ash insoluble in dilute HCL on dry basis not more than 1.5% by weight
8 Colouring powder expressed as curcuminoid content on dry basis not less than 2% by weight
9 Total starch not more than 60% by weight
10 Test for lead chromate Negative
11 Date of manufacture
12 Expiry Date
13 Batch Number
Palm oil
1 It should be clear, free from rancidity, suspended or other foreign matter, separated water, added colouring and flavouring substances or mineral oil
2 Butyro-refractometer reading at 50°C should be 35.5–44.0, or refractive index at 50°C should be 1.4491–1.4452
3 Melting point not more than 37 degree celsius
4 Iodine value: 45–56
5 Saponification value – 195–205
6 Unsaponification value not more than 1.2%
7 Acid value not more than 10%
8 Date of manufacturing

(Continued)

Table A8.15 (Continued)   Raw Material Specifications

9 Expiry Date
10 Batch Number
Moongdal/Peas dal/Urad dal/Gram dal
1 It should be sound, dry, sweet wholesome and free from unwholesome substance
2 It should be free from mold, living and dead insects, insect's fragmentation, rodent contamination
3 Moisture not more than 14% by weight
4 Extraneous matter not more than 1% by weight.
RAW MATERIAL SPECIFICATION
5 Other edible grains not more than 4% by weight.
6 Damaged grains not more than 5% by weight.
7 Weevilled grains not more than 3% by weight,
8 Uric acid not more than 100 mg per kg.
9 Aflotoxin not more than 30 micrograms per kg.
10 The product should be free from added color, mineral oil and any other harmful substances.
Vegetables
Tomato: Should be of Orange Gold color in semi-ripened form, should be free from infestation, should not have Transit damage injuries. The lot should be in sorted form.
Carrot: should be raw, should be carrot pink, should be free from insects, fungus, mud and dust, should be having growth till root.
Potato: should be semi-ripened in condition, should be free from insects, fungal growth mud and dust, should not have tuber/root growth on the skin of the vegetable, should not have transit damage injury.
Quality Assurance Officer Procurement Manager BT Team Operation.

Source: TAPF, September 2010.

Assistance towards Funding the Meal

TAPF received subsidy from the GoG. As of June 2010, this was ₹2.00 + 100 grams of grains (a total equivalent of ₹2.42) per capita for students in standard I to VII. During 2009–2010, the average cost of a meal provided by TAPF was ₹4.69. Hence, there was a gap of ₹2.27. In addition, there was capital expenditure including, setting up of the kitchen. (Refer to Table A8.16 for the breakup of the average cost of a meal and capital expenditure in 2009–2010.)

To cover the gap, TAPF raised donations from the general public and organisations towards the meal costs of the 1.3 lakh children. Donations were raised through five main channels: ISKCON, preachers' efforts, TAPF India marketing effort, Trustees, and abroad (primarily, TAPF, USA). (Refer to Table A8.17 for the donations received for Gandhinagar.)

Augmenting Employment Opportunities

The minimum wage policy was enforced in the Gandhinagar and Vadodara units. There was also an attempt to ensure retention. Minimum wages for unskilled workers started at ₹3,513 per month and for skilled labour at ₹3,578 per month. The Gandhinagar unit had 260 workers: 28 women and 232 men. The women worked from 10 am to 6:30 pm and engaged in vessel washing, vegetable cutting, *dal* cleaning etc. The men worked in multiple shifts in administration, production, and housekeeping departments. The workers were trained to function in accordance with the quality requirements.

Attrition was a major challenge. Workers would leave the job for better wages or to engage themselves in farming during the agricultural season. Some of the employees were addicted to tobacco and liquor consumption. However, the stringent culture at TAPF forbade them from this, resulting in turnover.

To address the situation, apart from the minimum wage policy, TAPF contributed an additional 12.5 per cent wages towards provident fund and insurance under the Provident Fund Act and Employees State Insurance Corporation. Even though the kitchen remained unutilised during the April to June vacation period, the workers were paid 50 per cent of their salary.

Concerns

Effective management solutions were implemented such as the deployment of Tally for purchase requisitions and payables, and biometric Employees State Insurance card to all the employees. Department-wise task lists and schedules were provided to every employee by the Human Resource Department. It also consisted of task tracking sheets which were filled out by the employees and reviewed every week. TAPF also maintained an operations dashboard which gave a comprehensive view of the activities. This was used for planning and monitoring. (Refer to Table A8.18

Table A8.16   Expenditure Details of TAPF, Gandhinagar in 2009–2010

Average Cost of a Meal in 2009–2010	
Particulars	Cost (₹)
Direct Costs	3.43
Cost of Food	2.93
Distribution Expenses	0.50
Indirect Costs	0.65
Administration Expenses	0.29
Fund Raising and Communication Expenses	0.05
Apportionment of HO Expenses	0.31
Notional Costs	0.61
Depreciation	0.61
Cost per Meal	**4.69**
Less: Subsidy Received	**2.42**
Cost Borne by TAPF	**2.27**

Capital Expenditure		
	Gross Block	
Assets	April 1, 2009	March 31, 2010
Land	–	–
Buildings		–
Lease Hold Improvements	20,54,841	32,05,730
Kitchen and Related Equipment	47,87,908	1,13,89,907
Office Equipment	–	–
Computer Equipment	2,11,750	3,37,761
Furniture and Fixtures	7,09,621	9,56,262
Vehicles	1,01,22,107	1,46,31,593
Distribution Vessels	30,59,017	49,02,517
Total	**2,09,45,244**	**3,54,23,769**

Source: TAPF, September 2010.

Table A8.17 Donations Received by TAPF

	ISKCON Contribution		Preachers' Efforts		Marketing Effort		Trustees' Efforts		Abroad		Others		Total	
	2009–2010	2010–2011	2009–2010	2010–2011	2009–2010	2010–2011	2009–2010	2010–2011	2009–2010	2010–2011	2009–2010	2010–2011	2009–2010	2010–2011
April 2009	–	–	12,58,366	9,49,089	–	14,400	–	–	–	–	–	–	12,58,366	9,63,489
May 2009	–	–	5,89,913	10,33,936	6,72,000	–	–	–	–	–	–	–	12,61,913	10,33,936
June 2009	–	–	4,87,069	12,78,836	600	10,30,000	–	–	–	4,017	–	–	4,87,669	23,12,853
July 2009	–	–	13,86,661	19,01,821	–	41,200	–	–	–	–	–	–	13,86,661	19,43,021
August 2009	–	–	11,39,653	14,20,123	–	–	–	–	7,94,808	–	–	–	19,34,461	14,20,123
September 2009	–	–	12,59,054	–	–	–	–	–	–	–	–	–	12,59,054	–
October 2009	–	–	9,47,723	–	–	–	–	–	–	–	–	–	9,47,723	–
Novemeber 2009	–	–	27,55,072	–	2,42,000	–	–	–	54,58,287	–	–	–	84,55,359	–
December 2009	–	–	12,91,821	–	94,976	–	–	–	–	–	–	–	13,86,797	–
January 2010	–	–	13,31,764	–	92,315	–	–	–	–	–	–	–	14,24,079	–
February 2010	–	–	22,80,058	–	10,200	–	–	–	–	–	–	–	22,90,258	–
March 2010	–	–	40,66,635	–	–	–	–	–	–	–	–	–	40,66,635	–
Total	–	–	**1,87,93,789**	**65,83,805**	**11,12,091**	**10,85,600**	–	–	**62,53,095**	**4,017**	–	–	**2,61,58,975**	**76,73,422**

Source: TAPF, September 2010.

Table A8.18   TAPF, Gandhinagar's Operations Dashboard for July 2010–August 2010

Capacity Details	July 2010	August 2010
Installed Kitchen Capacity (Number in meals)	1,08,000	1,08,000
Kitchen Capacity Utilization (in %)	83	83
Installed Logistics Capacity (Number in meals)	90,000	90,000
Logistics Capacity Utilization (in %)	95	95
Key Statistics - Schools		
Total Beneficiaries (Beneficiaries per day * total working days)	2,411,591	2,046,042
Total Enrolment	131,227	132,031
Ratio of Beneficiaries - PS:UPS:HS	3:01:00	2:01:00
Overall Ratio Total Beneficiaries: Total Enrolment	0.68	0.65
No of Feeding Days	27	24
No of Schools	525	525
Cost Per Meal (In ₹)		
– Ingredient Costs	1.43	–
– Rice/Wheat (National) Costs	0.38	0.40
– Fuel (Cooking) Costs	0.26	0.31
– Direct Labour Costs	0.28	0.36
– Other Direct Costs	0.21	0.24
Total Direct Costs (Cash Only)	–	–
– Distribution Costs	0.51	0.49
– Administrative Costs	0.18	0.22
– Publicity & Promotion Costs	0.06	0.06
– Finance Charges	0.00	0.00
Total Indirect Cost (Cash Only)	–	–

(Continued)

Table A8.18 (*Continued*) TAPF, Gandhinagar's Operations Dashboard for July 2010–August 2010

Capacity Details	July 2010	August 2010
Total Cash Cost Per Meal	3.76	4.36
Less: Subsidies	2.00	2.00
– Grain Subsidy	0.38	0.40
– Cash Subsidy	2.00	2.00
Cash Cost per Meal - Deficit	1.76	2.36
Cash Deficit (in%)	47	54
Norms Compliance (In%)		
Primary School	–	–
– Rice/Wheat	83.0	84.19
– *Dals* & Pulses	100	103.75
– Vegetables & Fruits	78	80.56
– Oils & Fats	100	111.30
– Salt	–	–
Income & Expenditure (In ₹)		
Income	–	–
– Donations from Local Donors	1,026,349.00	1,051,558.00
– Donations from Corporate Donors	–	368,565.00
– Grants/Subsidies from MDM	4,823,182.00	4,091,084.00
– Other Income	–1,079.00	38,924.00
Expenditure (in lakhs)	–	–
– Cost of Food (MDM)	2.81	3.14
– Distribution Costs (MDM)	0.42	0.60
– Publicity & Promotion Costs (MDM)	0.06	0.06
– Administrative Costs (MDM)	0.18	0.22
Excess of Income over Expenditure	–	–

(Continued)

Table A8.18 (*Continued*)   TAPF, Gandhinagar's Operations Dashboard for July 2010–August 2010

Capacity Details	July 2010	August 2010
Cash Flow (in ₹)		
Opening Balance of Funds in Hand	10,385.00	8,025.00
Funds Received from (in lakhs)	–	–
– HO	0.00	3,000,000.00
– Local Donations (in lakhs)	10.26	10.51
– Committed Donors (in lakhs)	–	–
– Foreign Donations (in lakhs)	–	–
– Realization of Receivables (in lakhs)	–	–
Funds Received (for Capital expenditure) (in lakhs)	–	–
– HO (in lakhs)	–	–
– Local Donations (in lakhs)	–	3.68
– Committed Donors (in lakhs)	–	–
– Foreign Donations (in lakhs)	–	–
Payments towards Expenses (in lakhs)		
Investments/Capitalization	527,278.00	124,707.00
Inter Unit Transfers/Loans	–	–
Closing Balance of Funds in Hand	8,025.00	26,332.00
Funds Required from HO	–	–
Fund Raising (₹ in Lakhs)		
Total Local Donations Raised	14.01	14.19
Total Local Donations Committed	–	–
Excess/Deficit	–	–
No of New Donors Added	–	–
Donations from New Donors	–	–

(*Continued*)

Table A8.18 (*Continued*)   TAPF, Gandhinagar's Operations Dashboard for July 2010–August 2010

Capacity Details	July 2010	August 2010
Donations from Existing Donors	–	–
Top Donor of the Month	–	–
No of Donation Leads	–	–
Procurement		
Average Cost of Rice/Wheat per kg (market)	–	–
Average Cost of Dals & Pulses per kg	33.30	34.26
Average Cost of Vegetables per kg	26.02	25.75
Average Cost of Oils & Fats per kg	47.86	53.00
Average Cost of Groceries per kg	28.73	29.10
Average Cost of Spices per kg	135.73	138.96
Inventory		
Value of Inventory of Direct Material (₹ in lakhs)	19.02	23.88
No of Days of Inventory	9	9
Production		
Average Cook-to-Consume Time (in hrs)	5.20	3.06
Average Production Hours per day	5	6.00
Average Batches of Cooking per day	–	–
No of Deviations from MPS	4	NIL
No of Deviations from MRP	4	NIL
Logistics		
No of Routes	33	33
Average Length of the Route (in km)	101	101
Average No of Schools/Route	16	16
Average No of Beneficiaries/Route	2,707	2,707

(Continued)

Table A8.18 (*Continued*)   TAPF, Gandhinagar's Operations Dashboard for July 2010–August 2010

Capacity Details	July 2010	August 2010
Ratio of Owned: Hired Vehicles	21:14	21:12
Average Meals per Vehicle	2,552	2,695
Average Fuel Cost per Meal (in ₹)	0.10	0.14
Average Logistics Cost per km (in ₹)	8.19	10.00
No of Vehicle Accidents Reported	–	–
No of Beneficiaries affected due to delays	7,740	9,464
No of Vessel Account Discrepancies	14	12
Human Resources & Compliance		
Average Head Count	261	261
Total Wage Bill (in ₹)	1,145,963.00	1,149,255.60
Total Overtime (in hrs)	678.00	888.00
Average Staff Cost Per Meal (in ₹)	0.48	0.48
No of Statutes to which the unit has not complied with	2	2
Process Compliance		
Are you claiming subsidies based on beneficiaries and actual consumption	Beneficiaries	Beneficiaries
Adherence to Preventive Maintenance Schedule	Yes	Yes
Adherence to Planning Process (MPS/MRP)	Yes	Yes
Product Substitutions/New Products Introduced	Yes	No
Implementation of New Tally	Yes	Yes
Physical Stock Take Tallied	Yes	Yes

Source: TAPF, September 2010.

for an abridged version of the operations dashboard.) In order to make the process lean, they implemented *Kaizen* in the Vadodara kitchen.

TAPF was considered to be a role model due to its innovations in technology and processes, quality of food, variety provided in the meal, large scale production, and timely delivery. However, according to Dasa, there was scope for improvement in the supply chain ranging from procurement to distribution. Some of the specific concerns were:

- How much to produce was based on the estimate given by the school authorities every day. However, this still led to food wastage.
- Supplier selection and price fluctuations were major challenges for purchase decisions.
- Storage posed a concern because of space crunch. The raw materials had to be used up within two weeks.
- The *rotis* were prepared in one building while rice and *sabji* were prepared in another. This resulted in loading of containers at two places.
- *Rotis* cut in squares or triangles would have served to be more efficient than *rotis* cut in the classic round shape. However, especially in case of square *rotis,* the question of acceptability of the shape posed a doubt. With a square or triangle shape, the left over in one cycle of the *atta* sheet would reduce. This would save the recycling costs of the unused portion of the *atta* sheet.
- The width of the *atta* sheet being inserted into the *roti*-making machine was always aligned with the width of the rotary cutting die. Strips of the *atta* sheet were being continuously cut out from the corner to ensure alignment. These were also recycled.
- The cut *roti* in the *atta* sheet was not always round or had some folds in them. These were manually inspected, removed, and recycled. Due to this, instead of 40,000 *rotis* per hour, approximately 35,000 *rotis* were produced per hour.
- Undercooked or overcooked *rotis* were discarded by two workers stationed at the end of the *roti* machine. Since this was being done manually, some inappropriately cooked *rotis* went unnoticed and got packed into the containers.
- Breakdown in the equipment being used in the kitchen caused delay in the cooking time, thereby leading to late distribution of food to the schools. (For photos, refer to Figure A8.12.) There was not sufficient standby equipment, and even when there was, they were not always ready for quick and easy start up for use.
- The kitchen area where the vegetables were being cooked had an open section to facilitate the movement of the containers to the vehicles which could lead to contamination of the food. Air curtains were considered a possible solution. Also, the vegetables were being stirred in open containers. (For photos, refer to Figure A8.12.)

a)   b)

Figure A8.12   Operational concerns. a) Fixing the links to the chain. b) Cooking in open containers.

- In case the requirement of any school was below 600 *rotis*, the capacity of the medium-sized containers would not be utilized optimally, resulting in higher space occupancy in the vehicle.
- High temperatures in the kitchen owing to the steam caused fatigue among the workers.
- The routes were not optimized with respect to the maximum number of schools that could be covered. Global Positioning System was being proposed, given the General Packet Radio Service network connectivity.
- Food wastage at school was a critical concern. It was observed that if a few children in a group finished first, the rest would leave with their peers even if it meant throwing away the left over in the plate. In most schools, the leftover food was distributed to *goshalas*.

Fostering Growth

TAPF was being asked to increase the number of beneficiaries in the Ahmedabad and Gandhinagar schools. They found it difficult to do so due to the infrastructural limitations. The Gandhinagar kitchen, though starting as a temporary arrangement, had evolved as a permanent facility. However, it was less than adequate. A centralized kitchen required about 15 acres of land with a good approach road. TAPF wanted GoG to provide such land and fund the capital expenditure of the kitchen (possibly in collaboration with any corporate). TAPF were also looking for exemptions in sales tax/value-added tax on the procured items and motor vehicle tax on the vehicles. TAPF was included as a destination in the government conducted educational tours of Gujarat.

GoG felt that the need of the hour was to make people aware of the feeding programs and to reach out to a larger section of the needy population. Therefore, TAPF was getting ready to serve food to the patients in the Civil Hospitals by October 2010. Lunch was to be priced at ₹25 and dinner at ₹20. The menu would be designed in consultation with the doctors and the dieticians.

Discerning the exigency that needed to be addressed, Dasa reflected on how to expand the noble cause of the program. Enthusiastic about the possibilities, he hoped to erase the meaning of classroom hunger in the cities of Gujarat, aligned with the TAPF's vision 'No child in India shall be deprived of education because of hunger'.

Glossary

Atta – Whole wheat flour
Dal – Stew prepared from lentils, peas or beans
Goshalas – Protective shelters for cows in India
Kaizen – Refers to the practices of continuous improvement of processes
Khichdi – Mixture of rice and lentils
Mandi – Vegetable market
Moong mogar – A lentil prepared using green beans
Mutter dal – A stew prepared from peas
Prasadam – A religious offering, usually an eatable, that is first offered to the deity and then offered to the devotees
Roti – Unleavened flatbread made from whole wheat flour
Sabji – Cooked vegetables
Sukhadi – A popular sweet made from fortified *atta*
Taluka – Administrative and geographical block consisting of some villages
Tava – A name for the *roti* making equipment
Thepla – A *roti*, where the dough is mixed with greens

Abbreviations

ESIC – Employees State Insurance Corporation
GoG – Government of Gujarat
GoI – Government of India
GPRS – General Packet Radio Service
GPS – Global Positioning System

GSCSC – Gujarat State Civil Supplies Corporation
HS – Higher Secondary
INR – Indian Rupees
ISKCON – International Society for Krishna Consciousness
ISO – International Organisation for Standardization
FEFO – First Expiry First Out
FIFO – First In First Out
LIFO – Last In First Out
MDM – Mid-Day Meal
MoU – Memorandum of Understanding
NGO – Non-Governmental Organisation
NP-NSPE – National Programme-Nutritional Support to Primary Education
PS – Primary School
Std – Standard
TAPF – The Akshaya Patra Foundation
UPS – Upper Primary School

Acknowledgements

Prepared by Shravanti Mitra, Professors G. Raghuram, and Atanu Ghosh, Indian Institute of Management, Ahmedabad. Research assistance provided by Shivani Shukla is acknowledged.

The authors acknowledge the access, time, and help provided by TAPF and particularly thank Chanchalapathi Dasa, Shridhar Venkat, Jaganmohan Krishna Dasa, Adi Keshava Dasa, Sushant Shetty, Chirag Seth, Rohit Agrawal, Nilesh Mehta, Vinay Kumar, and Ashwin Trivedi.

Cases of the Indian Institute of Management, Ahmedabad, are prepared as a basis for class discussion. They are not designed to present illustrations of either correct or incorrect handling of administrative problems.

Index

W

Warehouse management system
      (WMS), 576–580
Warehousing facilities, 23
Wastage
   source, 169
   storage, 169–171
   transportation, 171
Weather-based crop insurance, 230
Web based auctions, 388–390
Websites, agribusiness, 338–339
Wholesalers
   direct marketing through, 163–167
   farm market
      informal financial system and
         intermediation, 80
      Mandi system, 79–80
      procurement by government agencies
         and cooperatives, 80–84
      structural features, 78–79
   infrastructure requirements, 265
   intermediation process
      banana, 94
      pomegranate, 91
WMS, *see* Warehouse management system
Woolworths Limited
   Australian retail industry
      five year analysis, 558–569
      Project Mercury, 555, 570
      Project Refresh, 553–555
      retail industry turnover, 550

state-wise population and retail income
      distribution, 551
   supermarket, 550–557
company overview, 543, 545–546
major events, 544–545
major participants and market shares,
      601–603
management of, 534–537
Network Model Project, 586–587
organizational structure, 591–593
peer comparisons, 603–604
project refresh, 538–542
retail activities
   General Merchandize, 547–550
   supermarkets, 546
   wholesale, 550
retail brand, 548–549
supply chain strategy
   replenishment system, 581–583
   TMS, 580–581
   WMS, 576–580
Workshops, agricultural communications
      network, 260
World agricultural trade, 198–199
World Trade Organization (WTO),
      1995, 336–337
World Trade Organization—Sanitary
      and Phytosanitary (WTO-
      SPS), 192
WTO, *see* World Trade Organization
WTO-SPS, *see* World Trade Organization—
      Sanitary and Phytosanitary